THE TRIPLET STATE

THE TRIPLET STATE

PROCEEDINGS OF AN INTERNATIONAL SYMPOSIUM HELD AT THE AMERICAN UNIVERSITY OF BEIRUT, LEBANON

14–19 February 1967

CAMBRIDGE
AT THE UNIVERSITY PRESS
1967

CAMBRIDGE UNIVERSITY PRESS
Cambridge, New York, Melbourne, Madrid, Cape Town, Singapore,
São Paulo, Delhi, Dubai, Tokyo

Cambridge University Press
The Edinburgh Building, Cambridge CB2 8RU, UK

Published in the United States of America by Cambridge University Press, New York

www.cambridge.org
Information on this title: www.cambridge.org/9780521126502

First published 1967
This digitally printed version 2009

A catalogue record for this publication is available from the British Library

ISBN 978-0-521-06892-5 Hardback
ISBN 978-0-521-12650-2 Paperback

PREFACE

Long-lived phosphorescence had been known for almost a century before G. N. Lewis and M. Kasha showed in 1944 that it was associated with an energy state which was paramagnetic. The paramagnetism of this state was investigated in a much more detailed way by C. A. Hutchison Jr. and B. W. Mangum in 1958 using the techniques of electron paramagnetic resonance. Also in 1958 R. Williams discovered delayed fluorescence. A number of scientists discovered that this strange fluorescence may be attributed to triplet-triplet annihilation, thus initiating still another new line of triplet state research the full exploration of which is still under way.

Thus, research into all aspects of the triplet state has been actively pursued for more than two decades. Even so, few of the basic questions posed concerning it have received definite answers. That there is still much to be done is indicated by many gaps in our knowledge. For example, accurate determinations of such basic quantities as singlet-triplet and triplet-triplet extinction coefficients have yet to be made. On the theoretical side an adequate analysis of such molecular properties as the intersystem crossing rate constant is still lacking.

The wide variety of techniques being applied to the study of the triplet state and the present rapid increase in the type of problems associated with it suggested that it might be useful to review our understanding of the triplet state and to perhaps chart a course for future investigation. To this end the Symposium on the Triplet State was held at the American University of Beirut in the Republic of Lebanon from 14 to 19 February 1967.

This volume contains most of the papers presented at the Symposium. Those papers received after the publisher's deadline (which was quite early in an attempt to make this volume generally available as soon after the Symposium as possible) are represented by abstract only.

This Symposium was one of the events held in celebration of the Centennial of the American University of Beirut, founded in 1866 as the Syrian Protestant College.

The Organising Committee wishes to acknowledge the financial help of the American University of Beirut and of the Research Corporation (U.S.A.). The co-operation of the Symposium participants, officials and faculty members of the University and of the Cambridge University Press contributed greatly to the success of this meeting and to this volume which records it.

Organising Committee

G. M. ANDROES
F. W. HEINEKEN
A. B. ZAHLAN

CONTENTS

† Italic indicates the author who delivered the paper, and to whom discussion was directed.

Section 6. Delayed fluorescence and phosphorescence

Section 7. Triplet state related to biology

LIST OF PARTICIPANTS

Chairmen (on a daily basis)

J. B. Birks
H. F. Hameka
F. R. Lipsett
C. A. Parker
A. Pullman
S. A. Rice
G. W. Robinson
B. Stevens
H. C. Wolf

Secretaries (on a daily basis)

T. Farzaneh
D. A. Frey
I. Khubeis
M. F. Mayahi
M. Y. Mentalecheta
R. I. Razouk
M. P. Stevens
F. Tittel
A. Zmerli

R. Acharya, Department of Physics, American University of Beirut, Beirut, Lebanon

H. H. Aly, Department of Physics, American University of Beirut, Beirut, Lebanon

G. Ameer, Department of Physics, American University of Beirut, Beirut, Lebanon

G. Androes, Department of Physics, American University of Beirut, Beirut, Lebanon

R. Astier, Laboratorie de Physique, Ecole Polytechnique, Paris, France

J. S. Avery, Department of Chemistry, Imperial College, London, S.W. 7, England

E. Awad, Department of Chemistry, American University of Beirut, Beirut, Lebanon

R. G. Badro, Department of Physics, American University of Beirut, Beirut, Lebanon

J. B. Birks, Atomic and Molecular Physics Group, The Physical Laboratories, The University, Manchester 13, England

J. Boutros, Department of Biology, American University of Beirut, Beirut, Lebanon

F. Bruin, Department of Physics, American University of Beirut, Beirut, Lebanon

P. Cervenka, Department of Physics, American University of Beirut, Beirut, Lebanon

R. B. Cundall, Department of Chemistry, University of Nottingham, University Park, Nottingham, England

M. G. Delacôte, Institut d'Etudes Nucléaires d'Alger, Conseil de la Recherche Scientifique, Blvd. Franz Fanon, Alger, Algerie

G. A. Derwish, Nuclear Research Center of the Iraqi Atomic Energy Commission, Baghdad, Iraq

D. L. Dexter, Department of Physics and Astronomy, University of Rochester, River Campus Station, Rochester, New York 14627, U.S.A.

R. Douglass, Department of Physics, American University of Beirut, Beirut, Lebanon

P. Douzou, Laboratorie de Bio-Spectroscopie, Institut de Biologie Physico-Chimique, 13, Rue Pierre Curie, Paris Ve, France

F. Dupre la Tour, S. J. Dean, Faculty of Medicine and Pharmacy, St Joseph University, P.O.B. 5076, Beirut, Lebanon.

T. B. El-Kareh, 74 Tübingen, Postfach 1403, W. Germany

T. Farzaneh, Department of Physics, Pahlavi University, Shiraz, Iran

G. Finger, Department of Physics, American University of Beirut, Beirut, Lebanon

D. A. Frey, Department of Physics, Robert College, Bebek Post Box 8, Istanbul, Turkey

R. Gautron, Saint-Gobain, B.P. No. 8, 92-Antony, France

F. H. Giles, Department of Physics, University of Baghdad, Baghdad, Iraq

S. A. Goudsmit, Physics Department, Brookhaven National Laboratory, Upton, L. I., N.Y., 11973, U.S.A.

M. Guéron, Bell Telephone Labs. IC 515, Murray Hill, New Jersey, U.S.A.

H. F. Hameka, Department of Chemistry, University of Pennsylvania, Philadelphia, Pa. 19104, U.S.A.

G. Hanania, Department of Chemistry, American University of Beirut, Beirut, Lebanon

F. W. Heineken, Department of Physics, American University of Beirut, Beirut, Lebanon

C. Hélène, Laboratoire de Biophysique-Museum, 61 Rue Buffon, Paris (5e), France

R. M. Hochstrasser, Department of Chemistry, University of Pennsylvania, Philadelphia, Pa. 19104, U.S.A.

J. Hofland, Department of Physics, Robert College, Bebek Post Box 8, Istanbul, Turkey

C. A. Hutchison Jr., Department of Chemistry, University of Chicago, Chicago, Illinois, U.S.A.

C. K. Jen, Applied Physics Laboratory, Johns Hopkins University, Silver Spring, Maryland, 20910, U.S.A.

J. Joussot-Dubien, Département de Chimie Physique, Faculté des Sciences de Bordeaux, 33, Talence, France

J. A. Katul, Department of Physics, American University of Beirut, Beirut, Lebanon

A. A. Kazzaz, Physics Department, College of Science, Adhamiyha, Iraq

A. Kellmann, Laboratoire de Chimie Physique, Faculté des Sciences, Orsay, France

R. E. Kellogg, Central Research Department, Experimental Station, E. I. du Pont de Nemours and Co., Wilmington, Delaware, U.S.A.

I. Khubeis, Department of Physics, Faculty of Science, Jordan University, Amman, Jordan

T. A. King, The Physical Laboratories, The University, Manchester 13, England

J. Kommandeur, Department of Chemistry, The University, Grönin-gen, The Netherlands

H. Lemaire, B.P. 269, Grenoble, France

J. M. Lhoste, Laboratoire de Biophysique du Muséum National d'Histoire Naturelle, Paris

L. Lindqvist, Laboratorie de Chimie Physique, Faculté des Sciences, Orsay, France

F. R. Lipsett, National Research Council, Ottawa, Canada

G. R. Luckhurst, Varian Laboratories, Klausstr. 43, 8008 Zurich, Switzerland

J. E. LuValle, Director of Research, Fairchild, 300 Robbins Lane, Syosset, L.I., N.Y. 11791, U.S.A.

H. Maria, Department of Physics, American University of Beirut, Beirut, Lebanon

M. F. Mayahi, Department of Chemistry, College of Science, Adhami-yha, Baghdad, Iraq

J. W. McClain, Department of Physics, American University of Beirut, Beirut, Lebanon

M. Y. Mentalecheta, Institut d'Etudes Nucléaires d'Alger, Conseil de la Recherche Scientifique, Blvd. Franz Fanon, Alger, Algerie

Y. H. Meyer, Laboratoire de Physique, Ecole Polytechnique, Paris, France

J. Mirhij, Department of Biology, American University of Beirut, Beirut, Lebanon

P. Mourad, Department of Physics, American University of Beirut, Beirut, Lebanon

C. S. Nassar, Beirut College for Women, Beirut, Lebanon

N. Shahin, Imm. Shahin, Rue Hamra, Beirut, Lebanon

J. Olmsted III, Department of Chemistry, American University of Beirut, Beirut, Lebanon

C. le Pair, Department of Physics, American University of Beirut, Beirut, Lebanon

C. A. Parker, Admiralty Materials Laboratory, Holton Heath, Poole, Dorset, England

C. S. Parmenter, Department of Chemistry, Indiana University, Bloomington, Indiana 47401, U.S.A.

B. Pouyet, Faculté des Sciences de Lyon, 41, Chemin des Petites Brosses, 69 Caluire, France

M. Ptak, Laboratoire de Biophysique, 61, Rue Buffon, Paris (5e), France

A. Pullman, Institut de Biologie Physico-Chimique, 13 Rue Pierre Curie, Paris (5e), France

A. Rassat, B. B. 269, Grenoble, France

R. I. Razouk, Department of Physical Sciences, The American University in Cairo, 113, Sharia Kasr-el-Aini, Cairo, U.A.R.

S. A. Rice, Department of Chemistry, University of Chicago, Chicago, Illinois 60637, U.S.A.

G. W. Robinson, Gates & Grellin Laboratories of Chemistry, California Institute of Technology, Pasadena, California 91109, U.S.A.

A. Sarkis, Department of Chemistry, American University of Beirut, Beirut, Lebanon

M. Schwoerer, 3. Physikalisches Institut der Technischen Hochschule, 7 Stuttgart, Azenbergstrasse 12, Germany

M. Sharnoff, Department of Physics, University of Delaware, Newark, Delaware 19711, U.S.A.

W. Siebrand, Division of Pure Chemistry, National Research Council, Ottawa 2, Canada

S. Siegel, Laboratory Operations, Aerospace Corporation, Box 95085, Los Angeles, California 90045, U.S.A.

M. Silver, Department of Physics, University of North Carolina, Chapel Hill, N.C., U.S.A.

R. A. P. Singh, Department of Physics, University of Baghdad, Baghdad, Iraq

B. Stevens, Department of Chemistry, University of Sheffield

M. P. Stevens, Department of Chemistry, Robert College, Bebek, Istanbul, Turkey
J. Tanaka, Department of Chemistry, Nagoya University, Chikusa, Nagoya, Japan
F. Tittel, Department of Physical Sciences, The American University in Cairo, 113, Sharia Kasr-el-Aini, Cairo, U.A.R.
J. S. Vincent, Department of Chemistry, University of California, Davis, California 95616, U.S.A.
A. Vondjidis, Department of Physics, American University of Beirut, Beirut, Lebanon
J. H. van der Waals, Koninklijke/Shell-Laboratorium, Amsterdam, The Netherlands
W. Welten, S.J., University St Joseph, Rue Damas, Beirut, Lebanon
M. W. Windsor, Chemical Sciences Department, TRW Systems, Redondo Beach, California 90278, U.S.A.
H. C. Wolf, 3. Physikalisches Institut der Technischen Hochschule, 7 Stuttgart, Azenbergstrasse 12, W. Germany
A. B. Zahlan, Department of Physics, American University of Beirut, Beirut, Lebanon
O. Zamani-Khamiri, Department of Chemistry, American University of Beirut, Beirut, Lebanon
W. Zantut, Rue Gemayel, Chiah, Beirut, Lebanon
A. Zmerli, Faculté des Sciences, University of Tunis, 8, Rue de Rome, Tunis, Tunisia

1 SPIN-ORBIT COUPLING AND INTERSYSTEM CROSSING

SPIN-ORBIT INTERACTIONS IN ORGANIC MOLECULES†

H. F. HAMEKA

Forbidden transitions

In spectroscopy the expression 'forbidden transition' is used to describe a transition with a probability that is much smaller than normal. We may indicate how strongly forbidden a transition is by giving the ratio between its probability and the probability of a normal transition; there are cases where this ratio is 10^{-10} or even less. The word 'forbidden' is thus used in a relative sense; in fact, all known forbidden transitions must necessarily have finite transition probabilities.

From a theoretical point of view a transition is forbidden if its probability is zero in zeroth-order approximation and if its magnitude must be evaluated from higher-order approximations. Let us first discuss this zeroth-order approximation. We consider an atom or a molecule containing N electrons; in the case of a molecule we introduce the variable R to denote symbolically the positions of the nuclei. Such a system is described in zeroth-order approximation by the Hamiltonian

$$H_0 = \sum_{j=1}^{N} \frac{p_j^2}{2m} + V(\mathbf{r}_1, \mathbf{r}_2, \ldots, \mathbf{r}_N; R). \tag{1}$$

We denote the eigenvalues of H_0 by E_n and the corresponding eigenfunctions by Ψ_n. These functions contain the position coordinates $(\mathbf{r}_1, \mathbf{r}_2, \ldots, \mathbf{r}_N)$ of the electrons, the electron spin variables $(s_1, s_2, \ldots, s_N)$ and also the nuclear coordinates R. It follows from the exclusion principle that the functions Ψ_n must be antisymmetric with respect to permutations of the electron coordinates and, consequently, that each

† Work supported by the Advanced Research Projects Agency and by the National Science Foundation.

function Ψ_n must be an eigenfunction of the operators S^2 and S_z. These two operators are defined with the aid of the vector

$$\mathbf{S} = \mathbf{S}_1 + \mathbf{S}_2 + \mathbf{S}_3 + \ldots + \mathbf{S}_N, \tag{2}$$

where the operators $\mathbf{S}_j$ define the spins of the individual electrons. Hence

$$S^2 = (\Sigma_j S_{jx})^2 + (\Sigma_j S_{jy})^2 + (\Sigma_j S_{jz})^2,$$
$$S_z = \Sigma_j S_{jz}. \tag{3}$$

In general

$$\left.\begin{aligned} S^2\Psi_n &= s(s+1)\hbar^2\Psi_n, \\ S_z\Psi_n &= m_s\hbar\Psi_n, \end{aligned}\right\} \tag{4}$$

where s and m_s are the quantum numbers that label the spin states. Their possible values are

$$\left.\begin{aligned} s &= 0, 1, 2, 3, \ldots, \text{etc.} \quad (N \text{ is even}), \\ s &= \tfrac{1}{2}, \tfrac{3}{2}, \tfrac{5}{2}, \ldots, \text{etc.} \quad (N \text{ is odd}), \\ m_s &= -s, -s+1, \ldots, s \end{aligned}\right\} \tag{5}$$

It is easily seen that for a given value of s the corresponding spin state is $(2s+1)$-fold degenerate. This number, $2s+1$, is called the multiplicity of the spin state. It is used to identify the eigenvalues of S^2 and is written as a superscript on the left-hand side of the eigenfunction. For example, ${}^1\Psi_n$ describes a singlet state with $s = 0$, ${}^2\Psi_{n,j}$ is a doublet state with $s = \frac{1}{2}$ and $m_s = j$, ${}^3\Psi_{n,j}$ is a triplet state with $s = 1$ and $m_s = j$, etc. Different eigenfunctions are orthogonal,

$$\langle {}^k\Psi_{n,i} | {}^l\Psi_{m,j} \rangle = \delta_{n,m}\delta_{k,l}\delta_{i,j}. \tag{6}$$

Also

$$\langle {}^k\Psi_{n,i} | G_{0p}(\mathbf{r}_1, \mathbf{r}_2, \ldots, \mathbf{r}_n, R) | {}^l\Psi_{m,j} \rangle = 0 \tag{7}$$

for different spin states, that is if $k \neq l$ or $i \neq j$, because the spin functions belonging to different spin states are orthogonal.

Let us now consider the calculation of the transition probability between two states (0) and (f). A plane monochromatic light wave may be represented by a vector potential $\mathbf{A}$, which is given as

$$\mathbf{A} = A_0 \mathbf{e}\{\gamma\, e^{-2\pi i\nu t} + \gamma^* e^{2\pi i\nu t}\} \tag{8}$$

with the abbreviation

$$\gamma = e^{i\boldsymbol{\sigma}\mathbf{r}}. \tag{9}$$

Here $\mathbf{e}$ is a unit vector, which represents the direction of polarisation of the light, $\boldsymbol{\sigma}$ is the wave vector which describes the direction of propagation and also the wavelength,

$$\sigma = 1/\lambda, \tag{10}$$

and ν is the frequency of the light. We have taken the scalar potential of the radiation field equal to zero; this is allowed if we take

$$\operatorname{div} \mathbf{A} = 0, \tag{11}$$

or

$$(\mathbf{e} \,.\, \boldsymbol{\sigma}) = 0. \tag{12}$$

The electric and magnetic field strengths of the light wave are given by

$$\mathbf{E} = -\frac{1}{c}\frac{\partial \mathbf{A}}{\partial t}, \quad \mathbf{H} = \operatorname{curl} \mathbf{A}. \tag{13}$$

Finally, A_0 is a constant, which is a measure of the energy density of the radiation field.

In the presence of an electromagnetic field the Hamiltonian (1) becomes

$$H_1 = \frac{1}{2m}\Sigma_j\left(\mathbf{p}_j + \frac{e}{c}\mathbf{A}_j\right)^2 + V - e\Sigma_j \Phi_j. \tag{14}$$

Here the electromagnetic field is represented by a scalar potential Φ and a vector potential $\mathbf{A}$; Φ_j and $\mathbf{A}_j$ are the potentials experienced by the j-th electron. The perturbation term which is responsible for optical transitions is the term of (14) which is linear in $\mathbf{A}$, that is

$$H_1' = \frac{e}{mc}\Sigma_j \mathbf{A}_j \,.\, \mathbf{p}_j. \tag{15}$$

Here and in (14) the charge of the electron is given as $(-e)$.

From (8) and (15) we derive [1] that the time-proportional transition probability between two states (0) and (f) is given by

$$W_{0 \to f} = \frac{e^2}{2\pi m^2 \nu_{0f}^2} H_{f0} H_{f0}^*, \tag{16}$$

with

$$\left.\begin{aligned} H_{f0} &= \langle \Psi_f | \Sigma_j \gamma_j (\mathbf{e} \,.\, \nabla_j) | \Psi_0 \rangle, \\ h\nu_{0f} &= E_f - E_0. \end{aligned}\right\} \tag{17}$$

Here E_f and E_0 are the eigenvalues and Ψ_f and Ψ_0 are the eigenfunctions of the states (f) and (0) respectively.

In atoms and molecules the dimensions of the systems are usually much smaller than the wavelength of the light and it is customary to assume that

$$\boldsymbol{\sigma} \,.\, \mathbf{r} \ll 1, \quad \gamma_j \approx 1, \tag{18}$$

may be substituted into (17). In that case (17) may be transformed to the customary expression

$$\left.\begin{aligned} H_{f0} &= -(2\pi m \nu_{0f}/e\hbar)(\mathbf{e} \,.\, \mathbf{P}_{f0}), \\ \mathbf{P}_{f0} &= \langle \Psi_f | e\Sigma_j \mathbf{r}_j | \Psi_0 \rangle, \end{aligned}\right\} \tag{19}$$

where $\mathbf{P}_{f0}$ is called the transition moment between the states (0) and (f).

We call a transition forbidden if the transition moment (19) is equal to $\mathbf{0}$. It is possible that $\mathbf{P}_{f0}$ is $\mathbf{0}$ by coincidence, but this is very unlikely and we do not consider this. Otherwise there are two possible reasons that can cause the transition moment to be zero. The first one is a difference in spin quantum numbers between the states (0) and (f); we speak then of spin-forbidden transitions. The second reason is that the integration over the electron position coordinates in (19) gives zero because of the symmetries of the states (0) and (f); in that case the transition is called symmetry-forbidden.

The above discussion is based on the zero-order approximation, and we mentioned already that the finite probabilities of forbidden transitions can be explained only by considering higher-order approximations. In the case of spin-forbidden transitions we ought to consider more accurate Hamiltonians than H_0 of (1) and H_1 of (14) in order to derive a satisfactory theory. In the case of symmetry-forbidden transitions we can arrive at finite transition probabilities by using (17), and there is no need to consider higher-order terms of the Hamiltonian. Here there are two mechanisms that make the transition possible. The first mechanism is derived by observing that the approximation (18) is inadequate and that instead we ought to substitute

$$\gamma_j \approx 1 + i(\boldsymbol{\sigma} \cdot \mathbf{r}_j) \tag{20}$$

into (17). We obtain then

$$H_{f0} = \langle \Psi_f | i\Sigma_j(\boldsymbol{\sigma} \cdot \mathbf{r}_j)(\mathbf{e} \cdot \nabla_j) | \Psi_0 \rangle \tag{21}$$

since the first term on the right-hand side of (20) gives zero if the transition is forbidden. Transitions that are described by (21) are called magnetic dipole or electric quadrupole transitions, as opposed to electric dipole transitions, which are described by equations (19).

In molecules there exists a second mechanism that can make a symmetry-forbidden transition possible, namely, the molecular vibrations. The symmetry rules for the electronic wavefunctions are derived on the assumption that the nuclei are fixed at their equilibrium positions and the small, but finite, motion of the nuclei has the effect of distorting the symmetry of the electronic wavefunctions and to make the transition possible. We speak here of a vibrationally-allowed forbidden transition.

In the following section we will discuss the theory of spin-forbidden transitions, and we will see that the general theory of forbidden transitions is exactly the same for atoms and for molecules. In practice, on the other hand, there are important differences between atoms and molecules because the orders of magnitude of the various effects that make forbidden transitions observable are quite different in the two cases. In an atom a symmetry-forbidden transition can occur only as an electric quadrupole or a magnetic dipole transition, and this means that the transition probability is at most 10^{-6} times the probability of an allowed transition. In a molecule a symmetry-forbidden transition can be made possible by vibrations or a number of other effects and the probability varies between 10^{-4} and 10^{-1} times a normal transition probability. Consequently the symmetry restrictions are much more rigid for atoms than for molecules.

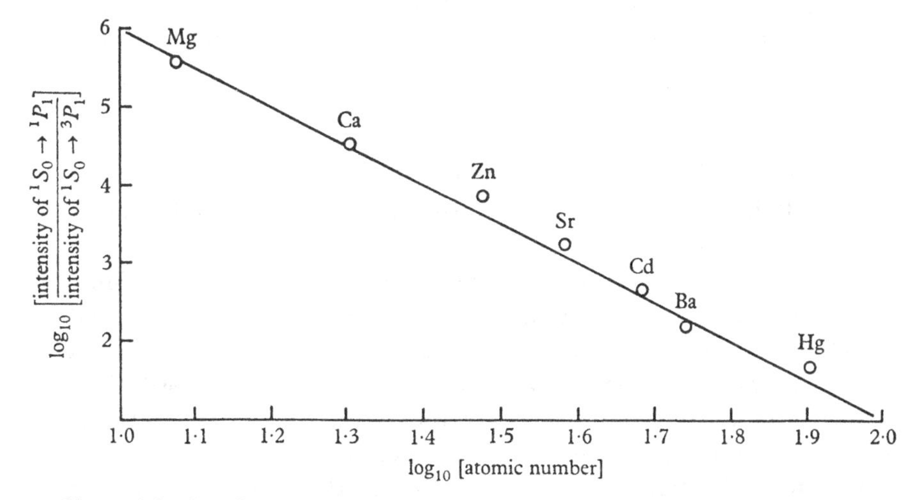

Fig. 1. A log-log plot illustrating the effect of atomic number on the intensity of a spin-forbidden transition.[2]

The general theory of spin-forbidden transitions is the same for atoms and molecules, but a very important difference between an atom and a molecule is due to the central symmetry of an atom as opposed to a molecule. As a consequence we are able to transform the theory of spin-forbidden transitions in atoms to a much simpler form than we can do for molecules. It is important to recognise that the detailed theoretical expressions for atoms cannot be applied to molecules since they were derived on the assumption that the system under consideration possesses central symmetry. A misappreciation of this point has led to some confusion in the literature. The probability, or intensity, of a spin-forbidden transition varies widely between different atoms. For example, in fig. 1 we have plotted[2] the intensity ratio between the ${}^1S_0 \rightarrow {}^3P_1$ and the ${}^1S_0 \rightarrow {}^1P_1$ transitions in a number of atoms; it is shown that this ratio varies between 10^{-6} and 10^{-1} and that it is inversely proportional to the fifth power of the atomic number Z. We usually consider spin-forbidden transitions in organic molecules only where the atoms are fairly light ($Z < 10$). Here the probabilities of spin-forbidden transitions vary between 10^{-5} and 10^{-10} times the probabilities of allowed transitions.

It follows from the above orders of magnitude that any molecular transition with a probability that is less than 10^{-5} times a normal transition is very likely spin-forbidden. In heavy atoms spin-forbidden transitions are almost as intense as normal transitions, and much more intense than symmetry-forbidden transitions. In light and intermediate atoms spin-forbidden and symmetry-forbidden transitions have comparable probabilities.

Spin interactions in the Hamiltonian

The finite probabilities of spin-forbidden transitions can be understood only if we base our considerations on Hamiltonians that are more accurate than the ones reported in equations (1) and (14). More specifically, we must derive the parts of the Hamiltonian that depend on the electron spin. The derivation of these spin-interaction terms is a straightforward problem, which has been solved satisfactorily in a number of different ways. However, there seems to be some confusion in the literature as to the correct form of the spin Hamiltonian for molecules. This confusion stems from the fact that in an atom the spin Hamiltonian can be transformed further because of the central symmetry and that attempts to use these atomic spin Hamiltonians for molecules led to inconsistent results. We have even encountered statements in the literature[3] which claim that the spin Hamiltonian for a molecule is not known. These statements are obviously incorrect since the general form of the spin Hamiltonian, valid for both atoms and molecules, has been known since the late twenties.

We mentioned already in the previous section that the angular momentum of a spinning electron is represented by an operator $\mathbf{S}$. For a single electron the quantum number s, defined by equation (4), is $\frac{1}{2}$ and the possible values of the other quantum number m_s are $m_s = \pm\frac{1}{2}$. We denote the two eigenfunctions of S_z by α and β. By drawing an analogy with the properties of the orbital angular momentum operator it may then be derived that

$$\left.\begin{aligned} S_x\alpha = \tfrac{1}{2}\hbar\beta, \quad S_y\alpha = \tfrac{1}{2}i\hbar\beta, \quad S_z\alpha = \tfrac{1}{2}\hbar\alpha, \\ S_x\beta = \tfrac{1}{2}\hbar\alpha, \quad S_y\beta = -\tfrac{1}{2}i\hbar\alpha, \quad S_z\beta = -\tfrac{1}{2}\hbar\beta. \end{aligned}\right\} \tag{22}$$

A spinning electron possesses also a magnetic moment $\boldsymbol{\mu}$, which is related to $\mathbf{S}$ by means of

$$\boldsymbol{\mu} = (-e/mc)\,\mathbf{S}. \tag{23}$$

This expression was first proposed by Goudsmit & Uhlenbeck[4] because it leads to a satisfactory explanation of the experimental data; it can be derived from the Dirac equation and it was also obtained by Kramers[5] from a non-quantum mechanical, relativistic, argument.

Let us first derive the spin Hamiltonian of an electron with a charge $(-e)$ that moves with a velocity $\mathbf{v}$ in a circular orbit around a nucleus with a charge Ze. We first observe that a magnetic moment $\boldsymbol{\mu}$ at rest in a magnetic field $\mathbf{H}$ has the energy

$$E_\mu = -\boldsymbol{\mu}\,.\,\mathbf{H}. \tag{24}$$

It follows from equations (23) and (24) that the spin energy of an electron at rest, in the presence of an electromagnetic field $(\mathbf{E}, \mathbf{H})$, is given by

$$E_s = (e/mc)\,(\mathbf{S}\,.\,\mathbf{H}). \tag{25}$$

If the electron moves with a velocity $\mathbf{v}$ with respect to the nucleus then it seems to an observer who is stationed on the electron that the nucleus moves with a velocity $-\mathbf{v}$. This motion gives rise to a magnetic field

$$\mathbf{H} = -\frac{Ze}{c}\frac{\mathbf{v}\times\mathbf{r}}{r^3}, \tag{26}$$

where $\mathbf{r}$ is the distance from the nucleus to the electron. According to this argument the spin energy of the electron becomes

$$E_s = \frac{-Ze^2}{mc^2}\frac{(\mathbf{v}\times\mathbf{r})\,.\,\mathbf{S}}{r^3}. \tag{27}$$

However, this result is incorrect. Equation (27) was derived from the point of view of an observer stationed on the electron, and the correct result should have been derived from the point of view of an observer stationed on the nucleus. This makes a difference because to an observer on the nucleus it appears that the coordinate system where the electron is at rest rotates with a frequency[6]

$$\boldsymbol{\omega}_T = (1/2c^2)\,(\mathbf{v}\times\mathbf{a}). \tag{28}$$

This is called the Thomas frequency. The acceleration **a** of the electron may be replaced by

$$\mathbf{a} = Ze^2\mathbf{r}/r^3m, \tag{29}$$

so that

$$\boldsymbol{\omega}_T = \frac{Ze^2}{2mc^2}\frac{\mathbf{v}\times\mathbf{r}}{r^3}. \tag{30}$$

This means that the electron experiences in addition to the magnetic field **H** of equation (26) a field

$$\mathbf{H}_T = \frac{Ze}{2c}\frac{\mathbf{v}\times\mathbf{r}}{r^3}. \tag{31}$$

Consequently it experiences a total magnetic field

$$\mathbf{H}' = \mathbf{H}+\mathbf{H}_T = -\frac{Ze}{2c}\frac{\mathbf{v}\times\mathbf{r}}{r^3} \tag{32}$$

and its energy is

$$E_s' = -\frac{Ze^2}{2mc^2}\frac{(\mathbf{v}\times\mathbf{r}).\mathbf{S}}{r^3}. \tag{33}$$

Equation (33) differs from equation (27) by a factor of one half, which is known as the Thomas factor. Finally we introduce the angular momentum

$$\mathbf{L} = m(\mathbf{r}\times\mathbf{v}) \tag{34}$$

of the electron and we write (33) as

$$E_s' = \frac{Ze^2}{2m^2c^2r^3}(\mathbf{L}.\mathbf{S}) = \xi(r)\,\mathbf{L}.\mathbf{S}. \tag{35}$$

This is the well-known expression for the spin-orbit coupling of an electron in a central force field. It is important to realise that equation (35) is neither valid for non-central force fields nor for many-electron systems.

It was shown by Thomas[6] and, subsequently, by Kramers[5] that the general expression for the spin energy of an electron in an electromagnetic field (**E**, **H**) is given by

$$E_s' = \frac{e}{mc}\mathbf{S}.\left\{\mathbf{H}+\frac{1}{2c}(\mathbf{E}\times\mathbf{v})\right\}. \tag{36}$$

This result is again consistent with the Dirac equation. We may combine this with equation (14) in order to obtain the total Hamiltonian H of an electron in an electromagnetic field. The field is given by the field strengths **E** and **H** or by the scalar potential Φ and the vector potential **A**. The Hamiltonian is

$$H = \frac{1}{2m}\left(\mathbf{p}+\frac{e}{c}\mathbf{A}\right)^2 - e\Phi + \frac{e}{mc}\mathbf{S}.\left\{\mathbf{H}+\frac{1}{2mc}(\mathbf{E}\times\mathbf{p})\right\}. \tag{37}$$

For example, this is the proper Hamiltonian for the hydrogen molecular ion if we take the electromagnetic field as the sum of the fields of the two nuclei and the exterior field.

Let us now generalise the Hamiltonian (37) to a molecule with N electrons, represented by the coordinates $\mathbf{r}_j$ and the momenta $\mathbf{p}_j$ and with N' nuclei, represented by the coordinates $\mathbf{R}_n$. We assume that the nuclei are at rest so that they give rise to an electric field $\mathbf{F}$ only. In addition, there is an exterior field, which is derived from the potentials Φ and $\mathbf{A}$ and which has an electric field strength $\mathbf{E}$ and a magnetic field strength $\mathbf{H}$. The total Hamiltonian for the molecule is then represented as

$$H = H_0 + H_{\text{int.}} + H_{\text{s.o.}} + H_{\text{s.s.}}. \tag{38}$$

Here H_0 is the zero-order Hamiltonian

$$H_0 = \frac{1}{2m}\sum_{i=1}^{N} p_i^2 + e^2\left[-\sum_{i=1}^{N}\sum_{n=1}^{N'}\frac{Z_n}{r_{n,i}} + \sum_{i>j}\frac{1}{r_{i,j}} + \sum_{n>k}\frac{Z_n Z_k}{R_{n,k}}\right]. \tag{39}$$

Here we have introduced

$$\mathbf{r}_{i,j} = \mathbf{r}_j - \mathbf{r}_i, \quad \mathbf{r}_{n,i} = \mathbf{r}_i - \mathbf{R}_n, \quad \mathbf{R}_{n,k} = \mathbf{R}_k - \mathbf{R}_n. \tag{40}$$

The electric charge on nucleus n is eZ_n.

The interaction between the orbital motion of the electrons and the exterior electromagnetic field is represented by

$$H_{\text{int.}} = \sum_j\left[\frac{e}{mc}\mathbf{A}_j\mathbf{p}_j + \frac{e^2}{2mc^2}A_j^2 - e\Phi_j\right]. \tag{41}$$

The spin-dependent parts of the Hamiltonian are $H_{\text{s.o.}}$, which is linear in the spin operators and which is called the spin-orbit interaction, and $H_{\text{s.s.}}$, which is quadratic in the spin operators and which is called the spin-spin interaction. We write the spin-orbit interaction as

$$H_{\text{s.o.}} = \frac{e}{mc}\sum_j \mathbf{S}_j \,.\, \mathbf{B}_j \tag{42}$$

with

$$\mathbf{B}_j = \mathbf{H}_j + \frac{1}{2mc}[\mathbf{E}_j \times \mathbf{p}_j] + \frac{1}{2mc}[\mathbf{F}_j \times \mathbf{p}_j] + \frac{e}{mc}\sum_{k\neq j} r_{j,k}^{-3}[(\mathbf{p}_k - \tfrac{1}{2}\mathbf{p}_j) \times \mathbf{r}_{j,k}]. \tag{43}$$

The term $H_{\text{s.s.}}$ is quadratic in the spin operators; it is called the spin-spin interaction since it represents the interactions between different electron spins. Its form is

$$H_{\text{s.s.}} = (e^2/m^2c^2)\sum_i\sum_{j>i} r_{i,j}^{-5}\{r_{i,j}^2(\mathbf{S}_i \,.\, \mathbf{S}_j) - 3(\mathbf{r}_{i,j} \,.\, \mathbf{S}_i)(\mathbf{r}_{i,j} \,.\, \mathbf{S}_j)\} - (8\pi e/3mc)\sum_i\sum_{j>i}(\mathbf{S}_i \,.\, \mathbf{S}_j)\,\delta(\mathbf{r}_{i,j}). \tag{44}$$

The second sum of equation (44) is called the Fermi contact potential; it can be derived from classical electromagnetic theory if the finite dimensions of the electrons are taken into consideration.

The Hamiltonian defined by (38) and subsequent equations has been derived also from quantum electrodynamics[7] by means of a relativistic expansion. Here it is assumed that the electron velocities v are small compared with the light velocity c so that it is possible to expand the Hamiltonian as a power series in (v/c). The first terms of this power series expansion constitute the Hamiltonian H of equation (38). Even though the Hamiltonian H is not complete, since there are always terms of higher order that should be added to it, there is no doubt that it is correct as long as $v \ll c$.

Most of the confusion in the literature with regard to the spin-orbit interaction stems from attempts to approximate (42) and (43) for atoms. We have seen in equation (35) that for the hydrogen atom the spin-orbit coupling can be expressed as the scalar product of **L** and **S**. This has led to the approximation

$$H_{\text{s.o.}} \approx \sum_{j=l}^{N} \xi(r_j)\, \mathbf{L}_j \,.\, \mathbf{S}_j \tag{45}$$

for an arbitrary atom. However, if we rewrite equation (43) for an atom, taking **E** and **H** equal to zero,

$$\mathbf{B}_j = (e/2mc)\Big[\Big(Zr_j^{-3} - \sum_{k \neq j} r_{j,k}^{-3}\Big)\mathbf{L}_j - 2\sum_{k \neq j} r_{j,k}^{-3}\mathbf{L}_k + \sum_{k \neq j} r_{j,k}^{-3}\{2(\mathbf{r}_j \times \mathbf{p}_k) - (\mathbf{r}_k \times \mathbf{p}_j)\}\Big] \tag{46}$$

we see what the shortcomings of this approximation are. Only the first term of (46) leads to an expression of the form (45), namely

$$H_{\text{s.o.}}^{0} = (e^2/2m^2c^2)\sum_j \Big(Zr_j^{-3} - \sum_{k \neq j} r_{j,k}^{-3}\Big)(\mathbf{L}_j \,.\, \mathbf{S}_j), \tag{47}$$

but it is not obvious why the other terms may be neglected. The second term of (46) gives a contribution

$$H_{\text{s.o.}} = (-e^2/m^2c^2)\sum_j \sum_{k \neq j} r_{j,k}^{-3}(\mathbf{L}_k \,.\, \mathbf{S}_j) \tag{48}$$

to the spin-orbit coupling. This is called the spin-other orbit coupling and it is sometimes taken into account in atomic theory. However $H_{\text{s.o.}}$ contains still an additional term

$$H_{\text{s.o.}}^{2} = (e^2/2m^2c^2)\sum_j \sum_{k \neq j} r_{j,k}^{-3}\{2(\mathbf{r}_j \times \mathbf{p}_k) - (\mathbf{r}_k \times \mathbf{p}_j)\} \,.\, \mathbf{S}_j, \tag{49}$$

which fits neither into (47) nor into (48).

In molecules the field **F** is not even central so that there is really no part of $H_{s.o.}$ that can be approximated by (45). Any attempt to reduce a molecular spin-orbit Hamiltonian to the customary atomic form is pointless and it will lead to unreliable results. We believe that such attempts are mainly responsible for the confusion as to the correct form of the molecular spin-orbit Hamiltonian.[3]

Singlet–triplet transitions

Spin-forbidden transitions have been observed in atoms, in ions and in molecules. We will limit the present discussion to organic, relatively large, molecules because atoms and diatomic molecules have been extensively reviewed. Apart from a few exceptions, the best known being O_2, all molecules are diamagnetic in their ground states; this means that these ground states must be singlet states. It may be derived from general theoretical considerations that some of the lower excited molecular states can be triplet states, but that the states with higher spin multiplicity have such high excitation energies that they fall outside the optical region. Consequently, the only spin-forbidden transitions that are observed in molecular spectroscopy are singlet-triplet (or triplet-singlet) transitions.

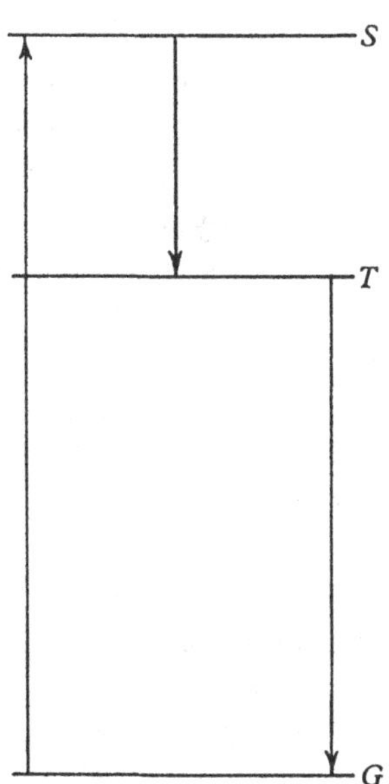

Fig. 2. Level diagram representing the mechanism of phosphorescent transitions in organic molecules.

The first observation of a triplet-singlet transition[8] in an organic molecule dates from as early as 1888. Wiedemann found at that time that a number of organic compounds (in particular quinine, aesculin and some dyes) in solid solution show a strong afterglow after excitation by a mercury lamp. Many similar observations followed and in 1944 Lewis & Kasha[9] made an extensive and systematic study of this type of phosphorescence and also offered an explanation for it, involving the molecular triplet state. The proposed mechanism is illustrated in fig. 2. Each molecule is originally in the singlet ground state G and under the influence of the ultraviolet mercury light it is excited to the singlet state S. A fraction of the molecules reaches the metastable state

T by means of a radiationless transition from the state S. Since the probability of the transition $T \rightarrow S$ is very small a considerable fraction of the molecules are in the state T. If we now shut off the mercury light we observe the transition $T \rightarrow G$ and we find that the intensity of this transition $I(t)$ decays exponentially as a function of time

$$I(t) = I(0)\, e^{-t/\tau}. \tag{50}$$

The time τ is known as the afterglow time of the phosphorescence and its inverse is the transition probability $w_{T \rightarrow S}$,

$$(1/\tau) = w_{T \rightarrow S}. \tag{51}$$

In aromatic hydrocarbons τ is of the order of 10 s; this means that the probability $w_{T \rightarrow S}$ is about 10^{10} smaller than a normal transition probability. Lewis & Kasha[9] proposed that the state T is the lowest excited triplet state of the molecule since only a spin-forbidden transition can have such a small probability in an organic molecule. This assumption was supported experimentally in 1955 when Evans[10] observed that the phosphorescent state of triphenylene is paramagnetic. The final confirmation was supplied by the electron spin measurements of Hutchison & Mangum[11] on the phosphorescent states of organic molecules.

It is necessary for all these experiments to be performed in solid solutions. Since the transition probabilities are so extremely small even small perturbations such as the intermolecular interactions between pairs of triplet molecules strongly affect the afterglow times and, consequently, the concentration of molecules in their phosphorescent state. Lewis & Kasha dissolved the phosphorescent compounds in EPA, that is a mixture of alcohol, ether and isopentane, at liquid air temperatures.

The values of triplet-singlet transition probabilities can be derived theoretically from the Hamiltonian (38). There are two mechanisms that lead to finite probabilities. The first one is usually negligible since it has been shown[12] to be too small even to account for the very small triplet-singlet intensities in aromatic hydrocarbons, but we will briefly mention it for the sake of completeness.

We consider again a radiation field that is described by the vector potential $\mathbf{A}$ of equation (8). It follows from equation (13) that the electric and magnetic field strengths of the radiation field are given by

$$\left.\begin{aligned} \mathbf{E} &= A_0 \mathbf{e}(2\pi i\nu/c)\{\gamma\, e^{-2\pi i\nu t} - \gamma^*\, e^{-2\pi i\nu t}\}, \\ \mathbf{H} &= A_0 i(\boldsymbol{\sigma} \times \mathbf{e})\{\gamma\, e^{-2\pi i\nu t} - \gamma^*\, e^{2\pi i\nu t}\}. \end{aligned}\right\} \tag{52}$$

By substituting this electromagnetic field into equation (43) we find that its interaction energy with the spins is given by

$$H'_s = (eA_0 i/mc) \sum_j \{\gamma_j e^{-2\pi i\nu t} - \gamma_j^* e^{2\pi i\nu t}\} \{(\boldsymbol{\sigma} \times \mathbf{e}) + (\pi\nu/mc^2)(\mathbf{e} \times \mathbf{p}_j)\} . \mathbf{S}_j. \tag{53}$$

The total interaction between the radiation field and the molecule is obtained as the sum of H'_s and of the operator H'_1 of equation (15). Since the operator H'_s contains the spin operators it has non-vanishing matrix elements between singlet and triplet functions, ${}^1\Psi_0$ and ${}^3\Psi_f$, respectively. However, these matrix elements are so small that they are usually negligible.[12]

Let us now discuss the second mechanism, which is generally regarded as being responsible for triplet-singlet transitions. We observe that the functions ${}^1\Psi_k$ and ${}^3\Psi_{n,i}$, which we have used so far for the calculations of transition probabilities, are eigenfunctions of the Hamiltonian H_0. However, a more accurate Hamiltonian is

$$H_m = H_0 + H_{\text{s.o.}} + H_{\text{s.s.}} = H_0 + H', \tag{54}$$

and for the calculation of triplet-singlet transition probabilities we ought to use the eigenfunctions of H_m instead of the eigenfunctions of H_0. Because the correction terms $H_{\text{s.o.}}$ and $H_{\text{s.s.}}$ are very small, we derive the eigenvalues ${}^1\epsilon_k$ and ${}^3\epsilon_{n,j}$ and the eigenfunctions ${}^1\Phi_k$ and ${}^3\Phi_{n,j}$ of H_m by means of perturbation theory from the eigenvalues ${}^1E_k$ and ${}^3E_n$ and the eigenfunctions ${}^1\Psi_k$ and ${}^3\Psi_{n,i}$ of the operator H_0.

The perturbed eigenfunction of the singlet state (0) is given by

$$\begin{aligned} {}^1\Phi_0 = {}^1\Psi_0 - \sum_{k \neq 0} ({}^1E_k - {}^1E_0)^{-1} \langle {}^1\Psi_k | H_{\text{s.o.}} + H_{\text{s.s.}} | {}^1\Psi_0 \rangle {}^1\Psi_k \\ - \sum_n \sum_i ({}^3E_n - {}^1E_0)^{-1} \langle {}^3\Psi_{n,i} | H_{\text{s.o.}} + H_{\text{s.s.}} | {}^1\Psi_0 \rangle {}^3\Psi_{n,i}. \end{aligned} \tag{55}$$

The derivation of the perturbed triplet functions is slightly more complicated because of the threefold degeneracy. It is convenient to make use of the variational method and to write the perturbed functions of the triplet state t as

$${}^3\Psi_t = \sum_i a_i \, {}^3\Psi_{t,i} + \sum_{n \neq t} \sum_j b_{n,j} \, {}^3\Psi_{n,j} + \sum_k c_k \, {}^1\Psi_k. \tag{56}$$

The variational equations for the coefficients are then

$$\begin{aligned} \sum_i (\langle {}^3\Psi_{t,i'} | H_m | {}^3\Psi_{t,i} \rangle - \epsilon \delta_{i,i'}) a_i + \sum_{n \neq t} \sum_j \langle {}^3\Psi_{t,i'} | H_m | {}^3\Psi_{n,j} \rangle b_{n\,j} \\ + \sum_k \langle {}^3\Psi_{t,i'} | H_m | {}^1\Psi_k \rangle c_k = 0 \quad (i' = 1, 2, 3), \end{aligned} \tag{57a}$$

$$\sum_i \langle {}^3\Psi_{m,j'}|H_m|{}^3\Psi_{t,i}\rangle a_i + \sum_{n\neq t}\sum_j (\langle {}^3\Psi_{m,j'}|H_m|{}^3\Psi_{n,j}\rangle - \epsilon\delta_{m,n}\delta_{j,j'})\, b_{n,j} + \sum_k \langle {}^3\Psi_{m,j'}|H_m|{}^1\Psi_k\rangle c_k = 0 \quad (m \neq t), \tag{57b}$$

$$\sum_i \langle {}^1\Psi_l|H_m|{}^3\Psi_{t,i}\rangle a_i + \sum_{n\neq t}\sum_j \langle {}^1\Psi_l|H_m|{}^3\Psi_{n,j}\rangle b_{n,j} + \sum_k (\langle {}^1\Psi_l|H_m|{}^1\Psi_k\rangle - \epsilon\delta_{l,k})\, c_k = 0. \tag{57c}$$

We may assume that ϵ is very close to ${}^3E_t$ and that the coefficients $b_{n,j}$ and c_k are much smaller than unity. Consequently, it may be derived from equations (57b) and (57c) that

$$\left.\begin{aligned} b_{n,j} &= -({}^3E_n - {}^3E_t)^{-1} \sum_i \langle {}^3\Psi_{n,j}|H'|{}^3\Psi_{t,i}\rangle a_i, \\ c_k &= -({}^1E_k - {}^3E_t)^{-1} \sum_i \langle {}^1\Psi_k|H'|{}^3\Psi_{t,i}\rangle a_i. \end{aligned}\right\} \tag{58}$$

We substitute this result into equation (57a) and we also replace ϵ by

$$\epsilon = {}^3E_t + \Delta. \tag{59}$$

The result is

$$\left.\begin{aligned} &\sum_i (\Omega_{i',i} - \Delta\delta_{i',i})\, a_i = 0, \\ \Omega_{i',i} = {} & \langle {}^3\Psi_{t,i'}|H'|{}^3\Psi_{t,i}\rangle - \sum_{n\neq t}\sum_j ({}^3E_n - {}^3E_t)^{-1} \\ & \times \langle {}^3\Psi_{t,i'}|H'|{}^3\Psi_{n,j}\rangle\langle {}^3\Psi_{n,j}|H'|{}^3\Psi_{t,i}\rangle - \sum_k ({}^1E_k - {}^3E_t)^{-1} \\ & \times \langle {}^3\Psi_{t,i'}|H'|{}^1\Psi_k\rangle\langle {}^1\Psi_k|H'|{}^3\Psi_{t,i}\rangle. \end{aligned}\right\} \tag{60}$$

We denote the eigenvalues of the Hermitian matrix $\Omega_{i',i}$ by Δ_σ and the corresponding eigenvectors by a_i^σ. The perturbed triplet eigenvalues and the corresponding eigenfunctions are then

$$\left.\begin{aligned} {}^3\epsilon_{t,\sigma} &= {}^3E_t + \Delta_\sigma, \\ {}^3\Phi_{t,\sigma} &= \sum_i a_i^\sigma\, {}^3\tilde{\Phi}_{t,i}, \\ {}^3\tilde{\Phi}_{t,i} &= \Psi_{t,i} - \sum_{n\neq t}\sum_j \frac{\langle {}^3\Psi_{n,j}|H'|{}^3\Psi_{t,i}\rangle\, {}^3\Psi_{n,j}}{{}^3E_n - {}^3E_t} \\ &\quad - \sum_k \frac{\langle {}^1\Psi_k|H'|{}^3\Psi_{t,i}\rangle\, {}^1\Psi_k}{{}^1E_k - {}^3E_t}. \end{aligned}\right\} \tag{61}$$

The probability of a triplet-singlet transition is evaluated by substituting the perturbed functions (55) and (61) into equation (17)

$$H_{t\sigma,0} = \langle {}^3\Phi_{t,\sigma}|\sum_j (\mathbf{e}\cdot\nabla_j)|{}^1\Phi_0\rangle; \tag{62}$$

where we have taken γ_j equal to unity. It follows from our perturbation treatment that the eigenfunctions of the operator H_m are no longer pure singlet or triplet functions. As a result of the perturbation small amounts of triplet functions are mixed with the singlet functions, and vice versa. The matrix element (62) is therefore different from zero and we obtain a finite triplet-singlet transition probability.

Usually it is difficult to distinguish experimentally between the three different triplet states and we calculate the three matrix elements

$$H'_{ti,0} = \langle {}^3\tilde{\Phi}_{t,i}|\sum_j (\mathbf{e}\,.\,\nabla_j)|{}^1\Phi_0\rangle. \tag{63}$$

It follows from equations (55) and (61) that

$$H'_{ti,0} = -\sum_n ({}^3E_n - {}^1E_0)^{-1}\langle {}^3\Psi_{n,i}|H'|{}^1\Psi_0\rangle\langle {}^3\Psi_{t,i}|\sum_j (\mathbf{e}\,.\,\nabla_j)|{}^3\Psi_{n,i}\rangle$$
$$-\sum_k ({}^1E_k - {}^3E_t)^{-1}\langle {}^1\Psi_k|H'|{}^3\Psi_{t,i}\rangle\langle {}^1\Psi_k|\sum_j (\mathbf{e}\,.\,\nabla_j)|{}^1\Phi_0\rangle. \tag{64}$$

We define

$$\left.\begin{aligned} {}^1\mathbf{P}_{k0} &= \langle {}^1\Psi_k|e\sum_j \mathbf{r}_j|{}^1\Psi_0\rangle, \\ {}^3\mathbf{P}_{tn} &= \langle {}^3\Psi_{t,i}|e\sum_j \mathbf{r}_j|{}^3\Psi_{n,i}\rangle \quad (i = -1, 0, 1). \end{aligned}\right\} \tag{65}$$

Furthermore, we note that in the matrix elements in (64) we may replace H' by $H_{\text{s.o.}}$ since the contributions of the spin-spin interactions are negligible. Hence, equation (64) may be written as

$$H'_{ti,0} = \frac{2\pi m}{e\hbar^2}\left[\sum_n \frac{{}^3E_t - {}^3E_n}{{}^3E_n - {}^1E_0}\langle {}^3\Psi_{n,i}|H_{\text{s.o.}}|{}^1\Psi_0\rangle(\mathbf{e}\,.\,{}^3\mathbf{P}_{tn})\right.$$
$$\left.+\sum_k \frac{{}^1E_k - {}^1E_0}{{}^1E_k - {}^3E_t}\langle {}^1\Psi_k|H_{\text{s.o.}}|{}^3\Psi_{t,i}\rangle^*(\mathbf{e}\,.\,{}^1\mathbf{P}_{k0})\right]. \tag{66}$$

The transition moments may be derived from the absorption spectrum of the molecule and the calculation of the triplet-singlet transition probability reduces to the evaluation of the matrix elements of the spin-orbit interaction. It is customary to take into account only a few states that have energies in the vicinity of ${}^1E_0$ and ${}^3E_t$ since it is practically impossible to consider all excited states.

The experimental values of the triplet-singlet transition probabilities of organic molecules are usually obtained from the phosphorescent afterglow times. It is very difficult to observe singlet-triplet transitions in absorption since the probabilities are so small that the absorption bands are usually obliterated by the broad singlet-singlet absorption bands in the vicinity.

Spin-orbit interaction in molecules

For a numerical evaluation of the matrix elements of $H_{\text{s.o.}}$ in a molecule we have to use approximate molecular eigenfunctions; we will assume that these approximate wavefunctions are antisymmetrised products of one-electron functions. In this approximation the non-degenerate singlet ground state of the molecule may be written as

$$\begin{aligned}{}^1\Psi_0 = \{(2N)!\}^{-\frac{1}{2}} \sum_P P\delta_P[&\eta_1(1)\,\eta_1(2)\,\alpha(1)\,\beta(2)\,\eta_2(3)\,\eta_2(4)\,\alpha(3)\,\beta(4)\\ &\ldots\eta_k(2k-1)\,\eta_k(2k)\,\alpha(2k-1)\,\beta(2k)\\ &\ldots\eta_N(2N-1)\,\eta_N(2N)\,\alpha(2N-1)\,\beta(2N)], \quad (67)\end{aligned}$$

where the one-electron functions η_k are known as molecular orbitals. The only excited states we consider are obtained by transferring an electron from an orbital η_l to an excited η_n. The excited singlet and triplet states that are obtained in this way have wavefunctions

$$\begin{aligned}{}^1\Psi_{l,n} = \{2(2N)!\}^{-\frac{1}{2}} \sum_P P\delta_P[&\prod_{k\neq l} \{\eta_k(2k-1)\,\eta_k(2k)\alpha(2k-1)\,\beta(2k)\}\\ &\times\{\eta_l(2l-1)\,\eta_n(2l)+\eta_n(2l-1)\,\eta_l(2l)\}\,\alpha(2l-1)\,\beta(2l)], \quad (68)\end{aligned}$$

$$\begin{aligned}{}^3\Psi_{l,n,i} = \{(2N)!\}^{-\frac{1}{2}} \sum_P P\delta_P[&\prod_{k\neq l} \{\eta_k(2k-1)\,\eta_k(2k)\alpha(2k-1)\,\beta(2k)\}\\ &\times \eta_l(2l-1)\,\eta_n(2l)\,{}^3\zeta_i(2l-1,2l)], \quad (69)\end{aligned}$$

with

$$\left.\begin{aligned}{}^3\zeta_1(l,n) &= \alpha(l)\,\alpha(n),\\ {}^3\zeta_0(l,n) &= 2^{-\frac{1}{2}}\{\alpha(l)\,\beta(n)+\beta(l)\,\alpha(n)\},\\ {}^3\zeta_{-1}(l,n) &= \beta(l)\,\beta(n).\end{aligned}\right\} \quad (70)$$

The evaluation of the spin-orbit matrix elements involves two steps: the first one is to express the matrix elements in terms of one- or two-electron integrals over the molecular orbitals, the second one is the calculation of these integrals. We will illustrate the first step for a simple case, a four-electron system. It will be seen that even for this small number of electrons we have to consider almost 50 000 terms at the beginning of the derivation and we have to be very systematic in order to cope with such a large number of terms and to reduce them to a convenient form.

As an example we calculate the matrix element of the spin-orbit interaction between a singlet state (0) with a wavefunction

$${}^1\Psi_0 = (24)^{-\frac{1}{2}} \sum_P P\delta_P[a_1\alpha_1 a_2\beta_2 b_3\alpha_3 b_4\beta_4], \quad (71)$$

and a triplet state t with a wavefunction

$$^{3}\Psi_{t,i} = (24)^{-\frac{1}{2}} \sum_{P} P\delta_P[a_1\alpha_1 a_2\beta_2 b_3 c_4 {}^{3}\zeta_{i,34}]. \tag{72}$$

For the sake of brevity we denote the electron coordinates by subscripts. The matrix elements are

$$\begin{aligned} H_{t,i} &= (24)^{-1}\langle \sum_{P} P\delta_P[a_1\alpha_1 a_2\beta_2 b_3 c_4 {}^{3}\zeta_{i,34}] \,|(e/mc) \sum_{j} \mathbf{S}_j \,.\, \mathbf{B}_j| \\ &\qquad \times \sum_{P} P\delta_P[a_1\alpha_1 a_2\beta_2 b_3\alpha_3 b_4\beta_4]\rangle \\ &= (e/mc)\,\langle \sum_{P} P\delta_P[a_1\alpha_1 a_2\beta_2 b_3 c_4 {}^{3}\zeta_{i,34}] \sum_{j} \mathbf{S}_j \,.\, \mathbf{B}_j\, |a_1\alpha_1 a_2\beta_2 b_3\alpha_3 b_4\beta_4\rangle. \end{aligned} \tag{73}$$

By means of the above transformation we have reduced the number of terms by a factor 24.

Let us consider first the case $i = 0$. Here we have exactly the same number of α and β spins on the left- and right-hand sides of the operator. Consequently, only the z components of the spin operators lead to a non-zero result and we may write

$$\begin{aligned} H_{t,0} &= (e/mc)\, 2^{-\frac{1}{2}}\, \langle \sum_{P} P\delta_P[a_1\alpha_1 a_2\beta_2 b_3 c_4(\alpha_3\beta_4 + \beta_3\alpha_4)] \\ &\qquad \times |S_{1z}B_{1z} + S_{2z}B_{2z} + S_{3z}B_{3z} + S_{4z}B_{4z}|\, a_1\alpha_1 a_2\beta_2 b_3\alpha_3 b_4\beta_4\rangle. \end{aligned} \tag{74}$$

The operation of the spins gives

$$\begin{aligned} H_{t,0} &= (e\hbar/2mc\surd 2)\,\langle \sum_{P} P\delta_P[a_1\alpha_1 a_2\beta_2 b_3 c_4(\alpha_3\beta_4 + \beta_3\alpha_4)] \\ &\qquad \times |B_{1z} - B_{2z} + B_{3z} - B_{4z}|\, a_1\alpha_1 a_2\beta_2 b_3\alpha_3 b_4\beta_4\rangle. \end{aligned} \tag{75}$$

We have to consider only those permutations of the triplet functions that do not cause the spin functions to become orthogonal, so that

$$\begin{aligned} H_{t,0} = (e\hbar/2mc\surd 2)\,\langle & a_1 a_2 b_3 c_4 - b_1 a_2 a_3 c_4 - a_1 c_2 b_3 a_4 + b_1 c_2 a_3 a_4 \\ & - a_1 a_2 c_3 b_4 + c_1 a_2 a_3 b_4 + a_1 b_2 c_3 a_4 \\ & - c_1 b_2 a_3 a_4\, |B_{1z} - B_{2z} + B_{3z} - B_{4z}|\, a_1 a_2 b_3 b_4\rangle. \end{aligned} \tag{76}$$

For further transformations we make use of the specific form of the operators B_{jz}. In agreement with equation (43) we write

$$\left.\begin{aligned} (e\hbar/2mc)\, B_{jz} &= L_j + \sum_{i \neq j} \Omega_{j\ i}, \\ L_j &= (e\hbar/2mc)\,\{H_{jz} + (1/2mc)\,(E_{jx}p_{jy} - E_{jy}p_{jx}) \\ &\qquad + (1/2mc)\,(F_{jx}p_{jy} - F_{jy}p_{jx})\}, \\ \Omega_{j,i} &= (e^2\hbar/2m^2c^2)\, r_{j,i}^{-3} \\ &\qquad \{(p_{ix} - \tfrac{1}{2}p_{jx})\,(y_i - y_j) - (p_{iy} - \tfrac{1}{2}p_{jy})\,(x_i - x_j)\}. \end{aligned}\right\} \tag{77}$$

We note here that L_j depends on one set of electron coordinates and that $\Omega_{j,i}$ depends on two sets. Consequently $H_{t,0}$ may be reduced to

$$H_{t,0} = \sqrt{(2)}\,[-\langle c|L|b\rangle + \langle b_1c_2|\Omega_{1,2}-\Omega_{2,1}|b_1b_2\rangle$$
$$-2\langle a_1c_2|\Omega_{2,1}|a_1b_2\rangle + \langle c_1a_2|\Omega_{1,2}+\Omega_{2,1}|a_1b_2\rangle]. \quad (78)$$

We consider next the case $i = 1$, the matrix element is then

$$H_{t,1} = (e/mc)\,\langle \sum_P P\delta_P[a_1\alpha_1a_2\beta_2b_3c_4\alpha_3\alpha_4]$$
$$\times \Sigma\mathbf{S}_j\,.\,\mathbf{B}_j\,|a_1\alpha_1a_2\beta_2b_3\alpha_3b_4\beta_4|\rangle. \quad (79)$$

We observe that on the left-hand side there are three α spins and one β spin and consequently we obtain a non-zero result only when an operator S_{jx} or S_{jy} works upon a β spin. Hence

$$H_{t,1} = (e\hbar/2mc)[\langle \sum_P P\delta_P\{a_1\alpha_1a_2\beta_2b_3c_4\alpha_3\alpha_4\}$$
$$\times|B_{2,x}-iB_{2,y}|\,a_1\alpha_2a_2\alpha_2b_3\alpha_3b_4\beta_4\rangle + \langle \sum_P P\delta_P\{a_1\alpha_1a_2\beta_2b_3c_4\alpha_3\alpha_4\}$$
$$\times|B_{4,x}-B_{4,y}|a_1\alpha_1a_2\beta_2b_3\alpha_3b_4\alpha_4\rangle], \quad (80)$$

or

$$H_{t,1} = (e\hbar/2mc)[\langle -a_4 \sum_P P\delta_P\{a_1c_2b_3\}\,|B_{2,x}-iB_{2,y}|\,a_1a_2b_3b_4\rangle$$
$$+\langle a_2 \sum_P P\delta_P\{a_1b_3c_4\}\,|B_{4,x}-iB_{4,y}|\,a_1a_2b_3b_4\rangle]. \quad (81)$$

We substitute

$$(e\hbar/2mc)\,(B_{j,x}-iB_{j,y}) = L'_j + \sum_{k\neq j} \Omega'_{j,k}, \quad (82)$$

where

$$\left.\begin{aligned} L'_j &= (e\hbar/2mc)\,[H_{jx}-iH_{jy}) + (1/2mc)\,\{(\mathbf{E}_j\times\mathbf{p}_j)_x \\ &\quad -i(\mathbf{E}_j\times\mathbf{p}_j)_y + (\mathbf{F}_j\times\mathbf{p}_j)_x - i(\mathbf{F}_j\times\mathbf{p}_j)_y\;, \\ \Omega'_{j,k} &= (e^2\hbar/2m^2c^2)\,r_{j,k}^{-3}[\{(\mathbf{p}_k-\tfrac{1}{2}\mathbf{p}_j) \\ &\quad \times(\mathbf{r}_k-\mathbf{r}_j)\}_x - i\{(\mathbf{p}_k-\tfrac{1}{2}\mathbf{p}_j)\times(\mathbf{r}_k\times\mathbf{r}_j)\}_y]. \end{aligned}\right\} \quad (83)$$

The result is

$$H_{t,1} = \langle c|\,L'\,|b\rangle - \langle b_1c_2|\,\Omega'_{1,2}-\Omega'_{2,1}\,|b_1b_2\rangle$$
$$+2\,\langle a_1c_2|\,\Omega'_{2,1}\,|a_1b_2\rangle - \langle c_1a_2|\,\Omega'_{2,1}+\Omega'_{1,2}\,|a_1b_2\rangle. \quad (84)$$

The matrix element $H_{t,-1}$ is also known because

$$H_{t,-1} = H^*_{t,1}. \quad (85)$$

The results (78) and (84) are easily generalised to the case of $2N$ electrons, where the wavefunctions are given by equations (67) and (69). We have

$$\begin{aligned}\langle {}^3\Psi_{l,n,0}|\, H_{\text{s.o.}}\, |{}^1\Psi_0\rangle \\ &= -\sqrt{(2)}\,[\langle\eta_n|\, L\,|\eta_l\rangle + 2\sum_{k\neq l}\langle\eta_k(1)\,\eta_n(2)|\,\Omega_{2,1}\,|\eta_k(1)\,\eta_l(2)\rangle \\ &\quad -\sum_{k\neq l}\langle\eta_n(1)\,\eta_k(2)|\,\Omega_{1,2}+\Omega_{2,1}\,|\eta_k(1)\,\eta_l(2)\rangle \\ &\quad -\langle\eta_l(1)\,\eta_n(2)|\,\Omega_{1,2}-\Omega_{2,1}\,|\eta_l(1)\,\eta_l(2)\rangle]. \qquad (86)\end{aligned}$$

The expressions for $\langle {}^3\Psi_{l,n,\pm 1}|\, H_{\text{s.o.}}\, |{}^1\Psi_0\rangle$ are derived in a similar fashion from equation (84).

Equation (86) can be simplified further. We assume that all molecular orbitals are real functions. In that case

$$\langle\eta_k(i)\,\eta_n(j)|\, r_{i,j}^{-3}\{p_{ix}(y_i-y_j)-p_{iy}(x_i-x_j)\}\,|\eta_k(i)\,\eta_l(j)\rangle = 0, \qquad (87)$$

$$\begin{aligned}\langle\eta_k(1)\,\eta_n(2)|\,\Omega_{2,1}\,|\eta_k(1)\,\eta_l(2)\rangle &= (e^2\hbar/4m^2c^2) \\ &\langle\eta_k(1)\,\eta_n(2)|\, r_{12}^{-3}\{(x_1-x_2)\,p_{2y}-(y_1-y_2)\,p_{2x}\}\,|\eta_k(1)\,\eta_l(2)\rangle. \qquad (88)\end{aligned}$$

We observe that the integral

$$\mathbf{F}^k(2) = e\int\eta_k^*(1)\,\eta_k(1)\,\{r_{12}^{-3}(\mathbf{r}_1-\mathbf{r}_2)\}\,d\mathbf{r}_1 \qquad (89)$$

represents just the electric field at the position of electron 2 due to an electron in the molecular orbital η_k. We substitute (89) into (88) and we obtain

$$\langle\eta_k(1)\,\eta_n(2)|\,\Omega_{2,1}\,|\eta_k(1)\,\eta_l(2)\rangle = (e\hbar/4m^2c^2)\,\langle\eta_n|\,(F_x^k p_y - F_y^k p_x)\,|\eta_l\rangle. \qquad (90)$$

The last term of equation (86) can be transformed in a similar fashion

$$\begin{aligned}-\langle\eta_l(1)\,\eta_n(2)|\,\Omega_{1,2}-\Omega_{2,1}\,|\eta_l(1)\,\eta_l(2)\rangle \\ &= (e^2\hbar/4m^2c^2)\,\langle\eta_l(1)\,\eta_n(2)|\, r_{12}^{-3}\{(\mathbf{p}_1+\mathbf{p}_2)\times(\mathbf{r}_1-\mathbf{r}_2)\}_z\,|\eta_l(1)\,\eta_n(2)\rangle \\ &= (e^2\hbar/4m^2c^2)\,\langle\eta_l(1)\,\eta_n(2)|\, r_{12}^{-3}\{\mathbf{p}_2\times(\mathbf{r}_1-\mathbf{r}_2)\}_z\,|\eta_l(1)\,\eta_n(2)\rangle \\ &= (e\hbar/4m^2c^2)\,\langle\eta_n|\, F_x^l p_y - F_y^l p_x\,|\eta_l\rangle. \qquad (91)\end{aligned}$$

By making use of equations (90) and (91) we transform equation (86) to

$$\begin{aligned}\langle {}^3\Psi_{l,n,0}|\, H_{\text{s.o.}}\, |{}^1\Psi_0\rangle \\ &= -\sqrt{2}(e\hbar/4m^2c^2)\,\langle\eta_n|\,[(\mathbf{F}+2\sum_{k\neq l}\mathbf{F}^k-\mathbf{F}^l)\times\mathbf{p}]_z\,|\eta_l\rangle \\ &\quad -\sum_{k\neq l}\langle\eta_n(1)\,\eta_k(2)|\,\Omega_{1,2}+\Omega_{2,1}\,|\eta_k(1)\,\eta_l(2)\rangle, \qquad (92)\end{aligned}$$

where we assume that the exterior electromagnetic field is zero,

$$\mathbf{E} = \mathbf{H} = 0.$$

The first term of equation (92) contains the effective electric field

$$\mathbf{F}_{\text{eff.}} = \mathbf{F} + 2 \sum_{k \neq l} \mathbf{F}_k \tag{93}$$

due to the nuclei and to all electron pairs which have opposite spins in the triplet state. The sum that occurs in equation (92) contains what may be called exchange terms. We see that the first term of equation (92) may be expressed in the form (45) and our result gives therefore some justification for equation (45) as long as the exchange terms are negligible. However, this latter assumption is not always valid.

Numerical evaluation of spin-orbit interactions

For a numerical calculation of the spin-orbit matrix elements we must know the molecular orbitals η_k that were introduced in the previous section. The customary approximation consists of taking η_k as a linear combination of atomic orbitals

$$\eta_k = \sum_{n=1}^{N'} a_{k,n} \phi_{j,n}, \tag{94}$$

where the summation is to be performed over all nuclei and the subscript j describes the type of atomic orbital. Substitution of (94) into (92) leads to a linear combination of one- and two-electron integrals, containing the atomic orbitals $\phi_{j,n}$. We distinguish between one-, two-, three-, and four-center integrals, depending on how many different nuclei the atomic orbitals in the integral are centered.

We believe that the most convenient approach to the evaluation of these integrals is by means of the Fourier convolution theorem. This method was first suggested by Prosser & Blanchard[13, 14] and it was used extensively by Geller.[15] We derive the theorem for a two-electron, two-center integral; from this derivation it follows easily how other types of integrals ought to be treated.

The integral

$$I(\mathbf{R}) = \iint f(\mathbf{r}_{1,a})\, g(\mathbf{r}_{1,2})\, h(\mathbf{r}_{2,b})\, d\mathbf{r}_1\, d\mathbf{r}_2 \tag{95}$$

with

$$\left.\begin{aligned} \mathbf{r}_{1,a} &= \mathbf{r}_1 - \mathbf{r}_a, & \mathbf{r}_{2,b} &= \mathbf{r}_2 - \mathbf{r}_b, \\ \mathbf{r}_{1,2} &= \mathbf{r}_2 - \mathbf{r}_1, & \mathbf{R} &= \mathbf{r}_b - \mathbf{r}_a, \end{aligned}\right\} \tag{96}$$

is first transformed to

$$I(\mathbf{R}) = \iint f(\mathbf{r}_1)\, g(\mathbf{r}_2 - \mathbf{r}_1)\, h(\mathbf{r}_2 - \mathbf{R})\, d\mathbf{r}_1\, d\mathbf{r}_2 \tag{97}$$

by taking the origins for the vectors $\mathbf{r}_1$ and $\mathbf{r}_2$ as $\mathbf{r}_a$. Here f, g and h are arbitrary three-dimensional functions, which have Fourier transforms F, G and H,

$$\left.\begin{aligned}\mathbf{F}(\mathbf{k}) &= \int f(\mathbf{s})\, e^{i\mathbf{k}.\mathbf{s}}\, d\mathbf{s},\\ \mathbf{G}(\mathbf{k}) &= \int g(\mathbf{s})\, e^{i\mathbf{k}.\mathbf{s}}\, d\mathbf{s},\\ \mathbf{H}(\mathbf{k}) &= \int h(\mathbf{s})\, e^{i\mathbf{k}.\mathbf{s}}\, d\mathbf{s}.\end{aligned}\right\} \tag{98}$$

It follows from the Fourier integral theorem that

$$\left.\begin{aligned}f(\mathbf{r}_1) &= (2\pi)^{-3}\int F(\mathbf{k})\exp\{-i(\mathbf{r}_1.\mathbf{k})\}\, d\mathbf{k},\\ g(\mathbf{r}_2-\mathbf{r}_1) &= (2\pi)^{-3}\int G(\mathbf{k})\exp\{-i(\mathbf{r}_2-\mathbf{r}_1).\mathbf{k}\}\, d\mathbf{k},\\ h(\mathbf{r}_2-\mathbf{R}) &= (2\pi)^{-3}\int H(\mathbf{k})\exp\{-i(\mathbf{r}_2-\mathbf{R}).\mathbf{k}\}\, d\mathbf{k}.\end{aligned}\right\} \tag{99}$$

Substitution of (99) into (97) gives

$$I(\mathbf{R}) = (2\pi)^{-9}\int\int\int F(\mathbf{k}_1)G(\mathbf{k}_2)H(\mathbf{k}_3)\, d\mathbf{k}_1 d\mathbf{k}_2 d\mathbf{k}_3 \int\int \exp\{-i\mathbf{r}_1.(\mathbf{k}_1+\mathbf{k}_2)\} \times \exp\{-i\mathbf{r}_2.(\mathbf{k}_2+\mathbf{k}_3)\}\exp\{i\mathbf{R}.\mathbf{k}_3\}\, d\mathbf{r}_1 d\mathbf{r}_2. \tag{100}$$

We observe now that the integrals over $\mathbf{r}_1$ and $\mathbf{r}_2$ are representations of the three-dimensional δ-function,

$$(2\pi)^{-3}\int \exp\{-i\mathbf{r}.(\mathbf{k}+\mathbf{k}')\}\, d\mathbf{r} = \delta(\mathbf{k}+\mathbf{k}') = \delta(k_x+k_x')\,\delta(k_y+k_y')\,\delta(k_z+k_z'). \tag{101}$$

Equation (100) may, therefore, be written as

$$I(\mathbf{R}) = (2\pi)^{-3}\int\int\int F(\mathbf{k}_1)G(\mathbf{k}_2)H(\mathbf{k}_3)\exp\{i\mathbf{R}.\mathbf{k}_3) \times \delta(\mathbf{k}_1+\mathbf{k}_2)\,\delta(\mathbf{k}_2+\mathbf{k}_3)\, d\mathbf{k}_1 d\mathbf{k}_2 d\mathbf{k}_3, \tag{102}$$

or as

$$I(\mathbf{R}) = (2\pi)^{-3}\int F(\mathbf{k})\, G(-\mathbf{k})\, H(\mathbf{k})\, e^{i\mathbf{k}.\mathbf{R}}\, d\mathbf{k}. \tag{103}$$

The above theorem was used for the evaluation of a large number of atomic integrals,[16, 17] it was found that the one-center and most of the two-center integrals can be obtained in analytical form. Tables of the necessary Fourier transforms $F(\mathbf{k})$, $G(\mathbf{k})$ and $H(\mathbf{k})$ have been published[16, 17] and the integrals of the type (103) are easily evaluated with the aid of an electronic computer.[18]

It is our experience that the one-center integrals are much larger than the two-center integrals and that the latter are much larger than the three- and four-center integrals. In general it may be predicted that the spin-orbit matrix element is very small if all the one-center atomic integrals that it contains are zero or cancel. It is therefore important to consider, in addition to the total symmetry of the molecule, the local symmetry at each atom in the molecule in order to make qualitative predictions about the magnitudes of the spin-orbit matrix elements. From the total molecular symmetry we derive whether or not a given matrix element is zero, but we cannot predict how large the matrix element should be if it is different from zero. The latter information can be obtained by considering the local symmetry.

Let us consider, for example, the aromatic hydrocarbons and the azines. In the case of an aromatic hydrocarbon we calculate the triplet-singlet transition probability from matrix elements between states that differ only in the occupancy of the π-orbitals. It follows from the local symmetry that all one- and two-center integrals that occur in the matrix elements are zero and, consequently, that the spin-orbit matrix elements are very small. In the case of an azine, for example, pyrazine, we calculate the triplet-singlet transition probability from a spin-orbit matrix element between a triplet state where the two unpaired electrons are both in π-orbitals and an excited singlet state that is obtained from the ground state by transferring an electron from a lone-pair orbital on the nitrogen to an excited π-orbital. Here it follows from the local symmetry that the spin-orbit matrix element contains in principle a non-zero one-center atomic integral. We conclude that the triplet-singlet transition probabilities are much larger in azines than in aromatic hydrocarbons. This agrees with the experimental observation that the phosphorescence lifetimes for aromatic hydrocarbons are a thousand to ten thousand times larger than the lifetimes for the azines.

The molecule whose spin-orbit interaction has been most extensively investigated[19, 20, 21] is benzene. Here the phosphorescent triplet state has some unusual properties that give rise to additional complications. The first difficulty is that the lowest triplet state cannot be represented by a single Slater determinant as its wavefunction, instead we have to take the sum or difference of two Slater determinants. Consequently equation (86) does not apply to benzene, however it can be amended fairly easily. The other difficulty in the case of benzene is of a more serious nature. If we recall that the triplet-singlet transition probability is given by equation (66) then we find that for benzene every singlet state k that is mixed with the triplet state t either has a zero transition moment ${}^1P_{k0}$ or a zero matrix element $\langle {}^1\Psi_k | H_{\text{s.o.}} | {}^3\Psi_{t,i} \rangle$ if vibrations are not taken into account. The same difficulty occurs with the triplet states (n, i) that are mixed with the singlet state 0. It follows thus that the triplet-singlet transition in benzene is both spin-forbidden and symmetry-forbidden and that we obtain a finite transition probability only if we correct equation (66) for the effect of molecular vibrations. In our calculation on benzene[21] we did not consider the effect of vibrations on the spin-orbit matrix elements and we substituted the experimental values for the symmetry-forbidden transition moments. However, if molecular vibrations play a role in the calculation of the transition moments they should be considered also in the calculation of the spin-orbit matrix elements. Albrecht[22] has given a very detailed analysis of all possibilities in mixing triplet and singlet states, both through vibrational interactions and through spin-orbit coupling. It seems to us that the theory of triplet-singlet transition probabilities is probably more complex for benzene than for any other molecule.

A comparison between the experimental and theoretical triplet-singlet transition probabilities is further complicated by the inadequacy of equation (51). There are two competing mechanisms that tend to depopulate the lowest triplet level of a molecule: the first is the radiative transition probability, which we have discussed so far, and in addition there is a radiationless transition from the lowest triplet level to the ground state. The latter mechanism is represented by a probability $w'_{T\to S}$ and we have, instead of equation (51),

$$(1/\tau) = w_{T\to S} + w'_{T\to S} = (1/\tau_{\text{rad.}}) + (1/\tau_{nr}) \qquad (104)$$

where $\tau_{\text{rad.}}$ and τ_{nr} are the inverse of $w_{T\to S}$ and $w'_{T\to S}$ respectively. It is not easy to determine τ_{nr} experimentally. Its value can be derived indirectly by measuring the quantum yield of the phosphorescence, that is the ratio between the number of photons that is emitted in the phosphorescence and the number of photons that is observed. In this manner it was found by Gilmore, Gibson & McClure[23] that for benzene $\tau = 7$ s and $\tau_{\text{rad.}} = 35$ s and the theoretical value of $w_{T\to S}$ should therefore be compared with 35 s rather than 7 s. More recently Kellogg & Bennett[24] performed a new series of measurements of phosphorescence quantum yields for aromatic molecules; their conclusion is that $\tau_{\text{rad.}}$ for most aromatic hydrocarbons is around 30 s within a variation of about 20 %. By comparison, the experimental values of t for aromatic hydrocarbons vary between 2·6 s for naphthalene and 16 s for triphenylene.

Another interesting phenomenon is the effect of isotopic substitution on the afterglow time τ of the phosphorescence. It was discovered by Hutchison & Mangum[25] that ordinary naphthalene, in a solid solution in ordinary durene, has an afterglow time of 2·1 s, whereas deuterated naphthalene in deuterated durene has a phosphorescence lifetime of 17 s. It is interesting that Siebrand[26] was able to account for the large changes in τ upon deuteration by assuming that the isotopic substitution affects only τ_{nr}. Following work by Robinson & Frosch[27] Siebrand evaluated τ_{nr} from Franck–Condon overlap integrals between the phosphorescent and the ground states. In this way he obtained reasonably good agreement with the experimental values of τ by taking $\tau_{\mathrm{rad.}}$ for all aromatic hydrocarbons equal to 30 s.

It seems that there are various aspects of the phosphorescence in aromatic hydrocarbons for which the previous calculations of spin-orbit coupling[19, 20, 21] offers no satisfactory explanation. Perhaps we should not limit our calculations only to states that differ in the occupancy of the π orbitals, but we should consider also mixing with states that are obtained by transferring a π electron to an anti-bonding C–H orbital or a C–H bonding electron to an empty π-orbital. These various possibilities are currently under investigation.

Spin-orbit interactions and zero-field splittings

Even though it has always been the most popular hypothesis, there was some doubt at one time as to whether or not the lowest excited phosphorescent states in some organic molecules are really triplet states. This doubt was caused by the failure of many attempts to measure the electron spin resonance of the triplet state. However, after the first successful electron spin resonance measurement of an aromatic hydrocarbon in its phosphorescent state[11] it became clear why the previous attempts had failed. In a molecule, namely, the three triplet levels are already split in the absence of a magnetic field due to the spin forces between the electrons. These 'zero-field splittings' are of the order of 0·1 cm^{-1} for aromatic hydrocarbons and of the order of 1 cm^{-1} for molecules like CH_2 and NH. The zero-field splittings are thus of the same order of magnitude as the splittings that are introduced in electron spin resonance by the exterior magnetic field. Consequently, the resonance frequency in electron spin resonance depends very strongly on the orientation of the molecule with respect to the exterior magnetic field. This causes such an extreme broadening of the resonance line for an assembly of randomly oriented molecules that the ESR signal cannot be observed. Hutchison & Mangum[11] avoided this difficulty by measuring the electron spin resonance of naphthalene, embedded in a durene crystal, so that the triplet molecules all have the same orientation. Recently the experimental techniques for measuring electron spin resonance of organic molecules in their triplet states have been developed further [18, 29] and experimental values of the zero-field splittings for a large number of molecules are now available.

The theoretical values of the zero-field splittings are derived from the matrix (60) by taking the differences of the eigenvalues. In the case of aromatic hydrocarbons it is customary to approximate the matrix elements $\Omega_{i,i'}$ of equation (60) as

$$\Omega_{i',i} \approx \langle {}^3\Psi_{t,i'} | H_{\text{s.s.}} | {}^3\Psi_{t,i} \rangle. \tag{105}$$

It is easily verified that this approximation is permissible. In the case of benzene, for example, the spin-orbit matrix elements are of the order of $1\,\text{cm}^{-1}$ and the energy differences $({}^1E_k - {}^3E_t)$ are of the order of $10\,000\,\text{cm}^{-1}$. The contribution of the infinite sum in equation (60) to the matrix element is of the order of $10^{-4}\,\text{cm}^{-1}$ and it may be neglected with respect to the zero-field splittings, which are of the order of $0{\cdot}1\,\text{cm}^{-1}$. Furthermore, it may be shown that

$$\langle {}^3\Psi_{t,i'} | H_{\text{s.o.}} | {}^3\Psi_{t,i} \rangle = a\delta_{i,i'}, \tag{106}$$

so that these spin-orbit matrix elements do not affect the zero-field splittings. Instead they shift all three triplet levels by an equal amount.

The above considerations are valid only for aromatic hydrocarbons, but unfortunately they led many people to believe that the approximation (105) is generally valid for all molecules. It is easily seen that this general belief has little justification. In the case of NH, for example, the zero-field splitting (ZFS) is around $2\,\text{cm}^{-1}$ and it may be anticipated that the matrix elements of the spin-orbit coupling are of the order of $100\,\text{cm}^{-1}$. If we estimate the energy difference between the triplet and singlet state as $10\,000\,\text{cm}^{-1}$ then we find that the infinite sum in equation (60) is of the order of $1\,\text{cm}^{-1}$, which is half the ZFS. We decided to investigate numerically for a number of molecules what the contribution of the second-order spin-orbit perturbation to the ZFS is. In the two cases which have been studied, namely CH_2 and NH,[17, 18] these contributions are 11 and 25 %, respectively, and it follows that in calculations of ZFS the second-order spin-orbit perturbation should be included in the calculations.

It has been argued that the second-order spin-orbit perturbation terms are always negligible because the g factors that are derived from the ESR results are the same for different molecules. However, it may be seen from our considerations that this argument is not necessarily correct. The changes in the g factor are determined by spin-orbit matrix elements with respect to a pair of triplet functions with identical spin multiplicities. On the other hand, the effect of spin-orbit coupling on the zero-field splitting depends on spin-orbit matrix elements between triplet and singlet functions. Consequently there is no connection between changes in the g factors and spin-orbit contributions to the zero-field splittings.

Final remarks

It was not until 1944 that the triplet state was first involved in an interpretation of chemical properties of organic molecules.[9] Since then its importance has grown rapidly, it plays a prominent role in spectroscopy, the zero-field splitting measurements are now used to obtain information on molecular structure, it is often used to explain the mechanisms of chemical reactions and it is even used to interpret biological properties. We have made an attempt to outline the general theoretical principles that describe the properties of the triplet state and of the effects of spin-orbit and spin-spin interactions. In particular, we have tried to clarify a few aspects of the theory that do not seem to be generally appreciated. It is not our intention to give a complete review of all the experimental and theoretical work on spin-orbit interactions in organic molecules, instead we have limited our considerations to a few major experimental break-throughs and to some general theoretical questions.

References

1 H. F. Hameka, *Advanced Quantum Chemistry* (Addison-Wesley Publishing Company, Inc., Reading, Mass., 1965), chapter 7.
2 This figure is a reproduction of figure 6-2 in R. M. Hochstrasser, *Behavior of Electrons in Atoms* (W. A. Benjamin, Inc., New York, N.Y., 1964), p. 103.
3 J. N. Murrell, *The Theory of the Electronic Spectra of Organic Molecules* (Methuen and Co. Ltd., London, 1963), p. 299.
4 G. E. Uhlenbeck & S. Goudsmit, *Naturwiss.* **13**, 953 (1925); *Nature*, **117**, 264 (1926).
5 H. A. Kramers, *Physica*, **1**, 825 (1934).
6 L. H. Thomas, *Nature*, **117**, 514 (1926); *Phil. Mag.* **3**, 1 (1927).
7 T. Itoh, *Rev. Mod. Phys.* **37**, 159 (1965).

8 E. Wiedemann, *Ann. Physik*, **34**, 446 (1888).
9 G. N. Lewis & M. Kasha, *J. Am. Chem. Soc.* **66**, 2100 (1944); **67**, 994 (1945).
10 D. F. Evans, *Nature*, **176**, 777 (1955).
11 C. A. Hutchison Jr. & B. W. Mangum, *J. Chem. Phys.* **29**, 952 (1958); **34**, 908 (1961).
12 H. F. Hameka, *J. Chem. Phys.* **37**, 328 (1962).
13 F. P. Prosser & C. H. Blanchard, *J. Chem. Phys.* **36**, 1112 (1962).
14 F. P. Prosser & C. H. Blanchard, *J. Chem. Phys.* **43**, 1086 (1965).
15 M. Geller, *J. Chem. Phys.* **39**, 84 (1963); **39**, 853 (1963).
16 S. J. Fogel & H. F. Hameka, *J. Chem. Phys.* **42**, 132 (1965).
17 J. W. McIver Jr. & H. F. Hameka, *J. Chem. Phys.* **45**, 767 (1966).
18 J. W. McIver Jr., Thesis, University of Pennsylvania (1966).
19 D. S. McClure, *J. Chem. Phys.* **17**, 905 (1949); **20**, 682 (1952).
20 M. Mizushima & S. Koide, *J. Chem. Phys.* **20**, 765 (1952).
21 H. F. Hameka & L. J. Oosterhoff, *Mol. Physics*, **1**, 358 (1958).
22 A. C. Albrecht, *J. Chem. Phys.* **38**, 354 (1963).
23 E. H. Gilmore, G. E. Gibson & D. S. McClure, *J. Chem. Phys.* **20**, 829 (1952).
24 R. E. Kellogg & R. G. Bennett, *J. Chem. Phys.* **41**, 3042 (1964).
25 C. A. Hutchison Jr. & B. W. Mangum, *J. Chem. Phys.* **32**, 1261 (1960).
26 W. Siebrand, *J. Chem. Phys.* **44**, 4055 (1966); also W. Siebrand, private communication.
27 G. W. Robinson & R. P. Frosch, *J. Chem. Phys.* **37**, 1962 (1962); **38**, 1187 (1963).
28 J. H. van der Waals & M. S. de Groot, *Mol. Physics*, **2**, 333 (1959); M. S. de Groot & J. H. van der Waals, *Mol. Physics*, **3**, 190 (1960).
29 W. A. Yager, E. Wasserman & R. M. R. Cramer, *J. Chem. Phys.* **37**, 1148 (1962).

SINGLET-TRIPLET TRANSITIONS IN ORGANIC MOLECULES

G. CASTRO, R. N. CLARKE, C. MARZZACCO,
M. SCHAFER, J. WHITEMAN AND
R. M. HOCHSTRASSER

Abstract

The spectroscopy of lower-lying triplet states of numerous types of organic molecules, and types of electronic states will be discussed in relation to low temperature, high resolution absorption studies of single crystals in polarised light, and the associated theoretical descriptions of the molecular states.

Orbital and spin-orbital characteristics of these triplet states, for example of carbonyl, nitrogen heterocyclics, and aromatic compounds, will be described leading to assignments of these triplet states. Zeeman effect studies of these singlet-triplet transitions will be shown to lead directly to the experimental evaluation of the relative magnitudes of the spin-orbit coupling matrix elements in the Cartesian framework of the molecule.

If time permits, the pertinence of some new single crystal results to the theory of triplet excitons will be discussed.

The materials whose singlet-triplet transitions will be discussed are as follows: anthracene, phenazine, benzophenone, tribromobenzene and hexabromobenzene.

This work was supported in part by the Advanced Research Projects Agency, Contract SD69, and in part by a National Institute of Health grant GM 12692.

TRIPLET DECAY AND INTERSYSTEM CROSSING IN AROMATIC HYDROCARBONS

W. SIEBRAND

Abstract

A theory is developed for radiationless transitions in polyatomic molecules. It applies when vibrational relaxation is fast and thus near-resonance vibronic interaction between electronic states governs the transition rate. This is the case for triplet-ground state transitions and presumably also for singlet-triplet transitions in aromatic hydrocarbons. It is assumed that for a specific transition in a class of related molecules, such as the aromatic hydrocarbons, the Franck–Condon factor of the transition is the main variable. This factor depends on the non-orthogonality of the vibrational wavefunctions in the two electronic states involved in the transition. It determines how the electronic energy is distributed among the normal modes in the final state. General expressions are developed for Franck–Condon factors of assemblies of harmonic oscillators. The case of isotopic substitution is treated separately and a simple isotope rule for radiationless transitions is derived. The theoretical expressions are compared with an empirical relation obtained for triplet decay rate constants in normal and deuterated aromatic hydrocarbons. It is shown that these rate constants are governed by anharmonic distortions of the CH and CD stretching modes. Comparison with the phosphorescence spectra leads to absolute values for the Franck–Condon factors. By virtue of this calibration it is possible to compare the vibronic interaction terms operative in triplet-ground state and singlet-triplet transitions. In the latter case higher triplet states are likely to be involved leading to Franck–Condon factors close to unity. It is concluded that the vibronic terms which govern this process are 1–2 orders of magnitude larger than those governing triplet-ground state transitions.

Introduction

Radiationless transitions play an important part in populating and depopulating the lowest triplet state of aromatic hydrocarbons. Of special interest here are transitions between triplet and singlet states. Two commonly observed transitions of this type, known as intersystem crossing, are shown in fig. 1. Rate constant κ refers to a transition from the first excited singlet state $|S\rangle$ to the triplet state $|T\rangle$, rate constant β to the radiationless transition from $|T\rangle$ to the ground state

$|0\rangle$. In many cases the former process competes successfully with fluorescence (rate constant a in fig. 1), so that the rate constant κ must be of the order of $10^7\,s^{-1}$. Very few accurate data are available, however. More is known about the rate constant β. From the observed (monomolecular) triplet lifetime τ_0 we can estimate β since $\beta \geqslant \tau_0^{-1}$. There is good evidence that the phosphorescence rate constant b (cf. fig. 1) is about $0{\cdot}03\,s^{-1}$ for all aromatic hydrocarbons.[1, 2] Since typical values for τ_0 range from 0·01 to 10 s, β must vary from less than 10^{-1} to $10^2\,s^{-1}$. If rate constant κ is included in the comparison we can say that the rate constants of intersystem crossing between the singlet and the triplet manifold vary between 10^{-1} and $10^7\,s^{-1}$. In the following we attempt to account for these wide variations and, as far as possible, for the observed rate constants.

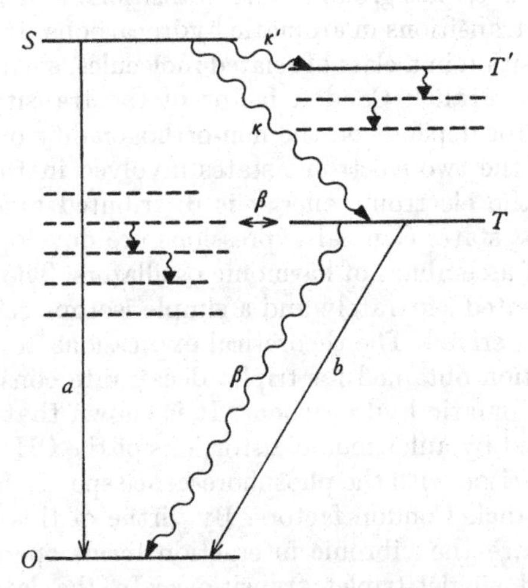

Fig. 1. Energy-level diagram of the lower electronic states of aromatic hydrocarbons. The broken lines indicate vibrational levels.

Radiationless transitions of the type under investigation involve near-resonance transfer between a singlet and triplet state followed by vibrational relaxation in the manifold of the final state. We shall only deal with the case where the relaxation process is very rapid and does not affect the observed kinetics. Recent work by various authors[2–4] indicates that this situation applies to aromatic hydrocarbons in solid matrices. The rate-determining step involves, at low temperatures, a transition from the lowest vibrational level of the initial electronic state to a vibrationally-excited level of the final state. This transition is due to two small interaction terms: spin-orbit coupling and vibronic coupling. The rate constant also depends strongly on the non-orthogonality of the vibrational wavefunctions in both states. If there is complete orthogonality only transitions between levels with the same vibrational quantum number (i.e. with $\Delta v = 0$) are allowed. Since in general radiationless transitions are multiphonon processes (Δv large) they are facilitated by an appreciable degree of non-orthogonality between corresponding vibrational wavefunctions in both states. The non-orthogonality is expressed in the Franck–Condon factor F of the transition. F is expected to be very sensitive to the vibrational quantum number v of the final state, and hence to the separation of the zero-point levels of the electronic states involved in the transition.

In view of the complexity of the problem no absolute calculation of β or κ will be attempted. Only the relative magnitudes of these quantities within the class of aromatic hydrocarbons will concern us here. We expect changes in β and in κ within this class to be mainly due to changes in F. As a first approximation we therefore assume all other parameters to be constant. This reduces the calculation of β or κ to the calculation of F. To calculate relative values of F advantage is taken of the fact that all aromatic hydrocarbons are built from essentially the same oscillators, so that a limited number of spectroscopic parameters determines F for all of them.

The relation between β (or κ) and F follows from time-dependent perturbation theory[2–4]

$$\beta = (2\pi\rho/\hbar)J_\beta^2 F \equiv C_\beta F. \tag{1}$$

Here ρ is the density of states in the final state, J_β is a vibronic matrix element and C_β is taken to be a constant proportionality factor. The Franck–Condon factor is

$$F = \sum_P P\left[\prod_{n=1}^{N} |\langle\chi_n^0(v_n)|\chi_n(0)\rangle|^2\right], \tag{2}$$

where χ_n and χ_n^0 are the vibrational wavefunctions of mode n in the initial and final state, respectively. The operator $\sum_P P$ permutes the vibrational quanta v_n among the N normal modes subject to the energy-conservation condition

$$E = \sum_n v_n \hbar\omega_n^0 = E_{00} \pm 1/2\rho, \tag{3}$$

where ω_n^0 is the angular frequency of mode n in the final state and E_{00} is the energy separation of the zero-point levels of both states. The density of states ρ, which is ultimately due to interactions with the matrix, is assumed so large that the vibrational fine-structure of F is washed out for large $\sum_n v_n$. Under these conditions F is expected to be a monotonically decreasing function of the continuous variable E.

Franck–Condon factors

To calculate F as a function of E we first assume that all oscillators are harmonic. This separates the molecular vibrations into (uncoupled) normal modes. In aromatic hydrocarbons these normal modes fall apart into groups of almost degenerate oscillators. Since we are assuming low spectral resolution it seems reasonable to neglect small differences between such oscillators and to treat them as degenerate. This reduces the problem of N normal modes to that of $M\,(\ll N)$ groups of quasidegenerate oscillators.

Let us first calculate $F(E)$ for a single non-degenerate oscillator. In general, the electronic transition will cause a change in equilibrium distance (Δq) and in frequency ($\Delta\omega$). If $\Delta\omega = 0$ (displaced oscillator) we obtain[5]

$$F(E) = e^{-\gamma}\gamma^v/v!; \quad \gamma = \tfrac{1}{2}k^0(\Delta q)^2/\hbar\omega^0, \tag{4}$$

where k^0 and ω^0 are the force constant and the angular frequency in the final state. This result is readily generalised to an N-fold degenerate displaced oscillator[2]

$$F(E) = e^{-N\gamma_n}(N\gamma_n)^v/v!, \tag{5}$$

which is of precisely the same form as (4). If we compare different aromatic hydrocarbons it is reasonable to assume that γ_n for a single oscillator decreases with an increasing number of quasidegenerate oscillators in the group. In the particular case that $N\gamma_n = \gamma =$ constant equation (5) reduces to (4). Thus the quasidegeneracy leaves the basic form of the $F(E)$ curve unaltered for displaced oscillators.

Consider now the case where $\Delta q = 0$ (distorted oscillator). For a non-degenerate oscillator one obtains[5]

$$F(E) = \frac{2(\omega^0\omega)^{\frac{1}{2}}}{\omega^0+\omega}\,\xi^v\,\frac{1.3.5\ldots(v-1)}{2.4.6\ldots v}; \quad v \text{ even}, \tag{6}$$

where $\xi = \Delta\omega/(\omega^0+\omega)$. If v is odd $F(E) = 0$. It is not difficult to extend this result to an N-fold degenerate oscillator:[2]

$$F(E) = \frac{2^N(\omega^0\omega)^{\frac{1}{2}N}}{(\omega^0+\omega)^N}\left(\frac{\xi}{N}\right)^v\frac{N(N+2)\ldots(N+v-2)}{2.4\ldots v}, \tag{7}$$

provided v is even. Here we have made the assumption[(2)] that

$$N\xi_n = \xi = \text{constant}$$

for all aromatic hydrocarbons. It is clear from (7) that degeneracy changes the form of the $F(E)$ curve for distorted oscillators.

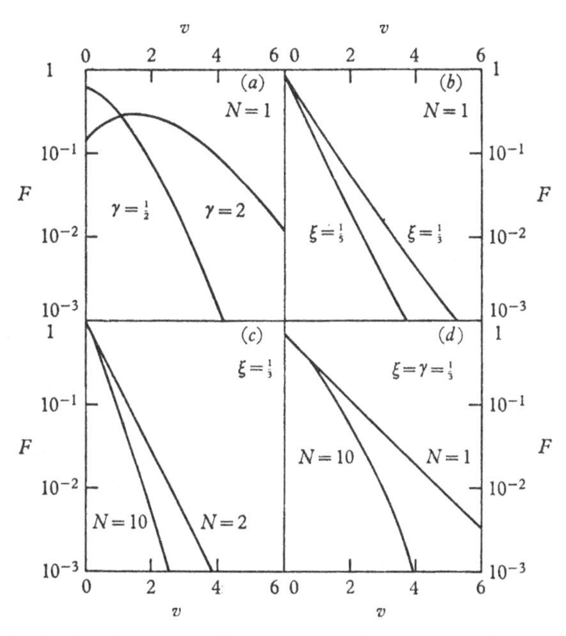

Fig. 2. Franck–Condon factors F of harmonic oscillators plotted as a function of the vibrational quantum number v for: (*a*) displaced oscillators; (*b*) non-degenerate distorted oscillators; (*c*) degenerate distorted oscillators; (*d*) displaced and distorted oscillators.

The difference between displaced and distorted oscillators is illustrated in fig. 2, where $\log F$ is plotted against v for representative values of γ, ξ and N. As explained above we treat v as a continuous variable under our conditions where the spectral resolution is low. Let us consider the logarithmic derivative of F

$$F'(E) \equiv [\partial \log F(E')/\partial E']_E. \tag{8}$$

For large enough v, $F'(E)$ is always negative. This means that the vibrational overlap decreases if the corresponding oscillator takes up energy. In general most of the electronic energy will flow into those oscillators for which $F'(E)$ is maximal, i.e. has the smallest negative value. Thus $F'(E)$ is a direct measure of the efficiency of the oscillators for taking up the energy released in the transition. Fig. 2 shows that for displaced oscillators $F'(E)$ decreases with increasing v, whereas for distorted oscillators $F'(E)$ increases for $N = 1$, remains constant for $N = 2$ and decreases slightly for $v > N > 2$. Only in the case that $v \ll N$ does $F'(E)$ for a distorted oscillator decrease as much as $F(E)$ for a displaced oscillator.

These considerations show that radiationless transitions follow a characteristic path. The energy to be disposed of first flows into the quasidegenerate group, α, with the largest $F'_\alpha(0)$. This is likely to be essentially a displaced oscillator, so that $F'_\alpha(E)$ decreases rapidly with increasing v_α. At a certain energy E_0 one reaches the point where

$$F'_\alpha(E_0) \equiv \left[\frac{\partial \log F_\alpha(E)}{\partial E}\right]_{E_0} = \left[\frac{\partial \log F_\beta(E-E_0)}{\partial E}\right]_{E_0}, \tag{9}$$

hence for $E > E_0$ group β rather than group α provides the major contribution to F. If β happens to be another displaced oscillator then a similar crossing will occur at a higher energy, above which a third group will take over, etc. If β is a non-degenerate distorted oscillator, however, $F'_\beta(E)$ does not decrease with increasing v_β and thus no other oscillator will be able to take over at higher energies. Hence in general a plot of $\log F$ against E for polyatomic molecules will show strong downward curvature similar to fig. 2 (*a*) for low energies and little or no curvature for high energies. The low-energy part applies to radiative, the high-energy part to non-radiative transitions. In most cases different oscillators will provide the major contributions in the two parts. Hence it is not possible to extrapolate the optical emission spectrum to the non-radiative limit. It is also incorrect to neglect frequency changes when dealing with radiationless transitions.

These conclusions indicate that we must consider the general case of a harmonic oscillator which is both displaced and distorted. In the absence of degeneracy ($N = 1$) one obtains[5]

$$F(E) = \frac{2(\omega^0\omega)^{\frac{1}{2}}}{\omega^0+\omega}\left(\frac{\xi}{2}\right)^v \frac{|H_v(ix)|^2}{v!}e^{-\gamma}, \tag{10}$$

where
$$x = [\gamma(1-\xi)/2\xi]^{\frac{1}{2}}, \tag{11}$$

and the $H_v(ix)$ are Hermite polynomials of an imaginary argument. For practical purposes we rewrite equation (10) in the form

$$F(E) = F(0)\,(z\xi)^v, \tag{12}$$

where $z = z(v; x)$. For $x < 1$, $v > 2$, z is a slowly varying function of v, as shown in fig. 3. Under these conditions, which should hold for a wide variety of radiationless transitions, $F'(E)$ is practically constant. Using (1) and the relation

$$v = (E - E_0)/\hbar\omega^0, \tag{13}$$

equation (12) gives rise to an exponential dependence of β on E. In the special case that $E_0 = 0$ the constant $F(0)$ is given by the v-independent term of (10), but in general E_0 denotes a crossing point as defined by equation (9), and then $F(0)$ is an unknown parameter.

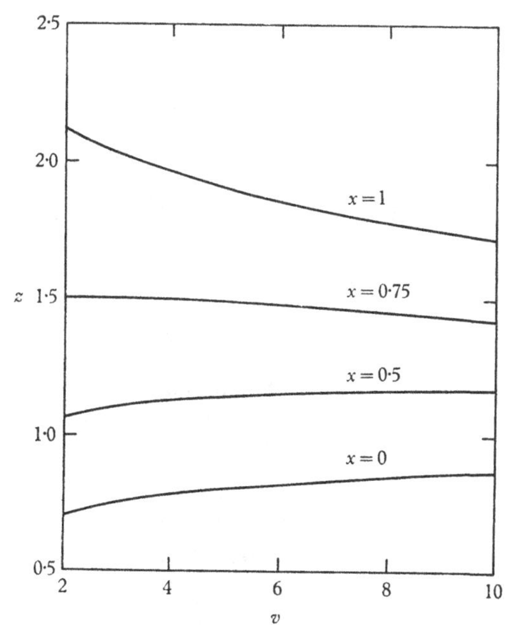

Fig. 3. Plot of $z(v; x)$ defined by equation (12). If z is independent of v in a certain range of v then $F(v)$ is an exponential function in this range.

If z varies sufficiently slowly with v it can be replaced by its average value $\bar{z}$. Then generalisation of (12) to a group of N degenerate oscillators can be achieved. First we assume that $N\xi_n = \xi$ and $N\gamma_n = \gamma$ are independent of N. Then it follows from equation (11) and fig. 3 that $\bar{z}$ is practically independent of N for small ξ, small x or large N. In these cases we have[2]

$$F(E) = F(0)\left(\frac{\bar{z}\xi}{N}\right)^v \frac{(N+v-1)!}{(N-1)!\,v!}. \tag{14}$$

This function decreases more rapidly with increasing E than $F(E)$ for a non-degenerate oscillator given by (12). An example is shown in fig. 2(*d*). If ξ and x are not small and N is not large $\bar{z}$ increases with increasing N and $F'(E)$ derived from equation (14) approaches $F'(E)$ derived from equation (12) more closely. Direct generalisation of this procedure to anharmonic oscillators seems to be impossible. However, anharmonicity can be introduced formally by means of the expansion

$$z\xi = z_0\xi_0 + \lambda v + \ldots \tag{15}$$

The definition of ξ suggests that (15) may be terminated after the linear term, provided z changes slowly with v.

Isotope rule

Since radiationless transitions depend on vibrational overlap they may be expected to be sensitive to isotopic substitution. It has indeed been found that, for example, the triplet lifetime of aromatic hydrocarbons is substantially longer for the deuterated species.[6] The magnitude of the isotope effect is readily derived (see above, p. 34 ff.) by substitution of the isotope factor

$$\sigma \equiv (\mu_i/\mu)^{\frac{1}{2}} = \omega^0/\omega_i^0, \tag{16}$$

where μ is the reduced mass and the subscript i labels the heavy isotope. For instance, equation (12) yields

$$F_i(E) = F_i(0)\,(z_i\xi)^{\sigma v}, \tag{17}$$

where $z_i(\sigma^{\frac{1}{2}}x;\sigma v) \approx z(x;v)$, whenever z is a slowly varying function of v. From this result the following isotope rule is obtained

$$F_i'(E) = \sigma F'(E). \tag{18}$$

The same rule can be derived from each of the equations (4)–(7) and (14), provided v is not too small.[2] Thus when comparing two (classes of) molecules which differ only by isotopic substitution one finds that the logarithmic derivatives of the Franck–Condon factors are proportional to the vibrational frequencies of the modes that are mainly responsible for the radiationless transition, in agreement with a suggestion of Robinson & Frosch.[3] This is perhaps the easiest way to determine the identity of these modes.

Triplet-ground state transitions

As a first application of the theory we consider radiationless transitions from the lowest triplet state to the ground state of normal and per-deuterated aromatic hydrocarbons. This process competes with phosphorescence, so that the non-radiative rate constant β follows from

$$\beta = \tau_0^{-1} - b, \tag{19}$$

where τ_0 is the observed triplet lifetime and b the radiative rate constant. Quantum-yield studies[1] indicate that for typical aromatic hydrocarbons $b \approx 30\,s^{-1}$; since more specific information is not available we adopt this value for all aromatic hydrocarbons. The Appendix contains a comprehensive list of the triplet lifetime data available from the literature, together with values of the triplet energy E_{TO}.

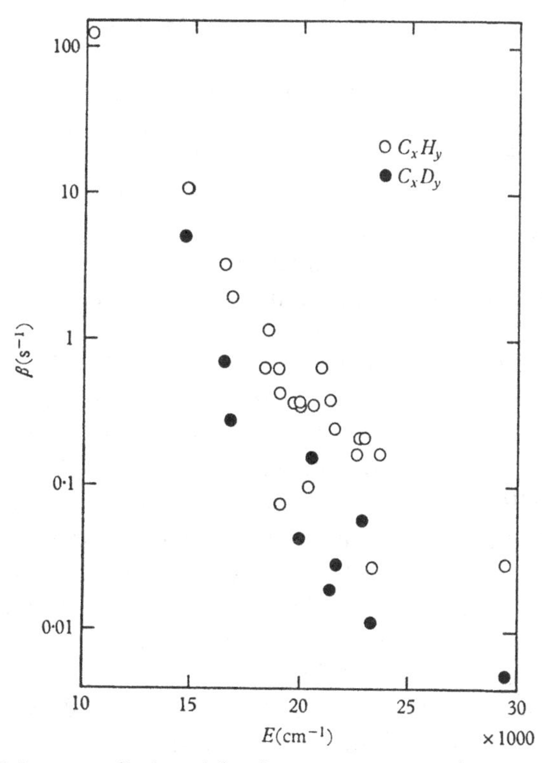

Fig. 4. Plot of the non-radiative triplet decay rate constant β against the energy of the lowest triplet state for normal and deuterated aromatic hydrocarbons.

In fig. 4 log β is plotted against E_{TO} for all aromatic hydrocarbons listed in the Appendix. This plot, which is analogous to fig. 2, shows that β and E are correlated, but not in a direct way. The decrease of β upon deuteration is evident, but does not seem to lead to a simple relation either. The correlation can be greatly improved, however, by allowing for structural differences between the hydrocarbons. Inspection of fig. 4 and the Appendix indicates that the relative number of hydrogen atoms, i.e. η for the hydrocarbon $C_{1-\eta}H_{\eta}$, is somehow involved in the correlation. All points corresponding to $\eta = 0{\cdot}400 \pm 0{\cdot}020$ fit a straight line reasonably well. Relative to this line all points for which $\eta > 0{\cdot}420$ and $\eta < 0{\cdot}380$ correspond to values of β that are too high and too low, respectively. A good fit of all points is obtained by plotting $\log\beta$ against E/η. Equation (13) suggests to generalise the latter quantity to $(E - E_0)/\eta$, where E_0 represents a crossing point defined by equation (9). The best correlation[2] is obtained by taking $E_0^{\mathrm{H}} = 4\,000\ \mathrm{cm}^{-1}$ for non-deuterated hydrocarbons. From (9) and (18) it follows that E_0^{D} (for deuterated hydrocarbons) must be greater than E_0^{H}. In this case there are not enough reliable points, however, to determine E_0^{D} unambiguously on the basis of the best correlation. By an indirect method to be explained later one can deduce a value $E_0^{\mathrm{D}} = 5\,500\ \mathrm{cm}^{-1}$. The corresponding plots of deuterated and non-deuterated hydrocarbons are shown in fig. 5. Note that the correlation becomes less good for $\beta < b$, in which case the calculation of β by means of equation (19) becomes inaccurate. However, the nature of the correlation confirms that the true radiative lifetime of aromatic hydrocarbons must be very close to 30 s.

The empirical relations between β and E shown in fig. 5 are sufficiently simple and precise to allow comparison with the theory. Application of the isotope rule (18) gives $\sigma = 1{\cdot}33$, which value is almost independent of the choice of E_0^{H} and E_0^{D}. Since $(\mu_{\mathrm{CD}}/\mu_{\mathrm{CH}})^{\frac{1}{2}} = 1{\cdot}362$ it follows that CH(CD) vibrations take up most of the energy in excess of $E_0^{\mathrm{H}}(E_0^{\mathrm{D}})$. Thus to a good approximation $F(E)$ for $E > E_0^{\mathrm{H}}$ can be described solely in terms of CH modes. The same conclusion can be drawn[1] from the appearance of the factor η in the correlation, since the extrapolation $\eta \to 0$, i.e. no CH modes, leads to $\beta \to 0$, i.e. no radiationless decay, for $E > E_0^{\mathrm{H}}$.

The behaviour of $F(E)$ for $E < E_0$ can be obtained from the phosphorescence spectrum. McCoy & Ross[7] have shown that for a number of hydrocarbons this function can be fairly accurately represented by (4) with $\hbar\omega^0 \approx 1\,400\ \mathrm{cm}^{-1}$ and $\gamma \approx 1{\cdot}5$. This part of $F(E)$ is mainly due to totally symmetric CC stretching vibrations and is insensitive to deuterium substitution. At the point E_0 where the two parts join they must have the same logarithmic derivative $F'(E_0)$ according to equation (9). This has been achieved in fig. 5 by drawing the empirical $F(E)$ curves for $E > E_0$ as tangents to the curve for $E < E_0$. This procedure leads to $E_0^{\mathrm{H}} \approx 4\,000\ \mathrm{cm}^{-1}$, in agreement with the value obtained above, and to $E_0^{\mathrm{D}} \approx 5\,500\ \mathrm{cm}^{-1}$. The connection between the radiative

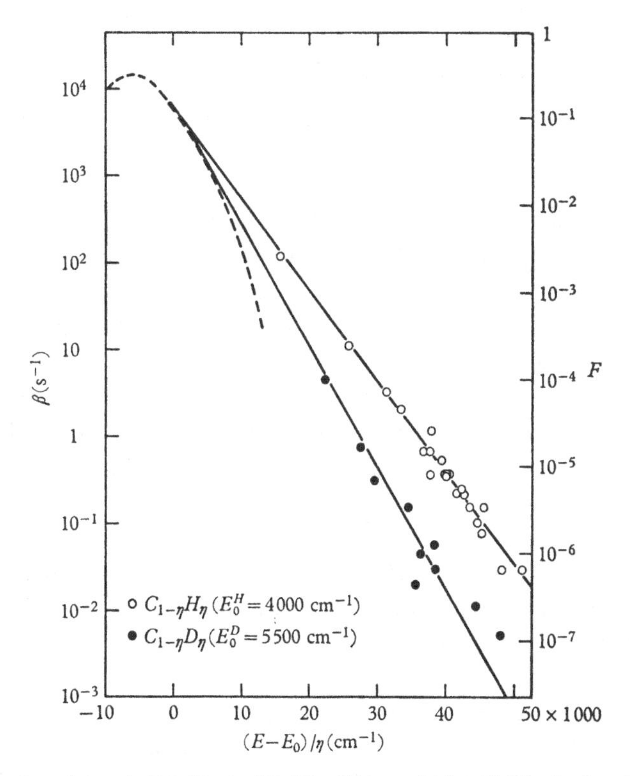

Fig. 5. Similar plot as in fig. 4 but with $(E-E_0)/\eta$ replacing E. The broken line represents F as derived from phosphorescence spectra. Drawing the two solid lines as tangents to this function determines the F-scale. The slopes of the two solid lines differ by a factor $1{\cdot}32 \approx (\mu_{\mathrm{CD}}/\mu_{\mathrm{CH}})^{\frac{1}{2}}$. [Note that points with $\beta \leqslant 10^{-2}\ \mathrm{s}^{-1}$ are inaccurate because of large corrections for the radiative process.]

and non-radiative parts of $F(E)$ calibrates the latter, empirical part. Specifically it leads to $C_\beta = 4 \times 10^4\ \mathrm{s}^{-1}$ and $\beta(0) = 10^4\ \mathrm{s}^{-1}$. These values are based on a strictly linear extrapolation of $F(E)$ for $E > E_0$ and are therefore somewhat uncertain.

Let us now compare fig. 5 with the theoretical expressions for $F(E)$ derived as shown above. From the linearity of the plot it follows at once that the oscillators involved must be essentially distorted so that (4) and (5) are not applicable. If we first try the assumption

$N = 1$ we can use (12) with $0{\cdot}5 < z < 2$. From (12), (13) and fig. 5 one obtains

$$\log z\xi = \frac{\hbar\omega^0}{\eta}\frac{\partial \log \beta(E)}{\partial(E-E_0)/\eta} = -0{\cdot}103\frac{\hbar\omega^0}{\eta}\times 10^{-3}. \qquad (20)$$

For $\eta = 0{\cdot}4$, which is an average value, this leads to the following results:

CH out-of-plane bending: $\hbar\omega^0 \approx 700\,\text{cm}^{-1}$, $z\xi = 0{\cdot}66$;

CH in-plane bending: $\hbar\omega^0 \approx 1\,100\,\text{cm}^{-1}$, $z\xi = 0{\cdot}52$;

CH stretching: $\hbar\omega^0 \approx 3\,000\,\text{cm}^{-1}$, $z\xi = 0{\cdot}17$.

No spectral features due to CH vibrations have so far been detected in phosphorescence spectra.[7] This rules out all except the last of the above alternatives. The result $z\xi = 0{\cdot}17$ for $\eta = 0{\cdot}4$ changes into $z\xi = 0{\cdot}12$ for $\eta = \frac{1}{3}$ (coronene) and into $z\xi = 0{\cdot}24$ for $\eta = \frac{1}{2}$ (benzene).

Comparison with the transition $|S\rangle \to |0\rangle$ in benzene, where $\xi \approx 0{\cdot}01$, indicates that these values are much too high. However, this spectroscopic value of ξ is not relevant here, since it is smaller than the anharmonicity of CH-stretching modes. For instance,

$$(1/\omega^0)\,\partial\omega^0/\partial v = 0{\cdot}04$$

in the ground state of the CH molecule, suggesting that $\partial\xi/\partial v \approx 0{\cdot}02$. Hence the linear rather than the constant term appears to dominate in the expansion (15), so that $F(E)$ is expected to be governed by anharmonic distortions.

On this basis a satisfactory formula for $F(E)$ can be obtained. Starting from equation (14) all CH-stretching modes are assumed quasidegenerate, so that the degeneracy equals the number of hydrogen atoms per molecule, N_{H}. The expression $\bar{z}\xi/N$ is replaced by $\lambda v/N_{\mathrm{C}}$, N_{C} being the number of carbon atoms. This number is used rather than N_{H} because the electronic rearrangement involves mainly π-electrons, which are associated with carbon atoms. Using these substitutions and representing v by (13), equation (14) assumes the form

$$F(E) = F(E_0)\,(\lambda v/N_{\mathrm{C}})^v\,(N_{\mathrm{H}}+v-1)!/(N_{\mathrm{H}}-1)!v!. \qquad (21)$$

This expression gives a good representation of figure 5, provided $\hbar\omega^0 \approx 3000\,\text{cm}^{-1}$ ($2250\,\text{cm}^{-1}$ for deuterated compounds). This is readily seen using the approximation

$$F(E) \approx F(E_0)\,(\mathrm{e}\lambda N_{\mathrm{H}}/N_{\mathrm{C}})^{(E-E_0)/\hbar\omega}. \qquad (22)$$

which is valid when $N_{\mathrm{H}} \gg v \gg 1$. Comparison with fig. 5 leads to $\lambda \approx 0{\cdot}1$, so that $\lambda/N_{\mathrm{C}} \approx 0{\cdot}02$ for benzene, in agreement with the above estimate. Thus we conclude that the rate constants of radiationless triplet decay are governed by anharmonic distortions of CH (or CD), stretching modes.

Singlet-triplet transitions

In conclusion we consider briefly radiationless transitions from the first excited singlet state $|S\rangle$ to the lowest triplet state $|T\rangle$. These transitions (rate constant κ) compete with fluorescence (rate constant a) and radiationless transitions from $|S\rangle$ to the ground state. The latter transitions are presumably much slower and will be neglected. In many cases a and κ are of the same order of magnitude and their precise values must be determined from quantum-yield measurements. Typical values for κ appear to be[1, 3, 9–11] 10^6–$10^8\,\mathrm{s}^{-1}$. The actual mechanism of this transition may be complicated by the presence of higher triplet states located between $|T\rangle$ and $|S\rangle$. For instance, in the situation represented by fig. 1 the rate constant κ' of the transition $|S\rangle \rightarrow |T'\rangle$ would contribute to the observed rate constant κ_0. In fact κ' is likely to outweigh the constant κ of the direct transition $|S\rangle \rightarrow |T\rangle$ because $E_{ST'} \ll E_{ST}$, so that one expects that $F(E_{ST'}) \gg F(E_{ST})$. This has been shown to be the case for anthracene[12–14], where[9, 13] $\kappa_0 \approx 4 \times 10^7\,\mathrm{s}^{-1}$. The fact that about the same value is observed in many other aromatic hydrocarbons, together with the absence of a deuterium effect[1, 9] suggests that a mechanism involving higher triplet states is common to most if not all of these systems.

If this conclusion is correct the corresponding Franck–Condon factors can be estimated to be $0{\cdot}01 < F(E_{ST'}) < 1$. Assuming $\kappa_0 = \kappa' = 10^7\,\mathrm{s}^{-1}$ and $F = 0{\cdot}1$ one obtains from (1) that $C_{\kappa'} = 10^8\,\mathrm{s}^{-1}$, which is to be compared with the result $C_\beta = 4 \times 10^4\,\mathrm{s}^{-1}$, derived above. This difference must be traced back to a difference in the matrix elements $J_{\kappa'}$ and J_β. Thus we conclude that in general $J_{\kappa'} \approx 50 J_\beta$.

Appendix

Energies and lifetimes of the lowest triplet state of aromatic hydrocarbons

Molecule	Formula	η	E_{TO} (cm^{-1})	τ_0 (s)	β (s^{-1})	Ref.
Benzene	C_6H_6	0·500	29500	16	0·029	a
	C_6D_6			26	0·0050	a
Naphthalene	$C_{10}H_8$	0·445	21300	2·4	0·39	b–o
	$C_{10}D_8$			19	0·020	d, g, i, k, n
Biphenylene	$C_{12}H_8$	0·400	19000	1·5	0·67	p
Biphenyl	$C_{12}H_{10}$	0·455	22900	4·0	0·22	e, g, h, k, o, q
	$C_{12}D_{10}$			11	0·057	g, h
Fluorene	$C_{13}H_{10}$	0·435	23800	5·0	0·17	b, c, e, m, q
Anthracene	$C_{14}H_{10}$	0·417	14700	0·045	22	d, o, r
	$C_{14}D_{10}$			0·14	7·1	r, s
Phenanthrene	$C_{14}H_{10}$	0·417	21600	3·5	0·25	b, c, e–i, k–m, t
	$C_{14}D_{10}$			16	0·029	g, i
Pyrene	$C_{16}H_{10}$	0·385	16900	0·5	2·0	b, c, g, t, u
	$C_{16}D_{10}$			3·2	0·29	g
Fluoranthene	$C_{16}H_{10}$	0·385	18500	0·85	1·2	e, t
Tetracene	$C_{18}H_{12}$	0·400	10300	0·008	130	v, w
Triphenylene	$C_{18}H_{12}$	0·400	23300	16	0·029	b, d, g, i, k, m, t
	$C_{18}D_{12}$			22	0·012	g, i
Chrysene	$C_{18}H_{12}$	0·400	20000	2·6	0·36	b, c, e, m, o, t
	$C_{18}D_{12}$			13	0·043	x
1,2-Benz-anthracene	$C_{18}H_{12}$	0·400	16500	0·3	3·3	b, c, f, t, u
	$C_{18}D_{12}$			1·7	0·59	s
3,4-Benzo-phenanthrene	$C_{18}H_{12}$	0·400	20000	2·5	0·37	t
p-Terphenyl	$C_{18}H_{14}$	0·438	20600	2·6	0·36	g, t
	$C_{18}D_{14}$			5·3	0·16	g
m-Terphenyl	$C_{18}H_{14}$	0·438	22700	4·1	0·21	y
1,2;5,6-Dibenzanthracene	$C_{22}H_{14}$	0·389	18300	1·5	0·67	b, c, t
3,4;5,6-Dibenzophenanthrene	$C_{22}H_{14}$	0·389	19800	2·5	0·37	t
Coronene	$C_{24}H_{12}$	0·333	19100	9·0	0·077	b, d, e
1,2;6,7-Dibenzopyrene	$C_{24}H_{14}$	0·368	20400	7·5	0·10	t
1,3,5-Triphenylbenzene	$C_{24}H_{18}$	0·428	22600	5·0	0·17	d, t, y
Hexahelicene	$C_{26}H_{16}$	0·380	19000	2·1	0·43	z

[a] M. R. Wright, R. P. Frosch & G. W. Robinson, *J. Chem. Phys.* **33**, 934 (1960).
[b] D. S. McClure, *J. Chem. Phys.* **17**, 905 (1949).
[c] D. P. Craig & I. G. Ross, *J. Chem. Soc.* p. 1589 (1954).
[d] M. S. de Groot & J. H. van der Waals, *Mol. Physics*, **3**, 190 (1960); **4**, 189 (1961).

[e] G. von Foerster, *Z. Naturforsch.* **18a**, 620 (1963).
[f] J. Czekalla, G. Briegleb, W. Herre & H. J. Vahlensieck, *Z. Elektrochem.* **63**, 1197 (1959).
[g] Ref. 6. [h] Ref. 11.
[i] E. C. Lim & J. D. Laposa, *J. Chem. Phys.* **41**, 3257 (1964).
[j] J. W. Hilpern, G. Porter & L. J. Stief, *Proc. Roy. Soc.* (*London*) A, **277**, 437 (1964).
[k] T. Azumi & S. P. McGlynn, *J. Chem. Phys.* **39**, 1186 (1963).
[l] D. Olness & H. Sponer, *J. Chem. Phys.* **38**, 1779 (1963).
[m] P. P. Dikun, A. A. Petrov & B. V. Sveshnikov, *Zh. Eks. Teor. Fiz.* **21**, 150 (1951).
[n] S. Siegel & H. S. Judeikis, *J. Chem. Phys.* **42**, 3060 (1965).
[o] G. N. Lewis & M. Kasha, *J. Am. Chem. Soc.* **66**, 2100 (1944).
[p] I. H. Munro, T. D. S. Hamilton, J. P. Ray & G. F. Moore, *Phys. Letters*, **20**, 386 (1966).
[q] V. V. Trusov & P. O. Teplyakov, *Opt. i Spektroskopiya*, **16**, 52 (1964).
[r] A. Beckett, *Nature*, **211**, 410 (1966).
[s] W. Siebrand & D. F. Williams, *J. Chem. Phys.* **46**, 403 (1967).
[t] E. Clar & M. Zander, *Chem. Ber.* **89**, 749 (1956).
[u] B. Stevens & M. S. Walker, *Proc. Roy. Soc.* (*London*), A, **281**, 420 (1964).
[v] S. P. McGlynn, M. R. Padhye & M. Kasha, *J. Chem. Phys.* **23**, 593 (1955).
[w] A. A. Lamola, W. G. Herkstroeter, J. C. Dalton & G. S. Hammond, *J. Chem. Phys.* **42**, 1715 (1965); β measured at 300 °K.
[x] Ref. 1.
[y] J. S. Brinen, J. G. Koren & W. G. Hodgson, *J. Chem. Phys.* **44**, 3095 (1966).
[z] W. Rhodes & M. F. A. El-Sayed, *J. Mol. Spectr.* **9**, 42 (1962).

References

1 R. E. Kellogg & R. G. Bennett, *J. Chem. Phys.* **41**, 3042 (1964).
2 W. Siebrand, *J. Chem. Phys.* **44**, 4055 (1966) and to be published.
3 G. W. Robinson, *J. Mol. Spectr.* **6**, 58 (1961); G. W. Robinson & R. P. Frosch, *J. Chem. Phys.* **37**, 1962 (1962); **38**, 1187 (1963).
4 G. R. Hunt, E. F. McCoy & I. G. Ross, *Aust. J. Chem.* **15**, 591 (1962); J. P. Byrne, E. F. McCoy & I. G. Ross, *ibid.* **18**, 1589 (1965).
5 E. Hutchisson, *Phys. Rev.* **36**, 410 (1930); C. Manneback, *Physica*, **17** 1001 (1951).
6 R. E. Kellogg & R. P. Schwenker, *J. Chem. Phys.* **41**, 2860 (1964).
7 E. F. McCoy & I. G. Ross, *Aust. J. Chem.* **15**, 573 (1962).
8 T. V. Ivanova & B. Y. Sveshnikov, *Opt. i Spektroskopiya*, **11**, 598 (1961) [English translation: *Opt. and Spectr.* **11**, 322 (1961)].
9 J. D. Laposa, E. C. Lim & R. E. Kellogg, *J. Chem. Phys.* **42**, 3025 (1965).
10 A. A. Lamola & G. S. Hammond, *J. Chem. Phys.* **43**, 2129 (1965).
11 V. L. Ermolaev, *Izv. Akad. Nauk SSSR, Ser. fiz.* **27**, 619 (1963).
12 R. E. Kellogg, *J. Chem. Phys.* **44**, 411 (1966).
13 W. H. Melhuish, *J. Phys. Chem.* **65**, 229 (1961).
14 D. F. Williams & G. Adolph (to be published in *J. Chem. Phys.*).

STATISTICAL ASPECTS OF RESONANCE ENERGY TRANSFER

J. S. AVERY AND J. C. PACKER

Abstract

The time-dependent luminescence of a randomly distributed collection of sensitiser and acceptor molecules is discussed using a Green's function method when the illuminating beam is an arbitrary function of time. An approximate 'cell model' of the system is introduced and the results are applied to the special case where the illuminating beam varies sinusoidally.

Introduction

Recent theoretical interest in sensitised phosphorescence and sensitised fluorescence[1–20] makes it desirable to study experimentally the exact R-dependence of transfer rate. Thus if we write the number of transitions per second in the form,[3, 4, 7]

$$1/\tau_{S-A} = (1/\tau_S).F(R), \tag{1}$$

where $1/\tau_S$ is the decay constant of the sensitiser excited state by photon emission and internal quenching and R is the distance between the sensitiser and the acceptor, then one wishes to determine $F(R)$ experimentally; however, the interpretation of experimental results is complicated by the fact that one usually observes the total phosphorescence or fluorescence of an ensemble of randomly distributed sensitiser and acceptor molecules in solution.[10, 11]

One is thus faced with the problem of relating the R-dependence of the transfer rate to properties of the ensemble. For example, experiments are performed to measure the fluorescence efficiency,[1, 2, 15, 16] and fluorescence lifetimes[21–25, 32, 33] of the sensitisers and acceptors as a function of acceptor concentration. A statistical treatment of this problem was introduced by Förster[2] and developed by others,[11] notably by Galanin,[6] and Inokuti & Hirayama.[10]

In this paper we would like to propose an alternative treatment which may have advantages for some applications such as phase fluorescence and cases where the behaviour of the acceptors is to be studied.[21–23]

A Green's function solution of the general equations

Let $S_j(t)$ represent the probability that at time t the j-th sensitiser is an excited state, and let $A_k(t)$ be the corresponding probability for the k-th acceptor. Let σ_S and σ_A represent the photon absorption cross-sections of the sensitiser and acceptor at the frequency of illumination and let $I(t)$ represent the intensity of illumination as a function of time.

The units of $I(t)$ are photons/cm^{2}s. So that the units of $\sigma_S I(t)$ and $\sigma_A I(t)$ are photons/s.

The probability that the j-th sensitiser is excited can change by several mechanisms: absorption of a photon from the monochromatic illuminating beam, spontaneous emission of a photon, and resonance transfer of the excitation energy to an acceptor.

Let us assume that these are the only competing mechanisms,[26, 27, 28] and rule out, for example, the possibility of a reverse resonance transfer of excitation energy back to the donor. This is a good approximation provided the Stokes shifts[28, 29] are such that overlap between the sensitiser absorption spectrum and the acceptor fluorescence spectrum are small. Then the rates of change of S_j and A_k are given by

$$\frac{dS_j}{dt} = -\frac{1}{\tau_S}\left[1+\sum_k F(R_{jk})\right] S_j + \sigma_S I(t) \tag{2}$$

and
$$\frac{dA_k}{dt} = -\frac{1}{\tau_A} A_k + \frac{1}{\tau_S} \sum_j F(R_{jk})\, S_j + \sigma_A I(t). \tag{3}$$

Here R_{jk} is the distance between the j-th sensitiser and the k-th acceptor. The term $-(1/\tau_S)\sum_k F(R_{jk})$ in equations (3) and (2) represents the rate of decay of S_j due to resonance transfer to one or other of the surrounding acceptors. In cases where the transfer proceeds by the Perrin–Förster mechanism[3, 30, 31]

$$F(R) = (R_0/R)^6. \tag{4}$$

We have assumed that the optical density of the system is small enough so that the illuminating beam is not greatly attenuated, and that diffusion is unimportant.[6, 16] Let us consider the function

$$S_j(t) = \int_{-\infty}^{\infty} dt'\, I(t') . G_j(t-t'), \tag{5}$$

where
$$G_j(t-t') = \sigma_S u(t-t') \exp\left[\frac{-(t-t')}{\tau_S}\left(1+\sum_k F(R_{jk})\right)\right] \tag{6}$$

and $u(t-t')$ is the step function

$$u(t-t') = \begin{Bmatrix} 0 & t < t', \\ 1 & t > t'. \end{Bmatrix} \tag{7}$$

The Green's function $(1/\sigma_S) G_j(t-t')$ can be thought of as the probability that the j-th sensitiser will emit a photon at time t if it is excited at time t'.

It is easy to see that the function $S_j(t)$, defined by equations (5)–(7) is a solution of (2), since

$$\left[\frac{d}{dt}+\frac{1}{\tau_S}\left\{1+\sum_k F(R_{jk})\right\}\right] G_j(t-t') = \sigma_S \delta(t-t'), \tag{8}$$

so that $\left[\frac{d}{dt}+\frac{1}{\tau_S}\left\{1+\sum_k F(R_{jk})\right\}\right] S_j(t)$

$$= \int_{-\infty}^{\infty} dt' I(t') \left[\frac{d}{dt}+\frac{1}{\tau_S}\left\{1+\sum_k F(R_{jk})\right\}\right] G_j(t-t')$$

$$= \int_{-\infty}^{\infty} dt' I(t') . \sigma_S \delta(t-t') = \sigma_S I(t). \tag{9}$$

Similarly, one can show that the function

$$A_k(t) = \int_{-\infty}^{\infty} dt' I(t') \, G_k(t-t'), \tag{10}$$

with

$$G_k(t-t') = u(t-t') \left[\sigma_A e^{-t/\tau_A} + \sigma_S \sum_j F(R_{jk}) \frac{(e^{-t/\tau_A} - e^{-t/\tau_S})\left\{1+\sum_k F(R_{jk})\right\}}{1-\frac{\tau_S}{\tau_A}+\sum_k F(R_{jk})} \right] \tag{11}$$

is a solution of (3). The function $\sum_j G_j(t-t')$ with $F(R) = (R_0/R)^6$ has been evaluated for computer-generated sets of random positions corresponding to various sensitiser and acceptor concentrations and the results are shown in fig. 1.

The cell model

We shall now introduce an approximate model of the system which will allow us to convert the sums into integrals.

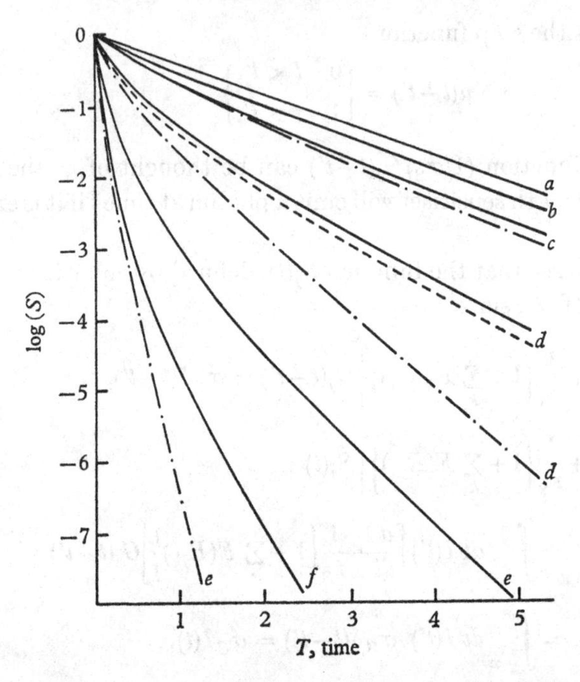

Fig. 1. The decay of sensitiser luminescence as a function of time. The abscissa represents time in units of τ_S, the ordinate the log of the excitation probability for an individual sensitiser excited at $T = 0$. The letters on the curves refer to the acceptor concentration C_A measured in units of C_0, $a = 0{\cdot}001$, $b = 0{\cdot}1$, $c = 0{\cdot}3$, $d = 1{\cdot}0$ $e = 2{\cdot}0$ and $f = 5{\cdot}0$. The solid lines show Förster's function, the short-dashed line the computer simulation

$$\sum_j G_S(t-t')/[u(t-t')\,\sigma_S],$$

and the dot-dashed curves show the cell-model function, $G_S(t-t')/[u(t-t')\,\sigma_S]$. At $C_A/C_0 \leqslant 0{\cdot}3$ the computer simulation and the Förster model give identical lines, with $C_A/C_0 \leqslant 0{\cdot}1$, all three functions are the same.

Consider an idealized system where the acceptors are regularly distributed throughout the solution. For example they might be arranged in a regular hexagonal lattice.[12] We shall also make the approximate assumption that the behaviour of the sensitisers within a sphere of radius R_1 about a particular acceptor is dominated by that acceptor. Thus the whole solution is thought of as being divided into approximately spherical cells. Finally we shall construct an average cell by superimposing all the cells in the solution and normalising the concentrations. This leads to a uniform continuous sensitiser distribution surrounding the central acceptor.

We can now try to evaluate the function $\sum_j G_j(t-t')$ where the sum is taken over all of the sensitisers in the averaged cell. Since there is only one acceptor to be considered we can write

$$\frac{1}{n_S}\sum_{j=1}^{n_S} G_j(t-t') = G_S(t-t')$$
$$\simeq \sigma_S u(t-t') \exp\left\{\frac{-(t-t')}{\tau_S}\right\} \frac{3}{R_1^3}\int_0^{R_1} R^2 dR \exp\left\{\frac{-(t-t')}{\tau_S} F(R)\right\}. \tag{12}$$

Here we have normalised the function

$$\sum_{j=1}^{n_S} G_j(t-t')$$

dividing it by n_S, the number of sensitisers in the cell. Fig. 1 shows

$$(1/\sigma_S)\, G_S(t-t')/u(t-t')$$

for the case where $F(R) = (R_0/R)^6$. This is to be compared with the corresponding function for the computer simulated system and with the results of Förster's statistical analysis.[2] Thus the total number of photons emitted per second by all sensitisers in the system is thus given by

$$\frac{N_S S(t)}{\tau_S} = \frac{N_S}{\tau_S}\int_{-\infty}^{\infty} dt'\, I(t')\, G_S(t-t'), \tag{13}$$

where $N_S = N_A n_S$ is the total number of sensitisers present and $G_S(t-t')$ is defined by equation (12). Analogously, the total number of photons per second emitted by the acceptors is

$$\frac{N_A A(t)}{\tau_A} = \frac{N_A}{\tau_A}\int_{-\infty}^{\infty} dt'\, I(t')\, G_A(t-t'), \tag{14}$$

where

$$G_A(t-t') = u(t-t')\Bigg[\sigma_A e^{-t/\tau_A}$$
$$+\frac{\sigma_S 3}{R_1^3}\int_0^{R_1} \frac{R^2 dR\, F(R)\,(e^{-t/\tau_A} - e^{-t/\tau_S}\{1+F(R)\})}{1-\tau_S/\tau_A + F(R)}\Bigg]. \tag{15}$$

The integral which forms part of equation (15) is shown in fig. 2 for various values of acceptor concentration in the case where

$$F(R) = (R_0/R)^6.$$

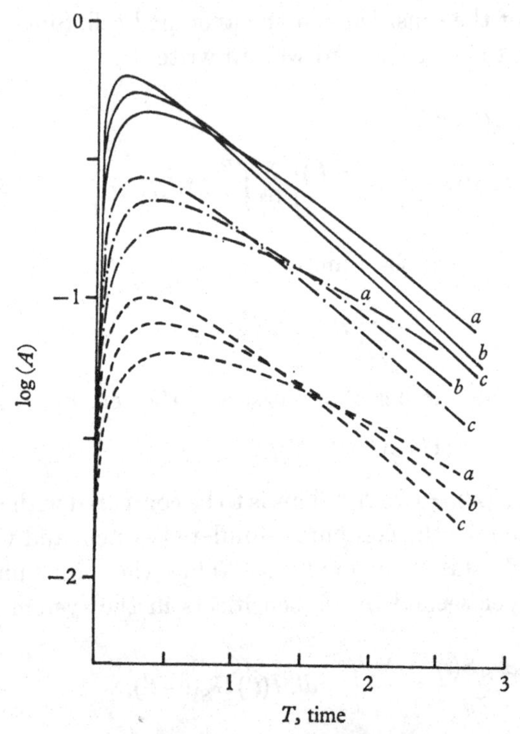

Fig. 2. Acceptor luminescence as a function of time. The ordinate represents the excitation probability of an acceptor initially in the ground state at $T = 0$. Time is measured in units of τ_A. The letters on the curves, $G_A((t-t'), \sigma_A = 0)/[u(t-t')\sigma_S]$, show the values of the parameters τ_A/τ_S used to evaluate equation (15), $a = 0{\cdot}5$, $b = 1{\cdot}0$, $c = 2{\cdot}0$. The solid lines show the function for $C_A/C_0 = 1{\cdot}0$, the dot-dashed line, $C_A/C_0 = 0{\cdot}3$, and the short-dashed line, $C_A/C_0 = 0{\cdot}1$.

Phase fluorimetry

A particularly interesting special case is that in which the intensity of the illuminating beam is a sinusoidally modulated function of time

$$I(t) = I_0 \cos \omega t + I_B = \operatorname{Re}\{I_0 e^{i\omega t} + I_B\}, \tag{16}$$

where I_B is the background intensity of illumination and Re stands for 'real part of'. Then the function $S(t)$ defined by equation (13) is given by

$$\begin{aligned} S(t) &= \operatorname{Re}\left\{\int_{-\infty}^{\infty} dt'(I_0 e^{i\omega t} + I_B)\, G_S(t-t')\right\} \\ &= \operatorname{Re}\left\{\int_{-\infty}^{\infty} dt'(I_0 e^{i\omega t} + I_B)\, \sigma_S \right. \\ &\quad \left. \times u(t-t') \exp\left\{\frac{-(t-t')}{\tau_S}\right\} \frac{3}{R_1^3} \int_0^{R_1} R^2\, dR \exp\left\{\frac{-(t-t')}{\tau_S} F(R)\right\}\right\}. \end{aligned} \tag{17}$$

Using the fact that

$$\int_{-\infty}^{\infty} dt' u(t-t') \exp\left\{\frac{-(t-t')}{\tau_S}\beta\right\} = \frac{\tau_S}{\beta}, \tag{18}$$

we have

$$S(t) = \sigma_S \tau_S \frac{3}{R_1^3} \int_0^{R_1} R^2 \, dR$$
$$\times \left\{ I_0 \left[\frac{(1+F(R)) \cos \omega t + \tau_S \omega \sin \omega t}{(1+F(R))^2 + (\omega \tau_S)^2}\right] + I_B/(1+F(R)) \right\}. \tag{19}$$

Analogously,

$$A(t) = I_B \left\{ \sigma_A \tau_A + \frac{3\sigma_S \tau_A}{R_1^3} \int_0^{R_1} \frac{R^2 \, dR \, F(R)}{1-\tau_S/\tau_A + F(R)} \left[1 - \frac{\tau_S/\tau_A}{1+F(R)}\right] \right\}$$
$$+ I_0 \tau_A \cos \omega t \left[\frac{\sigma_A}{1+(\omega\tau_A)^2} + \frac{3\sigma_S}{R_1^3} \int_0^{R_1} \frac{R^2 \, dR \, F(R)}{1-\tau_S/\tau_A + F(R)}\right.$$
$$\left.\times \left\{\frac{1}{1+(\omega\tau_A)^2} - \frac{\tau_S/\tau_A(1+F(R))}{(1+F(R))^2 + (\omega\tau_S)^2}\right\}\right] + \omega\tau_A^2 I_0 \sin \omega t \left[\frac{\sigma_A}{1+(\omega\tau_A)^2}\right.$$
$$\left.+ \frac{3\sigma_S}{R_1^3} \int_0^{R_1} \frac{R^2 \, dR F(R)}{1-\tau_S/\tau_A + F(R)} \left\{\frac{1}{1+(\omega\tau_A)^2} - \frac{(\tau_S/\tau_A)^2}{(1+F(R))^2 + (\omega\tau_S)^2}\right\}\right]. \tag{20}$$

In phase fluorimetry[21–23, 32, 33] one can measure the phase shift between the sinusoidally modulated illuminating beam and the luminescence of the sensitisers. From equation (19) it follows that the sensitiser phase shift ϕ_S is given by

$$\frac{\tan \phi_S}{\omega \tau_S} = \frac{\displaystyle\int_0^{R_1} \frac{R^2 \, dR}{(F(R)+1)^2 + (\omega\tau_S)^2}}{\displaystyle\int_0^{R_1} \frac{(F(R)+1) \, R^2 \, dR}{(F(R)+1)^2 + (\omega\tau_S)^2}}. \tag{21}$$

Similarly, for the acceptors the phase shift ϕ_A is given by

$$\frac{\tan \phi_A}{\omega \tau_A} = \left[\frac{\sigma_A}{1+(\tau_A\omega)^2} + \frac{3\sigma_S}{R_1^3} \int_0^{R_1} \frac{F(R) \, R^2 \, dR}{1-\tau_S/\tau_A + F(R)} \left\{\frac{1}{1+(\tau_A\omega)^2}\right.\right.$$
$$\left.\left. - \frac{(\tau_S/\tau_A)^2}{(1+F(R))^2 + (\tau_S\omega)^2}\right\}\right] \Big/ \left[\frac{\sigma_A}{1+(\tau_A\omega)^2} + \frac{3\sigma_S}{R_1^3}\right.$$
$$\left.\times \int_0^{R_1} \frac{F(R) \, R^2 \, dR}{1-\tau_S/\tau_A + F(R)} \left\{\frac{1}{1+(\tau_A\omega)^2} - \frac{\tau_S/\tau_A(1+F(R))}{(1+F(R))^2 + (\tau_S\omega)^2}\right\}\right]. \tag{22}$$

The functions $\tan \phi_S/\omega\tau_S$ and $\tan \phi_A/\omega\tau_A$ defined by equations (21) and (22) are shown in figs. 3 and 4.

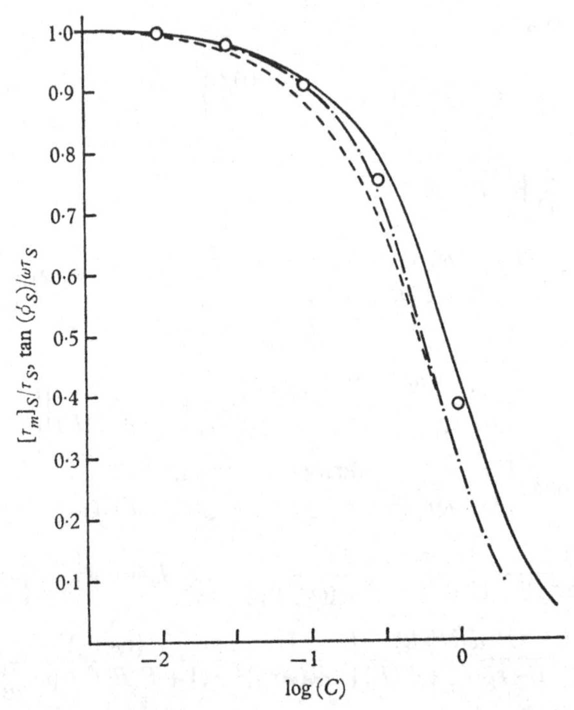

Fig. 3. The dependence of sensitiser lifetime and phase angle on acceptor concentration. The concentration is measured in units of C_0. The solid line shows $[\tau_m]_S/\tau_S$ obtained from Förster's decay function; the points were obtained from the computer simulation of $S(t)$. The dot-dashed line gives the function $\tan\phi_S/\omega\tau_S$ for $\omega\tau_S \leqslant 0{\cdot}04$ which is identical to $[\tau_m]_S/\tau_S$ derived from the cell-model. With $\omega\tau_S = 1{\cdot}33$ the form of $\tan\phi_S/\omega\tau_S$ is shown by the short-dashed line. At $C_A/C_0 \geqslant 1$ the functions $[\tau_m]_S/\tau_S$ and $\tan\phi_S/\omega\tau_S$ are the same, even with $\omega\tau \simeq 1$.

Luminescence efficiency and lifetimes

In the case of a time-independent illuminating beam, the results of the previous paragraph can be applied with $I_0 = 0$. One defines the sensitiser luminescence efficiency η/η_0 as the ratio of the sensitiser light output at a particular acceptor concentration to the corresponding output at zero acceptor concentration.(1–6) From equation (19) we find that

$$\left[\frac{\eta}{\eta_0}\right]_S = \frac{\int_0^\infty S(t, F(R))\,dt}{\int_0^\infty S(t, F(R) = 0)\,dt}$$

$$= \frac{3}{R_1^3}\int_0^{R_1} \frac{R^2\,dR}{1+F(R)}. \tag{23}$$

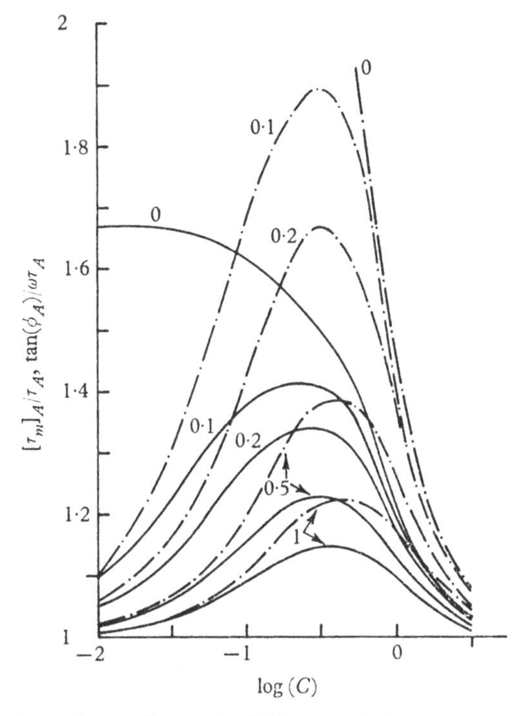

Fig. 4. The dependence of acceptor lifetime and phase angle on concentration. The solid lines show $[\tau_m]_A/\tau_A$ against reduced concentration C_A/C_0 obtained from equation (29) with $\tau_S/\tau_A = 1{\cdot}33$; the ratio τ_S/τ_A is shown on the curves. The dot-dashed lines are the function defined in equation (22), $\tan\phi_A/\omega\tau_A$, with $\tau_S/\tau_A = 1{\cdot}33$, $\tau_A/\omega = 1$, $\tau_S\omega = 1{\cdot}33$; σ_A/σ_S is shown on the curves. When $\omega\tau_S$, $\omega\tau_A \ll 1$, the functions $\tan\phi_A/\omega\tau_A$, $[\tau_A]_m/\tau_m$ become the same.

For the special case where $F(R) = [R_0/R]^6$

$$\left[\frac{\eta}{\eta_0}\right]_S = 1 - \frac{2C_A}{\sqrt{\pi}C_0}\tan^{-1}\left(\frac{\sqrt{\pi}C_0}{2C_A}\right), \tag{24}$$

where C_A is the acceptor concentration and C_0 is a 'critical concentration', defined by Förster[12]

$$C_0 = \frac{3}{2\sqrt{(\pi^3)}\,R_0^3}. \tag{25}$$

Hence the cell radius R_1 is given by

$$\left(\frac{R_1}{R_0}\right)^3 = \frac{\sqrt{\pi}}{2}\left[\frac{C_0}{C_A}\right]. \tag{26}$$

The interaction parameter, R_0, is the distance between an excited sensitiser and a ground state acceptor at which resonance energy transfer and spontaneous radiative decay of the sensitiser are equally probable. The function $[\eta/\eta_0]_S$ defined by equation (24) is shown in fig. 5 compared with the result of Förster's statistical analysis and the computer simulated system.

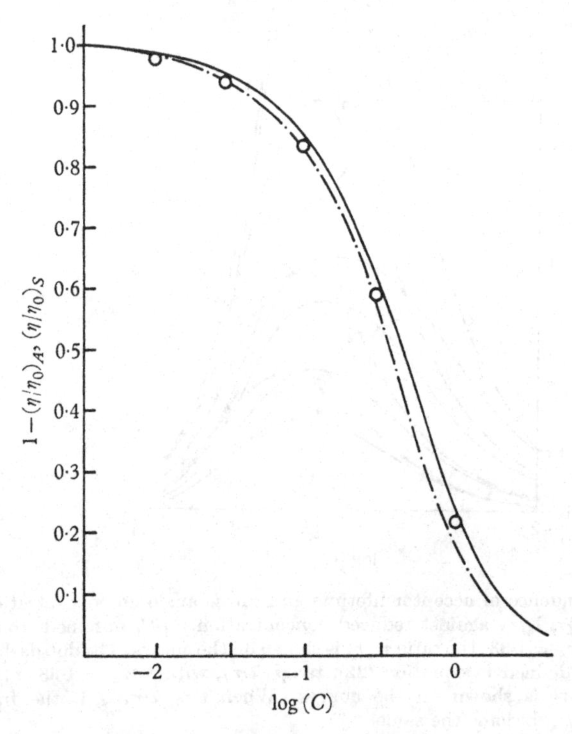

Fig. 5. Quantum efficiencies as a function of reduced acceptor concentration C_A/C_0. The ordinate shows the sensitiser quantum efficiency $[\eta/\eta_0]_S$; this is the same as $1-[\eta/\eta_0]_A$ where $[\eta/\eta_0]_A$ is the efficiency of the transfer. The solid line shows the results of Förster's statistics, the dot-dashed line is obtained from the cell-model, equation (24), and the points are from the computer simulation.

The mean lifetime of the sensitiser excited state in solution may be defined, following Inokuti,[10] as

$$[\tau_m]_S = \frac{\int_0^\infty t\,S(t)\,dt}{\int_0^\infty S(t)\,dt}, \tag{27}$$

$$\frac{[\tau_m]_S}{\tau_S} = \frac{\int_0^\infty t\,dt \int_0^{R_1} R^2\,dR\,e^{-t/\tau_S}(1+F(R))}{\tau_S \int_0^\infty dt \int_0^{R_1} R^2\,dR\,e^{-t/\tau_S}(1+F(R))}.$$

When $F(R) = [R_0/R]^6$

$$\frac{[\tau_m]_S}{\tau_S} = \frac{1+\dfrac{(C_A/C_0)^2}{2((C_A/C_0)^2+\frac{1}{4}\pi)} - \dfrac{3}{\sqrt{\pi}}\left(\dfrac{C_A}{C_0}\right)\tan^{-1}\left(\dfrac{\sqrt{\pi}}{2}\left(\dfrac{C_0}{C_A}\right)\right)}{1-\dfrac{2}{\sqrt{\pi}}\left(\dfrac{C_A}{C_0}\right)\tan^{-1}\left(\dfrac{\sqrt{\pi}}{2}\left(\dfrac{C_0}{C_A}\right)\right)}. \quad (28)$$

With $C_A/C_0 \ll 1$ equation (27) reduces to

$$[\tau_m]_S/\tau_S \simeq 1-\frac{\sqrt{\pi}}{2}\left(\frac{C_A}{C_0}\right)$$

as obtained by Galanin.[6] Similarly, for the acceptors we obtain the mean lifetime

$$\frac{[\tau_m]_A}{\tau_A} = 1+\frac{2}{\sqrt{\pi}}\left(\frac{C_A}{C_0}\right)\sigma_S\frac{\tau_S}{\tau_A}\left[\left\{\tfrac{1}{2}\tan^{-1}\frac{\sqrt{\pi}}{2}\left(\frac{C_0}{C_A}\right)\right.\right.$$
$$\left.\left.-\frac{\sqrt{(\pi)}\,(C_A/C_0)}{4(C_A/C_0)^2+\pi}\right\}\Bigg/\left\{\sigma_A+\frac{2}{\sqrt{\pi}}\left(\frac{C_A}{C_0}\right)\sigma_S\tan^{-1}\left(\frac{\sqrt{\pi}}{2}\left(\frac{C_0}{C_A}\right)\right)\right\}\right]. \quad (29)$$

Often in the interpretation of phase fluorimetry results the relation $\tan\phi = \omega\tau$ is used to relate lifetimes to observed phase shifts.[21–23, 32, 33] As may be seen from equations (21) and (28) and equations (22) and (29) this is not exactly correct in a sensitiser–acceptor system except with $\omega\tau_S \ll 1$ and $\omega\tau_A \ll 1$ when equation (21) reduces to equations (28) and (22) to (29) giving $\tan\phi/\omega$ as a direct measure of lifetimes as defined in equation (27). The functions $[\tau_m]_S/\tau_S$ and $[\tau_m]_A/\tau_A$ are shown in figs. 3 and 4.

Acknowledgements

We would like to thank Dr Hannah Gay, Mr R. S. Milner and Dr John Walkley for helpful discussions. Our thanks are also due to the Imperial College Computer Unit for access to their IBM 7090 Digital Computer.

References

1 T. Förster, *Z. Elektrochem.* **53**, 93 (1949).
2 T. Förster, *Z. Naturforsch.* **4***a*, 321 (1949).
3 T. Förster, *Ann. Physik*, **2**, 55 (1948).
4 T. Förster, *Disc. Faraday Soc.* **27**, 7 (1959).
5 T. Förster, Excitation transfer, in *Comparative Effects of Radiation*, M. Burton, J. S. Kirby-Smith and J. L. Magee, eds. (Wiley, New York, 1961), p. 300.
6 M. D. Galanin, *Soviet Phys.—JETP*, **1**, 317 (1955).
7 D. L. Dexter, *J. Chem. Phys.* **21**, 836 (1953).

8 V. L. Ermolaev & A. N. Terenin, *Pamiati S. I. Vavilova* (Moscow, 1952), p. 138. (English translation: NRC TT-540.)
9 A. N. Terenin & V. L. Ermolaev, *Dokl. Akad. Nauk. SSSR*, **85**, 547 (1952). (English translation: NRC TT-529.)
10 M. Inokuti & F. Hirayama, *J. Chem. Phys.* **43**, 1978 (1965).
11 K. B. Eisenthal & S. Siegel, *J. Chem. Phys.* **41**, 652 (1964).
12 J. S. Avery, Ph.D. Thesis, Imperial College, London University, 1965.
13 J. S. Avery, *Proc. Phys. Soc.* **88**, 1 (1966).
14 R. R. McLone & E. A. Power, *Proc. Roy. Soc.* (*London*) A, **286**, 573 (1965).
15 A. Kawski, *Ann. Physik*, **8**, 116 (1961).
16 J. Feitelson, *J. Chem. Phys.* **44**, 1497 (1966).
17 V. M. Agranovich, *Soviet Phys.—JETP* **10**, 307 (1960).
18 R. R. McLone & E. A. Power, *Mathematica*, **11**, 91 (1964).
19 W. T. Simpson, *Radiat. Res.* **20**, 87 (1963).
20 G. W. Robinson & R. P. Frosch, *J. Chem. Phys.* **38**, 1187 (1963).
21 A. Schmillen, *Z. Physik*, **135**, 294 (1953).
22 E. A. Bailey & G. K. Rollefson, *J. Chem. Phys.* **21**, 1315 (1953).
23 A. Schmillen, Decay times of organic crystals, in *Luminescence of Organic and Inorganic Materials*, H. P. Kallmann and G. M. Spruch, eds. (Wiley, New York, 1962), pp. 30–43.
24 M. Burton & H. Dreeskamp, *Disc. Faraday Soc.* **27**, 64 (1959).
25 R. K. Swank, L. J. Basile, *et al.*, *IRE Trans. Nucl. Sci.*, NS–**5**, 183 (1958).
26 H. P. Kallmann & M. Furst, *Phys. Rev.* **79**, 857 (1950).
27 D. F. McDonald, B. J. Dunn & J. V. Braddock, *IRE Trans. Nucl. Sci.*, NS–**7**, 17 (1960).
28 H. W. Leverenz, *An Introduction to Luminescence of Solids* (Wiley, New York, 1950).
29 A. Jablonski, *Z. Physik*, **73**, 460 (1931).
30 F. Perrin, *Ann. Phys.* (*Paris*), **17**, 283 (1932).
31 H. P. Kallmann & F. London, *Z. Phys. Chem.* B **2**, 207 (1929).
32 W. Kirchhoff, *Z. Physik*, **116**, 115 (1940).
33 W. Szymanowski, *Z. Physik*, **95**, 440 (1935).

DISCUSSION

HAMEKA

Van der Waals to Hameka: The vanishing of all one- and two-center integrals in the S(pin) O(rbit) C(oupling) Matrix elements for benzene because of 'local symmetry' only holds in the pure π-electron approximation. The very fact that the one-center contributions are so dominant in SOC for other molecules seems to indicate that intensity stealing from $\sigma\pi$ transitions is the major cause of phosphorescence in aromatic hydrocarbons. This belief is in line with the out-of-plane polarisation of the phosphorescence of polynuclear aromatic hydrocarbons and is also suggested by the mechanism of isotropic hyperfine interaction in aromatic radical ions. A calculation of SOC in naphthalene, including $\sigma\pi$ interaction leads to a radiative lifetime of the order of 10 s.

Hameka to van der Waals: This same idea occurred to us, and one of my collaborators investigated it for the case of benzene. In particular, we looked into the possibility that the triplet state would mix with excited singlet states, involving anti-bonding C—H or C—C orbitals. Unfortunately, we found that all spin-orbit matrix elements between the triplet state and the above described singlet states that we considered were zero. We did not investigate the case of naphthalene.

Rice to Hameka: Did you intend to imply that the frozen-core (electronic) approximation is adequate for the calculation of the zero-field splitting?

Hameka to Rice: No, this is not my intention. At first sight it would seem that the frozen-core approximation is totally inadequate in all cases. However, from a more detailed analysis it follows that the frozen-core approximation might work in certain cases. Therefore, I implied only that the frozen-core approximation might be adequate for certain molecules, but I would not want to make a more definite statement.

Robinson to Hameka: There is a point that may be pertinent here. If one takes the triplet radiative lifetimes of benzene, naphthalene, and anthracene to be all the same, which roughly speaking they seem to be, then the transition dipole squared goes in a ratio of one to two to three, respectively, in the three molecules. This implies that the complications associated with benzene, because of its high symmetry, must either be unimportant in the spin-orbit calculations or that such complications are also present in the other two molecules.

Robinson to Rice: The zero-field splitting that was observed in the singlet-triplet spectrum of gaseous formaldehyde, we thought, provided evidence that the frozen-core approximation was not valid for the low-lying excited states of this molecule. Uncertainties about the magnitude of spin-orbit contributions, however, caused us to have limited confidence in this interpretation.

Pullman & Robinson to Hameka: The inclusion of excited states involving σ electrons poses the problem of the accuracy of the σ representation. There is no such thing yet available for large molecules. What wavefunction did you use for the benzene calculation you mentioned?

Hameka to Pullman & Robinson: We used a very crude approximation to the wavefunction just to get some idea about the order of magnitude of the effect. The detailed form of the wavefunction became immaterial since we discovered that all terms we considered were zero because of symmetry considerations.

Sharnoff to Hameka: Might there not be a contribution to the spin-orbit coupling in aromatics which is analogous to the core polarisation observed in magnetic properties of transition metal ions?

Hameka to Sharnoff: This is a possibility. However, we considered only organic molecules with relatively light nuclei and for those systems the effects of core polarisation may be expected to be negligible.

HOCHSTRASSER

Birks to Hochstrasser: Do you have any new data on the phenomena which you attribute to triplet excimers?

Hochstrasser to Birks: All our observations on crystalline benzenes, up to this time, are consistent with the excimer interpretation of their emission spectra as published by Castro & Hochstrasser (*J. Chem. Phys.* **45**, 4352, 1966).

Tanaka to Hochstrasser: In what way do you explain the origin of the singlet-triplet absorption of benzophenone molecule, in which the direction of polarisation is determined to be tilted about 27° to the C=O axis?

Hochstrasser to Tanaka: We assume that this arises because the molecule utilises two components of the spin-orbit coupling operator, each mixing the triplet state in question with different singlet states, to different extents. Actually there is no reason, in general, why singlet-triplet transitions should be polarised along a molecular symmetry axis, or perpendicular to a symmetry plane; as I mentioned, only the mixed-in singlets are being observed. Admittedly, crystal field effects can also provide apparent rotations of the transition dipole.

SIEBRAND

Rice to Siebrand: I am impressed by the agreement between calculation and experiment, but I should like to ask about a question of principle. In your approach, first-order perturbation theory is used to describe a transition between a pure singlet state and a pure triplet state. How do you account for the conservation of angular momentum? Given that there are many ways of adding perturbations that will complement the pure electronic singlet-triplet transition and conserve the angular momentum, should not the presence of these perturbations also appear in the rate expression? I believe this is important in interpreting the mechanism of the transition and will supplement the considerations you have used, and may influence the conclusions drawn from the analysis.

Siebrand to Rice: My treatment is based on the assumption that the Franck–Condon factor is the only parameter which varies appreciably within the class of aromatic hydrocarbons. All other factors, including conservation of angular momentum, are lumped together in the constant C_β connecting the Franck–Condon factor and the radiationless rate constant. An *a priori* calculation of C_β is outside the scope of the present treatment. The good agreement between observed and calculated lifetimes indicates that it is reasonably constant for aromatic hydrocarbons. Thus it appears that an explicit consideration of the conservation of angular momentum, although important for an absolute calculation of triplet lifetimes, is not necessary when dealing with the relative rates of these lifetimes within the class of aromatic hydrocarbons.

Van der Waals to Siebrand: What is the physical significance of the fact that you may lump all vibrational modes together, irrespective of symmetry? Does it perhaps imply that the effect is mainly local in the C—H bonds?

Van der Waals to Rice: Spin angular momentum, of course, is conserved nearly perfectly. But when some 'leaks' out of the spin system into the orbital motion—as expressed by the $H_{s.o.}$ matrix elements of the present theory—it is no longer traceable. Not only is there very intimate coupling between the angular momenta of the electronic and nulcear motions (even in highly symmetrical non-linear molecules), but also inter-molecularly in condensed phase. In a gas the total angular momentum of a molecule would clearly be conserved by a change of the rotational angular momentum by one unit.

Siebrand to van der Waals: The radiationless transitions under consideration are multiphonon processes, so that restrictions due to symmetry are expected to be relatively unimportant. Thus whereas a transition $\Delta v = 1$ requires symmetric phonons, a transition $\Delta v = 2$ is allowed for any oscillator. Correspondingly, a transition involving many phonons and many oscillators will always be essentially allowed.

AVERY

Birks to Avery: At Manchester we are studying singlet energy transfer in liquid solutions using the Birks–Dyson phase and modulation fluorometer. If there are two exponentials involved in a luminescence decay and/or rise, the measurement of both phase and modulation at one frequency, or of either parameter at all frequencies is necessary and sufficient to obtain the two exponential time parameters. If the rise or decay function is non-exponential, it cannot be measured by phase fluorometry. It is then necessary to use flash (δ-function) excitation with observation of the luminescence time function by such techniques as pulse-sampling oscillography (Birks, Dyson, Munro, King) or photon-sampling analysis (Bollinger, Thomas, Koechlin, Lami, Birks, Smith-Saville).

Avery to Birks: δ-function excitation is especially interesting because the resulting luminescence is given directly by the Green's functions of the system, equations (12) and (15).

2 MAGNETIC RESONANCE AND MAGNETIC INTERACTIONS

MAGNETIC RESONANCE SPECTRA OF ORGANIC MOLECULES IN TRIPLET STATES IN SINGLE CRYSTALS†

CLYDE A. HUTCHISON JR.

1. Introduction

A salient feature of an organic molecule in a triplet electronic state is its magnetic moment arising from the almost pure electron spin angular momentum of its two so-called unpaired electrons. Magnetic methods and techniques have therefore played an important part not only in connection with the detection of triplet state species in a wide variety of organic chemical systems, but also in providing very detailed information on many fundamental aspects of a wide variety of problems related to various chemical and physical behaviours of such systems.

Interest in magnetic studies of aromatic organic molecules with two unpaired electron spins arose as early as 1935 in connection with the large biradicals such as Chichibabin's hydrocarbon.(1) The early susceptibility studies of Müller & Müller-Rodloff(1) showed no paramagnetism in such systems, but subsequent investigations by paramagnetic resonance methods(2) revealed very interesting paramagnetic properties. The relative roles of (*a*) triplet states well separated in energy from singlet states and of (*b*) states in which singlet and triplet are degenerate or the molecule is a 'double' radical is still not a completely decided matter.(3)

The main stimulus to studies of the magnetic properties of aromatic molecules in triplet states came much later from the suggestions of

† Parts of the work described in this paper which were done at the University of Chicago were financially supported by the United States Atomic Energy Commission, the United States Office of Naval Research and the National Science Foundation.

Electronic equipment used in experiments carried out at the University of Chicago was supplied by the United States Advanced Research Projects Agency.

Lewis, Lipkin & Magel[4] and of Terenin[5] that the long-lived phosphorescence of organic molecules in rigid media at low temperatures was produced by the slow decay of trapped triplet state molecules to their ground states, these triplet molecules having been produced by singlet-triplet intersystem crossings. The paramagnetism which should thus be an accompaniment of such phosphorescence was searched for by static field susceptibility methods and was observed and studied by Lewis, Calvin & Kasha,[6] by Frölich, Szalay & Szor,[7] by Evans,[8] and by Kortum,[9] beginning about 1955.

These static magnetic susceptibility measurements, together with the subsequent paramagnetic resonance investigations,[10, 11, 12] served both to establish the essential role of triplet states in organic phosphorescence and also to reveal magnetic techniques as extremely important methods for investigating organic molecules in triplet states and for studying many organic crystal processes in which triplet excitations were involved.

Since this first application to the study of photo-excited phosphorescing molecules, the scope of the applications of magnetic resonance techniques to triplet state organic molecules has been extended greatly. Many other triplet state systems of interest, e.g. organic crystal triplet excitons and organic molecules such as those of divalent carbon whose ground electronic states have triplet multiplicity, have been investigated using magnetic methods of various types.

This chapter will be mainly concerned with a summarisation of the results of the studies of the magnetic properties of organic molecules in triplet states in single organic crystals. Our subjects will be the magnetic resonance spectra and the information on single crystal systems obtainable from these spectra. Our illustrative materials will be drawn mainly from work done by various investigators in our laboratory at the University of Chicago.

2. Magnetic resonance spectra

The magnetic spectroscopy of those organic triplet state systems in which the triplet magnetic moment may be regarded as well trapped on individual molecules in the crystal may be conveniently summarised by giving the formal operator, i.e. spin Hamiltonian, whose eigenvalues in the triplet manifold of states (three in number) give term values which fit all the observed spectra. In table 1 an expression for this operator is written and approximate values of various terms are given for some exemplary aromatic molecule cases and for the conditions of a typical 3 cm wavelength microwave spectroscopy experiment.

The four terms of the spin Hamiltonian shown in table 1 summarise quite well the microwave spectroscopy observations, and so we will discuss them by talking about each term in turn.

Table 1

$$\mathscr{H} = \overset{(1)}{\mathbf{S}.\mathbf{T}.\mathbf{S}} + \overset{(2)}{|\beta_e|\,\mathbf{S}.\mathbf{g_e}.\mathbf{H}} - \overset{(3)}{\sum_k g_{nk}|\beta_n|\mathbf{H}.\mathbf{I}_k} + \overset{(4)}{\mathbf{S}.\sum_{i,k}\rho_i \mathbf{A}_{ik}.\mathbf{I}_k} \quad S = 1.$$

i indexes C's; k indexes H's or other nuclei

Term	$\lvert$Energy$\rvert/h$, mc s^{-1}		
	Photo-excited triplet	Substituted methylene	
(1)	3000	12000	
(2)	6000–12000	3000–22000	
(3)	8–16	4–35	for protons
(4)	0–20	0–8	for protons

2.1. S.T.S. *The fine structure term*

This term arises from those interactions which involve two angular momenta. In the photo-excited triplets it comes almost entirely from the magnetic dipole-dipole interaction between the two spins of the molecule which are locked parallel to each other in the triplet state. It is thus actually, as well as formally, a spin-spin interaction. The spin-orbit interactions of these electrons contribute to this term but to a smaller extent than the uncertainties associated with many of the spectroscopic experiments.[13] This spin-orbit contribution will always be expected to be quite small in long-lived phosphorescent molecules because, since it mixes singlet and triplet states, the long life implies its small value. In ground state triplet molecules it may be much larger but it is still small compared with the magnetic dipole-dipole interaction, i.e. 0·03 times the spin contribution.[14]

The fine structure term, arising from the tensor interaction between the two spins which are distributed over the aromatic framework, has a magnitude determined by the energies of interaction of two dipoles of strength $|\beta_e|$ at distances comparable with molecular dimensions. In the usual approximations, which represent the electrons' orbitals as linear combinations of single p_z orbitals on each carbon ($z \perp$ to plane of the flat aromatic molecule), this interaction cannot get larger than

$$|\beta_e|^2/(\text{aromatic nearest neighbour C—C distance})^3\, h = 5\,000\,\text{Mc s}^{-1}$$

because of the Pauli principle which allows only one 'odd' electron in a given orbital on a given C atom. In the substituted methylenes, whose ground electronic state is a triplet, the two electrons occupy orthogonal p orbitals, a $p\pi$ orbital and a $p\sigma$ orbital to a good approximation, on the same C atom, and thus the fine structure term may be much larger than in the photo-excited triplets, because the electrons are on the average much closer together and the energy of interaction varies as $\langle r_{12}^{-3}\rangle$. $\mathbf{r}_{12}$ is the position of one electron with respect to the other.

The dipole-dipole interaction between the two unpaired electrons is

$$+\frac{g_e^2\beta_e^2}{|\mathbf{r}_{12}|^3}\mathbf{S}_1\,.\,\mathbf{S}_2-3\frac{\mathbf{S}_1\,.\,\mathbf{r}_{12}\,\mathbf{S}_2\,.\,\mathbf{r}_{12}}{|\mathbf{r}_{12}|^2}.$$

The electrons are distributed over the aromatic molecular framework, and the spatial distribution is of course describable by an exact eigenfunction of the spinless many-electron Hamiltonian. In the photo-excited aromatic triplets the electrons are distributed over a large flattened nuclear structure leading of course to very different expectation values of the energy with direction of $\mathbf{S}$. In the ground state triplets the electrons are mainly in the toroidal $x\pm iy\,p$ distributions on the divalent carbon atom, again leading to a large anisotropy. The formulation of the calculation of the resulting magnetic dipole-dipole interaction energy in terms of the spinless components of the two electron density matrix in general form has been given.[15] If we make the approximation of considering only the two parallel spin electrons outside closed shells, use singly occupied orbitals to describe the outer two electrons, and expand these orbitals as linear combinations of atomic orbitals on the carbon atoms, the description is greatly simplified and is expressible as a sum of interactions between electrons in these atomic orbitals on the various carbon atoms.[16] If a and b are $2p_z$ orbitals on two carbon atoms in a photo-excited triplet, for example, the three triplet functions have the common orbital factor, $\psi_0[a(1)b(2)-b(1)a(2)]$, where the number in parentheses denotes the electron coordinates. Thus the contribution of these two carbon orbitals to the dipole-dipole interaction is represented in the triplet manifold,

$$\alpha\alpha\psi_0,$$

$$\frac{1}{\sqrt{2}}\{\alpha\beta+\beta\alpha\}\psi_0,$$

$$\beta\beta\psi_0,$$

by the matrix,

$$
\begin{array}{cccc}
 & \alpha\alpha & \frac{1}{\sqrt{2}}\{\alpha\beta+\beta\alpha\} & \beta\beta \\
\hline
\frac{3}{4}g_e^2|\beta_e|^2 & \multicolumn{3}{c}{\left(\begin{array}{ccc}
\iint \psi_0^2 \frac{r_{12}^2-3z_{12}^2}{r_{12}^2}\, d\tau_1\, d\tau_2 & 0 & \iint \psi_0^2 \frac{y_{12}^2-x_{12}^2}{r_{12}^5}\, d\tau_1\, d\tau_2 \\
0 & 0 & \\
\iint \psi_0^2 \frac{y_{12}^2-x_{12}^2}{r_{12}^5}\, d\tau_1\, d\tau_2 & 0 & \iint \psi_0^2 \frac{r_{12}^2-3z_{12}^2}{r_{12}^5}\, d\tau_1\, d\tau_2
\end{array}\right)}
\end{array},
$$

when principal magnetic axes, x, y, z, are used and the zero of energy is adjusted to be the energy of the lowest of the three states.

The same choice of molecular axes, x, y, z, diagonalises T in the spin Hamiltonian. Then the most general form in which the $\mathbf{S}.\mathbf{T}.\mathbf{S}$ term may be written is $DS_z^2+E(S_x^2-S_y^2)$, whose matrix in the strong field manifold, $|1\rangle$, $|0\rangle$, $|\bar{1}\rangle$ of triplet states is

$\lvert 1\rangle$	$\lvert 0\rangle$	$\lvert \bar{1}\rangle$
D	0	E
0	0	0
E	0	D

Thus we see that the two carbon atoms' contribution to the principal values of the fine structure tensor are related to the spin distributions on the two carbon atoms by

$$D = 3\frac{g_e^2|\beta_e|^2}{4} \iint \psi_0^2 \frac{r_{12}^2-3z_{12}^2}{r_{12}^5}\, d\tau_1\, d\tau_2,$$

$$E = 3\frac{g_e^2|\beta_e|^2}{4} \iint \psi_0^2 \frac{y_{12}^2-x_{12}^2}{r_{12}^2}\, d\tau_1\, d\tau_2.$$

In the simplest approximation, the D and the E in the spin Hamiltonian may be considered to be the sum of all such two center contributions. The essential features, however, are clear from this simple discussion. For example, it will be seen that for a distribution of spin which is oblate with respect to the z axis, as is the case for planar aromatic molecules such as those shown in fig. 1, the value of $\langle x^2+y^2-2z^2\rangle$ is positive, whereas for a prolate distribution it would be negative. Also, if the z axis is an axis of threefold or higher symmetry as in the case of benzene, for example, then $\langle y^2-x^2\rangle$, and therefore E, is zero. This very simple picture may be, and has been, extended to give much more sophisticated estimates of the numerical values of the contributions of the $\mathbf{S}.\mathbf{T}.\mathbf{S}$ term of the spin Hamiltonian.(17)

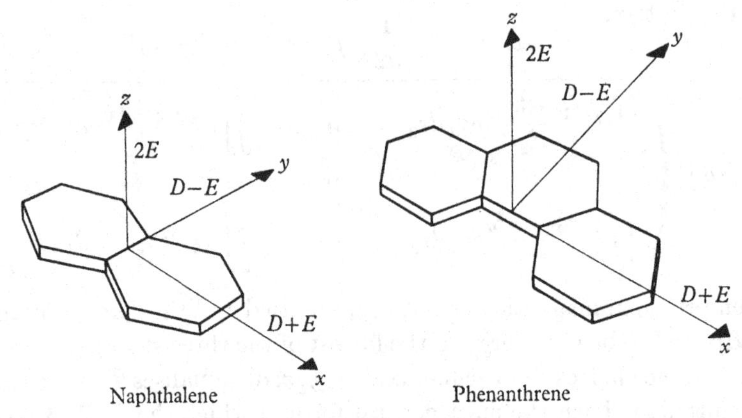

Fig. 1. Axis systems and polarisations.

Important interrelated consequences of this **S.T.S** term are the following:

(1) The spectra as usually obtained in magnetic fields of several thousand gauss are very anisotropic,[10, 18, 19] as shown in figs. 2 and 3. Plots of energy versus field strength (actually versus the proton magnetic fluxmeter frequency) such as those shown in the heavy lines of fig. 4 display the effects of terms (1) and (2) in the spin Hamiltonian,

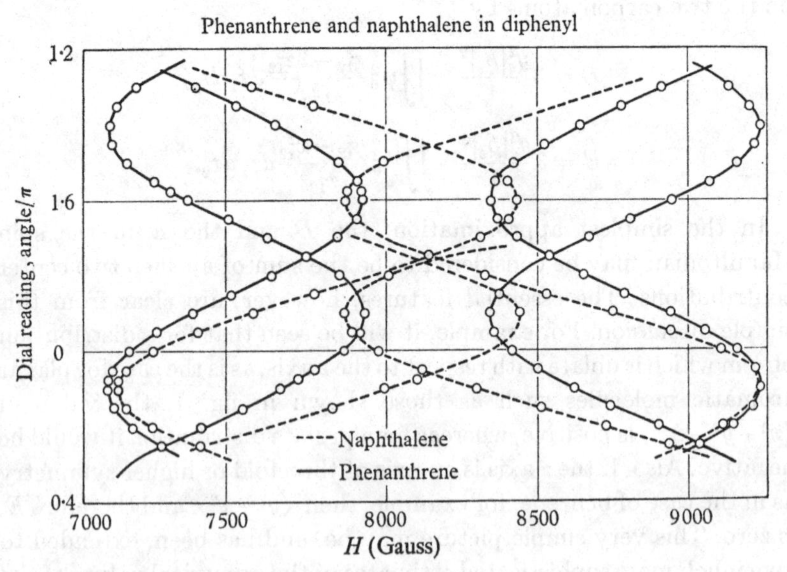

Fig. 2. $|H|$ for resonance against angle of H in crystallographic bc plane of diphenyl. Dial reading is angle/π.[18]

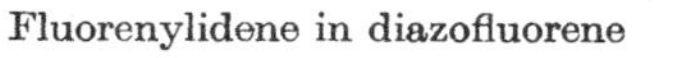

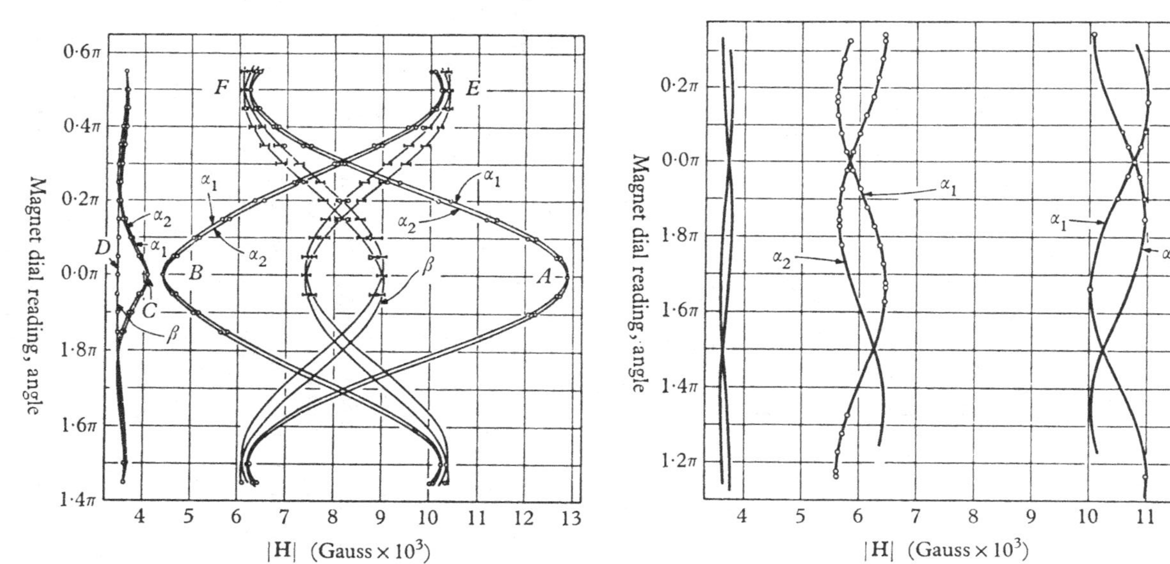

Fig. 3. $|\mathbf{H}|$ for resonance *vs* angle of **H** (left side) in *bc* plane of diazofluorene crystal and (right-side) in *xy* plane of fluorenylidene molecule on the left, $0{\cdot}0\pi$ denotes a direction $\parallel$ *c* axis on the right, $0{\cdot}0\pi$ denotes a direction $\perp$ *b* axis of the diazofluorene crystal.(196)

that is, they show the competition between the laboratory magnetic field and the molecular fields in the orientation of the spins, and thus clearly display the origins of the anisotropies of the EPR absorption lines.

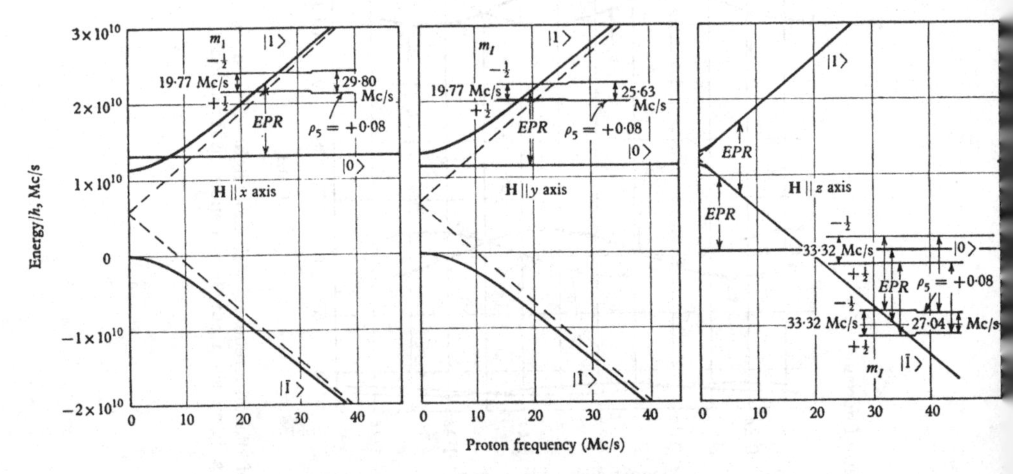

Fig. 4. Energy level diagram for triplet state molecule.[46]

(2) At zero external magnetic field, i.e. when term (2) (table 1) in the spin Hamiltonian is zero, the eigenvalues of term (1) (table 1) of the spin Hamiltonian are $D+E$, $D-E$, and 0. These energy splittings, as shown in table 1, correspond to transition frequencies in the microwave region. Thus it is possible to use a frequency scanning zero-field magnetic resonance spectrometer,[20] without laboratory magnet, for observation of paramagnetic resonance absorption in the triplet state molecules. Such an absorption obtained in a zero-field spectrometer is shown in fig. 5. Such direct measurements of zero-field splittings give the most accurate and precise numerical values of the D and E parameters, but of course they may also be obtained from the high-field measurements. With the zero-field spectrometer the values of the zero-field splittings may be measured to such a precision that environmental effects (organic crystal field effects) on the intramolecular electron spin distribution are much larger than the errors. For example, changing the host crystal to one composed of a different chemical substance changes the splittings by an amount which is two orders of magnitude larger than experimental errors. A change of 2 °K in temperature of the host crystal at 77 °K produces a change in zero-field splittings in naphthalene just equal to experimental error.

The zero-field magnetic dipole absorptions are of course predicted to be strongly polarised as shown in fig. 1. These polarisations have been observed and verified experimentally.[18]

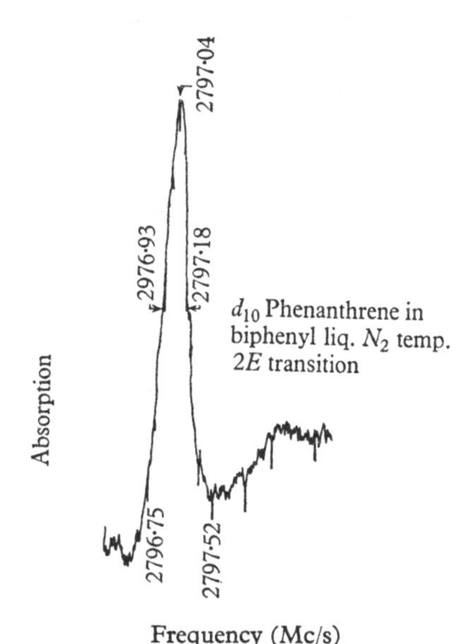

Fig. 5. Zero-field magnetic resonance signal.

(3) In conventional magnetic resonance experiments there is strong mixing of the high magnetic field eigenstates, i.e. the eigenstates of term (2) of the spin Hamiltonian, by term (1), the spin-spin term, because these two terms are comparable in size in such experiments.[10] Thus, for example, for a general orientation of the molecule in the magnetic field, and with the microwave magnetic field polarised perpendicular to the static laboratory magnetic field, all three transitions between the split three states of the triplet are observed. This is seen in fig. 3.

The eigenstates of the sum of terms (1) and (2) are given by

$$\cos\theta\,|1\rangle + \sin\theta\,|\bar{1}\rangle,$$
$$|0\rangle,$$
$$\sin\theta\,|1\rangle - \cos\theta\,|\bar{1}\rangle,$$

for the case when $\mathbf{H} \parallel z$ axis, where $\tan 2\theta = (E/g_{zz})\,|\beta_e|\,|\mathbf{H}|$, $|1\rangle$, $|0\rangle$ and $|\bar{1}\rangle$ are the eigenstates of term (2), and similar expressions hold for other directions. As a result of this state mixing we have the consequences that the expectation value of the spin in the direction of the field, $\langle S_H \rangle$, for that state which becomes $|1\rangle$ as $|\mathbf{H}|$ is increased, may fall far below 1 and that the angle between **S** and **H** may be very large.

(4) Because of the fact that when **H** is along the axes which diagonalise the **T** of the spin Hamiltonian, the values of **H** for magnetic resonance are stationary with respect to angle of **H** in the crystallographic axis system, as shown in fig. 3, and because of the fact that such axes are determined by the symmetry and structure of the molecular framework, one may, from such studies of this fine structure anisotropy, determine precisely the orientation of triplet state molecules in organic host crystals.[10, 18, 19] An illustration of this is given in table 2.

Table 2. *Direction cosines and their standard deviations in the diazofluorene crystallographic axis a', b, c system, of the x, y, z axes of the α and β fine-structure tensors of the triplet state species produced by irradiation of a pure diazofluorene crystal*

Fine-structure axis		Crystal axis a'		b		c	
		cos	$\sigma\{\cos\}$	cos	$\sigma\{\cos\}$	cos	$\sigma\{\cos\}$
α	x	+0·868	0·022	−0·494	0·039	+0·051	0·016
	y	+0·492	0·039	+0·869	0·022	+0·053	0·011
	z	−0·071	0·018	−0·021	0·007	+0·997	0·001
β	x	+0·554	0·025	−0·627	0·028	+0·548	0·025
	y	+0·424	0·025	+0·779	0·023	+0·463	0·026
	z	−0·717	0·023	−0·024	0·012	+0·697	0·023

(5) When a paramagnetic triplet state species, such as a photoexcited molecule or a substituted methylene, is created in an organic crystal and substitutes for a normal molecule in a position formerly occupied by such a normal molecule, its orientation is determined by the structure of the host which surrounds it. Many of the features of the crystal structure are thus reflected in this orienting of the guest and in particular the symmetry of the environment displays itself through the symmetry of the observed anistropic magnetic spectra. Thus inferences concerning the structure of the host crystal may be made from these spectra.[18, 19, 21] An example is shown in fig. 6 in which case it was deduced from the magnetic spectra that the host crystal was not centrosymmetric as later confirmed by X-ray diffraction studies.

A second example is shown in fig. 7 in which case it was deduced from the magnetic fine structure observations that in this crystal all

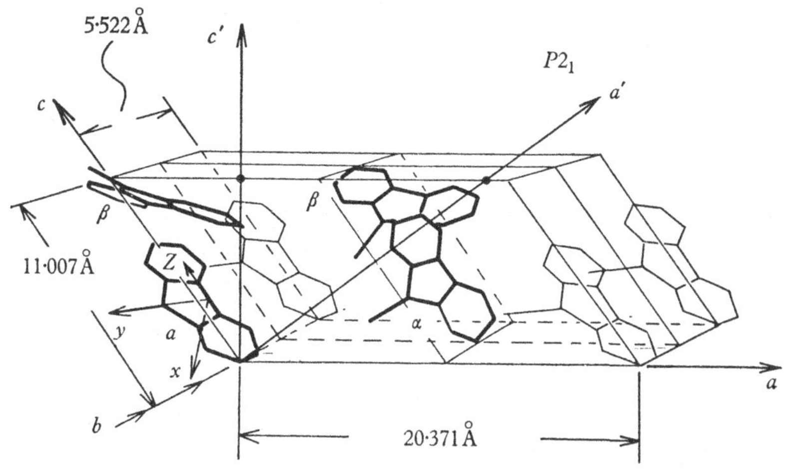

Fig. 6. Schematic diagram of proposed crystal structure of diazofluorene which is consistent with magnetic resonance studies on fluorenylidene in diazofluorene.[19b]

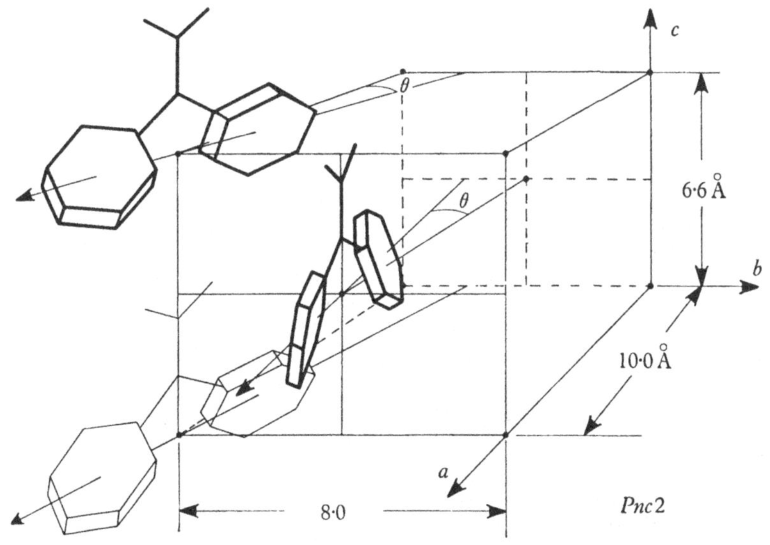

Fig. 7. Possible structure of 1,1′-diphenylethylene based on magnetic resonance, polarised optical absorption and X-ray data.[47]

the molecules were aligned with their principal axes almost parallel, and thus that this would be a good crystal for the study of polarised optical absorption.[21]

(6) The fine structure patterns are strikingly different for different triplet state molecules as shown in fig. 2 where two different photo-excited triplet state molecules coexist in the same host. It is easy to

distinguish the signals from the different guest species and thus to observe the transfer of energy from one such species to another.[22]

In table 3 are summarised measurements of principal values of **T** for a photo-excited triplet state molecule, together with theoretical calculations of them.

Table 3. *Experimental and theoretical values of spin Hamiltonian parameters for photo-excited triplet states*

	Experimental phenanthrene in diphenyl at 77 °K	Theoretical: Boorstein & Gouterman[17] Hückel orbitals	Theoretical: Boorstein & Gouterman[17] Hoffman orbitals	Theoretical: van der Waals & ter Maten[17] with $\sigma-\pi$ correction
D/hc (cm^{-1})	$+0{\cdot}100430$ $\pm 0{\cdot}000010$	$+0{\cdot}0731$	$+0{\cdot}0851$	$+0{\cdot}105$
E/hc (cm^{-1})	$-0{\cdot}046576$ $\pm 0{\cdot}000009$	$-0{\cdot}0269$	$-0{\cdot}0304$	$-0{\cdot}057$
g_{zz}	$+2{\cdot}00209$ $\pm 0{\cdot}00005$	—	—	—
g_{xy}	$+2{\cdot}00279$ $\pm 0{\cdot}00005$	—	—	—
g_{xx}	$+2{\cdot}0041$	—	—	—

	Experimental naphthalene in durene at 77 °K	Experimental naphthalene in diphenyl at 77 °K	Theoretical: Godfrey, Kern & Karplus,[17] Pariser wave function	Theoretical: Boorstein & Gouterman,[17] Hückel orbital	Theoretical: van der Waals & ter Maten,[17] with $\sigma-\pi$ correction
D/hc (cm^{-1})	$+0{\cdot}10119$	$+0{\cdot}09921$ $\pm 0{\cdot}00004$	$+0{\cdot}1081$	$+0{\cdot}1003$	$0{\cdot}101$
E/hc (cm^{-1})	$-0{\cdot}01411$	$-0{\cdot}01548$ $\pm 0{\cdot}00004$	$-0{\cdot}0093$	$-0{\cdot}0133$	$-0{\cdot}028$
g_{zz}	$+2{\cdot}0029$ $\pm 0{\cdot}0004$	—	—	—	—
g_{yy}	$+2{\cdot}0030$ $\pm 0{\cdot}0012$	—	—	—	—
g_{xx}	$+2{\cdot}0030$ $\pm 0{\cdot}0004$	—	—	—	—

Experimental values for a typical ground state triplet molecule are shown in table 4.[19, 23]

The measured values of the fine structure interactions in the case of the ground state triplets in single crystals have given some interesting insights into questions of molecular shapes and structures. For example, diphenylmethylene may be created in a host benzophenone

Table 4. *Value of spin Hamiltonian parameters for diphenylmethylene*

Parameter	Value in 1, 1′-diphenyl-ethylene	σ	Value in benzophenone	σ
D/hc	$\pm 0{\cdot}3964$	0·0002	$\pm 0{\cdot}4051$	0·0002
E/hc	$\mp 0{\cdot}01492$	0·0009	$\mp 0{\cdot}0192$	0·0001
g_{zz}	2·0022	0·0008	2·0025	0·0005
g_{xx}	2·0029	0·0002	2·0045	0·0005
g_{yy}	2·0021	0·0007	2·0043	0·0005

crystal by photodissociation of diphenylmethylene incorporated in the crystal.(19) The resulting magnetic resonance signal remains indefinitely at 77 °K and shows only Curie law intensity changes down to 4 °K so comes from either a ground state species or from one in a state very near its ground state. We may visualise the methylene as having one or another of the shapes shown in fig. 8. As stated before,

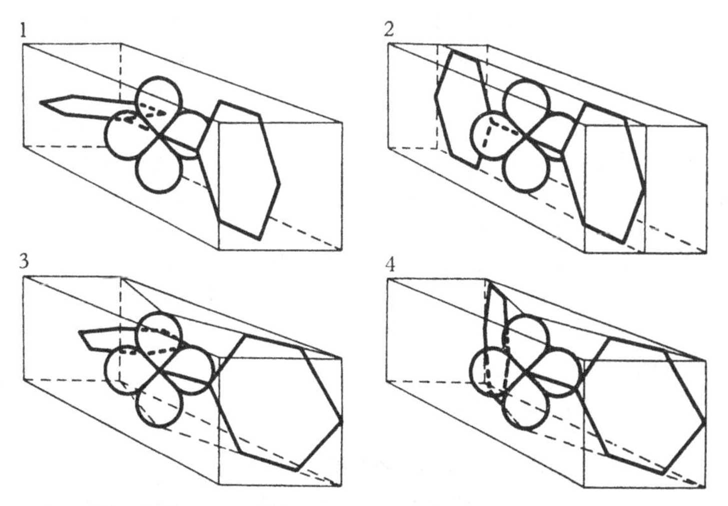

Fig. 8. Some possible structures of diphenylmethylene.

the D term arises mainly from the one center interaction between the two parallel spin electrons in the two orthogonal p orbitals on the central divalent C atom. The large value of D found in table 4 shows that the two electrons are on the average much closer than in the photo-excited triplets with their relatively small D's and thus identifies the formed species as a methylene. The best estimate of the size of D/hc for two such interacting p electrons on a C atom is $+0{\cdot}9055\ \mathrm{cm}^{-1}$.(24) The z axis is of course chosen normal to the plane of the two orthogonal p orbitals. The p distribution is very oblate with the large positive D.

The fact that D/hc is only 0·405 cm^{-1} shows that the p electron contribution from the divalent C is only about one-half of the expected value and one or both of the following may be true: (*a*) the spin may be delocalised from the central C and spread over the rings; (*b*) the molecule may be bent in one of the ways shown in fig. 8 so that the central C orbitals may be best described as having some s character and not as pure p orbitals. Distribution of spin on the rings can greatly reduce D both because the average interelectron distance can become much greater and because the distribution is more prolate giving negative contribution.

The bending of the molecule with implied s character of the central C orbitals drops the value of D inasmuch as the dipole-dipole interaction between a p and an s electron on the same C corresponds to a D/hc with the value $+0{\cdot}4803$ cm^{-1}.[25] Another point to be mentioned in connection with the fine structure interactions is the finite experimental size of E. Model 1 of fig. 8, which might be expected to be a stable configuration because it maximises overlap of the carbon p orbitals with the π electrons of the aromatic rings, cannot be a correct description because, as we have seen, the symmetry about the z axis (axis through ring centers) requires E to be zero. It should also be noted that as long as there are only pure p occupied orbitals on the central C, delocalisation of spin by different amounts from these p orbitals still gives no one center contribution to E. The two center contributions from spin on the rings can, however, give a finite E provided the molecule has one of the structures other than the one labelled 1. We defer further discussion of these points until we have discussed the other terms of the spin Hamiltonian.

2.2. $|\beta_e|\,\mathbf{S}\cdot\mathbf{g_e}\cdot\mathbf{H}$. *The electron Zeeman term*

It has already been stated that the electronic angular momentum in the organic triplet state molecules is almost pure spin. Inasmuch as the spin orbit coupling is so very small in these systems[13] the tensor describing the direct coupling of this spin to the laboratory magnetic field is nearly isotropic and is almost equal in numerical value to that for a free electron spin. Numerical values obtained by fitting the spin Hamiltonian to observed absorption are shown in table 3 for photo-excited triplets and in table 4 for a ground state triplet. Thus the only important way in which the anisotropy arises is through the correlation in electron position, over the very aspherical molecular framework, arising from the magnetic dipole-dipole interaction as described above.

2.3. $-\sum_k g_{nk}|\beta_n|\mathbf{H}.\mathbf{I}_k$. *The nuclear Zeeman term*

Although it is true that in general in systems with electron Zeeman interactions and also electron nucleus interactions we may find terms that are of first power in **H** and first power in **I** of just the form of term (3) of the spin Hamiltonian[26] (e.g. nuclear resonance chemical shifts due to admixture of higher paramagnetic states by the laboratory field), but which do not arise from the nuclear Zeeman interaction, in the case of organic triplets this term arises almost entirely from this latter source, for all other paramagnetic states are at much higher energies. Thus in the case of very large laboratory fields, in which both the electron spin and the nuclear spins, say proton spins in an aromatic molecule, are well quantised in the directions of the external magnetic field, this term makes a contribution to the energy which expressed as a frequency is of course just the nuclear resonance frequency of the protons in the water sample of the nuclear fluxmeter being used to measure the field for the triplet state resonance. Also, in such a situation, this term will correspond to an isotropic contribution to the energy and so this term has been written as a scalar product.

In the ordinary electron paramagnetic resonance spectrum, in this very high field case, this term will not manifest itself because the observed transitions occur between states of the same nuclear quantum number. Thus no splittings of spectral lines or shifts of line position are produced by it. The right-hand part of fig. 4 makes this point clear. Note that in fig. 4 the nuclear Zeeman level intervals are magnified by a factor of 100 with respect to the electronic level intervals.

2.4. $S\,.\sum_{i,k}\rho_i\mathbf{A}_{ik}.\mathbf{I}_k$. *The electron nucleus interaction term or hyperfine structure term (i labels the carbons and k labels the protons in the case of proton hyperfine interaction)*

In the organic triplet state molecules in single crystals this term arises from two sources; from the magnetic dipole-dipole interaction between the triplet state π electrons and the nuclei with magnetic moments and from the mixing of the π and σ systems of electrons by exchange interaction so that the predominantly π triplet-state electrons actually contact the nuclei of the aromatic molecules. The dipole-dipole part of the interaction is very anisotropic and the contact part is of course isotropic. (These interactions have been discussed in great detail and are well understood in the case of organic doublet radicals.[27])

The dipole-dipole part of the interaction is of the same form as that given previously for the interaction which leads to the fine structure. We may write for the energy of electron nucleus interaction the expression

$$-\frac{g_e g_n |\beta_e||\beta_n|}{|\mathbf{r}_{12}|^3}\left\{\mathbf{S}.\mathbf{I}-3\frac{\mathbf{S}.\mathbf{r}_{12}\mathbf{I}.\mathbf{r}_{12}}{|\mathbf{r}_{12}|^2}\right\},$$

where now $\mathbf{r}_{12}$ is the position of the nucleus with respect to the electron, β_n is the nuclear magneton and $\mathbf{I}$ is the nuclear spin. The electron g_e is regarded as a scalar in this expression and we have seen that this is nearly the case for triplet organic molecules. This expression may be conveniently rewritten as[28]

$$-g_e g_n |\beta_e||\beta_n|\,\mathbf{S}.\mathbf{A_d}.\mathbf{I},$$

where
$$\mathbf{A_d} \equiv \{\mathscr{I}-3\hat{\mathbf{r}}_{12}\hat{\mathbf{r}}_{12}\}/|\mathbf{r}_{12}|^3,$$

$\hat{\mathbf{r}}_{12}$ being a dimensionless unit vector in the direction of $\mathbf{r}_{12}$ and $\mathscr{I}$ being the unit dyadic operator. For the very high field case for which $\mathbf{S}$ and $\mathbf{I}$ are both well quantised along the external field $\mathbf{H}$, we may then write

$$-g_e g_n |\beta_e||\beta_n| M_S M_I \hat{\mathbf{H}}.\mathbf{A_d}.\hat{\mathbf{I}}$$

for this contribution to the energy.

We may further simplify this discussion for the case of spin angular momentum distributed over an aromatic molecular framework by approximating the molecular orbitals as linear combinations of carbon atomic orbitals and writing this dipole-dipole contribution in the form

$$-g_e g_n |\beta_e||\beta_n|\,\mathbf{S}.\mathrm{tr}\,(\rho\mathbf{A_d}).\mathbf{I},$$

where $\mathrm{tr}\,(\rho\mathbf{A_d})$ means
$$\sum_{ij}\rho_{ij}\langle\phi_j|\,\mathbf{A_d}\,|\phi_i\rangle,$$

the ϕ's are the carbon orbitals and ρ_{ij} is the product of coefficients of the corresponding ϕ's in the linear combination. For our case of triplet state molecules in which (*a*) the strong field state mixing by the fine structure interactions is appreciable and has large effects on the expectation value of the spin and the direction of the spin and in which (*b*) external fields of such size are employed that the deviations from perfect quantisation of the nuclear spin (Breit Rabi problem) must be considered we must diagonalise

$$\sum_{i,k}\{-g_e g_n|\beta_e||\beta_n|\,\langle\mathbf{S}\rangle.\mathrm{tr}\,(\rho_i\mathbf{A}_{\mathbf{d}ik}).\mathbf{I}_k-g_{nk}|\beta_n|\,\mathbf{H}.\mathbf{I}_k\},$$

the sum of the hyperfine term and the previously discussed term (3).

In discussions of the hyperfine interactions of protons, for example, with the triplet state electrons, in this approximation we may neglect the off diagonal elements of the matrix, ρ, but it is important to consider all diagonal elements, i.e. to consider the effect, on a given proton, of the spin distribution on all carbon atoms and not just the spin on the carbon adjacent to it as in the case of the isotropic hyperfine interactions[27] for radicals in solution. In many cases the effect of the spin on nonadjacent carbon atoms may be much larger than that of the spin on the adjacent carbon atom. The diagonal element of ρ will be called the spin density on the corresponding carbon atom. The isotropic component of the proton hyperfine interaction may be assumed to be proportional to the spin in the $p\pi$ orbital on only the carbon atom adjacent to the proton in question. This interaction will not be discussed in detail in this paper and will be presumed to arise from interactions similar to those which produce it in the case of doublet state radicals.[27]

These hyperfine interactions manifest themselves in the ordinary electron paramagnetic resonance spectra as splittings of the absorption lines of the spectrum. As stated before, the transitions which are observed are those between states with the same nuclear spin quantum number. Hence the Zeeman levels are shifted by the hyperfine interaction and thus the lines are split. This situation is made clear by the diagram on the right side of fig. 4. Remember that in fig. 4 the nuclear Zeeman and hyperfine level intervals are magnified by a factor of 100 with respect to the electron level intervals.

It should also be remarked that when one is considering the relation of the size of the splitting produced by the hyperfine interaction to the size of the interaction and the size of the electron spin density responsible for the interaction, that the interactions per unit spin density to which one is accustomed in the case of doublet radicals are transferable to this triplet situation when the total spin density of the molecule is normalised to 1, *not* to 2. When the spin of 1 flips from one orientation in the field to one differing in M_s by 1, then one may naïvely say that each of the 2 spins has flipped only halfway over and so has only half the effect of one spin in a doublet. More accurately, whereas both of the electron Zeeman levels, $+\frac{1}{2}$ and $-\frac{1}{2}$, in the case of $S = \frac{1}{2}$ are split by a given amount by the hyperfine interaction, in the case of $S = 1$, although one of them is split by twice this same amount, the other one, the state $|0\rangle$, is not split at all because the expectation value of the electron spin is zero in the direction of $\mathbf{H}$. So the net splitting of lines is the same as for the doublet case (see fig. 4).

Typical hyperfine patterns for a microwave paramagnetic-resonance absorption for photo-excited phenanthrene molecules,[29] which have five different pairs of equivalent protons, are shown in fig. 9. The proton hyperfine lines of such spectra are always found to be broadened sufficiently that not all the lines expected from the number of protons present are observed. Such broadening presumably comes mainly

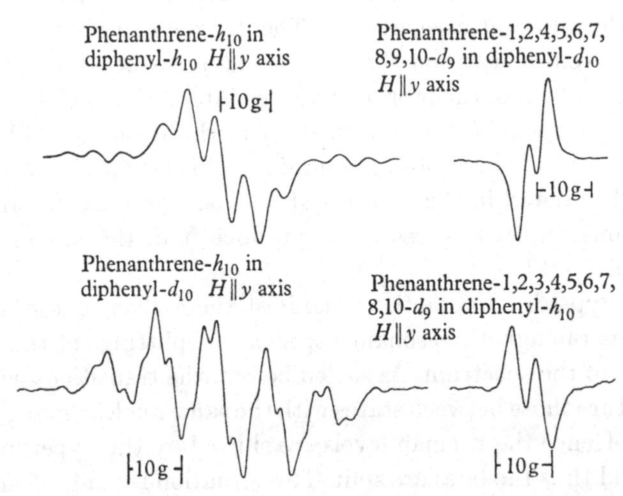

Fig. 9. Hyperfine patterns of photo-excited phenanthrene in diphenyl.

from dipole-dipole interactions with protons of neighbouring host molecules, from dipole-dipole interactions with other triplet state molecules in the vicinity and from anisotropy broadening and crystal field broadening arising from the imperfections in the structure of the host crystal. Hyperfine structure patterns for several partially deuterated species are helpful in the interpretation of hyperfine patterns in terms of the density of spin on the various C atoms. In the case of naphthalene,[30] for example, deuterating the β position gives a pattern arising almost completely from the 4α protons because of the small gyromagnetic ratio of the deuteron relative to the proton.

In the case of photo-excited naphthalene, spin densities on the three types of carbon atoms, the α and β C's and the ones without protons attached, which we will call the γ C's, have been determined from the experimental hyperfine patterns employing the following model.[30] Fig. 10 is helpful for visualising the model.

(1) The isotropic part of the hyperfine interaction arises only from separate interactions within the eight individual C—H fragments,

four of them in equivalent α positions and four in equivalent β positions in the ring. The contribution to the isotropic interaction with the proton adjacent to the i-th C atom is proportional to the spin density in the p_z orbital on the i-th C atom.

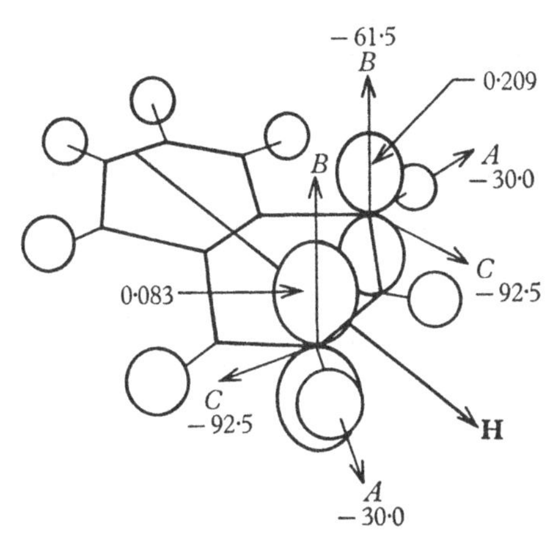

Fig. 10. Proton hyperfine interactions from adjacent carbon p orbitals in naphthalene. (Interaction energies/h per unit spin density are given in Mc sec^{-1}.)[48]

(2) The contribution of the spin density in the p_z orbital on the i-th C atom to the anisotropic dipole-dipole interaction with the proton adjacent to the i-th atom is the same, per unit spin density, as in the case of the central C—H fragment of the malonic acid radical.[34, 45]

(3) The contribution of the p_z orbital on the i-th C atom to the dipole-dipole interaction with the adjacent proton is proportional to the spin density in that p_z orbital.

(4) The proportionality constants for the interactions are the same for both the α and β positions.

(5) The contribution of the spin density ρ_j, in the p_z orbital on the j-th C atom to the interaction with the proton on the i-th C atom, $i \neq j$, is equal to that of two spin densities, each of size $1/2\rho_j$, located at points equal distances above and below (with respect to the aromatic plane) the j-th C atom. These distances are taken to be the most probable distances from the C atom for an electron in a Slater C orbital.

(6) The naphthalene molecule possesses orthorhombic symmetry. The α C—H bonds are all parallel to the y axis of the molecule. The β C—H bonds all make angles of magnitude $2\pi/6$ with the x axis of the molecule. All C—H bonds have the same length. The two rings are regular hexagons.

Numerical values drawn from various sources were employed as the values of the model parameters. The spin densities were obtained by an iterative procedure in which an assumed initial set of spin densities was used in conjunction with measured hyperfine interactions and the parameters of the model to obtain a refined set of spin densities. Table 5 summarises the results to which this process converged and compares the results with a number of theoretical calculations.

Table 5. *Spin densities for triplet state naphthalene*

	Results of present work			
	From anisotropic part (1)	From isotropic part $Q/h = -63{\cdot}08 \times 10^6$ s^{-1} See 4·3(1) (2)	From isotropic part $Q/h = -66{\cdot}50 \times 10^6$ s^{-1} (3)	Calculation by Amos† for lowest triplet state (4)
ρ_α	+0·219	+0·232	+0·220	+0·235
ρ_β	+0·062	—	—	+0·048
ρ_γ	−0·063	—	—	−0·066
ρ_α/ρ_β	+3·52	—	—	+4·89

	Single Hückel orbital, (5, 5′) configuration (5)	Pariser‡ calculation for lowest triplet state (6)	Hoyland & Goodman§ calculation for lowest triplet state (7)	Experimental value of Atherton & Weissman‖ for negative ion, $Q/h = -63{\cdot}06 \times 10^6$ s^{-1} (8)
ρ_α	+0·181	+0·168	+0·198	+0·220
ρ_β	+0·069	+0·074	+0·052	+0·083
ρ_γ	0·000	+0·015	0·000	−0·106
ρ_α/ρ_β	+2·62	+2·27	+3·81	+2·65

† See ref. (30). ‡ See ref. (31).
§ L. Goodman & J. R. Hoyland, *J. Chem. Phys.* **39**, 1068 (1963).
‖ N. M. Atherton & S. I. Weissman, *J. Am. Chem. Soc.* **83**, 1330 (1961).

It is, of course, well known that to a very good approximation the negative ion, the positive ion, and the singly-excited configuration for the lowest triplet will all give the same spin density distribution because of the pairing theorem.[33] It is only when mutual correlations of (*a*) the excited electron, and of (*b*) the hole which it leaves in the closed shell are taken into account that one gets a new effect which gives some expected differences between the spin distribution for triplets and doublets.[33]

The work which was just discussed indeed bears out the theoretical prediction that the spin density distribution for the lowest triplet state of naphthalene should closely resemble that of the negative ion. Experimentally the α spin densities are essentially identical in the two cases. The value of ρ_α/ρ_β is larger for the triplet molecule than for the negative ion by the factor 1·33. The negative spin density on the central C's is 1·7 times as large for the negative ion as for the triplet. It is difficult to see that the observed small differences between the doublet and triplet states bear any close resemblance to the computed differences except for the fact that they are relatively small.

The best values obtained, as described above, by the iterative calculation of the spin densities from the experimental hyperfine splittings may be used to obtain the principal values of the dipole-dipole component of the hyperfine interaction of the α proton with the p_z orbital on the adjacent C atom only (subtracting interaction with the p_z orbitals on not nearest C atoms). This gives, per unit spin density,

$$A'_{d\alpha}/h = +34{\cdot}7 \times 10^6\,\mathrm{s}^{-1},$$

$$B'_{d\alpha}/h = +0{\cdot}9 \times 10^6\,\mathrm{s}^{-1},$$

$$C'_{d\alpha}/h = -35{\cdot}5 \times 10^6\,\mathrm{s}^{-1}.$$

This iterative calculation of course uses the experimental parameters for the malonic acid radical, namely,

$$A_d/h = +35{\cdot}1 \times 10^6\,\mathrm{s}^{-1},$$

$$B_d/h = -0{\cdot}2 \times 10^6\,\mathrm{s}^{-1},$$

$$C_d/h = -35{\cdot}0 \times 10^6\,\mathrm{s}^{-1},$$

per unit spin density, to yield the spin density at each stage of the iteration.

The isotropic part of the α hyperfine interaction, $-14{\cdot}66 \times 10^6\,\mathrm{s}^{-1}$, divided by the α spin density, $+0{\cdot}219$, obtained in our calculation gives $-66{\cdot}9 \times 10^6\,\mathrm{s}^{-1}$ as the isotropic interaction per unit spin. This is quite close to the experimental value for the malonic acid radical which is $-68{\cdot}7 \times 10^6\,\mathrm{s}^{-1}$ per unit spin if we take the spin density on the central C to be 0·892. The calculation forces no agreement in the case of this isotropic part.

It is thus seen that our results for the hyperfine interactions of the α proton in naphthalene in its lowest triplet state are in agreement with the assumption that these interactions, both isotropic and anisotropic, are exceedingly close to those of the proton of the central C—H fragment of the malonic acid radical.

Table 6. ^{13}C *hyperfine interaction frequencies* A_{zz}/h, A_{xx}/h, A_{yy}/h *for the spin Hamiltonian* $\mathbf{I}.\mathbf{A}.\mathbf{S}$ *for ground state triplets in single crystals*

	Diphenylmethylene			Fluorenylidene		
	$i = z$	$i = x$	$i = y$	$i = z$	$i = x$	$i = y$
A_{ii}/h, 10^6 s^{-1}	115·4	189·6	214·8	203·4	280·0	307·3
$\sigma\{A_{ii}/h\}$, 10^6 s^{-1}	1·3	0·7	0·4	0·5	1·4	4·4
Isotropic component $(1/3h) \sum_{i=x,y,z} A_{ii}$, 10^6 s^{-1}		173·3			263·6	
σ\{isotropic component\}, 10^6 s^{-1}		0·5			1·55	
Anisotropic component $(1/h)A_{ii} - 1/3h \sum_{i=x,y,z} A_{ii}$, 10^6 s^{-1}	−57·8	+16·3	+41·5	−60·2	+16·4	+43·8
σ\{anisotropic component\}, 10^6 s^{-1}	1·4	0·9	0·6	1·6	2·1	4·7

In the case of organic ground state triplet molecules in single crystals although the hyperfine interactions with protons have not been observed in ordinary EPR spectra, the hyperfine interactions with the central C nucleus (divalent C nucleus) have been observed and measured.[19] The experimental results are given in table 6 for two such ground state triplet molecules, enriched in ^{13}C with $I = \frac{1}{2}$ in the central C position only. The anisotropic hyperfine interaction is traceless and thus the isotropic and anisotropic components may be separated as described in table 6. These hyperfine results give information on delocalisation of spin from the central C position in addition to that obtained from the fine structure measurements which was described previously. For example, consider the second molecule of table 6, fluorenylidene, whose structure is given in fig. 11. In fig. 12 is shown in the top two figures our expectations for the magnitudes of the ^{13}C hyperfine interactions. We may take the value 178 Mc s^{-1} as the expected value for the frequency corresponding to this interaction when **H** is along the axis of a C p orbital, based on available wave functions for the C atom. The anisotropic hyperfine interactions to be expected for not delocalised pure p triplet electrons on the divalent C are then shown in the second figure of fig. 12. The z axis is of course again normal to the plane of the p orbitals. The 2's occur in the arithmetic because of the requirement that the spin density be normalised to 1, not to 2, as discussed in detail earlier in this chapter. From the measured anisotropic components of the hyperfine interaction we get a unique solution for the delocalisation from each of the orbitals, the

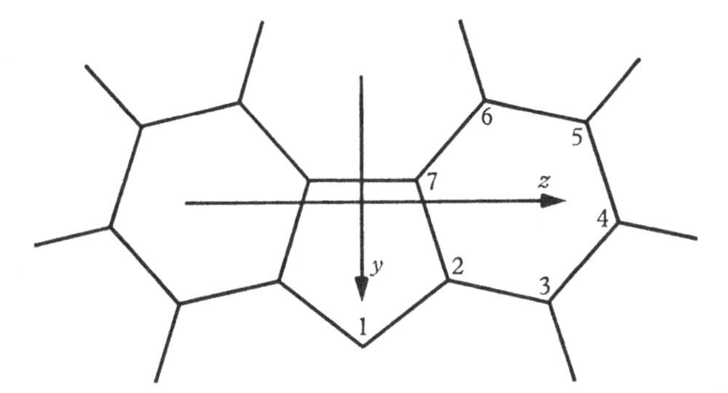

Fig. 11. Fluorenylidene molecule.[35]

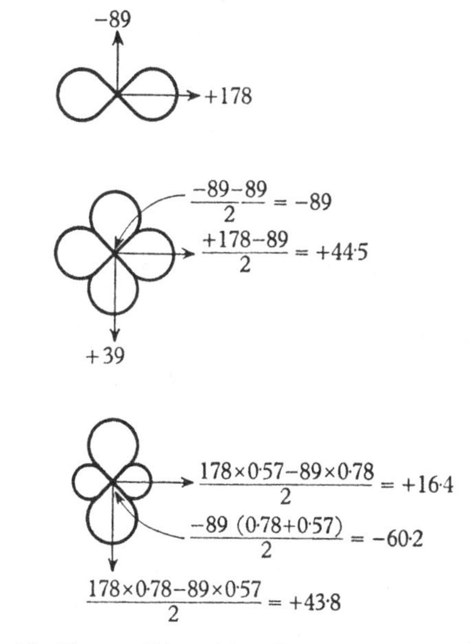

Fig. 12. Fluorenylidene hyperfine splittings (Mc/s).[46]

p_x orbital or π orbital (with respect to the aromatic plane) and the p_y or σ orbital. We can do this because we know from our fine structure information on both fluorenylidene and diphenylmethylene (this molecule as previously discussed is qualitatively very similar to fluorenylidene in all respects) the orientation of the fine structure principal axes. In the case of diphenylmethylene we know enough about the structure of the host to know just about how the molecules are oriented and thus to distinguish the directions in the plane of the central C p orbitals, that is, in the plane normal to the line of phenyl ring centers. Thus we are able to say in this fluorenylidene case that the orbital with 0·57 electron spin is the p_x or $p\pi$ orbital and the one with a 0·78 is the $p\sigma$ orbital, in the plane of the molecule, and therefore not expected to be as delocalised as the π electron which latter is a part of the π system of the whole molecule. If we take 3342 Mc s^{-1} as the size of the hyperfine interaction of a $2s$ electron on a C atom with the ^{13}C nucleus[34] we find that the measured isotropic component of the ^{13}C hyperfine interaction in fluorenylidene corresponds to 0·16 electron spin in the s orbital. If we assume that this s character arises from the fact that the —C— angle is not π,[25] that s admixture in the inplane orbitals gives sp hybrid orbitals with maxima in the direction of these bonds, and that one of our two triplet electrons occupies the non-bonding one of the three sp hybrids, then we find that our singly occupied $p\sigma$ orbital may be described as

$$0{\cdot}397\phi_{2s}+0{\cdot}918\phi_{2p}.$$

The angle of the —C— bond would then be 2·386 (137°).

It is possible to observe the hyperfine interaction of the protons in the aromatic rings of such a ground state triplet molecule when one uses electron nuclear double resonance methods. These experiments have been performed on the fluorenylidene molecule shown in fig. 11 with interesting results.[35] In this type of experiment one adjusts the microwave frequency and the external laboratory field **H** until one of the ordinary EPR absorptions shown in fig. 4 is obtained. Then another oscillating field at a second frequency is introduced which matches the small energy level intervals corresponding to the sum of the nuclear Zeeman interaction plus the nuclear hyperfine interaction with the triplet state electrons. Remember that in fig. 4 the nuclear Zeeman and hyperfine level intervals are magnified by a factor of 100 with respect to the electron level intervals. As this second frequency is varied, when the nuclear resonance condition is reached, the population of one of the two levels associated with the EPR transition to which the spectrometer is tuned is changed, but of course the level at the other end is not involved, its population is not so altered, and thus the intensity of the EPR line is changed. This intensity change is a very convenient indicator of the nuclear resonance condition and the nuclear frequencies may be measured in this way. The previously mentioned broadening of the ordinary EPR lines is just about that expected from the dipole-dipole interaction with the large number of magnetic nuclei of the host in the vicinity of the ground state triplet species. If this magnetic interaction is the source of the broadening, then the broadening of the proton nuclear resonance lines will be expected to be of the order of 1000 times less because the nuclear magneton is that much smaller than the Bohr magneton for the electron. In any event a reduction in broadening by just about this amount is observed and the nuclear resonances of all of the protons are observed and clearly resolved.

This double resonance technique is a powerful one for learning the spin distribution in the aromatic rings of such a ground state triplet. In the first place, since from the fine structure studies we know the molecule orientations, we not only resolve the proton structure but we can assign each resonance to a particular proton from observation of its variation with angle. We know from the fine structure information approximately the direction in the crystallographic axis system of each C—H bond and of course for a proton whose adjacent C has a relatively very large spin density, the nuclear resonance frequency will be stationary with angle as **H** goes through the C—H direction. We need to bear in mind that dipole-dipole effects from the spin on other C's are by no means negligible and that, moreover, we may have the large angle between **S** of the triplet electron and **H** as described previously. But it should be clear that enough is known that the excellent resolution of this double resonance method is utilisable through the assignability of each absorption to a particular nucleus. In the second place it is easily seen from fig. 4 that the proton nuclear resonance detected by the EPR spectrometer is shifted in frequency from the proton nuclear resonance detected by a proton resonance fluxmeter measuring the field by an amount which depends on the local field of the triplet electrons in its molecule (term (4) of the spin Hamiltonian) and with a sign which depends on the sign of this hyperfine interaction. Thus if the electron state is approximately $|+1\rangle$, then a positive spin density on the neighboring C will shift the proton frequency to value higher than that of the proton in the water sample of the fluxmeter, whereas if the spin density is negative the frequency will be shifted to lower values. The shifts are just the opposite of those described when the electron state is $|-1\rangle$. For an electron state described by $|0\rangle$ there will be no shift. Since we can tell which fine structure transition we are observing, we can measure the signs of the spins directly in the experiments.

In fig. 13 are shown some typical double resonance spectra. In fig. 14

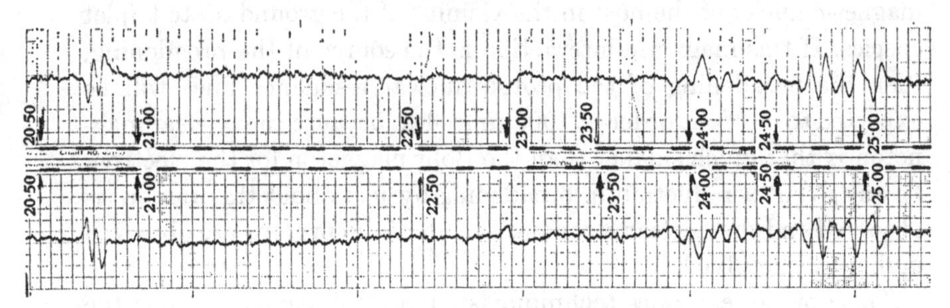

A. Proton *NMR* frequency = 24·510 mc/s

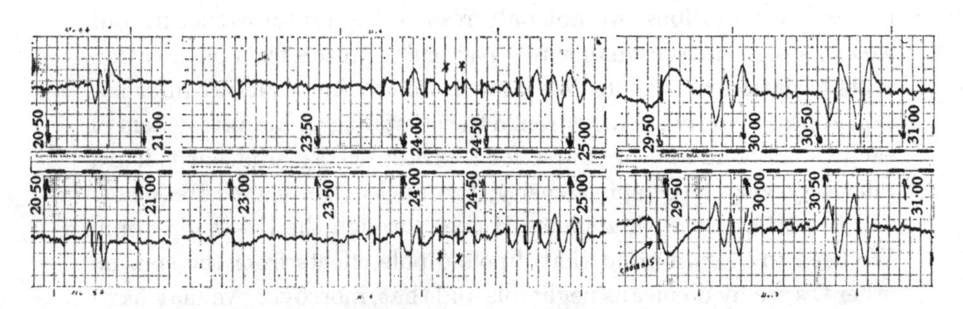

B. Proton *NMR* frequency = 24·507 mc/s

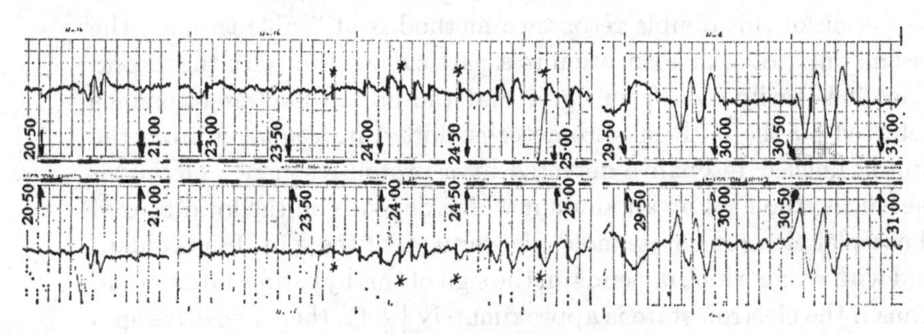

C. Proton *NMR* frequency = 24·497 mc/s

Fig. 13.[46]

are shown the shifts of these nuclear resonances for fluorenylidene from those for protons in water in the same field as a function of the angle which the field makes with the molecular axes in approximately the xy plane; **H** is going around in a plane normal to the line of ring centers. The assignments to particular ones of the four equivalent pairs

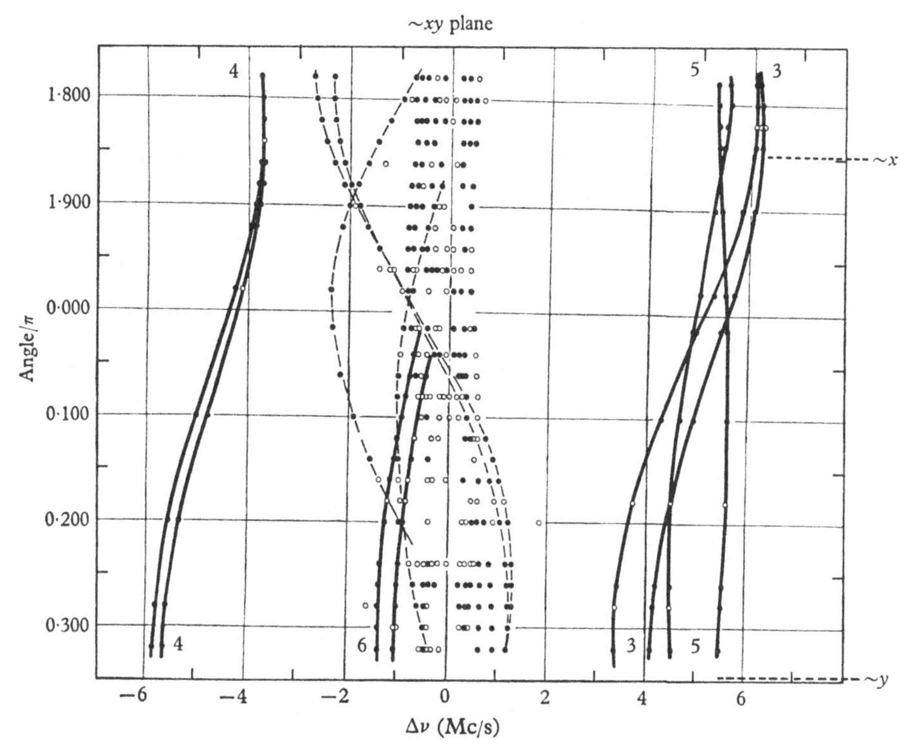

Fig. 14. Electron nuclear double resonance frequency shifts against angle with **H** rotating approximately in the xy plane of a fluorenylidene molecule.[46]

of protons (see fig. 11) contained in this molecule are indicated. A double line is seen in each case because the plane of scan was not exactly the xy plane. The EPR transition was the one shown on the left two diagrams of fig. 4.

It is very interesting that in fig. 14, in addition to the nuclear signals from the proton of the triplet molecule, one can see the resonances of many others from neighbors. It is easy to distinguish the distant ones from the local ones because the local ones have contact interactions and hence the interaction has a finite trace, that is, the sum of the interactions in three mutually perpendicular directions does not vanish. For the distant protons with only dipole-dipole interactions this sum vanishes.

After the components of the tensor $\mathbf{A}$ of term (4) of the spin Hamiltonian have been determined for each proton by this double resonance method, then spin densities may be found such that all of these interactions are satisfactorily accounted for. For this purpose a model must be assumed. This involves an assumed geometry (a fluorene-like model has been used for fluorenylidene[36]) for calculating the dipole-dipole interactions with spin in p orbitals of distant C atoms; an assumed set of interactions with nearest neighbor C p orbitals (those found for triplet naphthalene have been used); and assumptions about spin distributions on the central C (those found from ^{13}C hyperfine studies have been used) among others. There are three anisotropic components and one isotropic component of the interaction for each of the four C atoms of fluorenylidene giving sixteen measured quantities each of which may, using the model, be given as a linear expression in seven spin densities on the seven carbon atoms. The least squares best set of seven spin densities for fitting the measurements may thus be found. This work is not yet completed but a tentative set is tabulated

$\rho_1 = 0{\cdot}7$	$\sigma = 0{\cdot}1$
$\rho_2 =$ ______	
$\rho_3 = +0{\cdot}070$	$\sigma = 0{\cdot}008$
$\rho_4 = -0{\cdot}071$	$\sigma = 0{\cdot}007$
$\rho_5 = +0{\cdot}082$	$\sigma = 0{\cdot}014$
$\rho_6 =$ ______	
$\rho_7 = +0{\cdot}14$	$\sigma = 0{\cdot}14$

where the σ's are standard deviations

Fig. 15. Tentative spin density values from electron nuclear double resonance measurements on fluorenylidene.

in fig. 15. Dr E. C. Snyder[36] kindly supplied us with the results of his self-consistent field calculations of the spin densities in the π system of this fluorenylidene molecule which after dividing by 2 to put $\frac{1}{2}$ total spin in the π system and placing $\frac{1}{2}$ spin in the σ electron on the central C, making the total spin equal to 1 for reasons described previously, give us the results:

Carbon	1	2	3	4	5	6	7
Spin density	0·774	−0·019	+0·059	−0·014	+0·047	−0·010	+0·050

which may be compared with our measurements for the triplet state.

The fine structure studies, the ^{13}C hyperfine studies, and the electron proton double resonance studies complement each other nicely in this fluorenylidene case. From the ^{13}C hyperfine interaction we get the distribution of spin between $p\pi$, $p\sigma$ and s on the central C, 0·57, 0·78 and 0·16 respectively (total spin is 2). The double resonance results, with a rather large standard error for this C atom, are in good agreement with and confirm these results. With the value previously given for the D/hc for two triplet state spins in two orthogonal p orbitals and with $0 \cdot 4803 \text{ cm}^{-1}$ as the value[25] for the case of one spin in a p and the other in an s orbital, we would have a central C D/hc contribution of $0 \cdot 446 \text{ cm}^{-1}$. From the double resonance results we can assess the D/hc contribution from the rings and it is very small and about $-0 \cdot 026 \text{ cm}^{-1}$. We thus find from ^{13}C and proton hyperfine studies the value, $+0 \cdot 420 \text{ cm}^{-1}$, for D/hc which is to be compared with $+0 \cdot 409 \text{ cm}^{-1}$ given by the fine structure measurements. The matter of the angle of the —C— bond of the fluorenylidene molecule is interesting.[19,23] The isotropic contribution to ^{13}C hyperfine interaction gave an angle 2·386 (137°). The fluorene angle (in fluorene there are two H's on the central C) is 1·843 (106°)[36] and the triplet state molecule would not be expected to have a central C position in the molecular framework corresponding to an angle much different from 1·843. Thus the hyperfine data have been taken to indicate that the sp hybrids of the central C are not 'directed' toward the neighboring C's.[23] It has also been observed in connection with the fluorenylidene model used to get the spin densities from the double resonance results, that the standard deviations of spin densities are smallest in the vicinity of the fluorene angle if the angle of the model is varied. This argues for a molecular framework close to that of fluorene.[36]

3. Some applications

This completes our discussion of the essential features of the magnetic properties of organic molecules in triplet states. I wish to conclude this paper by mentioning very briefly some applications of magnetic resonance methods to a chemical kinetics problem in organic single crystals and to the study of the transfer of triplet excitation in an organic single crystal as specific illustration of the possibility of using these magnetic properties as tools in a variety of situations where organic triplet state molecules occur.

3.1. *Reaction kinetics in single crystals*[38]

When EPR spectra are produced at 77 °K by optical irradiation of single crystals of 1,1-diphenylethylene (see fig. 7) containing diphenyldiazomethane, the magnetic resonance signals initially increase in intensity, but after about 1 min of irradiation with $\sim 3\%$ of the light from an A-H 6 high-pressure Hg arc through 5 cm of an aqueous solution containing 250 g $CuSO_4$ per liter the signals weaken and disappear. If optical irradiation is stopped at 77 °K, while magnetic signals are present in high intensity, the signals are quite stable but they disappear on warming to approximately 100 °K. On subsequent cooling no signals are present, but further optical irradiation will produce them again.

The kinetics of this interesting chemistry in single crystals has been investigated in this and other similar cases by measurements of the rates of growth and decay of the magnetic resonance signals at various temperatures.

All of the observations are consistent with the following assumptions. When the crystal containing the ground state triplet methylene, produced by the optical irradiation, is warmed, the reaction of the methylene with the host ethylene, probably to produce a diamagnetic cyclopropane, is accelerated. The magnetic signal thus disappears and is not present on cooling to 77 °K. But further optical irradiation at 77 °K produces more triplet state species from unused diazo compound regenerating the signal. Not only does the irradiation of the parent diazo compound destroy it, creating paramagnetic methylene, but also optical irradiation of the methylene eventually completely destroys it. Polarised optical absorption spectra[39, 40] obtained in these crystals have confirmed these assumptions and the optical frequencies effective for the different processes have been ascertained.

An apparatus for control of the temperature of the crystal during an EPR measurement has been constructed which permits measurement of the reaction rates by observation of the growth and decay of EPR signals. Preliminary measurements of the rates of disappearance of EPR signals produced by approximately 20 s irradiations of 1,1′-diphenyldiazomethane with $\sim 1\%$ of the light from a 500 W Hg–Xe lamp have been measured as a function of temperature. The light is passed through 5 cm of an H_2O solution containing 500 g of $NiSO_4$ per liter. This solution transmits light at $\sim$5000 A in the visible and also in the range from 3 100–2 300 A in the ultraviolet. Such an irradiation produces approximately 3×10^{17} triplet state species when the light falls on 0·25 cm^2 of the cleavage plane of a crystal grown from the melt as judged by intensity comparison with a standard Mn^{+2} sample EPR signal. Polarised optical absorption[39, 40] studies also showed essentially the same results based on absolute determinations of absorbances of the diazo compound and assuming that initially every diazo molecule destroyed produced a methylene.

The decay of the EPR signal is very closely exponential. Preliminary observations give first-order rate constants at various temperatures as follows:

T (°K)	Rate constant (s^{-1})
77·2	$0{\cdot}002 \times 10^{-2}$
88·5	$0{\cdot}34 \times 10^{-2}$
96	$0{\cdot}91 \times 10^{-2}$
97	$1{\cdot}33 \times 10^{-2}$

A straight line, least squares fitted to a plot of logs of these rate constants *vs* $1/T$ gives a slope which interpreted as $-\Delta E/R$ gives $\Delta E \cong 5$ kcal mole^{-1}.

3.2. *Transfer of energy in organic single crystals*[41, 42, 43]

It has been shown by electron magnetic resonance studies that energy is transferred in single crystals of diphenyl, containing low concentrations of phenanthrene and naphthalene molecules, from optically-excited phenanthrene to unexcited naphthalene, where it is trapped, thus creating triplet state naphthalene molecules. The lowest naphthalene triplet state is ~ 300 cm^{-1} lower than the lowest phenanthrene triplet state. In fig. 2 we see the characteristic fine structures of the phenanthrene and naphthalene triplet magnetic resonance spectra. The naphthalene spectrum was observed under conditions when only the phenanthrene could absorb light, proving that the transfer occurred.[41] High precision measurements of D and E, using the zero-field spectrometer as described previously, showed that the transfer occurred through the host structure, and not because there were a large number of complexes between phenanthrene and naphthalene in the crystal, or a larger than statistical number of naphthalene phenanthrene pairs occupying neighboring sites in the crystal. In these last cases the high precision measurements would have detected the environmental effects on D and such effects were found to be absent.

Very interesting results were found[42] when diphenyl crystals containing small amounts of deuterophenanthrene and ordinary naphthalene were irradiated with light from a high-pressure A–H 6 Hg arc until steady-state conditions obtained, at which time the irradiation was terminated and the decay times of the paramagnetic resonance signals were measured. What was found was that from ~ 4 to ~ 85 °K the decays of the phenanthrene and naphthalene signals were exponential with mean times of $\sim 10{\cdot}0$ and $\sim 2{\cdot}2$ s respectively. Then, in a relatively narrow temperature range, between 85 °K and $\sim$115–120 °K, as the temperature for the measurement is raised, the phenanthrene magnetic resonance signal decay deviates from exponential behavior

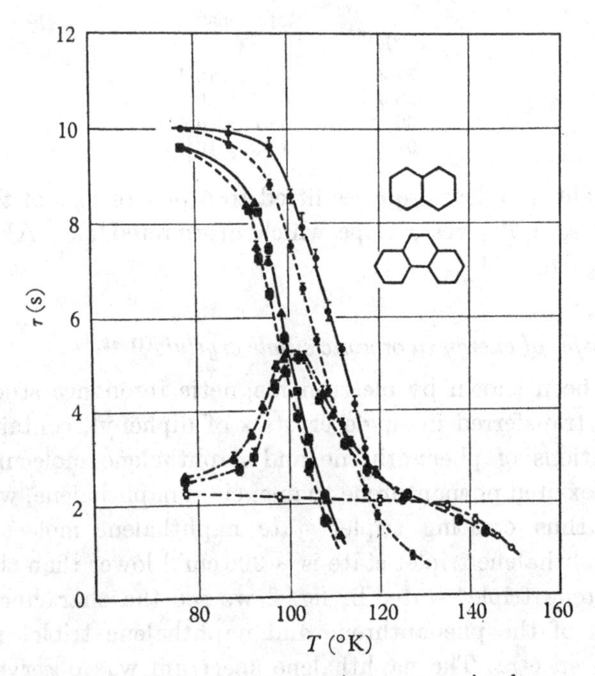

Fig. 16. $1/e$ times of paramagnetic resonance signals as a function of temperature.[42]

and its $1/e$ time falls to zero within the uncertainty of the measurement. The naphthalene signal's $1/e$ time goes through a maximum of $\sim 5{\cdot}5$ s and falls again to its low temperature value of $\sim 2{\cdot}2$ s. This behavior is described graphically in fig. 16. All curves which tend to ~ 10 s at low T are phenanthrene $1/e$ signal times and all those tending to ~ 2 s at low T are for naphthalene. The phenanthrene curves occur in broken and solid line pairs because in the range of the large drop the decay is not strictly exponential, and the solid lines give the time for the first $1/e$ of the signal to decay and the broken line the second $1/e$. The higher T phenanthrene curve pair is for phenanthrene only, in diphenyl ($\sim 0{\cdot}2$ mole %). The lower T phenanthrene curve pair is for the same concentration of phenanthrene with a small amount of naphthalene present ($\sim 0{\cdot}05$ mole %). The phenanthrene concentration was approximately doubled with little effect (squares denote $\sim 0{\cdot}24$ mole %

and triangles $\sim 0{\cdot}45$). The dramatic drop in the phenanthrene magnetic resonance signal lifetime, produced by the addition of the naphthalene, should be noted. For example, at 100 °K the time falls from ~ 9 to ~ 5 s when naphthalene is added. The naphthalene signal lifetime behavior is equally interesting. The broken line without a maximum is for naphthalene by itself in the diphenyl crystal, whereas the two peaked curves, nearly the same, are for the naphthalene in the presence of the two different concentrations of phenanthrene as mentioned above. The sharp rise in naphthalene signal lifetime by more than a factor of two at the same temperature range as the drop in phenanthrene time is to be noted.

This temperature-dependent lifetime behavior has been satisfactorily interpreted as being a result of a temperature-dependent transfer of triplet excitation energy from the phenanthrene to the naphthalene. The lowest triplet state of a diphenyl molecule is $\sim 1\,600\ \mathrm{cm}^{-1}$ higher than the lowest triplet state of phenanthrene. Thus the proposed mechanism, which seems to account for all the observations, involves vibrational (thermal) excitation of the triplet phenanthrene to an energy about equal to that of the triplet exciton band of the host diphenyl which would be approximately $1\,600\ \mathrm{cm}^{-1}$ with respect to the

$$(1)\quad P_T \xrightarrow{k_1} P_S \qquad\qquad (4)\quad D_T + P_T \xrightarrow{k_4} D_S + P_S$$

$$(2)\quad P_T \underset{k_{\bar{2}}}{\overset{k_2}{\rightleftarrows}} P_{T*} \qquad\qquad (5)\quad D_T + N_S \xrightarrow{k_5} N_T + D_S$$

$$(3)\quad P_{T*} + D_S \underset{k_{\bar{3}}}{\overset{k_3}{\rightleftarrows}} D_T + P_S$$

Fig. 17. Process involved in triplet energy transfer from phenanthrene-d_{10} to naphthalene in a diphenyl single crystal.

unexcited phenanthrene. The excitation then is transmitted to the deeper naphthalene traps via the exciton band. Fig. 17 describes the processes involved. Processes (2) and (3) occur at rates which are very large compared with those of the intersystem conversion process (1) or the triplet-triplet annihilation process (4).

We thus assume that over a very short period of time, relative to the intersystem conversion time, the rates of creation and destruction of P_T by (2)→ and (2)← become equal and the rates of creation and destruction of D_T by (3)→ and by (3)← become equal. A constant ratio $[P_{T*}]/[P_T]$ is thus established, where [ ]'s denote concentrations; this ratio is assumed to have the thermal equilibrium value, $e^{-\Delta E/kT}$, where

ΔE is the difference in energies of P_{T*} and P_T. A steady-state value of $[D_T]$, namely

$$[D_T] = k_3[P_{T*}][D_S]/k_{\bar{3}}[P_S]$$

is also established. We now add the relatively slow process of intersystem conversion (1) and triplet-triplet annihilation (4) which relatively slowly alter $[P_T]$ leading to the first expression in fig. 17 for the

$$d[P_T]/dt = -k_1[P_T] - k_4(k_3[P_T]^2[D_S]/k_{\bar{3}}[P_S])\exp(-\Delta E/kT)$$

$$\begin{aligned} d[P_T]/dt &= -k_1[P_T] - k_4[P_T][D_T] - k_5[D_T][N_S] \\ &= \left\{-k_1 - \frac{k_5 k_3}{k_{\bar{3}}}\frac{[D_S][N_S]}{P_S}\exp(-\Delta E/kT)\right\}[P_T] \\ &\quad - \left\{\frac{k_4 k_3[D_S]}{k_{\bar{3}}[P_S]}\exp(-\Delta E/kT)\right\}[P_T]^2 \end{aligned}$$

Fig. 18. Kinetic equations for triplet energy transfer from phenanthrene-d_{10} to naphthalene in a diphenyl single crystal.

decay of $[P_T]$ with time. For the three-component system containing naphthalene we add process (5) of fig. 17. This process removes D_T's which are replaced by the very fast process (3), being created from P_{T*}'s as described previously. The P_{T*}'s are in turn very rapidly generated by the excitation (2)→ from P_T's. Thus we have a third contribution to the decay of $[P_T]$ in addition to the intersystem conversion and the triplet-triplet annihilation discussed for the two-component system. This leads to the second expression shown in fig. 18.

This mechanism leads to a satisfactory explanation of the essential features of the magnetic resonance signal decay rates as a function of temperature. At low enough T's both of the equations of fig. 18 lead to a first-order T independent rate of decay of the population of triplet state phenanthrene molecules. At higher T's the first equation of fig. 18 leads to a decay rate for phenanthrene in diphenyl which is the sum of the first-order lower T decay plus a second-order T independent decay. At higher T's the second equation which applies to the three-component phenanthrene naphthalene diphenyl system leads to a decay rate which is a sum of a first-order and a second-order decay but this time the first-order one also depends on T. The detailed magnetic resonance studies which have been made show just these details of behaviour.

Moreover, when the magnetic results are analysed in such a way as to find a ΔE for the equations such that they give a best fit to the results, the value found for $\Delta E/hc$ is in the range $\sim 1\,400$—$1\,550\ \text{cm}^{-1}$. The remarkable agreement with the probable $1\,600\ \text{cm}^{-1}$ position of the host exciton band with respect to phenanthrene's lowest triplet state confirms the assumption that the activation process is an excitation to this level.

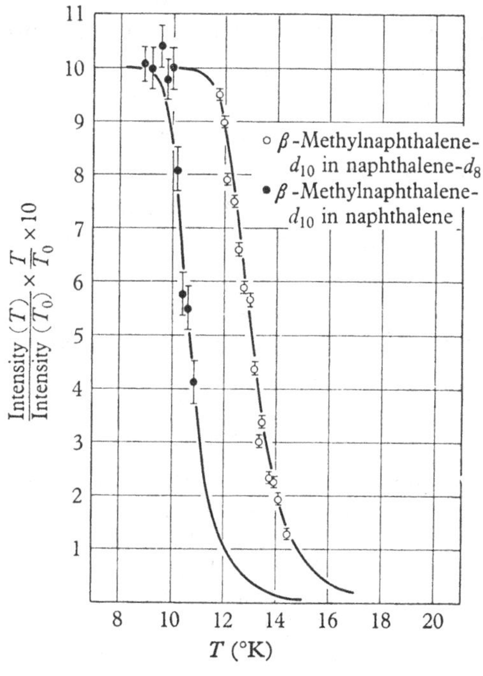

Fig. 19. Magnetic resonance signal intensities with constant illumination for β-methylnaphthalene-d_{10} in naphthalene-d_8 and in ordinary naphthalene.[42]

These experiments have recently been greatly extended to a large variety of similar systems.[43] Deuteration of guest and host may be employed to examine the effects of the resulting shifts in energy levels on the rates of signal decay and the effects are always just those predicted. A related example is shown in fig. 19 for β-methylnaphthalene-d_{10}, in naphthalene-d_8 and in naphthalene.[44] Here the trap depth on the methylated compound relative to the triplet state of ordinary naphthalene is probably $\sim 140\ \text{cm}^{-1}$. The triplet state of ordinary naphthalene is $\sim 130\ \text{cm}^{-1}$ lower than that of deuteronaphthalene. Thus we would expect to need to go to higher T's in the deuterated host to get the same drop in signal lifetime or decrease in signal intensity under constant illumination caused by excitation to the host band and subsequent triplet-triplet annihilation, as found in the ordinary host. This is indeed seen to be qualitatively the case and the quantitative aspects are in agreement with the model. Fig. 19 shows signal intensities at constant illumination.

Conclusion

I have, in this lecture, surveyed the essential features of the magnetic resonance spectra of organic molecules in single crystals. I have indicated the types of information concerning electronic states and molecular structures which may be deduced from the observations and measurements. In addition, I have described a couple of applications in which magnetic resonance methods are used as tools for the investigation of some problems involving triplet states in single crystal systems.

I gratefully acknowledge the cooperation of my many colleagues and students at the University of Chicago and elsewhere in making possible these investigations.

References

1 E. Müller & I. Müller-Rodloff, *Ann. Phys.* **517**, 134 (1935).
E. Müller, *Fortschr. Chem. Forsch.* **1**, 325 (1949).

2 C. A. Hutchison Jr., A. Kowalsky, R. C. Pastor & G. W. Wheland, *J. Chem. Phys.* **20**, 1485 (1952); H. S. Jarrett, G. J. Sloan & W. R. Vaughan, *J. Chem. Phys.* **25**, 697 (1956); D. C. Reitz & S. I. Weissman, *J. Chem. Phys.* **33**, 700 (1960).

3 H. M. McConnell, *J. Chem. Phys.* **33**, 1868 (1960).

4 G. N. Lewis, D. Lipkin & T. T. Magel, *J. Am. Chem. Soc.* **63**, 3005 (1941); G. N. Lewis & M. Kasha, *J. Am. Chem. Soc.* **66**, 2100 (1944); G. N. Lewis & M. Kasha, *J. Am. Chem. Soc.* **67**, 994 (1945).

5 A. Terenin, *Acta Physicochim. SSSR*, **18**, 210 (1943).

6 G. N. Lewis & M. Calvin, *J. Am. Chem. Soc.* **67**, 1232 (1945); G. N. Lewis, M. Calvin & M. Kasha, *J. Chem. Phys.* **17**, 804 (1949).

7 P. Frolich, L. Szalay & P. Szor, *Acta Chem. Phys. Univ. Szeged (Hungary)*, **2**, 96 (1948).

8 D. F. Evans, *Nature*, **176**, 777 (1955).

9 G. Kortüm, *Angew. Chem.* **70**, 14 (1958).

10 C. A. Hutchison Jr. & B. W. Mangum, *J. Chem. Phys.* **29**, 952 (1958); **34**, 908 (1961).

11 J. H. van der Waals & M. S. de Groot, *Mol. Physics* **2**, 333 (1959); **3**, 190 (1960).

12 W. A. Yager, E. Wasserman & R. M. R. Cramer, *J. Chem. Phys.* **37**, 1148 (1962).

13 D. S. McClure, *J. Chem. Phys.* **20**, 682 (1952); M. Mizushima & S. Koide, *J. Chem. Phys.* **20**, 765 (1952); H. F. Hameka & L. J. Oosterhoff, *Mol. Physics* **1**, 358 (1958); D. P. Craig, J. M. Hollas & G. W. King, *J. Chem. Phys.* **29**, 974 (1958).
14 S. H. Glarum, *J. Chem. Phys.* **39**, 3141 (1963); S. J. Fogel & H. F. Hameka, *J. Chem. Phys.* **42**, 132 (1965).
15 R. McWeeny, *J. Chem. Phys.* **34**, 399 (1961).
16 M. Gouterman & W. Moffitt, *J. Chem. Phys.* **30**, 1107, 1369 (1959).
17 S. A. Boorstein & M. Gouterman, *J. Chem. Phys.* **39**, 2443 (1963); **41**, 2776 (1964); Y.-N. Chiu, *J. Chem. Phys.* **39**, 2736 (1963); H. F. Hameka, *J. Chem. Phys.* **31**, 315 (1959); M. Geller, *J. Chem. Phys.* **39**, 853 (1963); M. Geller & R. W. Griffith *ibid.* **40**, 2309 (1964); J. H. van der Waals & G. ter Maten, *Mol. Physics* **8**, 301 (1964); M. Godfrey, C. W. Kern & M. Karplus, *J. Chem. Phys.* **44**, 4459 (1966).
18 R. W. Brandon, R. E. Gerkin & C. A. Hutchison Jr., *J. Chem. Phys.* **41**, 3717 (1964).
19 (*a*) R. W. Brandon, G. L. Closs & C. A. Hutchison Jr., *J. Chem. Phys.* **37**, 1878 (1962).
(*b*) R. W. Brandon, G. L. Closs, C. E. Davoust, C. A. Hutchison Jr. B. E. Köhler & R. Silbey, *J. Chem. Phys.* **43**, 2006 (1965).
20 L. E. Erickson, *Phys. Rev.* **143**, 295 (1966).
21 G. Closs, C. A. Hutchison Jr. & B. E. Köhler, *J. Chem. Phys.* **44**, 413, (1966); also unpublished work.
22 R. W. Brandon, R. E. Gerkin & C. A. Hutchison Jr., *J. Chem. Phys.* **37**, 447 (1962).
23 R. W. Murray, A. M. Trozzolo, E. Wasserman & W. A. Yager, *J. Am. Chem. Soc.* **84**, 3213 (1962); G. Smolinsky, E. Wasserman & W. A. Yager, *J. Am. Chem. Soc.* **84**, 3220 (1962).
24 J. Higuchi, *J. Chem. Phys.* **38**, 1237 (1963).
25 J. Higuchi, *J. Chem. Phys.* **39**, 1339 (1963).
26 D. Halford, C. A. Hutchison Jr. & P. M. Llewellyn, *Phys. Rev.* **110**, 284 (1958).
27 H. M. McConnell, *J. Chem. Phys.* **24**, 764 (1956); S. I. Weissman, *J. Chem. Phys.* **25**, 890 (1956); R. Bersohn, *J. Chem. Phys.* **24**, 1066 (1956); H. S. Jarrett, *J. Chem. Phys.* **25**, 1289 (1956).
28 H. M. McConnell & J. Strathdee, *Mol. Physics* **2**, 129 (1959).
29 A. M. Ponte Goncalves & C. A. Hutchison Jr., unpublished work.
30 N. Hirota, C. A. Hutchison Jr. & P. Palmer, *J. Chem. Phys.* **40**, 3717 (1964).
31 A. T. Amos, *Mol. Physics* **5**, 91 (1962).
32 R. Pariser, *J. Chem. Phys.* **24**, 250 (1956).
33 A. D. McLachlan, *Mol. Physics* **5**, 51 (1962).
34 M. Karplus & G. K. Fraenkel, *J. Chem. Phys.* **35**, 1312 (1961).
35 C. A. Hutchison Jr. & G. A. Pearson, *J. Chem. Phys.* **43**, 2545 (1965).
36 D. M. Burns & J. Iball, *Proc. Roy. Soc.* (*London*) A, **227**, 200 (1955).
37 L. C. Snyder, personal communication.

38 D. C. Doetschman & C. A. Hutchison Jr., unpublished work.
39 G. L. Closs, C. A. Hutchison Jr. & B. E. Köhler, *J. Chem. Phys.* **44**, 413 (1966).
40 G. L. Closs, C. A. Hutchison Jr. & B. E. Köhler, unpublished work.
41 R. W. Brandon, R. E. Gerkin & C. A. Hutchison Jr., *J. Chem. Phys.* **37**, 447 (1962).
42 N. Hirota & C. A. Hutchison Jr., *J. Chem. Phys.* **42**, 2869 (1965).
43 N. Hirota, *J. Chem. Phys.* **43**, 3354 (1965).
44 N. A. Ashford & C. A. Hutchison Jr., unpublished work.
45 T. Cole, C. Heller & H. M. McConnell, *Proc. Natl. Acad. Sci. U.S.* **45**, 525 (1959); H. M. McConnell, C. Heller, T. Cole & R. W. Fessenden, *J. Am. Chem. Soc.* **82**, 766 (1960).
46 C. A. Hutchison Jr. & G. A. Pearson, *J. Chem. Phys.* **47**, 15 July (1967)
47 C. A. Hutchison Jr., *J. Phys. Chem.* **71**, 203 (1967).
48 C. A. Hutchison Jr., *Record of Chemical Progress*, **24**, 105 (1963).

MAGNETIC INTERACTIONS RELATED TO PHOSPHORESCENCE

J. H. VAN DER WAALS AND M. S. DE GROOT

Abstract

A simple review is presented of spin-spin and spin-orbit interaction on the basis of group theory. Selection rules are given for the different multiplicities, orbital symmetries and spin symmetries of the states that yield non-vanishing matrix elements for these interactions, with special emphasis on molecules of symmetry C_{2v} or higher.

Subsequently, the paths of excitation into and radiative decay from the phosphorescent lowest triplet state are considered with reference to quinoxaline as an example. It is made plausible that entry into the triplet manifold from the lowest excited singlet state is, in general, accompanied by spin polarisation. Three different methods to detect such polarisation at liquid helium temperature are suggested and some preliminary results presented. The results indicate that flash excitation of quinoxaline at 1·56 °K leads to (almost) quantitative polarisation of spin angular momentum in the molecular plane.

Introduction and basis of the description

The purpose of the present paper is twofold. First, the different types of magnetic interaction with which one is concerned when studying the triplet state are reviewed. A central problem here is the selection rules according to which the lowest, metastable triplet state of phosphorescent organic molecules can be populated and depopulated. Although the experimental and theoretical work of many investigators has led to an overall pattern from which, in Kasha's terminology,[1] certain preferred 'paths of molecular excitation' stand out, the detailed mechanism remains to be uncovered.

Secondly, suggestions are made as to how such preferred paths of excitation should manifest themselves in the magnetic and optical properties of phosphorescent molecules at liquid helium temperature. In particular it is predicted that cross-over into the phosphorescent state following absorption of light in the singlet system should lead to spin polarisation. Some results on the effect of magnetic fields on phosphorescence decay at 1·5 °K are presented as a first experimental verification of the above suggestions.

Our starting-point is the familiar Jablonski, Lewis & Kasha diagram[2] which has been reproduced in fig. 1 for quinoxaline as an example. On the left are the singlet ground state and the three lower excited singlet states, on the right the lowest (phosphorescent) triplet state and the approximate positions of some of the higher triplets.

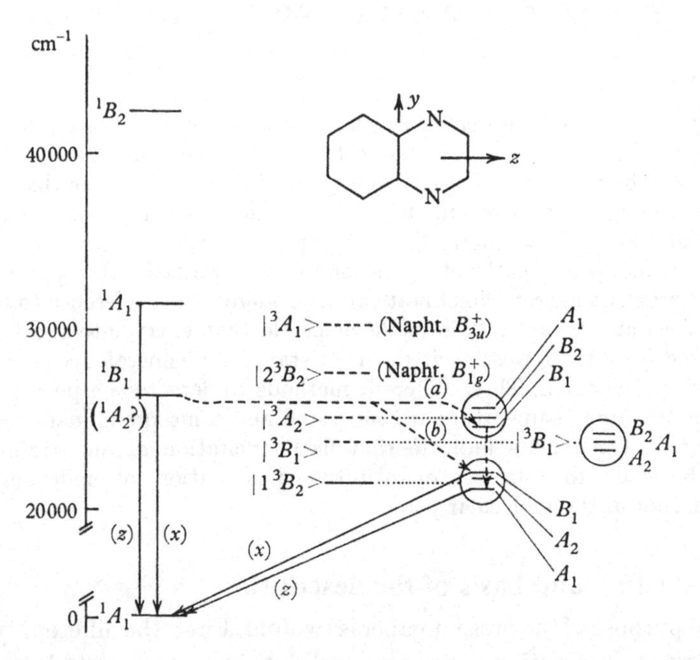

Fig. 1. The lower electronic states of quinoxaline. The fully drawn levels have been observed,[3] the dotted levels have been taken from calculations.[4,46] Labels in kets are orbital symmetries in C_{2v}, other labels are total symmetries. The individual spin components of the three lower triplets are shown in the circles in the order τ_z, τ_y, τ_x, from top to bottom. The polarisation of the two symmetry-allowed components of phosphorescence are indicated together with two singlet-singlet transitions from which they derive intensity; (*a*) and (*b*) constitute possible paths of excitation.

Our first task will be to review spin-spin and spin-orbit coupling and to examine in what manner these interactions may cause phosphorescence through singlet-triplet mixing, and split the three magnetic components of each triplet.

The molecules to be considered are all planar aromatic systems and in order to keep things simple we may think of them in terms of a molecular orbital (m.o.) picture, although many of the conclusions have wider validity. If the molecular plane is designated as $x = 0$, there are a number of π electron m.o.'s arising through interaction of $2p_x$

atomic orbitals (a.o.'s) and a group of σ orbitals having their maximum density in the plane $x = 0$. The precise form of the m.o.'s need not be known, but they should be so formulated that they belong to irreducible representations of the symmetry group of the molecule concerned. Later we shall return to the example of quinoxaline and indicate how fig. 1 is to be understood in terms of a molecular orbital picture (see below, p. 118).

Table 1. *Character table of the group C_{2v} and two-electron spin functions*

C_{2v}			E	C_2	σ_y	σ_x
σ	z	A_1	1	1	1	1
τ_z		A_2	1	1	-1	-1
τ_y	x	B_1	1	-1	1	-1
τ_x	y	B_2	1	-1	-1	1

We consider molecules that can be regarded as isolated systems, dispersed in a rigid medium (glass or host crystal) at low temperature, and in first instance neglect vibrations and rotations. It is to be noted, however, that for certain properties of interest it may not really be permissible to regard the solute molecules in a host crystal as isolated species. The very weak electric dipole transition of phosphorescence, in particular, is liable to severe alteration by the host.[5, 6]

With the above approximations the phenomena to be studied are governed by a Hamiltonian of the general form

$$H = H_0 + H'. \tag{1}$$

Here H_0 represents the kinetic energy of the electrons and the Coulomb energy of electrons and nuclei. The second term, H', represents the magnetic interactions. Although H' is a very small perturbation relative to H_0 it is of crucial importance, being that part of the Hamiltonian which contains spin-dependent operators and thus lifts the degeneracy of the triplet states and allows them to interact with singlets. For instance, the singlet-triplet splitting of a given electron configuration (e.g. the distance between the ${}^1B_2$ and lower ${}^3B_2$ state in fig. 1) is due to a difference in electrostatic repulsion of the electrons and of the order of $10^4\,\text{cm}^{-1}$, whereas the zero-field splitting is a manifestation of H' and only of the order of $0{\cdot}1\,\text{cm}^{-1}$.

Neglecting H', we choose as zero-order wavefunctions for a given electronic state the corresponding eigenfunctions of H_0, which can always be written as an antisymmetrised linear combination of products of orbital functions ϕ_j and spin functions χ_j,[7]

$$\mathscr{A}\phi_j([\mathbf{r}])\chi_j([s]) = |m^{\mu}\Gamma_0, \gamma_s\rangle. \tag{2}$$

Here the positional coordinates of all electrons are collectively designated by $[\mathbf{r}]$ and the spin coordinates by $[s]$; $\mathscr{A}$ is the antisymmetriser (cf. (5) below for a specific example).

The zero-order wavefunctions are specified by the notation on the right, which determines the behaviour under coordinate transformation of the factors ϕ_j and χ_j. The label Γ_0 refers to orbital symmetry and denotes the irreducible representation of the molecular point group to which the ϕ_j belong, μ is the multiplicity and m a running index that numbers the states with identical $^{\mu}\Gamma_0$ in order of increasing energy. (In cases where no ambiguity can arise m will be omitted.)

Each electronic state $|m^{\mu}\Gamma_0\rangle$ has $\mu = 2S+1$ spin components. For a specific spin component all χ_i in (2) not only are eigenfunctions of $\mathbf{S}^2$ with eigenvalue $S(S+1)$, but they also must all belong to the same irreducible representation Γ_s of the molecular point group. In order to understand how the label γ_s is related to Γ_s a few words may first be said about the transformation properties of spin functions (for an extensive discussion of this subject cf. Wigner).[7]

The operations of a molecular symmetry group can all be written as products of pure rotations and the inversion. When looking first at the effect of the former, the spin functions of a *single* electron transform like the $D^{(\frac{1}{2})}$ representation of the continuous rotation group. As a consequence, the systems with an even number of electrons in which we are interested possess singlet spin functions σ corresponding to $D^{(0)}$ and invariant under rotation, triplet spin functions τ_u transforming like the three-dimensional representation $D^{(1)}$, etc. Secondly, we have to consider the effect of the inversion. For a single electron this leads to multiplication by a factor which, because of the two-valuedness of $D^{(\frac{1}{2})}$, may be chosen as either $+1$ or -1 in a (unique) representation of a point group. Accordingly, any spin function of an *even* number of electrons must necessarily be invariant (*gerade*) under inversion.

In atomic theory it is customary to write the three triplet spin functions in a representation in which they are diagonal in S_z, where S_z is the component of the total spin angular momentum operator along an arbitrary direction in space. Such a description is specifically adapted to situations where the physical problem possesses cylindrical symmetry about an axis, which is then taken to be the z axis, as in the Zeeman effect for an atom. However, in the case of a triplet state of non-linear molecules, cylindrical symmetry does not apply—not even when a field is present—and it is then often more illuminating to work in a representation adapted to the molecular symmetry group. In the following we shall mainly deal with molecules possessing at least one plane of symmetry in addition to the molecular plane; the point group of such a molecule then contains C_{2v} as a subgroup. Accordingly, we choose triplet functions that belong to different irreducible representations of C_{2v}; for the simple case of two electrons we thus have:[8]

$$\begin{array}{llll} & & \multicolumn{2}{c}{\text{Representation}} \\ & & C_{2v} & D_{2h} \\ \sigma(1,2) = 2^{-\frac{1}{2}}[\alpha_1\beta_2 - \beta_1\alpha_2] & \mathbf{S}\sigma = 0 & A_1 & A_g \\ \tau_x(1,2) = 2^{-\frac{1}{2}}[\beta_1\beta_2 - \alpha_1\alpha_2] & S_x\tau_x = 0 & B_2 & B_{3g} \\ \tau_y(1,2) = 2^{-\frac{1}{2}}i[\beta_1\beta_2 + \alpha_1\alpha_2] & S_y\tau_y = 0 & B_1 & B_{2g} \\ \tau_z(1,2) = 2^{-\frac{1}{2}}[\alpha_1\beta_2 + \beta_1\alpha_2] & S_z\tau_z = 0 & A_2 & B_{1g} \end{array} \tag{3}$$

with

$$S_x\tau_y = -S_y\tau_x = i\tau_z, \text{ etc.}\dagger \tag{3a}$$

The symmetry labels in the last two columns of equation (3) indicate the irreducible representations of the groups C_{2v} and D_{2h} to which these functions belong (Note: the assignment of axes for quinoxaline (C_{2v}) differs from that normally used for naphthalene (D_{2h})[4]. In C_{2v} the two fold axis is labelled z (cf. table), whereas for naphthalene the z-axis is chosen perpendicular to the molecular plane). Since the singlet spin function is invariant under any rotation or inversion, it must necessarily be totally symmetric: A_1 or A_g. The symmetries of the triplet functions as given in equation (3) then follow from the identity

$$s_{1u}\sigma(1,2) = \tfrac{1}{2}\tau_u(1,2). \tag{4}$$

Here s_{1u} is the u component ($u = x$, y, or z) of the spin angular momentum operator acting on electron 1; as σ is totally symmetric, τ_u must necessarily transform as the u component of angular momentum.[9] Hence, the functions τ_x, τ_y, τ_z belong to the same representations of C_{2v} (or any other group) as the three components of angular momentum.

† Spin angular momentum operators are expressed in units of $\hbar$. Further $\mathbf{S}$ refers to total spin angular momentum and $\mathbf{s}_i$ to that of electron i.

In equation (3) we have further indicated that every τ_u is an eigenfunction of S_u with eigenvalue 0. Hence the three triplet functions correspond to situations in which the spin angular momentum vector lies in one of the three coordinate planes. Whenever these planes are predetermined by molecular symmetry, this is a most natural description.

In the above example two-electron spin functions were considered, but this is no limitation. An n-electron system will in general have a number of singlet spin functions $\sigma(1 \ldots n)$. One can always generate sets of three $D^{(1)}$ type functions out of each of them by the same trick as used before (equation (4)),

$$s_{iu}\,\sigma(1 \ldots n) = \tfrac{1}{2}\tau_u(1 \ldots n).$$

The resulting functions—many of which will be linearly dependent—must be the triplet functions, and, by the same argument as before, transform like angular momenta.

As we are exclusively interested in singlet and triplet states, we shall henceforth designate in the kets

$$\gamma_s = \sigma \quad \text{for the singlet,}$$

$$\gamma_s = \tau_x, \tau_y, \tau_z \quad \text{for the three triplets.}$$

Clearly, the total symmetry of the triplet component $|m^3\Gamma_0, \tau_u\rangle$ is then obtained as the direct product of Γ_0 with the representation Γ_u to which a rotation around the u axis belongs.

To conclude the discussion on wavefunctions we may write down equation (2) by way of example for the upper spin component of phosphorescent naphthalene. In the usual π electron molecular orbital description one then has

$$\mathscr{A}a(1)\,a(2)\,b(3)\,b(4)\,c(5)\,c(6)\,d(7)\,d(8)\,e(9)\,e'(10)$$
$$\times\,\sigma(1,2)\,\sigma(3,4)\,\sigma(5,6)\,\sigma(7,8)\,\tau_x(9,10) = |1\,^3B_{2u'}, \tau_x\rangle. \quad (5)$$

The $a \ldots e$ are the bonding m.o.'s in order of increasing energy and e' the lowest anti-bonding m.o.[4] Clearly, the total symmetry of the state (5) is $B_{2u} \otimes B_{3g} = B_{1u}$, cf. equation (3) for D_{2h}.

The spin Hamiltonian and its matrix elements

According to quantum mechanics an electron with spin angular momentum $\mathbf{s}_i$ has a magnetic moment $g\beta\mathbf{s}_i$, where β is the Bohr magneton and $g = 2{\cdot}0023$ a proportionality constant. We must now examine the various terms in H' that arise owing to interactions involving this magnetic moment. The most important of these, at least for the present purpose, are spin-spin and spin-orbit coupling. Since we are interested in general conclusions rather than specific calculations, the discussion will be in terms of symmetry properties; for more detailed expressions the reader is referred to other papers in the Symposium.

Before considering the separate terms in H', an almost trivial remark may be made: in the absence of external fields the orientation of the molecule in space is physically irrelevant. This implies the invariance of H', not only under the operations of the molecular symmetry group, but also under any rotation of the system of axes, provided its effect on both positional and spin coordinates is properly considered.

Spin-spin coupling

This is the magnetic dipole-dipole interaction between two electrons resulting from their spin. In classical theory the potential due to a dipole of moment $\boldsymbol{\mu}_1$ at a distance $\mathbf{r}$ is equal to

$$\boldsymbol{\mu}_1 . \nabla \frac{1}{r}.$$

If a second dipole $\boldsymbol{\mu}_2$ is now placed at a distance $\mathbf{r}_{12}$ from the first, the energy becomes

$$\boldsymbol{\mu}_2 . \nabla_2(\boldsymbol{\mu}_1 . \nabla_1)\, 1/r_{12} = -\boldsymbol{\mu}_2 . \nabla_1(\boldsymbol{\mu}_1 . \nabla_1)\, 1/r_{12},$$

because $\nabla_2 r_{12} = -\nabla_1 r_{12}$, and we can further drop the subscripts to the gradient operator.

In quantum mechanics the corresponding Hamiltonian for spin-spin interaction in an assembly of electrons becomes[10, 23]

$$\begin{aligned} H_{\mathrm{s.s.}} &= -(g\beta)^2 \sum_{i<j} \mathbf{s}_j . \nabla(\mathbf{s}_i . \nabla)\, 1/r_{ij} \\ &= -(g\beta)^2 \sum_{i<j} \sum_{u,\, v=x,\, y,\, z} \frac{\partial^2 1/r_{ij}}{\partial u_{ij}\, \partial v_{ij}}\, s_{iu} s_{jv}. \end{aligned} \tag{6}$$

Now $r^3\partial^2 r^{-1}/\partial u\, \partial v$, with $u, v = x, y, z$, is the generating function for the spherical harmonics of the second degree.[11] As for instance noted by Van Vleck,[12] this implies that the second sum in equation (6) can be arranged in five products of an orbital and a spin factor, so that the orbital factors form a basis for the irreducible respresentation $D^{(2)}$ of the three-dimensional rotation group. (Or, in other words, they are proportional to the angular parts of the wavefunctions for d electrons.) But, since $H_{\text{s.s.}}$ must be invariant under rotation, the spin factors must transform contragradiently to the orbital factors, i.e. also like $D^{(2)}$. Accordingly, one has on rearranging the terms in equation (6)

$$H_{\text{s.s.}} = -\tfrac{3}{2}(g\beta)^2 \sum_{i<j} r_{ij}^{-5} [(3z_{ij}^2 - r_{ij}^2)(s_{iz}s_{jz} - \tfrac{1}{3}\mathbf{s}_i \cdot \mathbf{s}_j)$$
$$+ (y_{ij}^2 - x_{ij}^2)(s_{iy}s_{jy} - s_{ix}s_{jx}) + \sum_{u \neq v = x,y,z} u_{ij}v_{ij}(s_{iu}s_{jv} + s_{iv}s_{ju})]. \quad (7)$$

The further reduction of equation (7) and the calculation of its matrix elements is based on a number of considerations which we shall stepwise discuss. Let us first investigate what multiplicities are connected by $H_{\text{s.s.}}$ irrespective of molecular symmetry. In the integral

$$\langle \mathscr{A}\phi_j \chi_j | H_{\text{s.s.}} | \mathscr{A}\phi_j' \chi_j' \rangle, \quad (8)$$

the spin-dependent part of the operator was found to transform like the representation $D^{(2)}$ of the continuous rotation group. This implies that it can only be different from zero if the direct product of the representations to which χ_j and χ_j' belong also contains $D^{(2)}$. Since singlet and triplet spin functions correspond to $D^{(0)}$ and $D^{(1)}$, respectively, $H_{\text{s.s.}}$ cannot cause interaction between singlets ($D^{(0)} \otimes D^{(0)} = D^{(0)}$), nor between singlets and triplets ($D^{(0)} \otimes D^{(1)} = D^{(1)}$).† As far as rotational invariance is concerned, however, $H_{\text{s.s.}}$ must be expected to give non-vanishing elements amongst triplet spin functions, because $D^{(1)} \otimes D^{(1)} = D^{(2)} + D^{(1)} + D^{(0)}$. A side remark may here be made: from the fact that $H_{\text{s.s.}}$ does not cause singlets to interact with triplets, one should not conclude that it is incapable of mixing states of different multiplicities in general. For instance, it may well mix doublets with quartets.

Let us return to our symmetrical molecules. Because of the previous result, we need only consider matrix elements of $H_{\text{s.s.}}$ involving triplet spin functions. The important ones are those diagonal in the orbital configuration, which are responsible for the zero-field splitting of the triplet manifolds,

$$\langle m\,{}^3\Gamma_0, \gamma_s | H_{\text{s.s.}} | m\,{}^3\Gamma_0, \gamma_s' \rangle. \quad (9)$$

† This conclusion also holds for a *rotating* molecule. When referring the spins to a set of axes moving with the molecule, the triplet spin functions are simply transformed amongst one another[13] and our reasoning remains valid in the rotating coordinate system. Or in physical language: the coupling between spin dipolar interaction and rotational motion cannot cause a shortening of the lifetime of the phosphorescence state.

Although spin-spin coupling between different electron configurations must also occur, we shall not consider it since it leads to very small perturbations not amenable to experimental investigation.

As the orbital factor is the same on both sides of the integral (9), only those parts of the operator that are totally symmetric in the coordinates can contribute. Thus far the orientation of the coordinate system in $H_{\text{s.s.}}$ was irrelevant, but by assigning, say, $x = 0$ and $y = 0$ to the two molecular symmetry planes, only the first two terms of equation (7) are of the proper symmetry in the orbital coordinates. The last term of equation (7), containing the three products $x_{ij}y_{ij}$, $y_{ij}z_{ij}$, and $z_{ij}x_{ij}$, may be discarded when calculating matrix elements diagonal in Γ_0, because these products belong to non-totally symmetric representations. We thus are left with

$$H_{\text{s.s.}} = -\tfrac{3}{2}(g\beta)^2 \sum_{i<j} r_{ij}^{-5}[(3z_{ij}^2 - r_{ij}^2)(s_{iz}s_{jz} - \tfrac{1}{3}\mathbf{s}_i \cdot \mathbf{s}_j) + (y_{ij}^2 - x_{ij}^2)(s_{iy}s_{jy} - s_{ix}s_{jx})]. \quad (10)$$

With the present choice of axes the orbital and spin factors in equation (10) separately belong to the totally symmetric representation of C_{2v}. But as the three triplet spin functions have been chosen to belong to different irreducible representations of C_{2v}, the only matrix elements that occur in equation (9) are those diagonal in the spin functions: $\gamma_s' = \gamma_s$.

We thus arrive at the important conclusion that each of the three eigenstates of H' for the components of a triplet in the absence of an external field corresponds to a location of the spin in one of the three 'preferred' planes of a symmetrical molecule. Since H' commutes with the molecular symmetry group and the three triplet spin functions belong to different representations of the group C_{2v} contained in it, they diagonalise H' independent of any degree of approximation, for instance whether spin-orbit coupling gives a higher order contribution to the splitting (it cannot in first order, cf. below).

Because of the symmetries of the three triplet functions τ_x, τ_y, τ_z, it is apparent that magnetic dipole transitions are allowed between all three components, with polarisations determined by equation (3a).[14] However, such zero-field transitions cannot show the familiar hyperfine splitting of ordinary ESR spectra, taken with a magnetic field. Namely, the operator for dipole-dipole interaction between the spin of nucleus K and that of electron i, by analogy with equation (7), is

$$H_{\text{s.I.}} = \mathbf{s}_i \cdot \mathbf{A}_{iK} \cdot \mathbf{I}_K,$$

where $\mathbf{I}$ is the operator for nuclear spin and the elements of the tensor $\mathbf{A}_{iK}$ are integrals over the electronic and nuclear coordinates. However, the part of $H_{\text{s.I.}}$ that acts on the electron spin coordinates transforms like an angular momentum in C_{2v} and since the zero-field eigenstates do so too—even for an unsymmetrical molecule in its principal axis system—$H_{\text{s.I.}}$ can only give matrix elements between *different* zero-field states. As $H_{\text{s.I.}}$ is smaller than $H_{\text{s.s.}}$ by a factor of 1 000, the second-order shift that remains should be small. Zero-field transitions have recently been observed by Brandon, Gerkin & Hutchison[15] and no hyperfine splitting was reported.

The Hamiltonian (10) is that used to calculate the energies of the three components of a triplet from an assumed electron distribution in the molecule; the first work of this kind was carried out by Gouterman & Moffitt[16] and by Hameka.[17] Now Van Vleck has pointed out[12] how a 'microscopic' Hamiltonian such as (7) or (10), in which the spin operators act on individual electrons, can be replaced by a phenomenological 'spin Hamiltonian' in which only operators for the *total* spin angular momentum occur, because the matrix elements of $(s_{iz}s_{jz}-\frac{1}{3}\mathbf{s}_i\,.\,\mathbf{s}_j)$, $(s_{iy}s_{jy}-s_{ix}s_{jx})$, etc., *within a given multiplet* $|m^{\mu}\Gamma_0\rangle$ are proportional to those of $(S_z^2-\frac{1}{3}\mathbf{S}^2)$, $(S_y^2-S_x^2)$, etc. This interesting property is a consequence of the Wigner–Eckart theorem for tensor operators and holds only when the operators are grouped together so as to form an irreducible representation $D^{(n)}$ of the three-dimensional rotation group. By operating on the functions (3) it can easily be verified that the above proportionality indeed holds within the two-electron triplet multiplet, but not between triplets and singlets.

Following Van Vleck, equation (10) may thus be rewritten as

$$H_{\text{s.s.}} = D(S_z^2-\tfrac{1}{3}\mathbf{S}^2)+E(S_x^2-S_y^2). \tag{11}$$

Here D and E are parameters characteristic for the particular molecule and triplet state concerned, which result from integration over the orbital coordinates. The actual expressions for these integrals in terms of the electron configuration can be obtained by calculating the energies of the three components of the triplet from both equations (10) and (11) and equating the results (cf. formulae (8) and (9) of ref.[18]).

The grouping of terms in equation (7), in conformity with the representation $D^{(2)}$ of the three-dimensional rotation group, allowed us to make some far-reaching conclusions and led to a considerable simplification of $H_{\text{s.s.}}$. In doing so, however, the basic equivalence of the three spin components of a triplet has been lost. This equivalence can be regained by expanding $\mathbf{S}^2$ in equation (11) into its component products and rearranging terms; one finds

$$H_{\text{s.s.}} = -(XS_x^2 + YS_y^2 + ZS_z^2). \tag{12}$$

The new parameters

$$X = \tfrac{1}{3}D - E, \quad Y = \tfrac{1}{3}D + E, \quad Z = -\tfrac{2}{3}D$$

have a very simple meaning: they are equal to the energies of the three components in the absence of a magnetic field, $X + Y + Z = 0$.

By choosing three triplet spin functions belonging to different representations of the group C_{2v}, we were able to diagonalise the Hamiltonian and arrived at the simple form (12). However, the zero-field splitting pattern will reflect the full molecular symmetry. If the components of angular momentum do not belong to degenerate representations of the molecular symmetry group, then X, Y, Z will in general all be different and the degeneracy of the triplet completely lifted, as in phosphorescent naphthalene. If, on the contrary, S_x and S_y, say, belong to the same two-dimensional representation of the molecular symmetry group, one has by symmetry $X = Y = -\tfrac{1}{2}Z$. This is the situation encountered in the phosphorescent state of triphenylene (D_{3h}) and in the $^3\Sigma$ ground state of the oxygen molecule.

A few words may be in order about molecules of low symmetry. If the molecular symmetry group does not contain C_{2v} as a subgroup, the three spin components to be taken for any given triplet state are no longer known *a priori*, but nevertheless almost all the preceding reasoning still applies. The transition from the full spin Hamiltonian (7) to the truncated form (10) then requires a coordinate transformation to principal axes that causes the sum containing the mixed functions $x_{ij}y_{ij}/r_{ij}^5$, etc., to vanish. Clearly, the three preferred spin planes for the eigenstates of $H_{\text{s.s.}}$ now are perpendicular to these principal axes. An unpleasant feature of such unsymmetrical molecules is that the directions of the principal axes of equation (7) vary with the orbital state $|m\Gamma_0\rangle$ considered and that there is no reason for coincidence with other preferred directions in the molecule, such as electronic transition moments.

The determination of zero-field splittings from electron resonance experiments is treated by Hutchison in another part of the Symposium. For the phosphorescent state of aromatic hydrocarbons and related molecules, in which the lowest triplet state corresponds to excitation within the π-electron system and the two electrons with parallel spin can never be on the same atomic center, the order of magnitude of the splitting is $0{\cdot}1\,cm^{-1}$. In situations where the two triplet electrons can simultaneously be in different orbitals on the same center the triplet splittings are considerably larger. For instance, for the triplet ground state of phenylmethylene $|D| = 0{\cdot}518$, $|E| = 0{\cdot}024\,cm^{-1}$,[19] and for that of oxygen $D = 3{\cdot}96\,cm^{-1}$, $E = 0$.[20] Theoretical estimates of spin-spin coupling are in agreement with these orders of magnitude[16–18, 20–22] and even give a reasonable semi-empirical account of the variation of D and E within a series of phosphorescent aromatic hydrocarbons.[18]

From the extensive work on the splitting of the $^3\Sigma_g^-$ state of O_2—where agreement between theory[20, 22] and experiment[20] is only moderate—two things stand out, however. First, the results of the calculations of the spin-spin interaction are very sensitive to the detailed form of the molecular wavefunction. Secondly, the second-order contribution to the splitting due to spin-orbit coupling has been estimated to amount to some 40% of the first-order term involving $H_{s.s.}$.[22 b] The symmetry of an aromatic hydrocarbon entails an extraordinarily weak spin-orbit coupling, but as soon as heteroatoms are present one should also consider the second-order spin-orbit contribution and not only the spin-spin coupling in calculating the splitting of a triplet. However, from the discussion following equation (16), it is clear that the second-order contribution of $H_{s.o.}$ is not to be limited to the interaction of the triplet with singlets (as is sometimes suggested), but the interaction with other triplets and possibly even quintets should also be considered.

Before concluding this section, a few words must be said about the Fermi contact interaction between the electrons. The formulation of equation (6), by analogy with the classical formula for two interacting dipoles, does not lead to the correct expression for $H_{s.s.}$ at the singular point $r_{ij} \to 0$. In fact, a term

$$-\sum_{i<j} \frac{32\pi}{3} \beta^2 \delta(r_{ij})\, \mathbf{s}_i \cdot \mathbf{s}_j, \tag{13}$$

where $\delta(r_{ij})$ is a Dirac delta function, has been neglected in equation (6).[23] However, by the Wigner–Eckart theorem, the matrix elements of the scalar product $\mathbf{s}_i \cdot \mathbf{s}_j$ within a multiplet are proportional to those of $\mathbf{S}^2$, which is a constant ($\mathbf{S}^2 = 2$ for a triplet state). Hence, equation (13) can at most cause a uniform (and very small) shift of all three components of a triplet, and all our previous conclusions remain valid. Moreover, in a two-electron description of the triplet state the contribution of the additional term (13) vanishes[14, 24] as the orbital parts of its matrix elements goes to zero when $r_{12} \to 0$, because of the antisymmetric character of a two-electron spatial wavefunction relative to electron interchange. But it is to be noted that in a many-electron description this is no longer true.

Spin-orbit coupling

This is the interaction of the magnetic moment of the spinning electron with the internal magnetic fields produced by the relative motion of electrons and nuclei. It can be written as

$$H_{\text{s.o.}} = g\beta \sum_i \mathbf{h}_i([\mathbf{r}], [\mathbf{p}]) \cdot \mathbf{s}_i. \tag{14}$$

Here $\mathbf{h}_i$ is the magnetic field acting on electron i, which is a function of the positions $\mathbf{r}_j$ and momenta $\mathbf{p}_j = -i\nabla_j$ of all electrons and further depends on the nuclear conformation. Because of the invariance of $H_{\text{s.o.}}$ under any rotation, the factors $\mathbf{h}_i$ and $\mathbf{s}_i$ in (14) have contragradient transformation properties, and since $\mathbf{s}_i$ transforms like an angular momentum $\mathbf{h}_i$ must do so also.

We shall not go into the details of spin-orbit coupling, but only give a general outline with special reference to those aspects of importance for what is to follow. For more explicit formulae and computational matters we refer to Hameka's review of the subject.

Although we hardly need it, it is illustrative to give the expression relating the local field $\mathbf{h}_i$ to the positions and momenta of the particles in the system. For a molecule fixed in space and in the absence of external fields, one has for the product $\mathbf{h}_i \cdot \mathbf{s}_i$ in equation (14)[10, 23]

$$\mathbf{h}_i([\mathbf{r}], [\mathbf{p}]) \cdot \mathbf{s}_i = \frac{e}{2mc}\left[\sum_K \frac{Z_K}{r_{Ki}^3}\mathbf{l}_{iK} + \sum_{j \neq i} \frac{1}{r_{ij}^3}(2\mathbf{p}_j - \mathbf{p}_i) \wedge \mathbf{r}_{ij}\right] \cdot \mathbf{s}_i. \tag{15}$$

Capital subscripts refer to nuclei and lower case subscripts to electrons. The first sum is a one-electron operator representing the contribution to the local field due to the motion of electron i in the Coulomb field of nucleus K with charge Z_K; $\mathbf{l}_{iK}$ is the angular momentum of electron i about nucleus K: $\mathbf{l}_{iK} = \mathbf{r}_{Ki} \wedge \mathbf{p}_i$, and $\mathbf{r}_{ij} = \mathbf{r}_j - \mathbf{r}_i$. The second term is a sum of two-electron operators and takes care of the contribution due to the motion of the electrons relative to one another.

Equation (15) is rigorous when $v/c \ll 1$. In actual calculations on molecules, where one no longer has the convenience of spherical symmetry of atomic theory—not even locally near the nuclei—the second sum presents a serious problem. In molecular orbital calculations on any but the smallest molecules, where only some of the electrons are being explicitly considered (e.g. the ten π electrons in naphthalene in equation (5)), the only practicable way out is to lump the other 'core' electrons together with the nuclei, use an effective nuclear charge $Z_K^{\text{eff.}}(r_i) < Z_K$ in the first term and take the second term only over the electrons in the m.o.'s considered.

The pioneering work on spin-orbit interaction in organic molecules was done by McClure.[25] He restricted himself to the one-electron operators in equation (15) and took $Z_K^{\text{eff.}}$ values from atomic theory to allow for the screening by other electrons. Calculations made since with this simple quasi-central field approach[26] have given a reasonable account of the large difference in the intensity of spin-orbit interaction (as manifested by phosphorescence lifetimes) between aromatic hydrocarbons on the one hand, and aza-aromatics on the other. Also, the polarisations of phosphorescence as experimentally observed can be properly understood on this basis. More elaborate calculations on polyatomic molecules, in which two-electron terms were considered, have so far only been made for benzene[8, 27, 28] and acetone.[8]

Let us now return to (14) and look at the general form of its matrix elements

$$\langle m^\mu \Gamma_0, \gamma_s | H_{\text{s.o.}} | n^\nu \Gamma_0', \gamma_s' \rangle. \tag{16}$$

As regards the multiplicities connected by $H_{\text{s.o.}}$, it is clear that, because the spin part of it transforms like $D^{(1)}$, a given multiplicity μ may be combined with $\nu = \mu + 2$, μ, $\mu - 2$ (only positive values allowed). Hence spin-orbit coupling causes triplet states to mix with singlets, triplets and quintets. Because of its importance for phosphorescence we shall mainly consider the interaction between singlets and triplets.

When next considering the integration over the orbital coordinates, the fact that the $\mathbf{h}_i$ transform as angular momentum implies—as first noted by McClure[25] and Weissman[9]—that the elements (16) vanish unless the direct product $\Gamma_0 \otimes \Gamma_0'$ contains a representation to which a component of angular momentum belongs. An important implication of this is that, when excluding orbitally degenerate states,† spin-orbit coupling in symmetrical molecules can never give matrix

† Because of great sensitivity to Jahn Teller instability, a satisfactory treatment of spin-orbit or spin-spin coupling in orbitally degenerate states should include vibronic coupling. In connection with the line shape of ESR spectra of aromatic radical-ions of high symmetry, a certain amount of work has been done on spin-orbit coupling in orbitally degenerate states.[29] The situation is by no means clear yet, and we exclude orbitally degenerate states from our present considerations.

elements diagonal in the electron configuration. Thus, it cannot contribute in first order to the zero-field splitting of a (orbitally non-degenerate) triplet state, nor cause interaction between the singlet and triplet states derived from the same electron configuration.

Let us next consider the integration over the spin coordinates. Here we have, just as with the orbital part, that the integrals

$$\langle \gamma_s | \mathbf{s}_i | \gamma_s' \rangle$$

vanish, unless the product $\Gamma_{\gamma_s} \otimes \Gamma_{\gamma_s'}$ transforms like one of the components of angular momentum. Thus, taking the triplet sublevel τ_x as an example, $H_{\text{s.o.}}$ allows it to interact with singlet states via the x components s_{ix} of the spin operators and with the τ_y and τ_z sublevels of other triplets through the components s_{iz}, s_{iy}, respectively. When combining the previous results it is clear that—at least for orbitally non-degenerate states—the interaction between a given singlet $|^1\Gamma_0, \sigma\rangle$ and triplet $|^3\Gamma_0', \tau_u\rangle$ can only occur through the *one* component τ_u for which $\Gamma_{\tau_u} = \Gamma_0 \otimes \Gamma_0'$.

It is physically illuminating to consider the interaction in a two-electron description. The spin-orbit coupling, on substituting the expressions (2), (14) and (15) into equation (16), then leads to integrals of the type

$$\langle \mathscr{A}\phi(1,2)\,\chi(1,2) |\, \mathbf{h}_1 . \mathbf{s}_1 + \mathbf{h}_2 . \mathbf{s}_2 \,| \mathscr{A}\phi'(1,2)\,\chi'(1,2) \rangle,$$

where χ and χ' are one of the functions (3). This can be rewritten as

$$\langle \mathscr{A}\phi(1,2)\,\chi(1,2) |\, \tfrac{1}{2}(\mathbf{h}_1 - \mathbf{h}_2)\,(\mathbf{s}_1 - \mathbf{s}_2) \,| \mathscr{A}\phi'(1,2)\,\chi'(1,2) \rangle$$
$$+ \langle \mathscr{A}\phi(1,2)\,\chi(1,2) |\, \tfrac{1}{2}(\mathbf{h}_1 + \mathbf{h}_2)\,(\mathbf{s}_1 + \mathbf{s}_2) \,| \mathscr{A}\phi'(1,2)\,\chi'(1,2) \rangle. \quad (17)$$

Since the spin function $\sigma(1,2)$ is *antisymmetric* and $\tau_u(1,2)$ *symmetric* with respect to interchange of the two spins, it follows on integration over the spin coordinates that the first integral in the expression (17) causes singlets to mix with triplets, whereas the second integral is responsible for the interaction of different components of different triplets, or, in formulae (cf. (3) and (4))

$$\left.\begin{aligned} |\langle \sigma(1,2) |\, \mathbf{s}_1 - \mathbf{s}_2 \,| \tau_u(1,2) \rangle| &= 1, \\ |\langle \tau_u(1,2) |\, \mathbf{s}_1 + \mathbf{s}_2 \,| \tau_v(1,2) \rangle| &= 1 \text{ for } v = u \mp 1 \text{ (cyclic)}, \end{aligned}\right\} \quad (18)$$

all other combinations being zero.

Apparently, singlet-triplet mixing—which is the turning over of one spin relative to the other—is proportional to the instantaneous field inhomogeneity ($\mathbf{h}_1 - \mathbf{h}_2$). A trivial corollary of this is that homogeneous external fields do not cause singlets to mix with triplets.

Thus far we have examined the requirements for singlet-triplet mixing on the basis of the overall symmetry of the molecule. However, from the work by Hameka & Oosterhoff[8] and from a number of other studies[26] it transpired that the striking difference in the extent of singlet-triplet mixing between aromatic hydrocarbons (phosphorescence lifetimes of the order of seconds) and ketones (lifetimes of milliseconds) is a consequence of a difference in local orbital symmetry. This can, qualitatively, be seen as follows.

If we work out matrix elements of the type (16), pick out the proper components by integrating over the spin coordinates and then expand the m.o.'s in linear combinations of a.o.'s, we end up with a sum of terms of the type

$$\langle a_A(1)\, b_B(2)|\, \mathbf{h}_1\, |c_C(1)\, d_D(2)\rangle, \tag{19}$$

where the a, b, c, d are a.o.'s centered on the nuclei A, B, C, D, not necessarily all different.

First of all, let us look at an aromatic hydrocarbon in the π electron description with $x = 0$ as the molecular plane. Then the product of four $2p_x$ type a.o.'s in (19) is symmetric about $x = 0$, and spin-orbit coupling can only occur, through the term $h_{ix} s_{ix}$ in $\mathbf{h}_i . \mathbf{s}_i$, as h_{iy} and h_{iz} are antisymmetric about $x = 0$. Consequently singlet-triplet mixing is here limited to coupling of the τ_x triplet components with singlets. However, that is not all. Since h_{ix} is antisymmetric about the two planes $y = 0$, $z = 0$, and the orientation of these planes for the purpose of calculating a specific integral (19) may be chosen to coincide with any symmetry planes inherent in the atomic orbital pattern of this integral, all one-center and two-center integrals must vanish by symmetry. Hence—as already noted by McClure[25]—in a π electron model singlet-triplet mixing can only be caused by very small three- and four-center integrals. Explicit calculation of a number of elements (16) between the lower singlet and triplet states of benzene has yielded values of 0·2–0·4 cm^{-1}.[8, 28] Thus, the orbital symmetry characteristic of a π electron description of aromatic hydrocarbons causes the matrix elements of spin-orbit coupling to be only about 1 % of the corresponding element between the lowest ^1S and ^3P states of the carbon atom!

In a ketone or azine, on the other hand, the situation is altogether different.[8, 26] Here excitations of modest energy occur from an in-plane

σ orbital to an out-of-plane π^* orbital involving, respectively, p_y and p_x a.o.'s on each of the nitrogen atoms of, say, quinoxaline in fig. 1. Some of the integrals (19) then take the form

$$\langle p_{xN}(1)p_{xN}(2)|\,\mathbf{h}_1\,|p_{yN}(1)p_{xN}(2)\rangle, \tag{20}$$

and these one-center integrals—via the h_z component which belongs to the same irreducible representation as the product $p_{xN}p_{yN}$—have values of the same order of magnitude as found for the N atom (50 cm^{-1}). Or, in other words, despite the 'quenching' of orbital angular momentum in a non-linear molecule, the local electron configuration around the nitrogen (or oxygen) nucleus in the molecules considered contributes an effective leak of angular momentum between the spin system and the orbital motion in the strong field near that nucleus, similar to that in a free atom.

Phosphorescence occurs because singlet and triplet states of equal total symmetry are mixed and intensity is thus 'stolen' from allowed transitions.[30] In a π electron model the only transitions considered are of π–π^* type and so the phosphorescence of an aromatic hydrocarbon might be expected to be in-plane polarised. During the last few years extensive experimental material has become available from which it is clear that this expectation is erroneous: the phosphorescence of a large number of aromatic hydrocarbons (with the possible exception of benzene[31]) is strongly out-of-plane polarised.[32]

Apparently, the dominant mechanism of spin-orbit coupling in, say, naphthalene is analogous to that in quinoxaline. In both cases the intensity of the phosphorescence emitted by the π–π^* lowest triplet state appears to be stolen from an out-of-plane transition in which an electron is transferred between a σ- and a π-type orbital. In quinoxaline the latter transition clearly involves the lone-pair electrons on the N atoms, in the hydrocarbon it must be (a) transition(s) between a π orbital and anti-bonding C—C, C—H, or possibly Rydberg orbitals.† The relation between phosphorescence polarisation and the mechanism of spin-orbit coupling is illustrated in the next section when the electronic energy level diagram of quinoxaline is discussed in some detail.

† The near-constancy of the *radiative* lifetimes of the phosphorescent states of aromatic molecules ($\approx$ 30 s)[33] is probably related to the fact that the emission is derived from such $\sigma^* \leftrightarrow \pi$ transitions, which would seem rather insensitive to detailed molecular structure.

Low temperature experiments for tracing the path of molecular excitation and de-excitation

Electronic energy levels of quinoxaline and the polarisation of its phosphorescence

Let us examine the example of the energy level diagram of quinoxaline (fig. 1), since this is the molecule which we have studied in our experiments; for a review of the electronic levels of aza-aromatics cf. Kasha.[34]

In quinoxaline two types of molecular orbitals have to be considered. First, the bonding (π) and anti-bonding (π^*) m.o.'s built up of $2p_x$ a.o.'s centered on the carbon and nitrogen atoms, which are closely related to those of naphthalene.[4] Secondly, two m.o.'s in which the in-plane lone pair orbitals on the nitrogens are taken together in a symmetrical (n_+), or antisymmetrical (n_-) combination.

The lowest excited singlet state in fig. 1 is an $n\pi^*$ state.[3] When the present work was started the belief prevailed that the $n\pi^*$ states observed in *p*-diaza-aromatic molecules actually were narrow doublets of an $n_-\pi^*$ and an $n_+\pi^*$ state. Recent work on pyrazine (*p*-diazabenzene), however, has destroyed this conviction. First the gas phase work of Innes, Simmons & Tilford[35] revealed that the first excited singlet state of pyrazine is not a doublet and of $n_+\pi^*$ symmetry. Secondly, Clementi[36] in an all-electron SCF calculation has found that the energy of the n_- orbital is, in fact, some 2·5 eV lower than that of the n_+ orbital, thus supporting Innes's experiments. Since the first absorption band of quinoxaline in the vapour does not show an apparent doubling either, we assume it to be of $n_+\pi^*$ type, leading to ${}^1B_1$ symmetry. Although this is a firm belief, it is not a rigorous assignment and we have indicated the alternative possibility $n_-\pi^*$ (${}^1A_2$) in brackets in fig. 1.

The next singlet up, ${}^1A_1$, is a $\pi\pi^*$ state, shifted some 1 000 cm^{-1} to the red relative to the corresponding state in naphthalene.[3] (From the change in intensity of the triplet ESR signal of quinoxaline in a durene single crystal with the direction of the beam of light exciting it through this second singlet-singlet absorption band, it was confirmed that the absorption is indeed z polarised, as in naphthalene.) The second $\pi\pi^*$ state[3] lies at a considerably higher energy and is thought to correspond to the ${}^1B^+_{2u}$ state of naphthalene, strongly shifted to the blue.

The lower triplet states of quinoxaline are drawn at the right in fig. 1. The separate components of some of them are shown on an enlarged scale in small circles. The total symmetries (printed at the right) have been obtained as the direct products of the representation of the orbital part (given in kets at the left) and that of the spin part (cf. (3)). Except for the lower triplet state,[38] the zero-field splittings are not known. The order being irrelevant for our purpose, we have arbitrarily assumed the energy always to increase in the order τ_x, τ_y, τ_z.

The lowest triplet state is a $\pi\pi^*$ state of B_2 orbital symmetry, which closely resembles the phosphorescent state of naphthalene as regards energy[3] and spin distribution.[38] Phosphorescence occurs because the sublevels of the lowest triplet are mixed by spin-orbit coupling with singlet states of equal total symmetry. When neglecting the effect of molecular vibrations, depletion of the population of the phosphorescent state may thus occur via τ_x with z polarised emission, or via τ_z with x polarised emission; two of the transitions from which intensity can be so stolen are indicated on the left, but they are by no means the only ones. (There can also be a contribution to the intensity of the emission from triplet character mixed into the ground state,[26c] but this would not modify our arguments.) In addition, one has to consider depletion of the phosphorescent state through radiationless return to the ground state, for which the selection rules, if any, are not known; for a *p*-diazine such as quinoxaline the probability of this process may be low.[1]

As we saw earlier (pp. 116–117) local symmetry causes spin-orbit coupling between $\pi\pi^*$ and $n\pi^*$ states to be at least an order of magnitude stronger than amongst pure π states. Consequently, phosphorescence in quinoxaline is expected to originate from the upper component (τ_z), e.g. by mixing with the near-by ${}^1B_1$ $n\pi^*$ triplet state. The lower component τ_x, on the other hand, can only weakly interact with ${}^1A_1$ $\pi\pi^*$ states. (Remember that the weakness of the latter interaction is a consequence of symmetry; it is conceivable that it would be enhanced considerably by an out-of-plane deformation of the molecule.) El-Sayed & Brewer confirmed that, in agreement with these ideas and the behaviour of $\pi\pi^*$ phosphorescent states in general, the emission of quinoxaline in a rigid glass at low temperature is predominantly x polarised.[39]

In some very recent work, however, it has been reported that in *durene* as a host crystal the phosphorescence of both naphthalene and quinoxaline is anomalous, since it is long axis polarised.[5] Whether this exceptional behaviour is due to a deformation of the solute molecules induced by a 'misfit' in the durene lattice, or perhaps by interaction with crystal states is not known. Clearly, one has to reckon with a distinct possibility that the radiative depopulation of quinoxaline when in a durene host, predominantly occurs through τ_x.

No experimental data are available on the higher triplets of quinoxaline, but there certainly must be one and perhaps two $n\pi^*$ triplet(s) below the corresponding singlet.[46] In the construction of the diagram, and in the discussion that follows, it is assumed that two $n\pi^*$ states but no further triplet states lie *below* the first excited singlet state. For rough guidance the two nearest $\pi\pi^*$ states as calculated by Pariser for naphthalene[4] have been drawn in. It is to be noted that simple perturbation theory predicts a marked shift to higher energy of the $|2^3B_2\rangle$ state of quinoxaline relative to the B_{1g}^+ state of naphthalene drawn in the figure, thus strengthening our assumption.

Paths for populating the phosphorescent state–spin polarisation

We have seen that spin-orbit coupling is a very selective interaction in the sense that the degree of coupling with the singlet system it affords varies widely for the different spin components of a given triplet. Experimentally, this is demonstrated by the fact that phosphorescence usually is strongly polarised, which implies that the radiative *depopulation* of the phosphorescent state occurs largely, or even entirely, through a single component. Now the *populating* of the phosphorescent state after excitation in the singlet system also occurs through spin-orbit coupling and there is little reason to believe that this process should not be selective relative to the different spin components. In fact, the same kind of matrix elements come into play and, again, a singlet may interact with a given triplet only through one of its zero-field components by the selection rules given.

Although progress has been made in the general understanding of intersystem crossing,[40] the detailed path in individual molecules is not known with any certainty. It is well established, however, that (barring one or two rare exceptions) the molecules of an aromatic substance, on excitation by light, in great majority first return to the lowest vibrational level of the first excited singlet state, $^1B_1$ in the present case. From there the molecule can either go to the ground state by fluorescence emission, or cross-over to a near-degenerate, vibrationally excited level of a triplet from which it then cascades down to the phosphorescent state, i.e. the ground vibrational level of the lowest triplet. In the present instance fluorescence has not been observed thus far, hence intersystem crossing to the triplet system must be fast with a rate $\geqslant 10^8\,s^{-1}$. The subsequent internal conversion to the phosphorescent state is even fastei and rates of the order of $10^{12}\,s^{-1}$ are plausible.[1,40]

If fig. 1 is correct (and we have little doubt it is), two 'allowed' routes for crossing are available: (*a*) via the τ_x component of the $n\pi^*$ triplet $|^3A_2\rangle$; (*b*) straight into a highly excited vibrational level of the $|1\,^3B_2\rangle$ state via its τ_z component, cross-over into the $|2\,^3B_2\rangle$ state is unlikely, but would also go through τ_z. In addition, crossing to the other components of the three lower triplets might occur by the intervention of asymmetrical vibrations. But, when intersystem crossing between a given singlet and triplet can go via an 'allowed' route, vibrational overlap would appear to favour this over 'forbidden' routes between these two states just as, in general, symmetry-allowed electric dipole transitions are appreciably more probable than forbidden ones.

In other words, for intersystem crossing to a specific triplet the invariance of $H_{\text{s.o.}}$ under the operations of the molecular symmetry group is likely to cause the molecule to go from a singlet spin function σ to *one* of the zero-field spin functions τ_u of the triplet. Thus, upon intersystem crossing the spin is expected to be quantised relative to the molecular frame. For instance, if in quinoxaline the upper route (*a*) is favoured in fig. 1, the majority of molecules should enter the triplet manifold via $|^3A_2, \tau_x\rangle$, with their spin 'polarised' in the plane $x = 0$. This initial polarisation is imposed by molecular symmetry and should still occur when an external magnetic field is present, since such fields were seen not to affect the spin-orbit coupling in first order.

Now it should be realised that, because of the weakness of magnetic interactions in non-linear molecules—where orbital angular momentum is quenched—the spin is only very loosely coupled to the electron distribution. Accordingly, reorientation of spin angular momentum—i.e. transition from one spin function to another—is a relatively improbable process in the absence of microwave radiation. Such 'relaxation' amongst the spin components of a given multiplet is produced by random modulation through thermal motion of the electron-electron and electron-nuclear spin dipolar couplings and is measured by the average 'spin-lattice' relaxation time, T_1. Values of T_1 can be inferred from saturation effects in ESR spectra; for quinoxaline in durene we estimate T_1 to be about 10^{-5} s at 80 °K and as much as a few tenths of a second at 1·6 °K.

In cascading down to the phosphorescent state the molecule changes its electron distribution and loses vibrational energy, but in neither of these processes forces specifically acting on the spin would seem to be operative.† Hence, as cascading occurs in a time short relative to T_1, there should only be a small probability for the spin to change from one preferred plane to another during this process, and *one would expect populating of the phosphorescent state to be accompanied by spin polarisation.* Obviously, thermal relaxation processes will tend to annihilate such polarisation and thus we may only hope to observe it at very low temperature where T_1 becomes of the order of, or even longer than, the phosphorescence lifetime (τ).

Effect of a magnetic field on the paths of excitation and de-excitation

Before coming to a discussion of some experimental aspects, the question should be answered to what extent an external magnetic field affects the above conclusions. As symmetrical arrangements yield the most information, we only consider orientations of the field along a molecular axis.

If the quinoxaline molecule is placed in a field H, the spin components of any of its triplets will be mixed by the Zeeman term in the Hamiltonian

$$H'' = g\beta \mathbf{H}.\mathbf{S} + H'.$$

For the phosphorescent state and a field $H = H_z$ along the z axis one has, for instance,

$$H'' = g\beta H S_z - [XS_x^2 + YS_y^2 + ZS_z^2] \quad (H = H_z). \tag{21}$$

The eigenenergies (relative to the center of the multiplet) and eigenfunctions of the components of the lowest triplet state then are[14]

$$\epsilon_0 = Z, \quad \epsilon_\pm = -\tfrac{1}{2}Z \pm \alpha_z \quad (H = H_z); \tag{22}$$

$$\left.\begin{aligned}
|{}^3B_2, +\rangle &= \frac{1}{\sqrt{2}}\left[1 - \frac{Y-X}{2\alpha_z}\right]^{\frac{1}{2}} |{}^3B_2, \tau_x\rangle \\
&\qquad + \frac{i}{\sqrt{2}}\left[1 + \frac{Y-X}{2\alpha_z}\right]^{\frac{1}{2}} |{}^3B_2, \tau_y\rangle \\
|{}^3B_2, 0\rangle &= |{}^3B_2, \tau_z\rangle \\
|{}^3B_2, -\rangle &= \frac{1}{\sqrt{2}}\left[1 + \frac{Y-X}{2\alpha_z}\right]^{\frac{1}{2}} |{}^3B_2, \tau_x\rangle \\
&\qquad - \frac{i}{\sqrt{2}}\left[1 - \frac{Y-X}{2\alpha_z}\right]^{\frac{1}{2}} |{}^3B_2, \tau_y\rangle
\end{aligned}\right\} \quad (H = H_z). \tag{23}$$

† Although the stationary spin eigenfunctions of $H_{s.s.}$ will change by a *steady* deformation of a molecule, they will remain as defined in (3) for vibrationally excited states of a symmetrical molecule.

Here +, 0, − denote the components in order of decreasing energy at high field, and α_z is a measure of the field strength,

$$\alpha_z = \{\tfrac{1}{4}(Y-X)^2 + (g\beta H_z)^2\}^{\frac{1}{2}}. \qquad (24)$$

In fields exceeding some 5 000 gauss the correction factors in square brackets tend to unity to within 1 %, and the functions (23) practically are the triplet eigenfunctions of S_z.

If the probability for emission from the phosphorescent state is denoted by k, then in the *usual random phase approximation*[41] k will be the sum of the probabilities for emission from the individual components,

$$k = \sum_i f_i k_i \,. \qquad (25)$$

Here f_i is the fraction of molecules being in the i-th spin component and k_i the probability of emission from that component; $i = x, y, z$ for the field free case and $i = +, 0, -$ when a field is present. The k_x, k_y, k_z are proportional to the square of the corresponding matrix elements for electric dipole radiation,

$$k_u \propto M_u^2, \quad \text{with} \quad M_u = |\mathbf{M}_u| = |\langle(^1A_1, \sigma)^{(1)}|\, e\mathbf{r}\, |(^3B_2, \tau_u)^{(1)}\rangle|. \qquad (26)$$

The superscript (1) indicates that the elements are not to be taken over the (spin-pure) zero-order wavefunctions, but over wavefunctions that include singlet-triplet interaction. For the specific situation of quinoxaline (fig. 1)

$$\left.\begin{aligned} M_x &= \langle(^1A_1, \sigma)^{(1)}|\, ez\, |(^3B_2, \tau_x)^{(1)}\rangle, \\ M_y &= 0, \\ M_z &= \langle(^1A_1, \sigma)^{(1)}|\, ex\, |(^3B_2, \tau_z)^{(1)}\rangle. \end{aligned}\right\} \qquad (27)$$

Now consider the situation in a strong field along the z axis, $H = H_z$. From the wavefunctions (23) and the transition moments (27) it follows that z polarised radiation may be emitted from the + and − components, and x polarised radiation from the central component. The probabilities are

$$k_0(H = H_z) = k_z \quad (x \text{ polarised}), \qquad (28)$$

$$k_\pm(H = H_z) = |\langle(^1A_1, \sigma)^{(1)}|\, e\mathbf{r}\, |(^3B_2, \pm)^{(1)}\rangle|^2,$$

which for a strong field is approximately equal to

$$k_\pm(H = H_z) \approx \left|\frac{1}{\sqrt{2}} M_x \pm \frac{i}{\sqrt{2}} M_y\right|^2 = \tfrac{1}{2} k_x \quad (z \text{ polarised}). \qquad (29)$$

Clearly, at relatively high temperature where the population of all three levels is very nearly equal and equilibrium between them is established in a time $T_1 \ll \tau \leqslant k^{-1}$, the total probability with or without the field is the same,

$$k = \tfrac{1}{3}k_x + \tfrac{1}{3}k_z. \tag{30}$$

If the temperature is very low, but equilibrium is still rapidly maintained, one would have to consider the appropriate Boltzmann factors.

From equations (23), (24) and some algebra it follows that the Boltzmann factors are

$$f_0 = [1+2\exp(\tfrac{3}{2}Z/kT)\cosh(\alpha_z/kT)]^{-1},$$

$$f_{\pm} = \text{ezp}(\mp\alpha_z/kT)[\exp(-\tfrac{3}{2}Z/kT)+2\cosh(\alpha_z/kT)]^{-1} \quad (H=H_z). \tag{31}$$

If it were not for the contribution of non-radiative decay and the predominance of relaxation effects (cf. below), these Boltzmann factors on substitution in (25) lead to a considerable magnetic anisotropy of phosphorescence decay whenever $g\beta H/kT \gtrsim 1$.

In fig. 2*a* we have given a schematic representation of the situation $H = H_z$ just considered; fig. 2*b* shows what happens for $H = H_y$. In the diagrams we have also indicated in which molecules the 'allowed' routes (*a*) and (*b*) of fig. 1 for populating the triplet state of quinoxaline terminate. However, before going over to experiment, we may sound a word of warning. The diagrams here shown have been obtained from a grossly idealised model. Molecular vibrations, and in particular the radiationless deactivation of the phosphorescent state, may open up additional routes to those shown in fig. 2, and experiment will ultimately decide the issue. (As before, one might expect radiationless deactivation to proceed mainly via the τ_x component, this being of A_1 overall symmetry.)

Experiments at liquid helium temperature (in collaboration with Mr I. A. M. Hesselmann)

When inspecting fig. 2 three types of experiment suggest themselves to detect spin polarisation on 'optical pumping' at liquid helium temperature, and to distinguish between the various paths of excitation and de-excitation.

(*a*) Populating of the phosphorescent state by flash-excitation and monitoring the decay as a function of time by photomultiplier (or possibly fast electron resonance), with and without magnetic field.

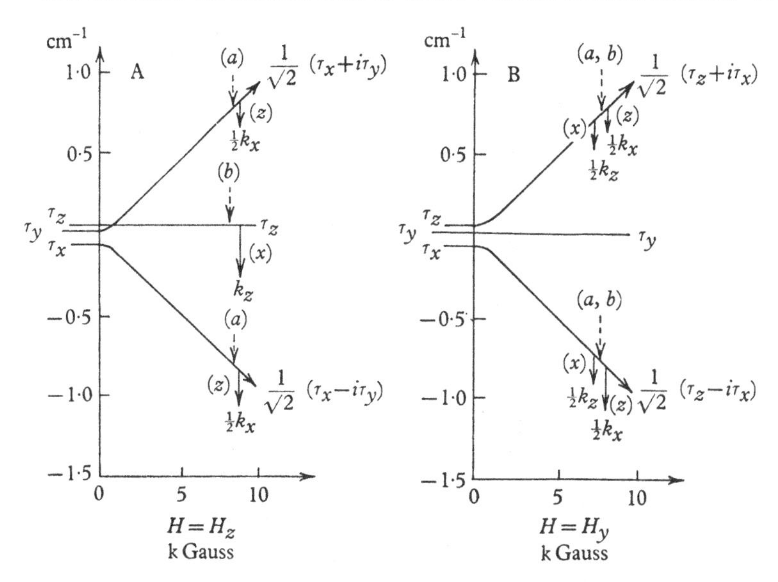

Fig. 2. Splitting of the components of the phosphorescent state of quinoxaline in an external magnetic field: A, field along molecular z axis, B, field along molecular y axis. The two components of symmetry-allowed radiative decay are given with their polarisation (in brackets); (a) and (b) are the paths of excitation taken from fig. 1.

(b) Electron resonance under continuous irradiation.

(c) Optical detection of electron resonance by measuring the effect of microwave saturation on phosphorescence output.

During the last few years we have tried all three types of experiment, using durene single crystals doped with quinoxaline or naphthalene. So far we have only been successful with method (a). However, the results obtained indicate that failure with the other methods is not of a fundamental nature, but due to the specific properties of the molecular systems chosen (largely the anomalous polarisation in durene,[5] (cf. below)). In fact, when the present paper was practically completed, it came to our notice that Schwoerer & Wolf have just reported[47] the detection of spin polarisation by method (b) in isotopically substituted naphthalene crystals. We finish with a review of the essence of the three types of experiment.

Phosphorescence decay at liquid helium temperature. The aspects of this method can best be illustrated by some of the results we obtained. (For some further details cf.[42].) Quinoxaline dissolved in a single crystal of durene was excited by an ultraviolet flash; as energy transfer from host to guest becomes very slow below the λ point excitation was done through a filter of a concentrated solution of durene in cyclohexane. At 4·2 °K phosphorescence decay is perfectly exponential with $\tau = 0{\cdot}237$ s, which is in agreement with the value reported at 77 °K ($0{\cdot}25 \pm 0{\cdot}02$ s).[38]

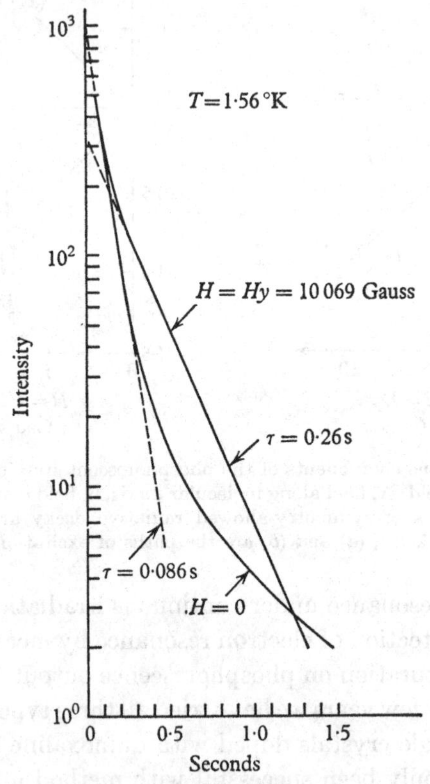

Fig. 3. Decay of the phosphorescence emission from quinoxaline in a durene host crystal following flash-excitation at 1·56 °K, in the absence of a field and in a strong field along the molecular *y* axis.

At lower temperatures, however, the situation becomes altogether different. In fig. 3 the decay is plotted, as log intensity versus time, without a field and with a field of 10 k Gauss along the molecular y axes (in durene the y axes of the two types of solute molecules in the unit cell practically coincide). In the absence of a field the beginning of the curve corresponds to near-exponential decay with a decay time of only 0·086 s, but when 90 % of the light has been emitted the curve begins to bend noticeably and the rate of decay drops off rapidly. With a field the initial intensity and rate of decay are roughly the same for about 0·1 s, after that the slope changes rapidly and then a long exponential part follows with $\tau = 0{\cdot}26$ s.

We suggest the following explanation. A scheme apparently holds where the molecules enter and leave the phosphorescent state through the same spin component τ_u; probably $\tau_u = \tau_x$ with entry via path (*a*) and decay via *z* polarised radiation and, possibly, some radiationless deactivation. For the sake of argument we shall further assume $\tau_u = \tau_x$, i.e. $k_y \approx k_z \approx 0$. (Although in agreement with the polarisation of phosphorescence in the present system, this needs verification.) For $H = 0$ the spin reorientation time T_1† considerably exceeds k_x^{-1}. Thus, initially $f_x \approx 1$, whilst the other two components are practically empty and decay occurs with a rate $k = k_x = (0{\cdot}086)^{-1}\,s^{-1}$ *which is three times the value at thermal equilibrium* (cf. equations 25 and 30). The non-exponential tail is caused by molecules that have gone over into one of the other spin components through relaxation before decaying straight from τ_x, and possibly to a small initial population of these other components.

When a field is present the energy separation between the components is increased some tenfold and spin reorientation becomes much faster, because the acoustic vibration spectrum has a higher density of modes at the higher frequencies involved.[43] Accordingly, the straight part of the curve for $H = H_y$ and $t \geqslant 0{\cdot}25$ s represents a situation where thermal equilibrium between the components is approximately maintained and decay occurs with the 'normal' rate of only $\frac{1}{3}k_x$.

The first part of the curve with a field represents a puzzling phenomenon, since for short times it exactly coincides with that for zero-field. Supposing populating is indeed by path (*a*), then in the field the molecules will initially be distributed over the + and − levels of fig. 2*b*. However, according to equation (29) the probability for radiative decay would then only be $\frac{1}{2}k_x$ and thus both *intensity* and *slope* should be smaller than in the field-free case. Although the phenomenon is somewhat reminiscent of the enhanced probability for emission associated with 'level crossing' in atomic spectroscopy,[41] it is unlikely that an analogous situation here arises because the splitting between the three spin components in the field would seem to be far greater than their natural line widths. Further experiments are planned in which the decay of the emission directly after the flash will be followed with improved instrumentation.

The present interpretation is supported by the fact that other

† Being interested in orders of magnitude, we hold the discussion in terms of a single relaxation time T_1. In actual fact relaxation may vary considerably between the three pairs of sublevels.

orientations of the field essentially give the same result. If entry and exit at zero-field were through different spin components, one would expect strongly orientational dependent phenomena as follows from fig. 2. Further decreasing the temperature to somewhat below 1·56 °K prolongs the region of coincidence of the two curves in fig. 3, apparently because spin re-orientation is slowed down still further.

Electron resonance under continuous irradiation. If the phosphorescence of quinoxaline in durene were x polarised—as originally expected—then entry into and exit from the phosphorescent state would go via different zero-field components (cf. fig. 1). When in such a situation the rate of spin re-orientation is reduced to below that of phosphorescence decay, the population of the 'exit' component must fall below its value at thermal equilibrium and stimulated emission of microwaves become a possibility. For instance, suppose that in fig. 2*a* $k_x = 0$ and $T_1 > k_z^{-1}$, then stimulated emission from the $|+\rangle$ to the $|0\rangle$ spin state may occur, if entry is via (*a*).

Unfortunately, our ESR experiments on quinoxaline in durene all turned out to be unsuccessful, and the decay experiments which have just been described revealed why. Probably because of the anomalous polarisation in durene as host, entry and exit are through the same zero-field component and thus population inversion cannot be attained by slowing down spin re-orientation. Nevertheless, the foregoing should illustrate the general principle of the method for detecting paths of excitation and de-excitation, and we are now applying it to other systems.

Very recently Schwoerer & Wolf reported the discovery of emission lines in the ESR spectrum of phosphorescent naphthalene $C_{10}H_8$ in a perdeutero naphthalene host crystal at 4·2 °K.[47] It is known[33] that in $C_{10}H_8$ more than 90 % of the population of the phosphorescent state decays through radiationless deactivation. From their experiments, in which emission was reported from the higher component containing τ_x for all three principal axis orientations of the field, it must be concluded that entry is through τ_x and exit (mostly non-radiative) at least through both τ_y and τ_z.† When examining the higher triplet states predicted by Pariser for naphthalene,[4] it is found that, because of the high symmetry of this molecule, none of those states lying between the phosphorescent ${}^3B_{2u}^+$ state at 2·7 eV and the fluorescent ${}^1B_{3u}^-$ state at 4·0 eV has a component of symmetry B_{3u} to which an 'allowed' intersystem crossing is possible. Thus there is no simple explanation why on optical pumping the spins are polarised in the component τ_x.

† Pariser's D_{2h} nomenclature with molecular plane $z=0$.

(The component $|^3B_{2u}, \tau_z\rangle$ of the phosphorescent state does have the proper symmetry, but this apparently lies too far below the fluorescent state to permit effective intersystem crossing.)[40]

Optical detection of electron resonance transitions. Several suggestions have been made to detect electron resonance absorption of phosphorescent triplets by an optical method similar to that first used by Brossel, Bitter & Kastler,[44] for the detection of transitions between the different Zeeman components of the fluorescent $6\,^3P_1$ state of mercury. According to the selection rules of the Zeeman effect for atoms, the polarisation of the fluorescence emitted by a mercury atom returning from the $6\,^3P_1$-excited state to its $6\,^1S$ ground state will vary with the value of m in the $^3P_1$ multiplet. If the external field is along the z axis, light emitted from the $m = 0$ component is z polarised, while that emitted from the $m = \pm 1$ levels is left or right circularly polarised in the xy plane. Transitions between the Zeeman components induced by an r.f. field may thus be detected as changes in the polarisation of the emitted light.

From the analysis given above (pp. 122–124) it follows that the above situation must not be expected to hold for phosphorescent aromatic molecules. Imagine for a moment that the phosphorescence of quinoxaline in zero-field were purely z polarised, then equation (23) and fig. 2 show that it must remain so in a strong magnetic field, because it can derive its intensity only from the τ_x part of the wavefunctions of the Zeeman components.

In the atomic case the value of m determines the orientation of the emitting dipoles relative to the magnetic field. In the present molecular case the emitting dipoles are fixed relative to the molecular frame. It is important to note that this conclusion stems from the low symmetry of a molecule as compared to an atom, and *not* from the *strength* of the dipolar interaction between the electron spins relative to the Zeeman energy. Only if two of the components in zero-field, say τ_y and τ_x, are bases for a degenerate representation of the symmetry group of the molecule, or if k_y and k_x happen to be equal accidentally, will there be circularly polarised emission for certain orientations of the molecule in a magnetic field, cf. equation (23).

In actual practice intermediate cases must arise. For instance, for naphthalene in zero-field a weak in-plane component probably occurs in addition to the strong out-of-plane component.[32] Hence, when placed in a field along the molecular y axis, the molecule will emit light from the top and bottom levels which is slightly elliptical.

As we have seen in the previous section, the probabilities for emission are in general different for the three Zeeman components of a phosphorescent molecule. If one wants to detect microwave transitions within the Zeeman multiplet by optical methods, one can best try and do this by measuring changes in the intensity of the phosphorescence produced by microwave saturation: the saturation will disturb the fractions f_i and thus alter the probability for radiative return to the ground state k as given by equation (25). As pointed out to us by Dr Wasserman, such an experiment will be similar in several respects to experiments carried out by Geschwind *et al.* on ruby.[45]

The authors are greatly indebted to Professor L. J. Oosterhoff for his advice and encouragement. In addition van der Waals would like to express his sincere thanks to Professor M. Kasha and all members of the Institute of Molecular Biophysics at Tallahassee, where a large part of this review was written.

References

1 M. Kasha, *Radiat Res. Suppl.* **2**, 243 (1960).
2 G. N. Lewis & M. Kasha, *J. Am. Chem. Soc.* **66**, 2100 (1944).
3 R. C. Hirt, F. T. King & J. C. Cavagnol, *J. Chem. Phys.* **25**, 574 (1956); A. Rousset, *J. Chim. Phys.* **61**, 1621 (1964).
4 R. Pariser, *J. Chem. Phys.* **24**, 250 (1956).
5 N. K. Chaudhuri & M. A. El-Sayed, *J. Chem. Phys.* **43**, 1423 (1965)*a*; **44**, 3728 (1966)*b*.
6 D. P. Craig & M. R. Philpott, *Proc. Roy. Soc.* (*London*) A, **290**, 583, 602 (1966); R. G. Body & I. G. Ross, *Aust. J. Chem.* **19**, 1 (1966).
7 E. P. Wigner, *Group Theory* (Academic Press, New York and London, 1959).
8 H. F. Hameka & L. J. Oosterhoff, *Mol. Physics* **1**, 358 (1958).
9 S. I. Weissman, *J. Chem. Phys.* **18**, 232 (1950).
10 W. Heisenberg, *Z. Physik*, **39**, 499 (1926).
11 E. Madelung, *Die Mathematischen Hilfsmittel des Physikers* (Springer, Berlin, Göttingen and Heidelberg, 1950), p. 107.
12 J. H. Van Vleck, *Rev. Mod. Phys.* **23**, 213 (1951).
13 W. Pauli, *Z. Physik*, **43**, 601 (1927); R. de L. Kronig, *Band Spectra and Molecular Structure* (Cambridge University Press, 1930).
14 J. H. van der Waals & M. S. de Groot, *Mol. Physics* **2**, 333 (1959).
15 R. W. Brandon, R. E. Gerkin & C. A. Hutchison Jr, *J. Chem. Phys.* **41**, 3717 (1964).
16 M. Gouterman & W. Moffitt, *J. Chem. Phys.* **30**, 1107 (1959); M. Gouterman, *J. Chem. Phys.* **30**, 1369 (1959).

17 H. F. Hameka, *J. Chem. Phys.* **31**, 315 (1959); R. M. Pitzer & H. F. Hameka, *J. Chem. Phys.* **37**, 2725 (1962).

18 J. H. van der Waals & G. ter Maten, *Mol. Physics* **8**, 301 (1964).

19 A. M. Trozzolo, R. W. Murray & E. Wasserman, *J. Am. Chem. Soc.* **84**, 4990 (1962).

20 M. Tinkham & M. W. P. Strandberg, *Phys. Rev.* **97**, 937 (1955).

21 J. Higuchi, *J. Chem. Phys.* **38**, 1237 (1963); S. A. Boorstein & M. Gouterman, *J. Chem. Phys.* **39**, 2443 (1963); Ying-Nan Chiu, *J. Chem. Phys.* **39**, 2736, 2749 (1963); M. Godfrey, C. W. Kern & M. Karplus, *J. Chem. Phys.* **44**, 4459 (1966), with further references to earlier work.

22 K. Kayama, *J. Chem. Phys.* **42**, 622 (1965)*a*; K. Kayama & J. C. Baird, *J. Chem. Phys.* **43**, 1082 (1965)*b*.

23 H. A. Bethe & E. E. Salpeter, *Handbuch der Physik*, S. Flügge, ed. (Springer, Berlin, Göttingen and Heidelberg, 1957), vol. xxxv, p. 267.

24 H. M. McConnell, *Proc. Natl. Acad. Sci. U.S.* **45**, 172 (1959).

25 D. S. McClure, *J. Chem. Phys.* **17**, 665 (1949); **20**, 682 (1952).

26 J. W. Sidman, *J. Mol. Spectr.* **2**, 333 (1958)*a*; E. Clementi & M. Kasha, *J. Mol. Spectr.* **2**, 297 (1958)*b*; L. Goodman & V. G. Krishna, *Rev. Mod. Phys.* **35**, 541 (1963)*c*.

27 M. Mizushima & S. Koide, *J. Chem. Phys.* **20**, 765 (1952).

28 E. Clementi, *J. Mol. Spectr.* **6**, 497 (1961).

29 H. M. McConnell, *J. Chem. Phys.* **34**, 13 (1961).

30 R. S. Mulliken, *J. Chem. Phys.* **7**, 14 (1939).

31 P. G. Russell & A. C. Albrecht, *J. Chem. Phys.* **41**, 2536 (1964).

32 T. Azumi & S. P. McGlynn, *J. Chem. Phys.* **37**, 2413 (1962); V. G. Krishna & L. Goodman, *J. Chem. Phys.* **37**, 912 (1962); M. A. El-Sayed, *Nature*, **197**, 481 (1963); F. Dörr, H. Gropper & N. Mika, *Ber. Bunsenges. Physik. Chem.* **67**, 46, 193 (1963).

33 R. E. Kellogg & R. G. Bennett, *J. Chem. Phys.* **41**, 3042 (1964).

34 M. Kasha, *Light and Life*, W. D. McElroy and B. Glass, eds. (The Johns Hopkins Press, 1961), p. 31.

35 K. K. Innes, J. D. Simmons & S. G. Tilford, *J. Mol. Spectr.* **11**, 257 (1963); G. W. Robinson, private communication.

36 E. Clementi, *J. Chem. Phys.* (to be published).

37 S. F. Mason, *J. Chem. Soc.* p. 1240 (1959).

38 J. S. Vincent & A. H. Maki, *J. Chem. Phys.* **39**, 3088 (1963).

39 M. A. El-Sayed & R. G. Brewer, *J. Chem. Phys.* **39**, 1623 (1963).

40 G. W. Robinson & R. P. Frosch, *J. Chem. Phys.* **37**, 1962 (1962); **38**, 1187 (1963); G. R. Hunt, E. F. McCoy & I. G. Ross, *Aust. J. Chem.* **15**, 591 (1962); J. P. Byrne, E. F. McCoy & I. G. Ross, *Aust. J. Chem.* **18**, 1589 (1965).

41 G. Breit, *Rev. Mod. Phys.* **5**, 91 (1933); P. A. Franken, *Phys. Rev.* **121**, 508 (1961); T. R. Carver & R. B. Partridge, *Am. J. Phys.* **34**, 339, (1966), with further references.

42 M. S. de Groot, I. A. M. Hesselmann & J. H. van der Waals, *Mol. Physics* **12**, 259 (1967)

43 G. E. Pake, *Paramagnetic Resonance* (Benjamin, New York, 1962); chapter 6, in particular fig. 6–4.

44 F. Bitter, *Phys. Rev.* **76**, 833 (1949); J. Brossel & A. Kastler, C.R. Acad. Sci. (Paris) **229**, 1213 (1949); J. Brossel & F. Bitter, *Phys. Rev.* **86**, 308 (1952).

45 S. Geschwind, G. E. Devlin, R. L. Cohen & S. R. Chinn, *Phys. Rev.* **137**, A 1087, (1965), with references to earlier work.

46 L. Goodman & R. W. Harrell, *J. Chem. Phys.* **30**, 1131 (1959).

47 M. Schwoerer & H. C. Wolf, Communicated at *XIVth Colloque Ampère*, Ljubljana, September 1966.

ESR INVESTIGATIONS OF NAPHTHA-LENE-d_8:NAPHTHALENE-h_8 MIXED CRYSTALS

M. SCHWOERER AND H. C. WOLF

Abstract

The ESR spectrum of pairs of naphthalene-h_8 molecules (naphthalene-h_8 unit cell) in a matrix of naphthalene-d_8 is analysed. From the width of the exchange narrowed line one can calculate the exchange energy between the two molecules, one of them in the lowest triplet state, as

$$J/hc = (5 \pm 0{\cdot}7)\ \mathrm{cm}^{-1}.$$

Evidence for ESR detection of optical spin polarisation is briefly discussed.

Introduction

Naphthalene-d_8: naphthalene-h_8 mixed crystals in their triplet state were investigated by ESR within the temperature range from 1·6° to 10° K. The concentration c of the undeuterated naphthalene (N-h_8) was varied from 0·05 to 20 %.

The low temperature was reached by immersing the crystal into liquid helium. The crystals were excited by the ultraviolet light of an Osram HBO 200 mercury lamp. The microwave power (9·39 Gc/s) could be varied from $2 \cdot 10^{-1}$ to $5 \cdot 10^{-9}$ W. The magnetic field was modulated by 100 kc/s.

Experimental results

Fine structure

The metastable ESR spectrum consists of six absorption lines (figs. 1, 2). All six lines vanish above 10 °K. Four of them ($A^{\pm}$, $B^{\pm}$ in figs. 1 and 2) are triplet lines of isolated N-h_8 molecules as described by the well-known spin Hamiltonian [1]

$$\hat{H}_s = \mu_B \vec{H}_0 \mathbf{g} \hat{\vec{S}} + D\hat{S}_z^2 + E(\hat{S}_x^2 - \hat{S}_y^2) \qquad (1)$$

with $S = 1$.

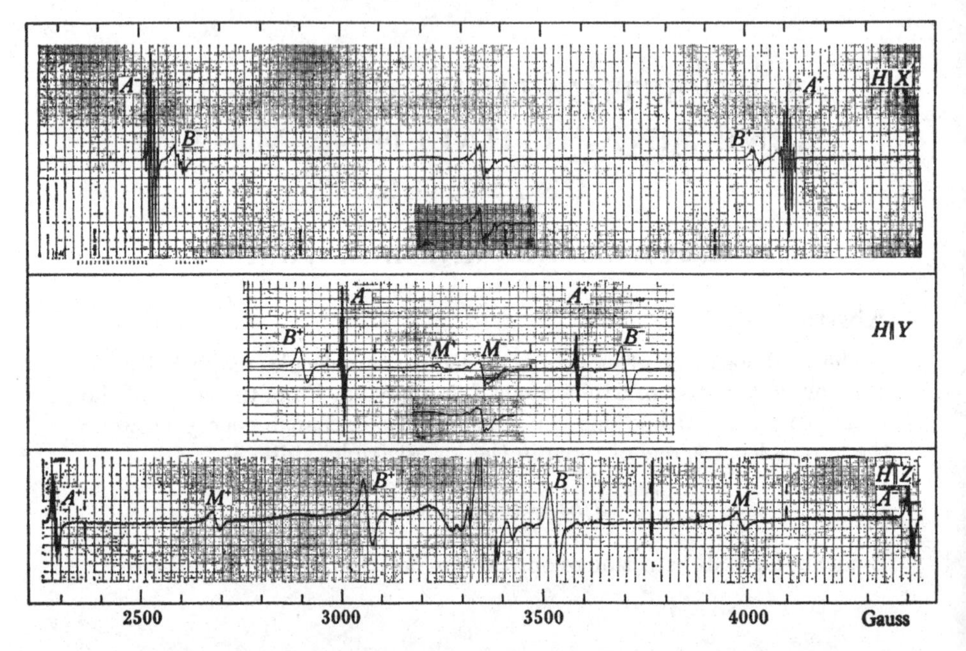

Fig. 1. ESR spectra of an N-d_8:2 % N-h_8 mixed crystal at LHeT during ultraviolet excitation. H_0 is along the x, y or z axis of one sort of the molecules (called A molecules). At the indicated field position (↓) the modulation amplitudes was switched to smaller and higher values respectively in order to avoid a smearing out of the hyperfine splitting. The line group at 3·350 Gauss is due to free radicals and is not metastable. The spectra below the principal spectra are measured without ultraviolet excitation.

The constants are as follows:

$$\left.\begin{aligned} g_{xx} &= 2{\cdot}0024 \pm 0{\cdot}0010, \\ g_{yy} &= 2{\cdot}0027 \pm 0{\cdot}0010, \\ g_{zz} &= 2{\cdot}0033 \pm 0{\cdot}0010, \end{aligned}\right\} \quad \begin{aligned} D/hc &= 0{\cdot}0994 \pm 0{\cdot}0003\,\text{cm}^{-1}, \\ E/hc &= -0{\cdot}0154 \pm 0{\cdot}0002\,\text{cm}^{-1}. \end{aligned}$$

The additional two lines ($M^\pm$ in figs. 1 and 2) are situated halfway between two regular lines, M^+ between A^+ and B^+, M^- between A^- and B^-. This field position is independent of the individual angle α between H_0 and the crystal axes (fig. 2).

The M lines are also described by the triplet state spin Hamiltonian:

$$\hat{H}_s^* = g^*\mu_B \cdot \vec{H}_0 \cdot \vec{\hat{S}} + D^*\hat{S}_{z^*}^2 + E^*(\hat{S}_{x^*}^2 - S_{y^*}^2), \tag{2}$$

with $S = 1$.

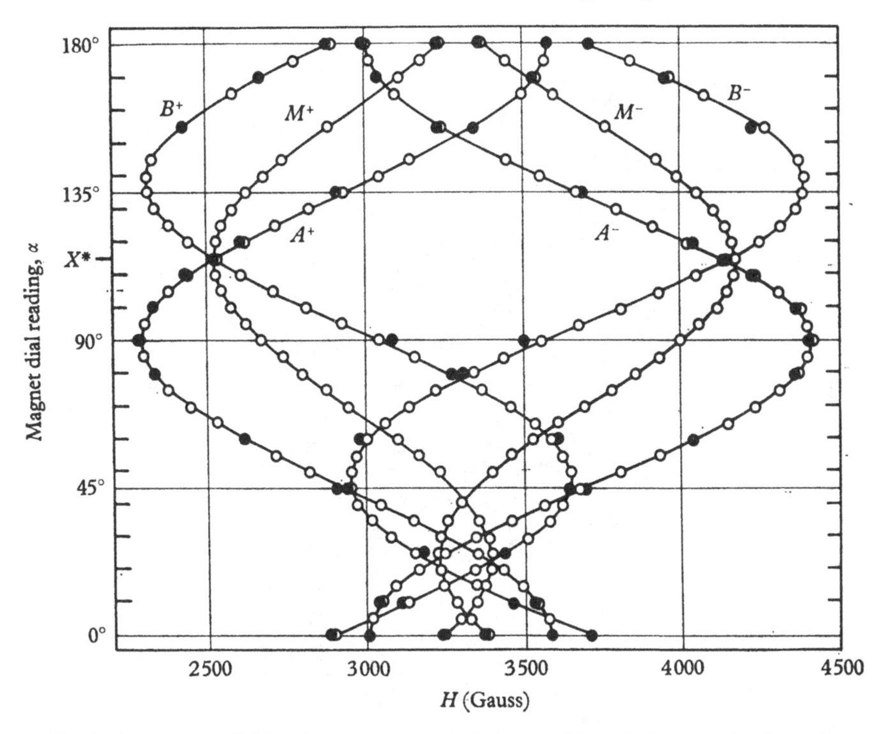

Fig. 2. Resonance field-values *vs.* angle α between H_0 and the y axis. Rotation in the y-z plane of the A molecules. Open circles: experimental; closed circles: theoretical points.

The constants are as follows:

$$g^* = 2{\cdot}0030 \pm 0{\cdot}001,$$

$$D^*/hc = -0{\cdot}0059 \pm 0{\cdot}0006\ \mathrm{cm}^{-1},$$

$$E^*/hc = +0{\cdot}0485 \pm 0{\cdot}0006\ \mathrm{cm}^{-1}.$$

The axes x^*, y^* and z^* are not the molecular axes but are related very closely to the crystal axes: z^* is identical with the b axis and x^* makes an angle of $+22{\cdot}4°$ with the a axis. The six-line pattern is observed at all concentrations between $c = 0{\cdot}05\%$ and $c = 10\%$ of N-h_8 in N-d_8.

Hyperfine splitting

The A and B lines are hyperfine split if H_0 is parallel to one of the three principal axes: if H_0 is parallel to the x axis the well-known five-line pattern [(1)] is observed. The intensity ratios of the components are about 1:4:6:4:1, their average distance $\delta = 8$ Gauss and their

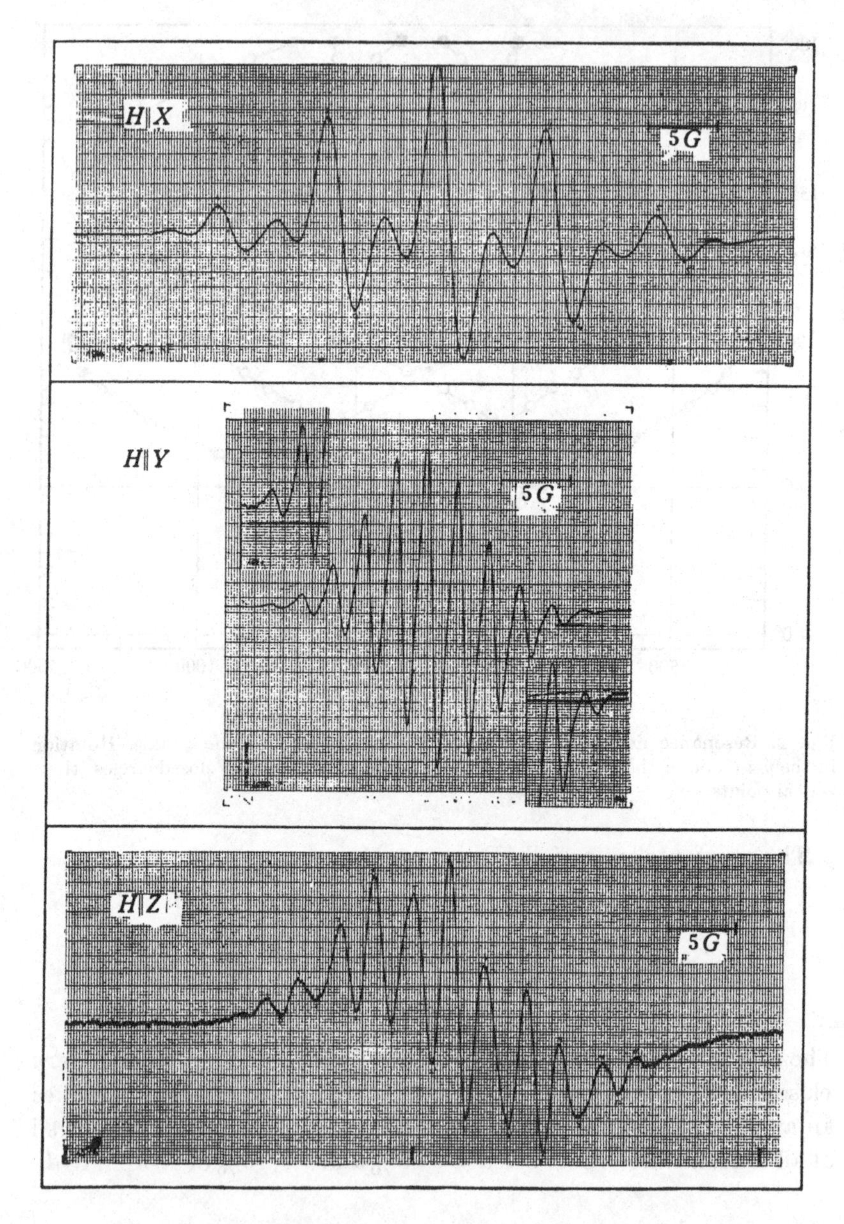

Fig. 3. Hyperfine structure of the low field A lines. H_0 is parallel to the x, y, and z axis respectively of the molecules.

maximum slope width $\Delta\omega = 2{\cdot}1$ Gauss. In addition the low field line shows six satellites between and outside the five principal hyperfine components as described by several authors.[2,3] If H_0 is parallel to the y axis the lines are split into eleven components ($\delta' = 2{\cdot}2$ Gauss, $\Delta\omega' = 0{\cdot}85$ Gauss). If H_0 is parallel to the z axis they are split into eleven components. All hyperfine splittings are, at least up to $c = 10\,\%$, independent of the N-h_8 concentration. The low field A lines for the orientation of H_0 parallel to the x, y and z axes respectively are shown in fig. 3.

Width of the M lines

The M lines are never hyperfine split and show saturation broadening. The shape is Lorentzian if the distance between the M lines and the corresponding A and B lines is large. The shape is nearly Gaussian for small distances.

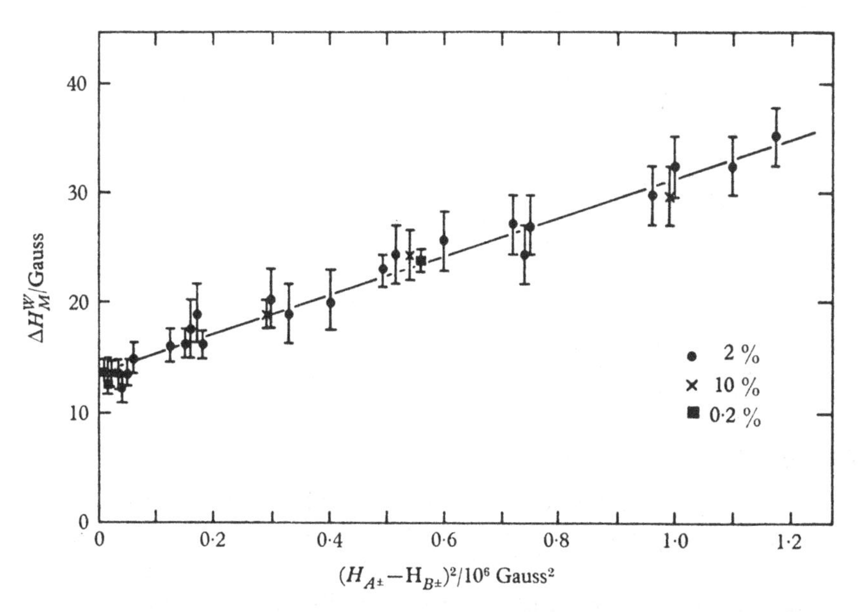

Fig. 4. M line width ΔH_M^w (distance between the points of maximum slope) *vs.* the square of the distance of the corresponding A and B lines.

The width of the M lines is dependent on the distance between the corresponding A and B lines (fig. 4). The width increases with the increasing A and B lines (fig. 4). The width increases with the increasing square of the $A - B$ distance.

Concentration dependence of intensities

The intensity ratio $I_M/I_{A,B} = P$ as a function of the N-h_8 concentration is shown in fig. 5. Up to 5 %, P is proportional to c. The limits of error are relatively large, because the A and B lines saturate much easier than the M lines, and because the saturation behaviour is concentration dependent.

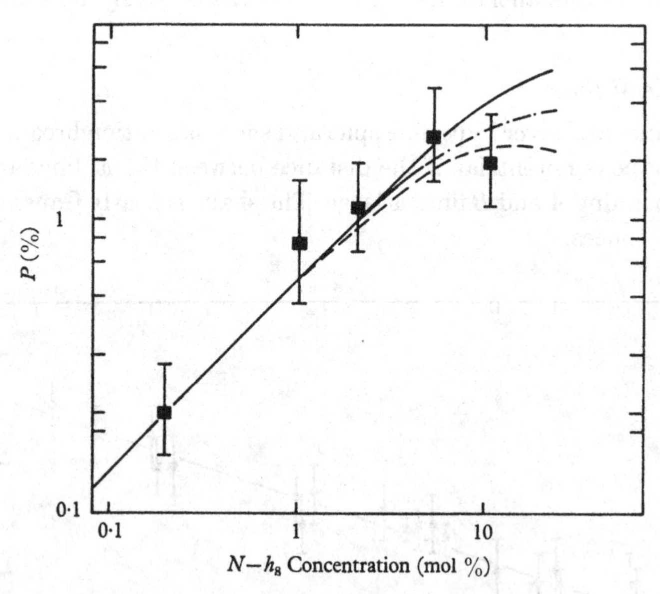

Fig. 5. $P = \frac{1}{2}(I_M/I_{A,B})(-I_M)$ is the intensity of the two M lines, $I_{A,B}$ the intensity of the four A and B lines. ■ experimental points. The drawn-out and dotted lines are calculated from statistics under slightly different assumptions.

Emissive lines

If the N-h_8 concentration is below 0·3 % always *one* of the two A or B lines (the $|0\rangle \rightarrow |+1\rangle$ or the $|-1\rangle \rightarrow |0\rangle$ transition) is *emissive*, the other *absorptive*. The transition depends on the direction of the magnetic field. If H_0 is parallel to the x or the z axis this holds *only during excitation*. After the exciting light has been switched off the emissive lines become absorptive in a time $T < \tau$, where τ is the triplet lifetime. In a dynamical experiment the emissive character can also be seen at concentrations $c > 1$ %.[4]

Decay time

The decay behaviour is the same for all the six lines between 1·6 °K and 4·2 °K. If $c < 1\%$ the decay is exponential and $\tau_{1/e} = (2{\cdot}6 \pm 0{\cdot}1)$ s. If $c > 1\%$ the decay is no longer exponential and $\tau_{1/e}$ decreases to 0·1 s at $c = 20\%$ and 4·2 °K. τ is independent of temperature below $c = 1\%$ and 4·2 °K but increases with decreasing temperature at $c > 1\%$.

Discussion of the M lines

The M lines are due to pairs of N-h_8 molecules as nearest neighbours in inequivalent positions in the lattice (naphthalene unit cell). One of the two molecules is excited into the triplet state. This excitation is rapidly exchanged between the two molecules. Therefore the M lines are exchange narrowed A and B lines.

This can be concluded by the agreement between the observed spin Hamiltonian (2) and the spin Hamiltonian calculated for this model.[5,6] The fine structure constants D^*, E^*, g^* and the axes x^*, y^*, z^* could be calculated from the fine structure of the isolated molecules and their orientations in the crystal.

The model is further proved by the lack of hyperfine splitting, by the observed line width and line shape and by the concentration dependence of the intensity.[5]

The theoretical width for an exchange narrowed line of pairs of molecules is given by the formula,[5,7]

$$\Delta\omega^{\frac{1}{2}}_M = (\Delta\omega^{\frac{1}{2}}_M)_0 + \frac{10/3[\frac{1}{2}(\omega_A - \omega_B)]^2}{J/\hbar}. \tag{3}$$

$\Delta\omega^{\frac{1}{2}}_M$ is the half width at half power of the M line,

$$(\Delta\omega^{\frac{1}{2}}_M)_0 = \frac{\Delta\omega^{\frac{1}{2}}_{A,B}}{\sqrt{2}}, \tag{4}$$

is the width of the inhomogeneously broadened A and B line in absence of exchange, $\omega_A - \omega_B$ is the distance between the corresponding A and B line ($A^+ - B^+$ or $A^- - B^-$). J is the exchange integral.

Application of this formula to our observations (fig. 4) gives the following experimental value for the exchange energy between two naphthalene molecules in the unit cell (distance $\frac{1}{2}(\vec{a} + \vec{b})$), one of them excited into the lowest triplet state:

$$J/hc = (5 \pm 0{\cdot}7)\,\text{cm}^{-1}.$$

This corresponds to a transfer time for the radiationless energy transfer of

$$\tau = \tfrac{1}{4}hJ^{-1} = 1{\cdot}7 \times 10^{-12}\,\mathrm{s}.$$

The experimental value J is in good agreement with the value calculated by Jortner *et al.*,[8] who obtained $J = 4{\cdot}5\,\mathrm{cm}^{-1}$.

Further discussion

The emissive character of the regular lines seems to be due to spin polarisation caused by strong selection rules within the intersystem crossing and the subsequent radiationless triplet-population together with a long T_1. Similar experiments are described by Tanimoto *et al.*[9]

The decrease of $\tau_{1/e}$ with increasing c and the temperature dependence of this effect indicates a thermally activated quenching of the triplet state in the crystal.

A more detailed paper is in preparation. The support of the Deutsche Forschungsgemeinschaft is gratefully acknowledged.

References

1 C. A. Hutchison Jr. & B. W. Mangum, *J. Chem. Phys.* **34**, 908 (1961).
2 J. S. Vincent & A. H. Maki, *J. Chem. Phys.* **39**, 3088 (1963).
3 A. W. Hornig & J. S. Hyde, *Mol. Physics* **6**, 33 (1963).
4 M. Schwoerer & H. C. Wolf, *XIVth Colloque Ampere, Ljubljana* 1966.
5 M. Schwoerer, Dissertation, 1967.
6 H. Sternlicht & H. M. McConnell, *J. Chem. Phys.* **35**, 1793 (1961).
7 P. W. Anderson, *J. Phys. Soc. Jap.* **9**, 316 (1954).
8 J. Jortner, S. A. Rice, J. L. Katz & S. I. Choi, *J. Chem. Phys.* **42**, 309 (1965).
9 D. H. Tanimoto, W. M. Ziniker & J. O. Kemp, *Phys. Rev. Letters* **14**, 645 (1965).

BIRADICALS AND POLYRADICALS IN THE NITROXIDE SERIES

H. LEMAIRE AND A. RASSAT

Abstract

Starting from nitroxide monoradicals, molecular systems containing two or more unpaired electrons have been prepared and investigated by electron spin resonance spectroscopy and static magnetic susceptibility measurements. Experimental results give information on the nature of the ground state of these molecular systems.

Within experimental error, the monoradical 2,2′,6,6′-tetramethyl-piperidine-4-one-1-oxy follows the Curie law to 1·9 °K. As does the biradical (2,2′,6,6′-tetramethylpiperidine-4-ol-1-oxy) carbonate for which the ESR results show that J is about 10^{-3} cm^{-1}.

The susceptibility of the monoradical 2,2,6′,6′-tetramethylpiperidine-4-ol-1-oxy increases as the temperature decreases, reaches a maximum at about 5 °K and decreases until 1·9 °K. Its Curie temperature is −6 °K. The extrapolated susceptibility for 0 °K is different from zero.

The biradical (2,2′,5,5′-tetramethylpyrrolidine-3-one-1-oxy)-3-azine shows the same behaviour except that its Curie temperature is −3 °K and that its extrapolated susceptibility may be zero at 0 °K.

The susceptibility of the biradical (2,2′,6,6′-tetramethylpiperidine-4-ol-1-oxy) suberate follows a Weiss-Curie law with a paramagnetic Curie temperature of +1 °K.

CHANGES INDUCED IN THE PHOSPHORESCENT RADIATION OF AROMATIC MOLECULES BY PARAMAGNETIC RESONANCE IN THEIR METASTABLE TRIPLET STATES†

C. K. JEN, L. C. AAMODT AND A. H. PIKSIS

Abstract

By using simple group theory, an analysis is made of the singlet state character in the individual Zeeman levels of the metastable triplet state of an aromatic molecule having D_{2h} symmetry embedded in a host crystal. An intra-molecular spin-orbit coupling mechanism is assumed to produce the singlet-triplet admixing. It is shown that the individual Zeeman levels have different transition probabilities for radiation to the ground state depending upon how the molecule is oriented in a steady magnetic field. Because of this, the non-Boltzmann population distribution produced among the triplet state Zeeman levels by paramagnetic resonance causes the total intensity of the observed phosphorescence to change. By noting this intensity change, paramagnetic resonance can be detected optically, and the polarised phosphorescence of the individual Zeeman levels can be studied separately. This information may prove useful in determining which singlet states are admixed into the triplet state, and may possibly offer some evidence concerning the nature of the singlet-triplet admixing mechanism. An experiment using a naphthalene molecule in a durene host crystal is being conducted at the present time in order to verify the theoretical predictions.

Introduction

In a pioneering work on the optical detection of paramagnetic resonance of the excited state of Mn^{4+} ion in Al_2O_3, Imbush & Geschwind[1] have demonstrated the feasibility of observing excited state electron spin resonance (ESR) by noting the change of emitted polarised light intensity in passing through magnetic resonance. It is of interest to inquire whether a similar scheme can be applied in detecting optically the effect of ESR on the triplet state of aromatic

† This work supported by the Bureau of Naval Weapons, U.S. Department of the Navy, under Contract NOw 62–0604–c.

molecules.[2] In comparing these two cases, a few differences become apparent. In the case of the atomic ion cited above, the radiative lifetime of the excited state (0·8 ms) is very much shorter than the spin-lattice relaxation time. On the other hand, the lifetime of the lowest triplet state of an aromatic molecule (approximately 2 s for naphthalene in durene) is several orders of magnitude longer than the spin-lattice relaxation time. Further differences also exist in the configuration and electronic symmetries of the active guest species embedded in the lattice of the host atoms or molecules. In addition, the host crystal (Al_2O_3) is uniaxial, whereas many molecular organic host crystals, such as durene, are biaxial. Some of these differences favor the optical detection of ESR in the atomic ion over this type of detection in organic molecules. In spite of the differences just mentioned, it seemed desirable to explore the possibilities of using optical detection of ESR in aromatic molecules as a probe for studying the nature of the interactions in the triplet-singlet transition. Our preliminary experiments have shown that the effect of ESR on the phosphorescent emission of naphthalene in durene is small and undetectable under our present experimental conditions. Nevertheless we believe that this problem is challenging and sufficiently worth while to warrant a progress report on this difficult investigation.

Theory of polarised phosphorescence[3,4,5] in a magnetic field[6]

A simplified energy level diagram[5,7] for the naphthalene molecule is given in fig. 1. Photo-excitation raises the energy of these molecules from the ${}^1A_{1g}$ ground state to the ${}^1B_{3u}$ state, or to some other higher excited singlet states. Then, by a series of radiationless transitions, some of these excited molecules cascade down to the metastable ${}^3B_{2u}$ triplet state. If this state were a pure triplet state, then a radiative transition back to the ground ${}^1A_{1g}$ state would be forbidden by the spin multiplicity selection rule. A weak phosphorescent transition is observed, however, indicating that either the triplet state contains a small admixture of singlet character or the ground singlet state contains some triplet character. In the following analysis, we will consider only the admixture of singlet state character into the excited triplet state, but it can be readily shown that similar results are obtained for a mixture of triplet state character into the ground state.

A number of workers have attributed the singlet-triplet state mixing to a spin-orbit interaction. If we assume such an interaction, then, by perturbation theory,

$$\psi(^3B_{2u}) = \psi^0(^3B_{2u}) + \sum_k \frac{\langle \psi(k)|\mathscr{H}_{\text{s.o.}}|\psi(^3B_{2u})\rangle}{E(^3B_{2u}) - E(k)} \psi(k), \tag{1}$$

where $\psi^0(^3B_{2u})$ is the pure triplet state wavefunction and $\psi(k)$ is the wavefunction of an excited singlet state having the energy $E(k)$.

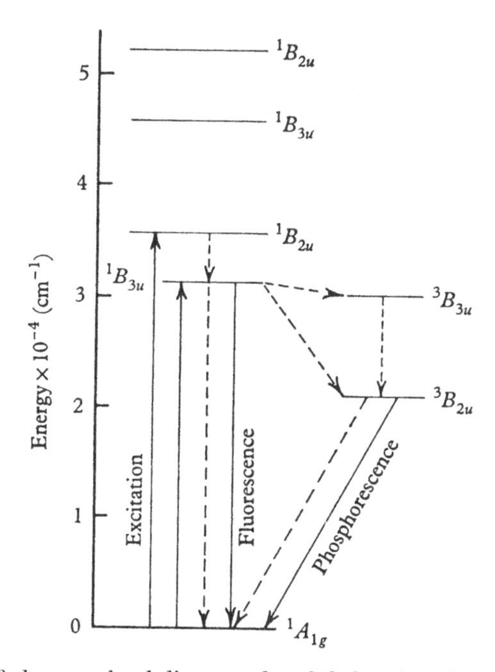

Fig. 1. A simplified energy level diagram of naphthalene in a durene single crystal. Level positions are shown approximately to scale. Dashed lines signify radiationless transitions.

In the presence of a sufficiently large steady magnetic field $\mathbf{H}_0$, the triplet $^3B_{2u}$ state Zeeman levels may be denoted by their magnetic quantum numbers $M(M = 1, 0, -1)$. For simplicity, suppose that the triplet-singlet admixture arises almost entirely from† one singlet state ^1X. Then we have, for each Zeeman level,

$$\psi_M(^3B_{2u}) = \psi_M^0(^3B_{2u}) + \alpha_M \psi(^1X), \tag{2}$$

where

$$\alpha_M = \frac{\langle \psi(^1X)|\, \mathscr{H}_{\text{s.o.}}|\, \psi_M^0(^3B_{2u})\rangle}{E(^3B_{2u}) - E(^1X)}. \tag{3}$$

† If several singlet states contribute, it is merely necessary to add the contributions from the individual states.

The radiative transition probability between the individual Zeeman levels and the ground state is determined by the electric dipole matrix element, which reduces to

$$\langle\psi(^1A_{1g})|\, e\mathbf{r}\,|\psi_M(^3B_{2u})\rangle = \alpha_M\langle\psi(^1A_{1g})\,|\, e\mathbf{r}\,|\psi(^1X)\rangle, \tag{4}$$

where $\mathbf{r}$ is the radius vector of the electron.

The selection rule governing equation (4) may be determined by a simple group theoretical treatment. The naphthalene molecule belongs to the $D_2 \times i = D_{2h}$ symmetry, where i is the inversion operator. The presence of the inversion operator establishes the selection rules $g \leftrightarrow g$ and $u \leftrightarrow u$ for the spin-orbit interaction and $g \leftrightarrow u$ for the electric dipole interaction. Thus, both α_M and $\langle\psi(^1A_{1g})|\, e\mathbf{r}\,|\psi(^1X)\rangle$ are non-zero only for those singlet states ^1X_u that have u symmetry.

The D_2 point group has four irreducible representations A_1, B_1, B_2, and B_3 which transform as r, z, y, and x respectively. (We choose our axes so that x is parallel to the long molecular axis, y is parallel to the short molecular axis, and z is perpendicular to the molecular plane.) If the ^1X state is assumed to transform according to the X irreducible representation in D_2, then since A_1 is the totally symmetric representation, the three matrix elements

$$\langle\psi(^1A_1)|\,x\,|\psi(^1X)\rangle,\ \langle\psi(^1A_1)|\,y\,|\psi(^1X)\rangle,\ \text{and}\ \langle\psi(^1A_1)|\,z\,|\psi(^1X)\rangle$$

transform as the direct products $B_3 \times X$, $B_2 \times X$, and $B_1 \times X$, respectively. For D_2 symmetry, only the direct products $B_i \times B_i$ $(i = 1, 2, 3)$ yield the totally symmetric representation and, therefore, the only possible non-vanishing electric dipole matrix elements on the right-hand side of equation (4) are

$$\langle\psi(^1A_{1g})|\,ex\,|\psi(^1B_{3u})\rangle,\quad \langle\psi(^1A_{1g})|\,ey\,|\psi(^1B_{2u})\rangle$$
$$\text{and}\quad \langle\psi(^1A_{1g})|\,ez\,|\psi(^1B_{1u})\rangle.$$

We determine the selection rule for α_M by considering the transformation properties of the various terms in a spin-orbit operator of the form

$$\mathscr{H}_{\text{s.o.}} = \frac{e\hbar}{2m^2c^2}\sum_j \mathbf{s}_j \cdot (\nabla U_j \times \mathbf{p}_j), \tag{5}$$

in which $\mathbf{s}_j$ is the spin of the j-th electron, $\mathbf{p}_j$ its momentum, and U_j is the electrostatic potential at this electron. Since the quantum mechanical equivalent of the momentum operator is $-i\hbar\nabla$, $\mathbf{p}_j$ transforms as ∇, which transforms as the radius vector $\mathbf{r}$. The quantity

∇U_j also transforms as ∇, and hence as $\mathbf{r}$, since U_j is totally symmetric in the D_2 symmetry. Thus the vector products

$$(\nabla U_j \times \mathbf{p}_j)_x, \quad (\nabla U_j \times \mathbf{p}_j)_y, \quad \text{and} \quad (\nabla U_j \times \mathbf{p}_j)_z$$

transform like the components of an axial vector. In D_2 symmetry, however, axial and polar vectors transform alike and, therefore,these three vector products transform according to the B_3, B_2, and B_1 representations, respectively.

If we now consider only the orbital part of the wavefunction, and restrict ourselves to those ${}^1X_u$ states that are allowed by the electric dipole selection rules (i.e., to the states ${}^1X_u = {}^1B_{1u}$, ${}^1B_{2u}$, and ${}^1B_{3u}$), we see that the matrix element $\langle \psi(X_u)|(\nabla U_j \times \mathbf{p}_j)|\psi_M(B_{2u})\rangle$ has only two non-vanishing components, namely,

$$\langle \psi(B_{1u})|(\nabla U_j \times \mathbf{p}_j)_x|\psi_M(B_{2u})\rangle \quad \text{and} \quad \langle \psi(B_{3u})|(\nabla U_j \times \mathbf{p}_j)_z|\psi_M(B_{2u})\rangle.$$

Thus, only ${}^1B_{1u}$ and ${}^1B_{3u}$ admixtures into the triplet state contribute to the phosphorescent emission. From the electric dipole selection rule we see that a ${}^1B_{1u}$ admixture produces z-polarised phosphorescence while a ${}^1B_{3u}$ admixture produces x-polarised phosphorescence. Regardless of the magnetic field orientation, no y-polarised radiation is allowed as long as the D_2 symmetry is strictly maintained.

Taking into account the spin part of the spin-orbit interaction, we see that the only triplet spin states that can phosphoresce are those coupled to ${}^1B_u$ admixing states by the x component of the spin operator and those coupled to ${}^1B_{3u}$ admixing states by the z component of the spin operator (the direction of each spin component is specified in the molecular coordinate system).

Which particular triplet Zeeman levels radiate depends upon the direction of the magnetic field relative to the molecular axes. To illustrate this dependence, consider the case where $H_0 \| y$, which was the experimental arrangement that we used in our attempt to detect ESR optically in the triplet state of naphthalene. Since the admixed states are singlets, they have only one magnetic state $M = 0$. Therefore, if we choose our triplet state spin functions such that the representation of the spin component along the magnetic field is diagonal, then s_y is the only component of the spin operator that can couple the $M = 0$ triplet level to the singlet spin state, while the $M = 1$ and $M = -1$ triplet levels are coupled to the admixed singlet state by both the x and z components of the spin operator. Hence, for $H_0 \| y$, the $M = 0$ triplet Zeeman level cannot radiate, while the $M = 1$ and

$M = -1$ levels radiate with x and z polarisations as long as admixtures of the proper symmetry are present.

By explicitly evaluating the spin matrix, the phase relationship between the x and z polarised light can be established. We have summarised these results in table 1 where we have also given the results corresponding to the other two principal magnetic field orientations.

Table 1. *Polarisation of the phosphorescent radiation of naphthalene for the three principal orientations of* $\mathbf{H}_0$.

M	$H_0 \parallel x$	$H_0 \parallel y$	$H_0 \parallel z$
1	x	$x+iz$	z
0	z	0	x
-1	x	$x-iz$	z

In table 1, only the direction of polarisation and the relative phases are indicated. The intensities of the differently polarised components depend upon the relative amounts of the $^1B_{1u}$ and the $^1B_{3u}$ character in the metastable triplet wavefunctions. If the admixtures are such that both x and z polarisations have equal intensity, then for $H_0 \parallel y$, right-hand circularly polarised light is emitted from the $M = 1$ Zeeman level, and left-hand circularly polarised light from the $M = -1$ level, both propagating in the y direction. If the two intensities are not equal (the general case), the emitted light is elliptically polarised. Whenever the singlet admixture has exclusively $^1B_{1u}$ or $^1B_{3u}$ symmetry, the phosphorescent emission becomes linearly polarised.

The results of table 1 are exhibited in figs. 2 and 3 for two principal magnetic field orientations.

Next, let us inquire how the earlier results will be affected if the assumption that the electron spin is perfectly quantised along the external magnetic field breaks down. Following Hutchison & Mangum,[8] we write the spin-Hamiltonian as

$$\mathscr{H} = g\beta H_0 S_\parallel + DS_z^2 + E(S_x^2 - S_y^2), \tag{6}$$

where $S_\parallel$ is the spin component along the magnetic field and S_x, S_y and S_z are the components of the spin along the principal molecular axes. If we orient the magnetic field along any one of the three principal molecular axes, then $S_\parallel$ equals one of the quantities S_x, S_y or S_z. In a spin representation where $S_\parallel$ is diagonal, the Zeeman term in the spin-Hamiltonian and one of the three quadratic spin terms will be diagonal, while the other two quadratic spin terms will have only off-diagonal matrix elements between the states $M = 1$ and $M = -1$.

Thus, when the magnetic field is parallel to any one of the three principal molecular axes, the $M = 0$ Zeeman level remains pure. Because of this, for $H_0 \| y$, our conclusion that there is no phosphorescent radiation from the $M = 0$ level remains valid, but some changes in polarisation of the radiation emitted from the individual $M = \pm 1$ levels result as H_0 is varied. This is the conclusion which serves as the basis for our attempt to detect ESR optically in the metastable triplet state of naphthalene. It should be noted, however, that this prediction is not strictly true unless $\mathbf{H}_0$ is accurately aligned parallel

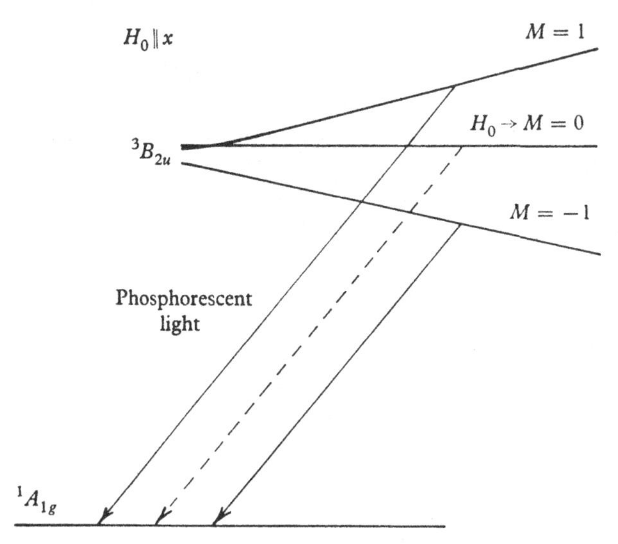

Fig. 2. The energy diagram of a naphthalene molecule in a durene crystal showing the polarisation of the phosphorescent emission from the triplet state to the ground state of $H_0 \| x$ axis, for x-polarised radiation ($^1B_{3u}$ admixture) (——) and for z-polarised radiation ($^1B_{1u}$ admixture) (- - - -).

to the y axis. Also, one must remember that all the results derived above are based upon the D_{2h} symmetry of the naphthalene molecule. Any perturbation which tends to destroy this symmetry weakens the invoked arguments.

Phosphorescent light change induced by ESR

A. *Population and light intensity changes*

In the phosphorescent triplet state of naphthalene the three Zeeman energy levels have radiative lifetimes of approximately 2 s, while the time required for the redistribution of the population among these levels is very short (spin-lattice relaxation time less than 10^{-4} s).

In the absence of microwave-induced transitions between the Zeeman levels, their relative populations are given by a Boltzmann distribution, but, when microwave power is present, forced transitions produce a non-Boltzmann distribution.

Let n_i and N_i represent the population of the ith Zeeman level when the microwave power is on and off, respectively, and let I represent the total intensity of the phosphorescent light emitted by all of the Zeeman levels. Since the phosphorescent emission from a given Zee-

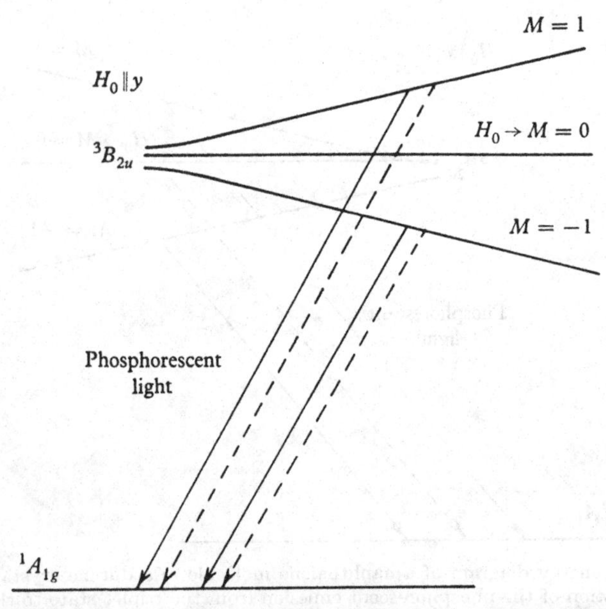

Fig. 3. The energy diagram of a naphthalene molecule in a durene crystal showing the polarisation of the phosphorescent emission from the triplet state to the ground state for $H_0 \| y$ axis, for x-polarised radiation ($^1B_{3u}$ admixture) (——) and for z-polarised radiation ($^1B_{1u}$ admixture) (– – – –).

man level is proportional to its population and to its radiative transition probability, the change in I caused by the application of microwave power is

$$\Delta I = c \sum_i (n_i - N_i)\, w_i, \tag{7}$$

where w_i stands for the probability of phosphorescent emission from the ith Zeeman level per unit time and c is a proportionality constant. From the results of the last section, it follows that w_i is a function of the orientation of the magnetic field relative to the molecular axes.

Under our experimental conditions where $H_0 \| y$, the $M = 0$ level does not radiate, while the $M = \pm 1$ levels radiate with the same transition probability. Consequently, for this magnetic field orientation,

$$\frac{\Delta I}{I} = \frac{n-N}{N}, \tag{8}$$

with $n = n_1 + n_{-1}$ and $N = N_1 + N_{-1}$. It is this fractional change in phosphorescent emission caused by ESR that we are studying. For the case of microwave saturation of the $M = 0 \leftrightarrow M = 1$ Zeeman transition, rate calculations show

$$\frac{n-N}{N} = \frac{h\nu}{4kT}, \tag{9}$$

provided $h\nu \ll kT$, where h denotes the Planck constant, ν the microwave frequency, and T the absolute temperature of the crystal.

The population change specified by equation (8) must be modified if the ESR line is 'inhomogeneously' broadened. In this case, the quantity on the right-hand side of equation (8) must be multiplied by a numerical factor η, which is less than unity. For $H_0 \| y$, an unresolved single ESR line is observed. Hutchison & Mangum[(8)] have shown that this line is a Gaussian and is composed of nine equally spaced hyperfine components. This situation represents a special form of inhomogeneous line broadening. Since the individual hyperfine components have slightly different resonant magnetic field (for the same microwave frequency), all of the hyperfine components are not equally excited at any instantaneous value of the applied magnetic field. At the center of the observed ESR line, the reduction factor η is about 0·5.

B. *Multiple molecular orientations*

Imagine the special case where all of the naphthalene molecules are similarly oriented in the crystal and hence emit phosphorescent light with the same polarisation pattern. In this case, if depolarisation effects are not too strong or can be evaluated, by using an optical analyser and viewing parallel to the y molecular axis, one can determine separately the amount of light polarised along the x and z molecular axes, and hence measure the amount of $^1B_{1u}$ and $^1B_{3u}$ admixture in the phosphorescent state. It is still impossible, however, to identify by optical means how many and which Zeeman levels are radiating with a particular polarisation since the Zeeman spectrum of the phosphorescence is not optically resolvable. Consequently, the singlet character of the individual Zeeman levels cannot be ascertained by using purely optical techniques. Such information can be obtained, however, if the ESR effect on the phosphorescent light can be measured. This is possible because ESR affects the various Zeeman levels differently. By varying the magnetic field orientation relative to the molecular axes and by choosing ESR transitions between different pairs of the Zeeman levels, we can, in principle, determine the nature of the polarised light emitted by each of the Zeeman levels separately.

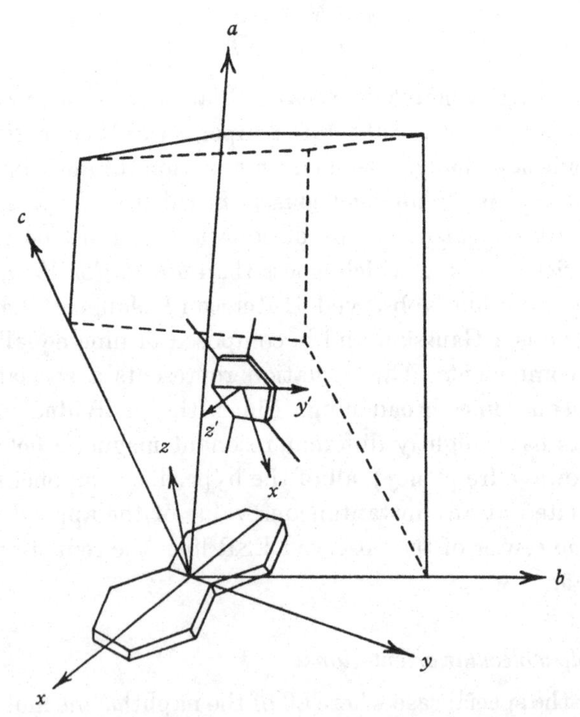

Fig. 4. A naphthalene molecule with x, y, z coordinate axes replacing one of the two durene molecules in a unit cell. The monoclinic[11] angle $\beta[= <(a, c)] = 113\,°18'$.

For the experimental case of naphthalene in durene, the molecular arrangement is more complicated. A naphthalene molecule embedded in a durene crystal can occupy either of two non-equivalent sites (fig. 4). In order to distinguish between these two different molecular orientations, we associate one orientation with an x, y, z coordinate system and the other with an x', y', z' coordinate system. For convenience, we will speak of a molecule having the first orientation as an 'unprimed' molecule and any molecule having the second orientation as a 'primed' molecule. It is important to note that, to a good approximation, $x \| z'$, $y \| y'$ and $z \| x'$. Thus the molecular planes of the differently oriented molecules are nearly mutually perpendicular and their short in-plane axes lie almost parallel.

Optically, both the primed and the unprimed molecules of naphthalene in durene contribute to the phosphorescence, and hence their individual polarisation patterns must be separated by analysis when purely optical data are being handled. For the ESR-induced phosphorescence measurements, however, the situation is simpler. This is due to the fact that with a given magnetic field orientation, a given microwave frequency, and a given ESR transition, the electron spin resonances of the two differently oriented types of molecules occur at different magnetic field intensities, except for the special circumstances where the three Zeeman levels are equally spaced in energy. Therefore, quite generally, optical detection of ESR permits separate excitation and hence separate study of the primed and unprimed molecules.

Experimental system

A block diagram of the experimental apparatus is shown in fig. 5. The excitation light from an a.c. operated GE A-H6 high-pressure mercury arc lamp is incident on the crystal in a direction parallel to the molecular z axis. This light is chopped by a rotating disc in order to allow the phosphorescent light to be observed free from other extraneous radiation during the excitation-off period. The phosphorescent radiation from the crystal passes through a narrow slit in the side wall of the cylindrical microwave cavity and leaves the cryostat through an unsilvered fused quartz tip to enter the aperture in a magnet pole-face and strike an RCA 1P21 photomultiplier. An optical analyser and an interference filter are placed in the light path within the axial hole of the electromagnet. The output of the photomultiplier is switched on and off in anti-synchronism with the excitation light by means of a relay device activated by photovoltaic sensors (fig. 6). Subsequently, the signal is fed into a low-pass filter having a 20 c/s cut-off.

The microwaves are generated by a Varian V-58 klystron, operated at a microwave cavity frequency of about 9320 Mc/sec. The microwaves are chopped at a 5 c/s rate by modulating the klystron reflector. The square-wave generator which provides the modulating voltage also supplies the reference signal to the phase sensitive lock-in detector. The chopped microwaves produce a corresponding variation in the population distribution among the triplet state Zeeman levels. This population variation modulates the phosphorescence output. The 5 c/s signal, after amplification, is phase-detected and recorded.

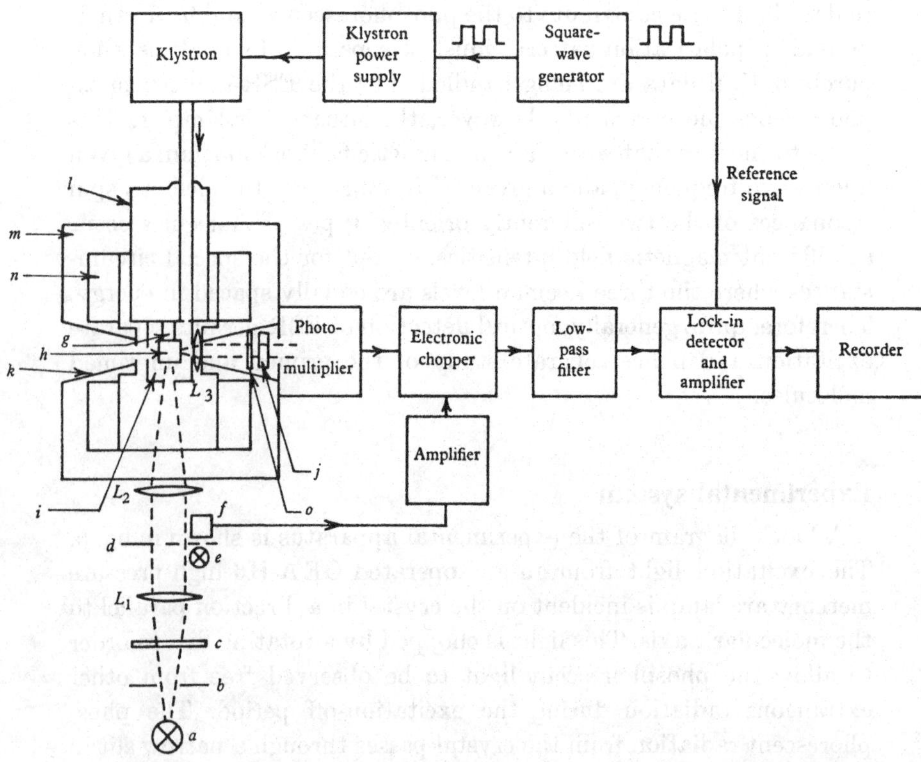

Fig. 5. A block diagram of the system for optical detection of ESR: (*a*) high-pressure A-H 6 mercury arc lamp, (*b*) variable stop, (*c*) Corning C. S. 7-54 glass filters, (*d*) mechanical light chopper, (*e*) small incandescent lamp, (*f*) photocell, (*g*) sapphire rod and crystal support wedge, (*h*) oriented single crystal, (*i*) singly-split cylindrical TE_{011} microwave cavity, (*j*) interference filter, (*k*) one of the two apertures in magnet pole-faces, (l) cryostat, (L_1), (L_2) fused quartz lenses, (L_3) crown glass lens, (*m*) electromagnet, (*n*) magnet current coil, and (*o*) optical analyser.

The single crystals of naphthalene in durene that were used contained an initial concentration of 0·4 mole % naphthalene and were grown from carefully purified (50 passes) melt in our laboratory by the modified Bridgman[9]–Stockbarger[10] method. We used only those test specimens which had excellent cleavage faces parallel to the *ab* crystallographic plane[11] and showed a high degree of perfection optically when examined with a polarising microscope. In all cases the crystal ingot grew with the *b* crystallographic axis parallel to the main temperature gradient in the Bridgman furnace. For the present experiment, the unpolished cleavage face of the crystal slab (typical dimensions 8 mm by 8 mm by 3·5 mm) was attached to a specially made sapphire wedge in such a manner that, when the wedge was mounted, the molecular plane of one of the two molecules of a unit cell was horizontal and parallel to the plane swept out by the field of the rotatable magnet. Depolarisation of the phosphorescent light by specular reflections was minimised by blackening the walls of the microwave cavity and all exposed surfaces of the sapphire wedge. Masks were not used around the crystal edges in order to avoid contaminating the crystal specimen and reducing the quality factor (Q) of the microwave cavity.

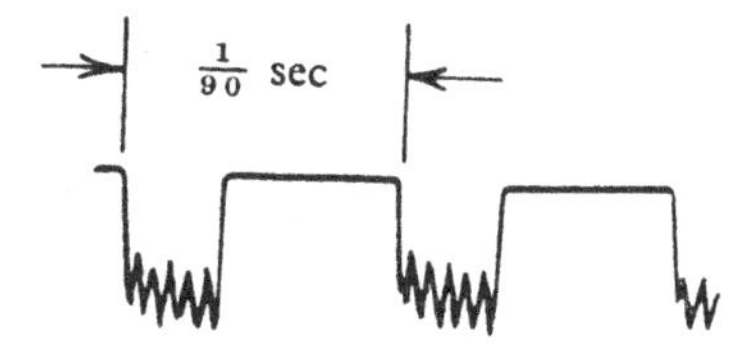

Fig. 6. The chopped phosphorescence voltage at the output of the photomultiplier as measured before the low-pass filter. The high-frequency noise that appears in the picture and the chopping frequency harmonics are eliminated by the low-pass filter.

Experimental results: naphthalene in durene

A. *Triplet ESR spectrum*

The ESR spectrum of the lowest triplet state of naphthalene in durene has been observed at both liquid nitrogen and liquid helium temperatures. The ESR results obtained in these experiments and the absence of spurious lines indicate that the samples used were good single crystals free from paramagnetic impurities. A typical ESR trace is shown in fig. 7. Our ESR results agree with those obtained by Hutchison & Mangum[8]. From intensity saturation results, we estimate that the spin-lattice relaxation time of the triplet state is 10^{-5} to 10^{-4} s.

B. *Polarised phosphorescence radiation*

The phosphorescent radiation from the naphthalene triplet state has been observed, and the relative intensities of its polarised components have been measured. The phosphorescence extends roughly from 4700 to 6300 Å, with its 0–0 transition[12] peaked at around 4746 Å. In this spectral region there is some conspicuous fluorescent radiation which decays very rapidly when the excitation is removed. It is about four times more intense than the 0–0 phosphorescence band of naphthalene. Because of this, all of our measurements have been made using an apparatus functioning effectively as a phosphoroscope (see fig. 5). The desired pass-band of phosphorescent wavelengths was

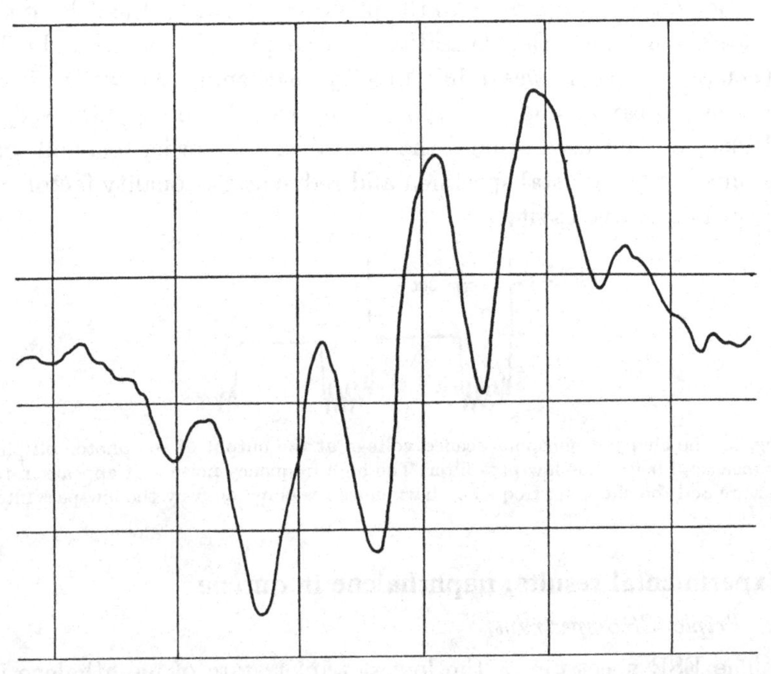

Fig. 7. The ESR spectrum of the triplet state of a naphthalene molecule in a durene crystal. $T = 77$ °K; $H_0 \| x$ axis; field modulation 1000 c/s; microwave frequency 9320 Mc.

attained with the aid of a Baird–Atomic interference filter having a transmission band width of 93 Å at one-half intensity maximum and a transmission peak at 4723 Å.

With the crystal mounted so that the molecular z axis is vertical in the laboratory (fig. 8), we have determined the values of the vertically and horizontally polarised phosphorescence intensities P_v and P_h as a function of the azimuthal angle in a horizontal plane. The polarisation ratio P_v/P_h is found to be 1·0 along the molecular y axis and 4·0 along the molecular x axis. Figure 8 shows the crystal geometry. In the y direction the light leaves the crystal approximately perpendicular to the crystal surface plane, while in the x direction it emerges from a narrower side face of the crystal, where some influence from the crystal substrate is felt. Because of this, the experimental polarisation ratio in the y direction must be regarded as much more reliable than that in the x direction. In either case, the interpretation of the observed polarisations is complicated by biaxial crystal optical properties and crystal surface effects such as double refraction and multiple reflections from the crystal faces. The obtained results, however, are in qualitative agreement with the theory developed above (see pp. 144–149) and with recent work by Chaudhuri &

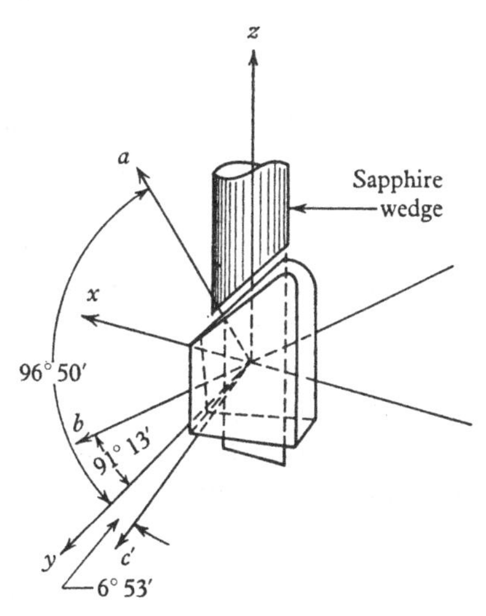

Fig. 8. A view of the crystal in its mounted position showing the molecular and crystal axes. By slightly wedging the outer crystal face relative to the cleavage (ab) plane, the y axis is made nearly perpendicular to the face viewed by the detector.

El-Sayed.[13,14] These authors state that the phosphorescence emission from $C_{10}D_8$ in durene is more than 70 % long axis polarised. Since our naphthalene molecule is not deuterated, the applicability of this latter fact based on $C_{10}D_8$ to our case is subject to closer experimental scrutiny.

C. *Optical detection of ESR*

To test the theory of optical detection of ESR, a number of preliminary experiments were conducted with the apparatus described above (see pp. 153–155). Each of the crystal samples that we used was so shaped and mounted that the light reaching the photomultiplier propagated perpendicular to its plane surface (fig. 8). For this reason and other reasons discussed above (pp. 147–149), we performed the experiments with both the magnetic field and the direction of observation along the molecular y axis.

The cryostat employed in this investigation was originally designed for another problem at liquid nitrogen temperature, but, with special modifications, it can be operated at temperatures near 10 °K. When the phosphorescent light intensity was measured experimentally, it became apparent that temperatures close to the temperature of boiling helium are required for detecting ESR optically in naphthalene. A new cryostat is being constructed that will permit reaching 4·2 °K and lower temperatures.

A semi-empirical signal-to-noise ratio (V_S/V_N) was calculated using the measured noise voltage at the output of the photomultiplier to determine V_N and equations (A. 1) and (A. 3) in the Appendix to compute the expected signal voltage V_S. The experimentally measured value of the photomultiplier current I_p was used in evaluating equation (A. 4). The measured noise voltage (V_N) was found to be about twice the theoretical noise voltage computed from equation (A. 2). The electronic circuitry following the photomultiplier has inherently low noise characteristics and does not make essential contribution to the noise level. The signal voltage (V_S) takes into account incomplete microwave saturation and the effect of inhomogeneous broadening as described above (see p. 151). The population change, given by equation (8) for complete microwave saturation, was reduced by approximately 50 % because of incomplete saturation under our experimental conditions. For an effective bandwidth of 0·1 c/s, the computed signal-to-noise ratio V_S/V_N equals 0·4 at 77 °K and 3 at 10 °K.

No signal can be expected at the liquid nitrogen temperature. Various tests were made at this temperature, however, in order to check the proper functioning of various parts of the experimental system. At the lower temperature of about 10 °K, the signal-to-noise ratio is marginal, and no signal has thus far been detected.

Discussion

Since the lowest triplet states of most aromatic molecules in molecular organic crystals have long lifetimes, at low temperatures they often have sufficient population differences between their Zeeman levels to allow direct ESR detection, but the relatively low phosphorescent emissions of these states and their wide bands of phosphorescent wavelengths make optical detection of ESR difficult. Because of this, when the latter technique is used exclusively as a means of observing ESR in the triplet state, it is considerably inferior to conventional ESR detection.

Optical detection of ESR is valuable, however, because this technique provides a means of obtaining information that cannot be obtained by purely optical methods. The advantages of optical detection of ESR are twofold. First, it allows the different Zeeman levels of the triplet state to be studied individually, and second, when the organic molecule can occupy more than one spatially nonequivalent site in the host crystal, it allows one to study the molecules located at each different site separately.

By inducing ESR transitions between different pairs of Zeeman levels at various orientations of the molecule relative to the magnetic field, one can determine the polarisation characteristics of the individual Zeeman levels of the triplet state. From this information it is possible to deduce the symmetry properties and relative amounts of admixture of the singlet state or states into the separate Zeeman components. Therefore, the optical detection of triplet state ESR may provide an additional clue for differentiating between possible triplet-singlet mixing mechanisms.

The possibility inherent in optical detection of ESR for separately studying the excited molecules that have different orientations in the host crystal is especially useful in cases, such as naphthalene in durene, where the molecular planes in the two orientations are almost mutually perpendicular and pairs of differently labelled, polarisation-active molecular axes align nearly parallel. In such cases, provided the molecules of both orientations are phosphorescing, it is difficult, if not impossible, to distinguish the two superimposed polarisation patterns of the differently oriented molecules by using purely optical techniques. Specifically, it is difficult to determine the direction of the transition dipole moments relative to the molecular axes.

The absence of an identifiable signal in our preliminary experimental attempt to detect optically ESR may be due to several factors. At the temperatures thus far used, the small calculated signal-to-noise ratio makes the possibility of detection marginal, and any degrading of the signal by factors not taken into consideration would make the signal too small to be detected. At lower temperatures, which will be provided by the new cryostat, the increased sensitivity will make possible a firmer test of the theory.

Theoretically, we have assumed that a π, π^* spin-orbit interaction is the dominant mechanism for producing the singlet admixture in the triplet phosphorescent state. We have also assumed that the excited molecule is sufficiently isolated in the crystal so that an oriented gas model, which neglects all intermolecular interactions of the real crystal, is a good approximation. If the spin-orbit interaction between the respective π, π^* states of the naphthalene molecule is unusually small, as concluded by McClure[15] for aromatic hydrocarbons, then, as he suggests, other mechanisms such as perturbations that destroy the symmetry plane of the molecule, spin-orbit interactions involving non-π-electron states of the molecule,[16] and spin-spin interactions could outweigh the spin-orbit interaction among the π-electron states. In this case the group theoretical prediction of these competing interactions would have to serve as a guide for experiment.

It is of interest to note that, for naphthalene-d_8 in a durene crystal, Chaudhuri & El-Sayed[13] conjecture that the singlet admixture into the lowest triplet state is increased through an intermolecular spin-orbit interaction mechanism that originates in the non-planarity of the naphthalene molecule, a perturbation produced by an intermolecular interaction. If this is the case for ordinary naphthalene, then the oriented gas model must be modified appropriately to take this interaction into consideration.

In conclusion, it should be noted that although the theory presented in this paper is based upon the spin-orbit interaction of our particular type and upon a D_{2h} molecular symmetry, the optical detection of ESR is more generally applicable. The feasibility of this technique does not depend upon any particular type of triplet-singlet mixing mechanism in the metastable triplet state nor does it require molecular D_{2h} symmetry.

Acknowledgements

We thank Dr F. J. Adrian and Dr A. N. Jette for their contributions to the group theoretical calculations, Dr S. N. Foner for his constant technical advice and encouragement, and Mr V. A. Bowers for his valuable technical help.

Appendix

Signal-to-noise ratio: optical detection of ESR

Let I_p be the photomultiplier output current produced by a phosphorescent light signal, large enough so that the photomultiplier dark current can be ignored, V_p be the voltage across the photomultiplier load resistance R due to I_p, and V_N be the r.m.s. noise voltage impressed across R due to statistical fluctuations in I_p. Then we have[17]

$$V_p = RI_p \tag{A. 1}$$

$$V_N = R(2e\Delta f\mu I_p)^{\frac{1}{2}}, \tag{A. 2}$$

where Δf is the effective bandwidth of the measuring apparatus, e is the magnitude of the electronic charge, and μ is the current amplification of the photomultiplier.

Using equation (8), the expected signal voltage V_S is

$$V_S = \eta' V_p \frac{n-N}{N}, \tag{A. 3}$$

where the numerical factor η' corrects for inhomogeneous broadening and for incomplete microwave saturation.

Combining equations (A. 1)–(A. 3), we have the signal-to-noise ratio

$$\frac{V_S}{V_N} = \eta' (I_p/2e\mu\Delta f)^{\frac{1}{2}} \frac{n-N}{N}. \tag{A. 4}$$

The last equation indicates that the signal-to-noise ratio depends upon several factors. It is directly proportional to the fractional population change caused by microwave-induced Zeeman transitions. Since this fractional population change is proportional to $h\nu/kT$, a low crystal temperature and a high microwave frequency (strong magnetic field) must be used. The signal-to-noise ratio is also proportional to $(I_p/\mu)^{\frac{1}{2}}$, where I_p/μ is the value of the photomultiplier current at the photocathode. Consequently, it is essential to operate with high phosphorescent light flux. Equation (A. 4) further shows the usual condition that small effective bandwidths are required for high sensitivity.

References

1 G. F. Imbush & S. Geschwind, *Phys. Rev. Letters* **18**, 109 (1965).
2 This topic was presented by C. K. Jen at the *Seventh International Symposium on Free Radicals*, Padua, Italy, 10 September 1965.
3 D. S. McClure, *J. Chem. Phys.* **17**, 665 (1949).
4 S. I. Weissman, *J. Chem. Phys.* **18**, 232 (1950).
5 R. Pariser, *J. Chem. Phys.* **24**, 250 (1956).
6 This presentation was substantially aided by private communications from Dr F. J. Adrian and Dr A. N. Jette.

7 See also the paper: G. R. Hunt, E. L. McCoy & I. G. Ross, *Aust. J. Chem.* **15**, 591 (1962).
8 C. A. Hutchison Jr. & B. W. Mangum, *J. Chem. Phys.* **34**, 908 (1961).
9 P. Bridgman, *Proc. Am. Acad. Arts Sci.* **60**, 305 (1925).
10 D. C. Stockbarger, *Rev. Sci. Instr.* **7**, 133 (1936).
11 J. M. Robertson, *Proc. Roy. Soc.* (*London*) A, **141**, 594 (1933).
12 T. N. Misra & S. P. McGlynn, *J. Chem. Phys.* **44**, 3816 (1966).
13 N. K. Chaudhuri & M. A. El-Sayed, *J. Chem. Phys.* **43**, 1423 (1965).
14 N. K. Chaudhuri & M. A. El-Sayed, *J. Chem. Phys.* **44**, 3728 (1966).
15 D. S. McClure, *J. Chem. Phys.* **20**, 682 (1952).
16 M. Levine, J. Jortner & A. Szöke, *J. Chem. Phys.* **45**, 1591 (1966).
17 R. W. Engstrom, *J. Opt. Soc. Am.* **37**, 420 (1947).

PARAMAGNETIC RESONANCE OF THE TRIPLET STATE OF TETRAMETHYLPYRAZINE

J. S. VINCENT

Abstract

EPR measurements are reported for solid solutions of tetramethylpyrazine (TMP) in durene under ultraviolet irradiation at 77 °K. The fine structure of the paramagnetic resonance spectrum observed at 9·2 and 35 GHz may be described by the spin Hamiltonian

$$\mathscr{H} = g\beta \mathbf{H}.\mathbf{S} + DS_z^2 + E(S_x^2 - S_y^2),$$

in which $S = 1$.

$$D/hc = \pm 0{\cdot}0990 \pm 0{\cdot}0005 \text{ cm}^{-1},$$

$$E/hc = \pm 0{\cdot}0043 \pm 0{\cdot}0007 \text{ cm}^{-1},$$

$$g(\text{isotropic}) = 2{\cdot}003 \pm 0{\cdot}001.$$

No hyperfine structure is observed in crystals doped with either normal TMP or perdeuterated TMP. The decrease in the line width upon deuteration and the variation of the line width upon rotation in the magnetic field is consistent with greater spin density on the ring carbon atoms than on the nitrogen atoms.

The triplet emission spectrum is also reported for TMP measured in an Argon and a hydrocarbon matrix at 4·2 °K. The O–O band of the $T \rightarrow S$ emission of TMP is at $26\,050 \pm 30$ cm^{-1} in both matrices. The emission decay constant is $0{\cdot}41 \pm 0{\cdot}03$ s in both matrices at 4·2 °K.

The EPR measurements and the emission studies indicate that the lowest triplet state of TMP is of π–π^* character. This is in contrast with the known n–π^* character of the triplet state of the parent compound, pyrazine. The extent of the conjugative perturbation necessary to depress the π–π^* level below the n–π^* level is investigated and compared with analogous systems.

ON MAGNETIC DIPOLE CONTRIBUTIONS TO THE INTRINSIC $S_0 \rightleftarrows T_1$ TRANSITION IN SIMPLE AROMATICS

MARK SHARNOFF

Abstract

Magnetic dipolar contributions to radiative $S_0 \rightleftarrows T_1$ processes in aromatic hydrocarbons are estimated by a semi-classical method in which spin-orbit coupling is intentionally neglected. For benzene and anthracene, the estimated contributions are 1 % and $\frac{1}{4}$% of the total radiative intensity.

In a classic experiment, Weissman & Lipkin demonstrated[1] in 1942 that the long-lived phosphorescence emission of fluorescein was electric dipole radiation. It has been a widely held belief since that time that the phosphorescence of all aromatic hydrocarbons is electric dipole in character. Radiation of this type is made possible by spin-orbit coupling of the singlet and triplet configurations within the shells of the heavy atoms in the molecule; McClure showed[2] that such a picture gave theoretical phosphorescence lifetimes in good agreement with the millisecond values observed in substituted aromatics and their derivatives. The much longer-lived (0·1–7 s) phosphorescence of the simple aromatics seemed anomalous; McClure[3] clarified this matter by proving that spin-orbit interaction in planar π–π^* systems vanished to high order. His paper concluded with the suggestion that residual spin-orbit and spin-spin couplings, involving very highly excited states of the molecule, were responsible for the (electric dipole) phosphorescence of these compounds. This suggestion seemed reasonable in view of the measured phosphorescence lifetimes of the day. Later quantum yield measurements by Kellogg & Bennett[4] showed, however, that the radiative lifetimes of the simple aromatics could not be smaller than 28 s, and attempts by Craig and co-workers[5] to observe the $S_0 \rightarrow T_1$ transition in absorption in carefully de-oxygenated benzene gave an upper limit of 7×10^{-12} for its oscillator strength, which would place the intrinsic radiative lifetime of the $T_1 \rightarrow S_0$ phosphorescence in excess of 300 s. Lifetimes of this magnitude are interestingly long, in view of the typical aromatic $S_1 \rightarrow S_0$ fluorescence lifetimes of 20–300 ns, and suggest that magnetic dipole processes may be of some importance in the intrinsic $S_0 \rightleftarrows T_1$ transition. In the present note we reopen the question of the nature of this transition, estimating by means of an elementary semi-classical calculation that magnetic dipole radiation by the electron spins may be a mechanism strong enough to contribute significantly to the intrinsic $S_0 \rightleftarrows T_1$ oscillator strength in benzene and perhaps in other simple aromatics.

For simplicity, we consider a system containing only two electrons, taking a single-particle viewpoint. The feature of the calculation which will refer specifically to the simple aromatics is the explicit neglect, in accord with McClure's analysis, of spin-orbitally mediated interaction between singlet and triplet configurations. Let Ψ be the molecular orbital occupied by the electrons when the system is in its ground state, and consider an orbitally excited state in which one of the electrons occupies an orbital χ. If the excited state is a triplet, the spatial wavefunction will be

$$[\,] \equiv (1/\sqrt{2})\,[\Psi(\mathbf{r}_1)\,\chi(\mathbf{r}_2) - \Psi(\mathbf{r}_2)\,\chi(\mathbf{r}_1)]; \tag{1}$$

if it is a singlet, the spatial function will be

$$\{\,\} \equiv (1/\sqrt{2})\,[\Psi(\mathbf{r}_1)\,\chi(\mathbf{r}_2) + \Psi(\mathbf{r}_2)\,\chi(\mathbf{r}_1)]. \tag{2}$$

The complete wavefunctions for the excited singlet and triplet states will be

$$|S_1\rangle = |\{\,\}; 0, 0\rangle \tag{3}$$

$$|T_1, m\rangle = |[\,]; 1, m\rangle \qquad (m = -1, 0, +1), \tag{4}$$

where the spin quantum numbers have been denoted in the usual way. The ground state is

$$|S_0\rangle = |\Psi(\mathbf{r}_1)\,\Psi(\mathbf{r}_2); 0, 0\rangle. \tag{5}$$

Let a plane electromagnetic wave be incident upon the system. The interaction, $J_\mu A_\mu$, of this wave with the molecule will contain electric multipole terms, which cannot, in absence of spin-orbit coupling, change the multiplicity of the system; and magnetic multipole terms, which can. The coupling of an electron spin to the wave has the form $2\beta\mathbf{s}\,.\,\mathbf{b}\exp(i\mathbf{k}\,.\,\mathbf{r})$, where β is the Bohr magneton, $eh/4\pi mc$, $\mathbf{s}$ is the spin of the electron and $\mathbf{r}$ its position, and where $\mathbf{k}$ is the propagation vector of the wave and $\mathbf{b}$ the amplitude of its magnetic induction. The semi-classical transition amplitude between the ground state and one of the components of the excited triplet is

$$2\beta\langle S_0|\,\mathbf{s}_1\,.\,\mathbf{b}\exp(i\mathbf{k}\,.\,\mathbf{r}_1) + \mathbf{s}_2\,.\,\mathbf{b}\exp(i\mathbf{k}\,.\,\mathbf{r}_2)\,|T_1, m\rangle. \tag{6}$$

This expression vanishes in lowest order in $\mathbf{k}\,.\,\mathbf{r}$. When the exponentials are expanded in power series, the lowest non-vanishing term reduces to

$$a(S_0 \rightleftarrows T_1, m) = 2(\sqrt{2})\,i\beta\langle 0, 0|\,\mathbf{b}\,.\,\mathbf{s}_2\,|1, m\rangle\langle\Psi|\,\mathbf{k}\,.\,\mathbf{r}\,|\chi\rangle. \tag{7}$$

Rather than trying to evaluate the orbital matrix element numerically, we exploit its connection with the electric dipole-induced transition from the ground state to the excited singlet. Consider a second electromagnetic wave, having electric field $\boldsymbol{\epsilon}'$ oriented in the direction of the propagation vector $\mathbf{k}$ of the first wave. The singlet-singlet transition amplitude of the second wave is

$$a(S_0 \rightleftarrows S_1) = (\sqrt{2})\, e\langle\Psi|\boldsymbol{\epsilon}'.\mathbf{r}|\chi\rangle = ((\sqrt{2})\, e\epsilon'/k)\langle\Psi|\mathbf{k}.\mathbf{r}|\chi\rangle \tag{8}$$

in lowest order. Now Maxwell's third equation requires that in a uniform plane wave, the magnitude of $\boldsymbol{\epsilon}'$ must be equal (in c.g.s. units) to the magnitude of the induction, $\mathbf{b}'$, divided by the index of refraction, n, of the medium. If the primed and unprimed waves have equal intensities, then $b = b' = n\epsilon'$, and the ratio of the transition amplitude will be

$$a(S_0 \rightleftarrows T_1, m)/a(S_0 \rightleftarrows S_1) = (2i\beta kn/eb)\langle 0,0|\mathbf{b}.\mathbf{s}_2|1,m\rangle. \tag{9}$$

The square of this quantity, averaged over all possible orientations of $\mathbf{b}$, is clearly the ratio of Einstein B-coefficients for the transitions.[17] Writing $k = 2\pi n/\lambda_\phi$, where λ_ϕ is the free-space wavelength of the $S_0 \rightarrow T_1$ transition, we find

$$\frac{B(S_0 \rightleftarrows T_1, m)}{B(S_0 \rightleftarrows S_1)} = \frac{1}{3}\left(\frac{n^2\lambda_c}{\lambda_\phi}\right)^2 |\langle 0,0|\mathbf{s}_2|1,m\rangle|^2, \tag{10}$$

where λ_c is the Compton wavelength of the electron, $h/mc = 0{\cdot}0243$ Å. Summation of this expression over the three spin states of the triplet gives the ratio of oscillator strengths predicted by neglecting the spin-orbit coupling between configurations:

$$\frac{f(S_0 \rightarrow T_1)}{f(S_0 \rightarrow S_1)} = \frac{1}{4}\left(\frac{n^2\lambda_c}{\lambda_\phi}\right)^2 \frac{\lambda_f}{\lambda_\phi}, \tag{11}$$

where λ_f is the free-space wavelength of the $S_0 \rightarrow S_1$ transition. With an average value of $\lambda_\phi = 3\,250$ Å for the $S_0 \rightarrow T_1$ system in benzene and with $n = 1{\cdot}50$, the ratio (11) becomes $5{\cdot}5 \times 10^{-11}$. Using Platt's measurement[6] of the oscillator strength of the benzene $S_0 \rightarrow S_1$ transition in conjunction with Craig's upper limit, we find

$$\frac{f(S_0 \rightarrow T_1)}{f(S_0 \rightarrow S_1)} < \frac{7 \times 10^{-12}}{2 \times 10^{-3}} = 3{\cdot}5 \times 10^{-9}. \tag{12}$$

On the basis of these rough calculations, then, it would seem as though magnetic dipole contributions amount to at least 1·5% of the intrinsic strength of the $S_0 \rightarrow T_1$ transition. For anthracene, on the other hand, equation (11) gives $1{\cdot}2 \times 10^{-11}$, while experiment[6, 7, 8] yields 4×10^{-8}. About 0·03 % of the $S_0 \rightarrow T_1$ oscillator strength might be attributed to magnetic dipole absorption.

Since there is nothing in the above argument which refers solely to absorption processes, we turn equation (10) around, using the well-known relation between the Einstein A- and B-coefficients, to derive a predicted ratio of phosphorescence and fluorescence radiative life-times:

$$\frac{\tau_\phi}{\tau_f} = 8\left(\frac{\lambda_\phi}{n^2\lambda_c}\right)^2\left(\frac{\lambda_\phi}{\lambda_f}\right)^3. \tag{13}$$

For anthracene and benzene, equation (13) gives 9×10^{11} and $6{\cdot}6 \times 10^{10}$, respectively.

Experimental values of τ_ϕ and T_f may be obtained by dividing[9] the observed fluorescence and phosphorescence lifetimes by appropriate quantum yields. Combining the data of Kepler *et al.*,[10] McClure,[2] Kellogg *et al.*,[11] Bowen & Williams,[12] and Smith[20] on anthracene, we find

$$(\tau_\phi/\tau_f)_{\text{expt.}} = (3 \pm 1) \times 10^9,$$

a value considerably higher than might have been expected from the absorption oscillator strengths.

For benzene, the data of Ivanova *et al.*[13] and Lim[14] give

$$(\tau_\phi/\tau_f)_{\text{expt.}} = 1{\cdot}2 \times 10^8,$$

a value lower than might have been predicted from the absorption oscillator strengths. None the less, the numbers confirm that about ¼ % of the phosphorescence might arise from magnetic dipole radiation.

The lack of reference, in the above calculations, to specific structural details (other than planarity[3]) of the aromatics constitutes a serious oversimplification. Vibronic effects, overlooked here, are important in the lower energy transitions of the aromatics, and it is not clear that the orbital matrix elements in equations (7) and (8) should actually be considered to be the same. The $S_0 \rightarrow S_1$ transition in benzene is allowed only vibronically, and it is known that the vibrational mechanisms of the $S_1 \rightarrow S_0$ and $T_1 \rightarrow S_0$ transitions in benzene are different.[15] Secondly, the retention of vibrational quantum numbers in the theory would invalidate[16] the use of the Einstein relations[17] in deriving equation (13), because the transitions seen in absorption involve the excited vibrational states of S_1 or T_1, whereas the emission involves the excited vibrational states of S_0. It should therefore be borne in mind that, in the absence of reliable determinations of the vibronic matrix elements, the theoretical estimates made here could be inaccurate by more than an order of magnitude. Since, then, it is possible that the effective magnetic dipole strengths are larger than indicated here, there would seem to be no basis at present for believing that magnetic dipole radiation by the electron spins does not play a significant role (2–20 %) in the radiative $S_0 \rightleftarrows T_1$ transitions of *any* of the aromatics. While the benzene radiative lifetime calculation of Hameka & Oosterhoff[19] inveighs such a thesis, a resuscitation of the method of Weissman & Lipkin would surely settle the matter.

Acknowledgement

This research was sponsored by the University of Delaware Research Foundation.

References

1 S. I. Weissman & D. Lipkin, *J. Am. Chem. Soc.* **64**, 1916 (1942).
2 D. S. McClure, *J. Chem. Phys.* **17**, 905 (1949).
3 D. S. McClure, *J. Chem. Phys.* **20**, 682 (1952).
4 R. E. Kellogg & R. G. Bennett, *J. Chem. Phys.* **41**, 3042 (1964).
5 D. P. Craig, J. M. Hollas & G. W. King, *J. Chem. Phys.* **29**, 974 (1958).
6 H. B. Klevens & J. R. Platt, *J. Chem. Phys.* **17**, 470 (1949).
7 P. Avakian, E. Abramson, R. G. Kepler & J. C. Caris, *J. Chem. Phys.* **39**, 1127 (1963).
8 M. R. Padhye, S. P. McGlynn & M. Kasha, *J. Chem. Phys.* **24**, 588 (1956).
9 See, for example, E. C. Lim & J. D. Laposa, *J. Chem. Phys.* **41**, 3257 (1964).
10 R. G. Kepler, J. C. Caris, P. Avakian & E. Abramson, *Phys. Rev. Letters* **10**, 400 (1963).
11 J. D. Laposa, E. C. Lim & R. E. Kellogg, *J. Chem. Phys.* **42**, 3025 (1965).
12 E. J. Bowen & A. H. Williams, *Trans. Faraday Soc.* **35**, 765 (1939).
13 T. V. Ivanova, P. I. Kudryashov & B. Y. Sveshnikov, *Dokl. Akad. Nauk SSSR*, **138**, 572 (1961) (English translation: *Soviet Phys.-Doklady*, **6**, 407 (1961).
14 E. C. Lim, *J. Chem. Phys.* **36**, 3497 (1962).
15 T. V. Ivanova & B. Y. Sveshnikov, *Opt. i Spektroskopiya*, **11**, 598 (1961) (English translation: *Opt. and Spectr.* **11**, 322 (1961).
16 Thus Robinson's criticism[18] of Craig's upper limit for $f(S_0 \rightarrow T_1)$ is without logical basis, for it is possible, without violation of the Einstein relations, that the phosphorescence radiative lifetime could be shorter than would be predicted from an oscillator strength measured from absorption spectra. The phosphorescence radiative

lifetime of anthracene is, for that matter, an order of magnitude longer than would be predicted from the $S_0 \rightarrow T_1$ oscillator strength. 'Anomalies' of this sort are well-known in inorganic substances, viz. in the optical properties of defects in the alkali halides.

17 E. C. Kemble, *The Fundamental Principles of Quantum Mechanics* (McGraw Hill Book Co., Inc., New York, 1937), pp. 448 ff.

18 M. R. Wright, R. P. Frosch & G. W. Robinson, *J. Chem. Phys.* **33**, 934 (1960).

19 H. F. Hameka & L. J. Oosterhoff, *Mol. Physics* **1**, 358 (1958).

20 G. C. Smith, *Bull. Am. Phys. Soc.* II, **11**, 777 (1966).

DISCUSSION

HUTCHISON

Schwoerer to Hutchison: Over what distance does the energy transfer take place?

Hutchison to Schwoerer: In the most dilute diphenyl crystal solutions in which we observed triplet excitation transfer, the phenanthrene and the naphthalene, if randomly distributed, were on the average approximately four diphenyl distances away from each other. The zero-field measurements, mentioned in my talk, indicate that the actual distribution was probably close to random in character.

Birks to Hutchison: Do you have any evidence (negative or positive) for the existence of the triplet state of excited complexes between similar aromatic molecules (triplet excimers) or different aromatic molecules (triplet π–π exciplexes)? Birks & King (*Phys. Letters* 1965) have proposed that excimers are dissociated in the triplet state, although Hochstrasser has suggested that triplet excimers exist.

Hutchison to Birks: We have never observed any triplet state magnetic resonance spectra which could be attributed to complexes between guest species in single crystals.

Hochstrasser to Hutchison: In what manner, and how confidently, can EPR spectroscopy of oriented triplets shed light on the nodal structure or obital character of the triplet state wavefunctions?

Van der Waals to Hutchison: As regards the question put forward by Professor Hochstrasser, the following can be said. ESR would seem to permit of making an unambiguous choice whenever the hyperfine structure can be resolved, as in Hutchison and Mangum's classical study of naphthalene. If one only has the zero-field splitting an assignment can, in general, still be made with reasonable confidence if one has satisfactory wavefunctions for the states concerned. We found the predicted splitting to be quite susceptible to the nodal pattern of the wavefunctions.

Wolf to Hutchison: You mentioned these very interesting ENDOR lines which belong to protons on neighbor molecules. Have you been able to identify and analyse such lines in detail?

Hutchison to Wolf: We have not yet attempted identification and analysis of such lines. As X-ray information on the crystal structures of the hosts which we are using becomes available, the analysis of these lines should be of considerable interest.

Wolf to Hutchison: If you compare EPR lines with compute fits, what do you take as linewidth for the individual hyperfine component? This is probably not the spin-spin relaxation time T_2.

Hutchison to Wolf: The linewidths used for the individual components in our Gaussian syntheses of phenanthrene EPR lines have in all cases (i.e. two monoprotonated cases and the completely protonated molecule) included hyperfine broadening. In the monoprotonated species we were fitting our observations with either a two-line spectrum (1 proton) or a six-line spectrum (1 proton, 1 deuteron in a nearby position). In the completely protonated species we were fitting our observations with only three proton splittings corresponding to the three protons which are in the three positions of highest spin density. Thus the widths which we used for individual components included hyperfine broadening from deuterons or protons and was not related to the spin-spin relaxation time. The observed hyperfine patterns are clearly too poorly resolved to infer anything quantitative about individual effects of protons in the low spin density positions, and so we cannot infer magnitudes of the fundamental linewidths, of individual hyperfine components from these data.

VAN DER WAALS

Hutchison to van der Waals: You stated that when $|\mathbf{H}| = 0$, the hyperfine splitting vanishes to first order. Is it not true that if the only perturbations are the electron-spin–electron-spin and electron-spin–nuclear-spin interactions, then the hyperfine splitting vanishes to all orders?

Van der Waals to Hutchison: Indeed, the zero-field states can never be split through interaction with a single proton (Kramer's degeneracy). However, very small second-order splittings of the zero-field transitions should result from the different ways in which a pair of protons (or a single nuclear spin with $I = 1$) can be coupled to the electron spin angular momentum.

Hameka to van der Waals: You have described the spin functions by means of the three-dimensional rotation group. I remember vaguely that it is necessary to exclude half of the rotations in doing so. Could you comment on this?

Van der Waals to Hameka: Since singlet and triplet spin functions involve the spin of *two* particles they must be invariant under inversion. Thus there is no need to exclude improper rotations (rotation followed by inversion).

Hochstrasser to van der Waals: With reference to your energy diagram, could you describe what would happen if the excitation were directly into the second $^3n\pi^*$ state. Would such excitation not uniquely populate one of the other spin states and provide you with a check on the assumption that intersystem crossing occurs to a specific zero-field state?

Furthermore, would not highly monochromatic (e.g. tuned laser) excitation of the $T_i \leftarrow S_0 (i = 2, 3 \text{ etc.})$ transitions provide fluctuations of polarisations in the emission spectrum assuming your discussion of the radiationless process is valid? Also selective excitation of vibronic bands of a given triplet should cause analogous fluctuations. It seems to me as if such experiments could be done and they may shed light on the internal conversion process within the triplet manifold.

Van der Waals to Hochstrasser: We have to work on dilute single crystals to avoid triplet-triplet interaction. The absorptivity of our small crystals for direct $T \leftarrow S$ absorption is far too low to realise these interesting suggestions. I think excitation into a specific spin component τ_u of an excited triplet state, by absorption of polarised light from the ground state, must subsequently lead to populating of the same spin state τ_u of the lowest triplet. (The idea is that spin symmetry tends to be conserved in internal conversion, where orbital symmetry is changed.)

Wolf to van der Waals: I have a comment and a question. The comment: optical spin polarisation seems to be a very general phenomenon for all triplet systems. It has been observed also for inorganic systems like M-centers in alkali halides.

The question: What do you mean when you say there is absence of energy transfer at 1·5 °K?

One more question: Have you evidence that the process responsible for

$$T_1 \approx \frac{1}{(\Delta\nu)^2}$$

is working? I have the impression that the T_1-process in durene as matrix is different from that in a naphthalene matrix because emissive lines in ESR can be observed only in the naphthalene matrix.

Van der Waals to Wolf: We find that light absorbed by the durene host at 1·6 °K is no longer transferred to the solute molecules.

I have no real proof for the direct dipole-dipole relaxation mechanism, but it seems the most plausible for the present system at very low temperature (cf. ref. 43 of the paper). We also have looked very carefully for microwave stimulated emission in naphthalene in durene and found nothing. One of the causes, we think, is the low value of T_1 in the durene host and possibly aspecific populating.

Sharnoff to van der Waals: I should like to comment upon Professor Wolf's question as to whether $T_1 \approx 1/(\Delta\nu)^2$. In the ESR-optical double quantum work I have done on naphthalene-d_8 in biphenyl, I find that at 1·8 °K the relaxation time of the $\Delta m = \pm 2$ transition is not longer than 10^{-2} s. This result supports Dr van der Waal's diminution of the phosphorescence lifetime of quinoxyline in durene which accompanies the application of an external magnetic field.

Birks to van der Waals: Is your theory consistent with the Azumi-McGlynn paper which shows that the relative populations of the triplet states are temperature-dependent and that the 'effective' multiplicity g of the triplet state which enters the lifetime equation differs from three?

Van der Waals to Birks: Differences in Boltzmann factors obviously are present at low temperature, but as demonstrated by the present experiments their effect may be overshadowed by relaxation effects.

SCHWOERER

Sharnoff to Schwoerer: I should like to ask whether Professor Robinson has any comment to make on the difference between the value of $J = 5 \mp 0{\cdot}7\ \text{cm}^{-1}$ found from these experiments and the value of $10\ \text{cm}^{-1}$ which he obtained from his studies of the $S_0 \rightarrow T_1$ absorption of crystalline naphthalene at 4·2 °K.

Robinson to Schwoerer: There seems to be a disagreement between our measured triplet Davydov splitting in *crystalline* naphthalene ($10\ \text{cm}^{-1}$) and your exchange coupling constant $J = 5\ \text{cm}^{-1}$ for *pairs*, but perhaps it arises because of different definitions.

Vincent to Schwoerer: In preliminary measurements on the tetramethyl pyrazine–durene mixed single crystal we see the exchanged M lines arising between the individual A and B lines when the latter lines are very close. These results are obtained at 77 °K in crystals which have approximately 2 % tetramethyl pyrazine incorporated into the durene crystal.

Schwoerer to Vincent: The M lines in $N\text{-}d_8$: $N\text{-}h_8$ become always very strong, when the corresponding A and B lines are close, because the M line width is small in this case.

Siebrand to Schwoerer: Are the decay times the same for the signal due to isolated molecules and the pair signal (M signal)?

Schwoerer to Siebrand: Yes, exactly.

Siebrand to Schwoerer: This would imply that there is no essential difference between the unimolecular triplet decay rate constant for isolated molecules (in an appropriate matrix) and molecular aggregates, including single crystals. In other words, the decrease in signal lifetime observed at high concentrations appears to be due to impurity quenching.

RASSAT

Kommandeur to Rassat: You seem to have measured the first radical, which will keep its susceptibility down to 1·9 °K. Can you fit your curves to a triplet-singlet excitation or singlet-triplet excitation function such as $\chi = C\{T(3+\exp[\pi/kT])\}^{-1}$? Is there any evidence for a phase transition at 5 °K (the temperature of the maximum)?

Rassat to Kommandeur: The susceptibility curve of 2,2′,5,5′-tetramethyl N-oxypyrrolidine-4-azine fits very well to such a singlet-triplet excitation curve, with a separation of about 5 °K. At this time, we only have measured the heat capacity C for 2,2′,6,6′-tetramethyl-4-piperidinol N-oxy; the curve $C(T)$ is anomalous by having a marked shoulder at 3·5 °K.

JEN

Sharnoff to Jen: Dr Jen has graciously informed the audience of my results with the ESR-optical double-quantum technique.

A melt-grown crystal of biphenyl containing 0·1 mol % naphthalene-d_8 was cleaved and placed in an X-band microwave cavity, where it was immersed in liquid helium maintained at 1·8 °K. The crystal was irradiated continuously with the ultraviolet light produced by an Osram HBO 200 mercury burner and filtered with a 5 cm. path of aqueous solution of $CoSO_4 . 7H_2O$ (84 g/l.) and $NiSO_4 . 6H_2O$ (276 g/l.). The sample's phosphorescence was monitored by a detector consisting

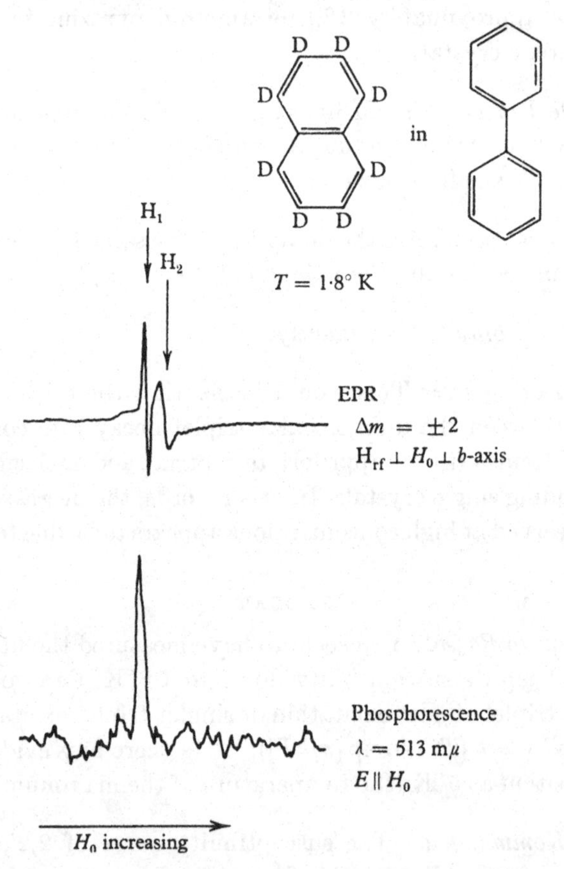

Fig. 1. ESR-optical double quantum results for naphthalene-d_8 and biphenyl, shown for a crystalline orientation in which all excited naphthalene molecules are magnetically equivalent. The phosphorescence was observed in the direction of the biphenyl b-axis. The narrow line at H_1 is a triplet state signal; the broader peak centered at H_2 is a background, $g = 4$, resonance whose intensity was independent of the ultraviolet illumination of the sample.

of a linear polariser, a low-resolution monochromator (64 Å/mm) and an RCA 7265 photomultiplier and was found to consist mainly, if not entirely, of the naphthalene-d_8 emission. Electron spin resonance of the triplet naphthalene-d_8 was observed both by conventional means and when the microwave power was square-wave modulated on and off at 40 Hz. The square wave from which this modulation was derived was also used to drive the reference channel of a phase-sensitive detector, whose input was fed by the phosphorescence signal from the photomultiplier. It was found that when an ESR transition was alternately saturated and allowed to relax, at the 40 Hz modulation frequency, the phosphorescence also became modulated.

The results of saturating the 'forbidden' $\Delta m = \pm 2$ transitions are shown in fig. 1, in which the phase-detected phosphorescence signal and the first derivative of the ESR are displayed simultaneously on a dual channel recorder. The number of triplet states observed in the ESR is approximately 3×10^{13}; about 2×10^{9} quanta per second ($\lambda = 513$ mμ) impinged upon the photocathode. The peak-to-peak modulation depth at 40 Hz of the phosphorescence was computed from the measured gain of the phase-sensitive detector and found to be 0·3 % of the total. Integrator time constants of 3 s were used in both channels.

The alteration of the intensity of the phosphorescence occurs as a result of the redistribution of the triplet sublevel populations which is produced by ESR saturation, which demonstrates that the radiative matrix elements connecting any triplet sublevel with the ground singlet electronic level are functions of the magnetic quantum numbers of that sublevel. For the phosphorescence, polarisation intensities of an ensemble of oriented triplets is a population-weighted average of the characteristic polarisation intensities of the individual sublevels and would not change with changing sublevel populations if the sublevels were radiatively equivalent.

Jen to Sharnoff: I wish to congratulate Dr Sharnoff on his success in observing a signal of phosphorescence intensity change at paramagnetic resonance in the triplet state. There are two factors in his experiment which, I think, are important: (1) the attainment of a temperature of 1·8 °K; (2) the use of naphthalene-d_8 in a biphenyl crystal. The effect of the first factor is obvious. The effect of the second factor may be due to the more 'accommodating' properties of biphenyl as a host than those of durene. I think these factors are largely contributory to his successful observation.

Sharnoff to Jen: I feel, on the basis of the preliminary investigation which I made for my experiment, that the effects will also be readily found for both light and heavy naphthalene in biphenyl and in durene.

Van der Waals to Sharnoff: How can you see the ESR absorption when saturating the transition?

Sharnoff to van der Waals: Total saturation cannot be produced, in principle, by finite microwave power. I estimate that the degree of saturation present in the data reported here is about 99 %.

Wolf to Sharnoff: Is there a difference in the observed effect for the different vibrionic bands of the phosphorescence spectrum?

Sharnoff to Wolff: I have observed the effect to be the same for both the 476 mμ band (0, 0) and the 513 mμ band. I have not investigated any of the other bands.

VINCENT

Van der Waals to Vincent: Can you understand the absence of resolved N-hyperfine splitting? Are your lines very weak and is the line perhaps temperature dependent? It is conceivable that the proximity of the π–π^* and n–π^* triplets introduces strong spin-orbit coupling shortening the lifetimes of the spin states.

Vincent to van der Waals: The lines are quite weak in comparison to the EPR of naphthalene or other π–π^* triplet states. I am not sure now why the lines show no hyperfine structure. A simulated spectrum with a normal component linewidth of 2–3 Gauss has structure with a rather wide range of proton and nitrogen coupling constants. The proximity of the n–π^* state to the π–π^* state might indeed lead to a relaxation process which broadens the line.

Schwoerer to Vincent: Is it possible that the line broadening is due to a T_1 process due to the rotation of the CH_3 groups?

Vincent to Schwoerer: Some line broadening may be due to a methyl group T_1 process, but I believe there may be other more important processes such as interaction with the slightly higher n–π^* state for broadening the line.

Siebrand to Vincent: I would like to make a general comment on the use of durene as a host crystal in triplet experiments. There is a good deal of evidence in the literature that durene, even when zone-refined, normally contains an appreciable concentration of a specific, unidentified impurity. This impurity is characterised by a triplet state some 3 000 cm^{-1} below the triplet state of durene and a triplet lifetime of the order of 0·1 s. In the past, the failure to recognise the presence of this impurity has led to erroneous conclusions on a number of occasions. I do not wish to suggest that this is also the case with the work reported here this morning, but I believe that the problem of the purity of the durene matrix deserves more attention than it appears to get at the moment.

Vincent to Siebrand: The host impurity problems in host-guest single crystal EPR measurements are not as critical as they are for emission or absorption measurements of pure crystals. The added guest is usually present at much higher concentrations than the host impurity and the EPR measurement primarily detects only the triplet state of the most concentrated impurity.

In an attempt to further purify durene, we have subjected extensively zone-refined durene to a sodium mirror in high vacuum for approximately one hour. This durene was then sublimed into the growing tube.

Robinson to Vincent: The purification procedure that we used for benzene and naphthalene using alkali metal purification probably will not work for alkylated hydrocarbons such as durene.

SHARNOFF

Hameka to Sharnoff: Your calculation is both similar and complementary to work which I published in the *J. Chem. Phys.* **37**, 328 (1962).

3 PHOTOCHEMISTRY

THE KINETICS OF ENERGY TRANSFER FROM THE TRIPLET STATE IN RIGID SOLUTIONS

R. E. KELLOGG

Abstract

Assuming the energy transfer is dependent on the inverse n-th power of the donor-acceptor distance ($1/R^n$), Inokuti and Hirayama derived an expression for the observed decay of donor emission, using flash excitation of a random solid solution:

$$I = I_0 \exp[-t/\tau - (2/\pi^{\frac{1}{2}})\Gamma(1-3/n)\gamma(t/\tau)^{3/n}],$$

where τ is the measured lifetime in the absence of an acceptor, γ is linear in acceptor concentration, and n is the distance dependence parameter. This reduces to the Förster result for a dipole-dipole interaction when $n = 6$. The extent to which we can determine n experimentally will allow us to measure the 'purity' of the dipole-dipole interaction presumed to induce the transfer.

To this effect, a detailed kinetic analysis of triplet-singlet energy transfer in rigid solutions has been carried out using triphenylene-d_{12} as the donor and Rhodamine B as the acceptor. A computer fit of the decay data to the above expression was carried out with a non-linear regression routine that determined I_0, γ, and n simultaneously using the least-squares criterion. Over the concentration range of $\gamma = 0{\cdot}26$ to $2{\cdot}1$, a value $n = 6{\cdot}05 \pm 0{\cdot}17$ was determined, indicating no measurable deviation from the behaviour predicted by Förster.

In principle, this procedure could be used to identify a quadrupole transition in the acceptor, where a value of $n = 8$ would signify a dipole-quadrupole interaction. A search for quadrupole transitions using this technique was fruitless.

TRIPLET STATES IN GAS-PHASE PHOTOCHEMISTRY

R. B. CUNDALL, A. S. DAVIES AND K. DUNNICLIFF

Abstract

Methods available for the identification and study of excited states of molecules in gas-phase photochemistry are briefly discussed. The importance of vibrational effects in determining the relative extents of internal conversion and intersystem crossing is indicated by studies which have been made using the *cis-trans* isomerisation of olefins technique. Triplet state yield data for toluene, monofluorobenzene, *o*-fluorotoluene, hexafluorobenzene and pyridine are given. Fluorination considerably reduces the non-radiative lifetime of the triplet state.

Introduction

Absorption of radiation generally produces excited molecules with the same multiplicity as the ground state, except when the absorption is very weak, as for a singlet-triplet transition, or when perturbing influences (spin-orbit coupling) operate. The fate of the excited state which results from the absorption of a photon may be

$$^1M \rightarrow {}^3M \text{(intersystem cross-over)} \tag{1}$$

$$^1M \rightarrow M + h\nu \text{ (fluorescence),} \tag{2}$$

$$^1M \rightarrow \text{ground state (internal conversion).} \tag{3}$$

Non-radiative transitions between states of different multiplicity are significant photochemical processes. Collision effects can influence all these possibilities markedly. (2) may be measured directly by observations of fluorescence decay, or measured relatively by studying fluorescence quenching. (1) has to be studied by one of the methods discussed below and knowledge of (3) has been dependent on data about (1) and (2). Simultaneous with these may be a process of chemical significance

$$^1M \rightarrow \text{dissociation or rearrangement.} \tag{4}$$

Vibrational energy is in very many cases a process-determining factor and wavelength dependence will be important as a consequence. The elucidation of many gas-phase photolyses mechanisms is complicated by the difficulty in experimentally separating the effects of vibrational deactivation and quenching of electronic excitation. Assignments of the role of specific excited states, and, in particular, of the triplet, are often uncertain.

Detection and estimation of excited states

The identification of excited states may be made in a number of ways, and instantaneous concentrations may be observed by a variety of physical techniques such as optical absorption, nuclear magnetic resonance, electron spin resonance, magnetic susceptibility, fluorescence and phosphorescence. They are all rarely applicable under normal photochemical conditions, particularly in the gas phase when concentrations are low. These techniques are usually used for demonstrating directly the existence of a particular state and obtaining data on the elementary processes in which such species participate.

Much useful information has been obtained from observations of emissions, both direct and sensitised. A generalised scheme for the technique is:

$$M + h\nu \rightarrow {}^1M, \tag{5}$$

$${}^1M + S \rightarrow {}^1S + M, \tag{6}$$

$${}^1M \rightarrow {}^3M, \tag{7}$$

$${}^1S \rightarrow {}^3S, \tag{8}$$

$${}^3M + S \rightarrow {}^3S + M, \tag{9}$$

$${}^1S \rightarrow S + h\nu'_s, \tag{10}$$

$${}^3S \rightarrow S + h\nu''_s. \tag{11}$$

Allowance must be made for the direct excitation.

$$S + h\nu \rightarrow {}^1S(\rightarrow {}^3S). \tag{12}$$

Ishikawa & Noyes[(1)] have used this technique to identify both singlet and triplet excited states and obtain estimates of rate constants in a number of systems. Biacetyl is unique in that it emits from both the singlet and triplet states, with the latter in high yield. The phosphorescence is not readily quenched except by paramagnetic species.[(2)]

The addition of agents which quench or perturb specific excited states is used, and kinetic analysis may allow the assignation of rate constants to specific processes. The use of oxygen as a diagnostic, if not unambiguous, test for triplet state is very often used. Heavy atom-containing molecules which perturb and modify the spin selection rule[3] by spin-orbit coupling effects have been used in solution photochemistry to enhance singlet-triplet crossover.[4] The triplet-triplet absorption increase which has been observed in flash photolysis arises from excited molecules which would otherwise have undergone either fluorescence or internal conversion from the singlet manifold. The technique is difficult to apply in the gas phase when either reaction or decomposition of sensitisers is more probable. A very high pressure of xenon might be used to convert singlet to triplet states in the gas phase. An extension of the emission technique has recently been made by Parker[5] in which measurements of delayed fluorescence are used to measure the efficiency of intersystem crossing. This may be applicable to gas-phase studies in some cases, particularly for aromatic hydrocarbons, but impurity effects are troublesome and any short-lived triplets would be difficult to detect. The high concentration of triplet states needed will be a limiting factor in many cases.

A purely chemical method for the detection and estimation of triplet states makes use of the *cis-trans* isomerisation of $\alpha\beta$-substituted olefins.[6] This technique arose out of studies on mercury photosensitisation[7] and involves transfer of electronic excitation energy from the triplet state of the sensitiser (the excited state under investigation) to the ground state of the olefin, which is then excited from the rigid planar ground state to the perpendicular excited form, capable of a ready interchange between the *cis*- and *trans*- configurations.[8] In the case of simple olefins, e.g. butene-2, there is no evidence of any barrier in the triplet state. Return to the ground state by a process which is very rapid because of overlap of the ground and triplet potential energy surfaces produces either the *cis*- or *trans*- isomer in fixed ratio (1·37:1 for butene-2). The advantages of the method are: (i) difficult emission measurements are not required, (ii) mono-olefins are not excited at wavelengths longer than 2100 Å, and in most conventional photochemical studies the sensitiser can be excited without effect upon the olefin; (iii) singlet-singlet transfer will not occur above 2100 Å. Disadvantages of the method are: (i) chemical interaction between olefin and sensitiser may occur, although there is no example in which the technique has been invalidated because of this; and (ii) the high energy of the olefin triplet. The olefin triplet energy is not well-established due to the exacting Franck–Condon conditions which apply to the electronic excitation of olefins by photon absorption. Evans,[9] by means of the high-pressure oxygen perturbation technique, has assigned a value of 82 kcal mole^{-1} to the 0–0 ground state-triplet separation. This is too high and photosensitisation studies would suggest that a value of 70–74 kcal mole^{-1} is more likely. Franck–Condon conditions are not restrictive during encounters at normal collision distances, particularly when excess of excitation energy is available for vibrational activation. A lower value for the energy of olefin triplets has been obtained by Kuppermann & Raff[10] who studied the excitation of ethylene in an electron beam of varying energy (*ca.* 70 kcal mole^{-1}) which agrees more closely with the chemical evidence.

It has been suggested that the isomerisation technique is not applicable to systems in which free radicals are involved.[11] However, this is not a serious limitation except in cases where a reversible addition to the double bond can occur as is the case when atoms of bromine and iodine interact with olefins.[8]

The *cis*- and *trans*- isomers of butene-2 are convenient for gas-phase work, both are readily available in a state of high purity and are easily analysed by gas chromatography.

Absolute values of triplet quantum yields can be obtained, but for the determination of rate constants the technique is comparative and other information, experimental or assumed, is needed before individual rate constants can be assigned.

Studies directed towards the characterisation of triplet states in the gas phase

1. *Benzene and simple benzene derivatives*

The biacetyl technique[1] and olefin isomerisation[12] methods have both been applied in detailed investigations of singlet and triplet states of benzene excited by absorption in the 2000–3000 Å region. Data obtained by the latter method show that the yield is 0·63[12, 13] when excitation is at 2537 Å. Noyes & Ishikawa[1] obtained a value of 0·78 by the biacetyl emission method but a detailed re-investigation of the fluorescence of benzene by Noyes, Mulac & Harter[14] brings the results of both methods into line and 0·63 (at 2537 Å) seems firmly established. Noyes & co-workers[14] find that both the singlet and triplet yields decrease with wavelength and some process other than fluorescence or intersystem crossing must compete for the excited singlet state ($^1B_{2u}$). For benzene-d_6 the triplet yield at 2537 Å is 0·55 and derived non-radiative triplet state lifetimes are 12·5 and 14·0 μs for C_6H_6 and C_6D_6 respectively.[13] These values are deduced by assuming that the rate constants for triplet state deactivation by *cis*-butene-2 is 2×10^9 l. mole^{-1} s^{-1}. This is based on values obtained for triplet quenching by oxygen obtained by Heicklen & Noyes[15] and Porter & West.[16] Butene-2 and oxygen are of identical efficiency in quenching triplet benzene[12] and the most consistent agreement with other published data is obtained if this value is assumed. After 2537 Å excitation about 19 % of the excited states undergo conversion to the gound state by some process(es) other than fluorescence or conversion to the triplet state.[14] The slight temperature decrease of both the fluorescence[17] and triplet yield[13] indicates that this process requires a small activation energy (E_3 (or possibly $E_3 - E_1$) $\approx$ 300–400 cal mole^{-1} at 2537 Å). This activation energy may tentatively be assigned to the need for population of an auxiliary vibrational mode which aids the $^1B_{2u} \rightarrow {}^1A_{1g}$ crossover. Pressure effects have been studied by Sigal,[18] and Kistiakowsky & Parmenter.[19]

This type of study has been extended to other aromatic molecules.

Experimental. The apparatus was essentially the same as that described elsewhere.[13] Quantum yields were measured relative to benzene (it was assumed that under all conditions for excitation by the 2537 Å mercury resonance line $\phi_{\text{triplet}} = 0{\cdot}63$) by comparison of rates of isomerisation when both benzene and sensitiser had the same optical density.

Results. In the course of the quantum yield estimations, it was found that, over a range of benzene pressures, the rate of isomerisation was directly proportional to the intensity of light absorbed. This confirms that, over the range of concentrations used, triplet yields are practically independent of the benzene concentration.

This does not exclude triplet quenching by ground state benzene but shows that this is unimportant under the experimental conditions.

All the experiments are at 25 °C. The *trans-cis* isomerisation of *cis*-butene-2 was studied in all experiments.

(i) *Toluene.* A mean quantum yield of $0{\cdot}58 \pm 0{\cdot}02$ was found for toluene concentrations between 1·75 and $3{\cdot}1 \times 10^{-4}$ moles l.$^{-1}$ and *cis*-butene-2 between 0·85 and $38{\cdot}5 \times 10^{-4}$ moles l.$^{-1}$. A fall off in the triplet yield was found below 0·3 torr of *cis*-butene-2.

(ii) *Monofluorobenzene.* The data are shown in fig. 1.

(iii) *o-Fluorotoluene.* As for monofluorobenzene, fluorination in-

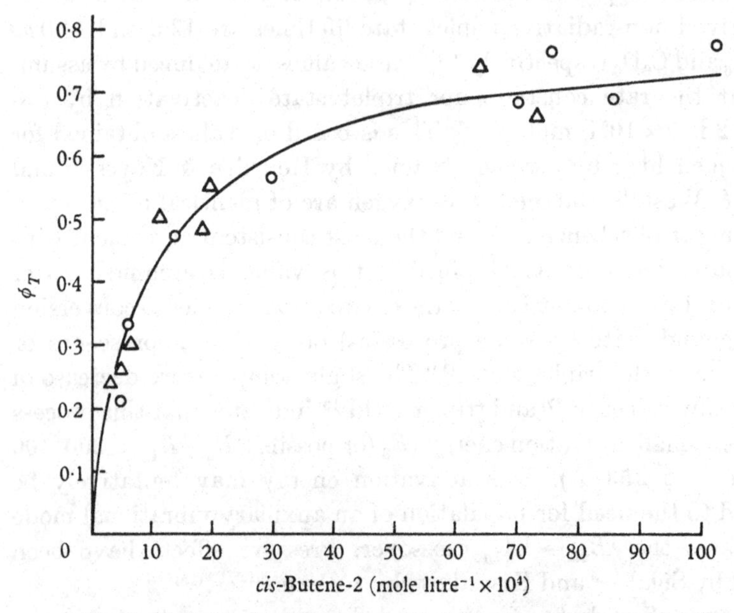

Fig. 1. Variation in observed triplet yield with olefin concentration. Monofluorobenzene: ○, 6×10^{-5} mole l.$^{-1}$; △, $1{\cdot}1 \times 10^{-4}$ mole l.$^{-1}$.

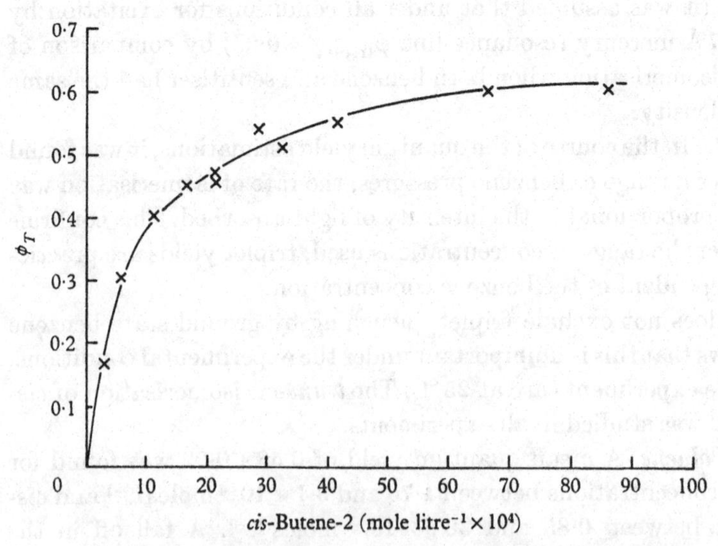

Fig. 2. Variation in observed triplet yield with olefin concentration. *o*-Fluorotoluene ($7{\cdot}5 \times 10^{-5}$ mole l.$^{-1}$).

creases the triplet yield above that for toluene and decreases the non-radiative triplet lifetime (fig. 2).

(iv) *Hexafluorobenzene.* A mean triplet yield of $0{\cdot}04 \pm 0{\cdot}02$ for $1{\cdot}75 \times 10^{-4}$ moles l.$^{-1}$ of sensitiser and $7{\cdot}8$–$37{\cdot}4 \times 10^{-4}$ moles l.$^{-1}$ of olefin. There was some indication of chemical reaction.

(v) *Pyridine.* $0{\cdot}03 \pm 0{\cdot}02$ was the triplet yield for pyridine ($2{\cdot}4 \times 10^{-4}$ moles l.$^{-1}$) for concentrations of *cis*-butene-2 between $10{\cdot}1$ and $22{\cdot}0 \times 10^{-4}$ moles l.$^{-1}$. This is a much lower value than that reported earlier.[12]

Discussion. A comparison of the triplet yields with the absorption spectra shows that the yield of triplet decreases with the extent of resolvable banded structure in the singlet-singlet absorption spectrum. Substitution affects the symmetry character of the molecule and can increase the possibility of vibrational overlap between excited states.

The triplet yield for fluorobenzene is less than that reported by Ungar[20] using the biacetyl method (*ca.* 0·9). A fluorine atom substituent reduces the non-radiative lifetime and may be expected to enhance intersystem crossing from the excited singlet state due to a slight heavy atom effect. In hexafluorobenzene it is possible that the low yield of triplet may be due to the very short lifetime of the latter due to the sixfold substitution. Very high olefin pressures would be needed to check this and the evidence available does not support this possibility.

Estimates of the lifetimes of the triplet states of monofluorobenzene and *o*-fluorotoluene have been made from the plots of

$$[O_c]/\phi_T \; vs. \; [O_c], \quad \text{where} \quad [O_c]$$

is the concentration of olefin shown in figs. 3 and 4. Kinetic analysis based on the scheme

$$B + h\nu \rightarrow B_s, \tag{13}$$

$$B_s \rightarrow B, \tag{14}$$

$$B_s \rightarrow B + h\nu_f, \tag{15}$$

$$B_s \rightarrow B_t, \tag{16}$$

$$B_t \rightarrow B, \tag{17}$$

$$B_t + O_c \rightarrow B + O_r, \tag{18}$$

$$O_r \rightarrow O_c, \tag{19}$$

$$\rightarrow O_t, \tag{20}$$

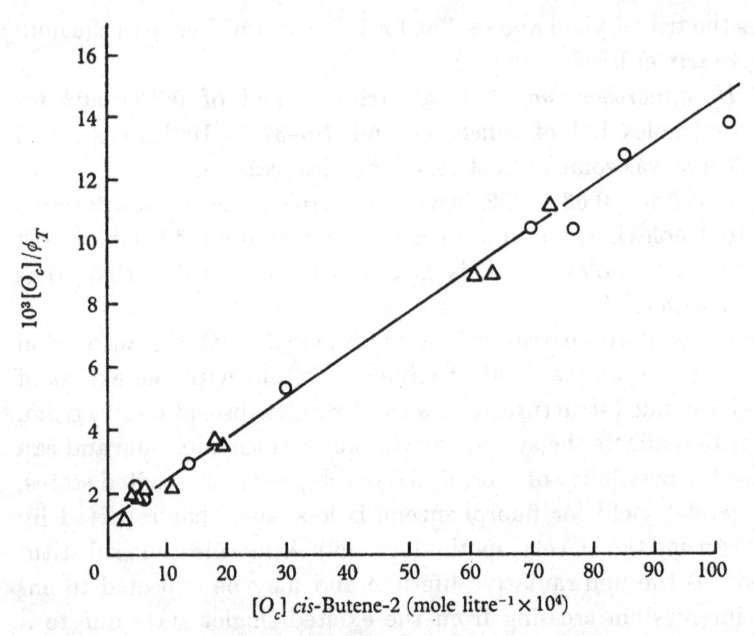

Fig. 3. $[O_c]/\phi_T$ against $[O_c]$. Monofluorobenzene: ○, 6×10^{-5} mole l.$^{-1}$; △, $1{\cdot}1\times 10^{-4}$ mole l.$^{-1}$.

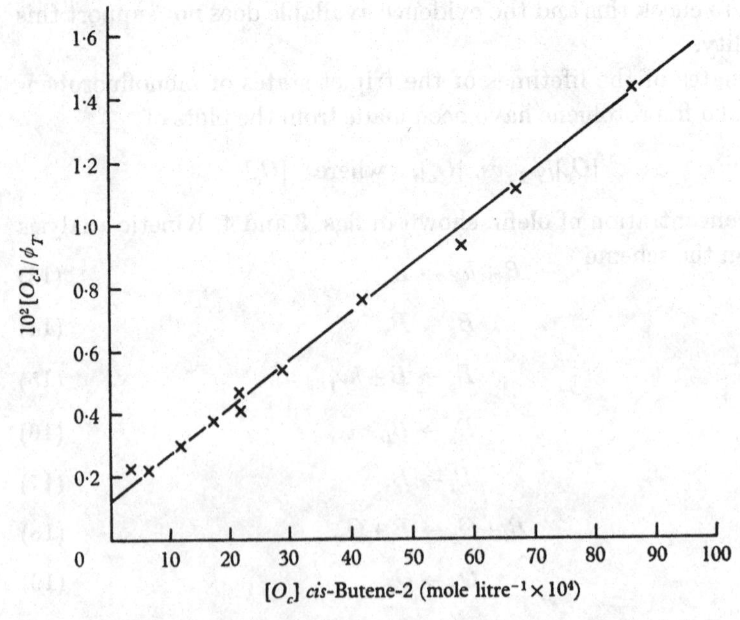

Fig. 4. $[O_c]/\phi_T$ against $[O_c]$. *o*-Fluorotoluene ($7{\cdot}5\times 10^{-5}$ mole l.$^{-1}$).

where O_t is the *trans* ground state, B the sensitiser and subscripts s or t indicate the singlet or triplet excited states, shows that the ratio of slope to intercept is k_{18}/k_{17}. If, as stated above, k_{18} is assumed, then k_{17} is the reciprocal of the non-radiative lifetime. 0·53 and 0·65 μs are thus obtained for monofluorobenzene and *o*-fluorotoluene. The lifetime of the toluene triplet is longer than that of benzene ($> 10^{-4}$ s).

No evidence has yet become available on the nature of the internal conversion processes. Isomerisation to a non-aromatic form is one possibility and experiments with suitably substituted derivatives should help clarify this problem. The evidence available indicates that it is facilitated by overlap of the vibrational levels. The process appears similar to a unimolecular reaction of the Rice Ramsperger Kassel type.

2. *Carbonyl compounds*

Remarkable similarity in photochemical behaviour is exhibited by compounds[21] of this class even when substitution and overall symmetry are very different. This has been explained in terms of 'total symmetry'.[22] Acetone[23] and acetaldehyde[24] photochemistry have both been examined by the butene-2 isomerisation method. Table 1 summarises the data for acetone photochemistry; the results for excited state decomposition being derived from kinetic analysis of the isomerisation data. In table 2 information for acetaldehyde and acetone are compared. The differences in behaviour in the primary stages are mainly due to the competitive processes in the scheme. A simple scheme which accounts for the behaviour is

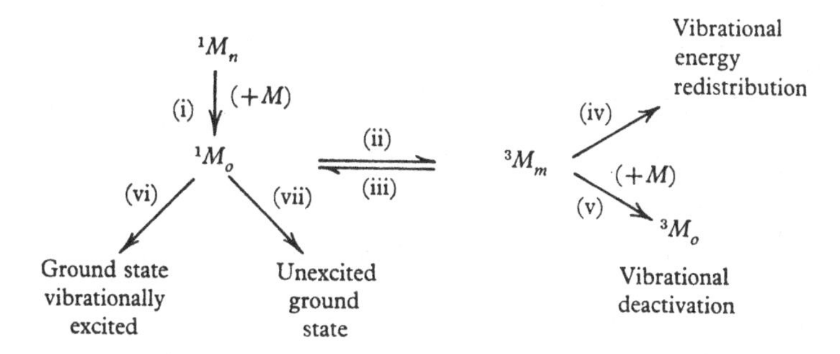

The production of triplet states is governed by the competition of (ii) with (vi) and (vii). The competition of the reverse (iii) with (iv) and (v) is not likely to be so important for polyatomic molecules. (vii) must be either electronic quenching or fluorescence. Enol formation

Table 1. *Effect of varying experimental parameters in the photolysis of acetone*[25]

	Incr. temp. (25–135 °C)	Incr. pressure of acetone	Exciting light: Incr. frequency	Exciting light: Incr. intensity
Steady state yield of				
Triplet	Reduced 1·0–0·27	Reduced slightly	Reduced	Reduced slightly
Singlet	Reduced	Reduced slightly	Reduced	Unaffected
Dissociation from				
Triplet	Increased 0·06–0·27	Reduced	Reduced	Reduced slightly
Singlet	Increased 0 → 0·73	Little effect	Increased	Unaffected
Fluorescence yield	Reduced	Reduced	Reduced	Unaffected
Phosphorescence yield	Reduced	Reduced	Reduced	Reduced slightly

Table 2. *Comparison of data for acetaldehyde and acetone at* 48 °C *with* 3130 Å *exciting light*[25]

Datum	Acetaldehyde	Acetone
(*a*) Triplet state yield	0·4	1·0
(*b*) Dissociation from triplet	0·2	0·15
apparent activation energy	*ca.* 5 kcal mole^{-1}	6·4 kcal mole^{-1}
rate constant at 48 °C	$6{\cdot}2\times10^5$ s^{-1}	$1{\cdot}25\times10^6$ s^{-1}
(*c*) Dissociation from singlet	Negligible at 48 °C	Negligible at 48 °C
(*d*) Non-radiative decay from triplet rate constant at 48 °K	0·1 + 0·2 4×10^5 s^{-1}	0·85 $1{\cdot}25\times10^5$ s^{-1}
(*e*) Internal conversion from singlet	*ca.* 0·6	0·0
(*f*) Fluorescence yield rate constant at 48 °C	5×10^{-3}	2×10^{-3} $3{\cdot}4\times10^5$ s^{-1}
(*g*) Phosphorescence yield rate constant at 48 °C	2×10^{-3} 8×10^2 s^{-1}	2×10^{-2} 5×10^3 s^{-1}
(*h*) Transfer of triplet energy to *cis*-butene-2		
Rate constants at 48 °C	$1-2\times10^8$ l mole^{-1} s^{-1}	2×10^7 to 2×10^9 l mole^{-1} s^{-1}

or other chemical reaction such as diradical formation for cyclic ketones may lead to disappearance of $^1M_0$.

Intersystem crossing can occur from any vibrational level between ^1M_n and $^1M_0$ and complete descriptions of photochemical behaviour must allow for this.

Detailed information[25] has been obtained on the behaviour of excited states as tables 1 and 2 show. For example, in table 2, item (*h*) provides a measure of the separation of the upper and lower vibrational levels of the donor triplet. A consequence of the difference in separation of energies between the two states and vibrational level structure is shown in fig. 5. Intersystem crossing is most favoured in acetone, an effect which may result from the closer spacing of vibrational levels partly due to the wider singlet-triplet energy gap. This is the opposite effect to that which occurs with the aromatic systems.

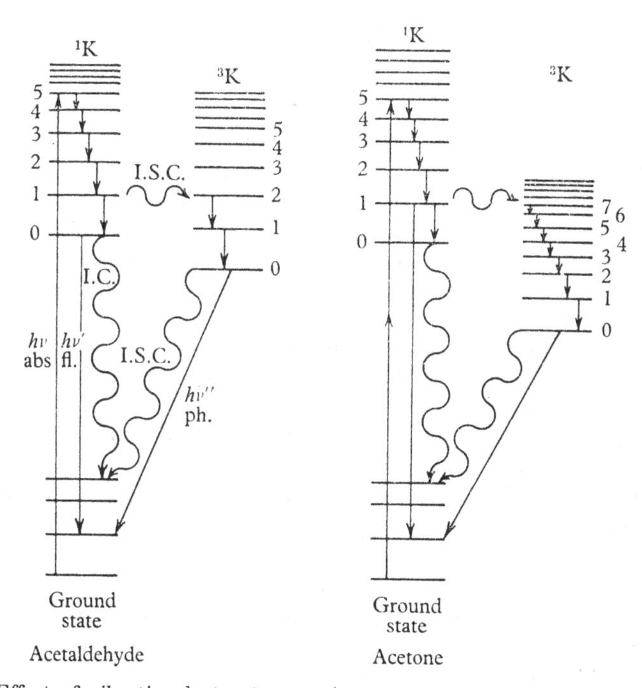

Fig. 5. Effect of vibrational structure on intersystem crossing $h\nu_{abs.}$ absorption of one quantum, $h\nu_{fl.}$ fluorescence, $h\nu_{ph.}$ phosphorescence. *Note.* I.S.C. shown from $^1K_1$ where matching with triplet levels is achieved; only requirement is that crossover occurs from low vibrational level.[25]

Conclusions

No firm conclusions are possible until much more experimental data are available. The rather limited data which are available would appear to emphasise the crucial role of vibrational excitation and structure in determining the fate of the singlet state. Internal conversion $S_1 \rightarrow S_0$ is not insignificant and the $T_1 \rightarrow S_0$ intersystem crossing is unexpectedly rapid in the gas phase.

The difference between the aromatic and carbonyl systems may be due to the fact that internal conversion of the aromatic excited singlet states proceeds through an isomeric form, whose production is favoured by excess vibrational energy, and facility of transfer between modes. A similar molecular modification is not available in the carbonyl compounds.

The authors are indebted to the Petroleum Research Fund of the American Chemical Society for support.

References

1 E. G. H. Ishikawa & W. A. Noyes Jr., *J. Chem. Phys.* **37**, 583 (1962).
2 G. B. Porter, *J. Chem. Phys.* **32**, 1587 (1960).
3 D. S. McClure, *J. Chem. Phys.* **17**, 905 (1949).
4 T. Medinger & F. Wilkinson, *Trans. Faraday Soc.* **61**, 620 (1965).
5 C. A. Parker & T. A. Joyce, *Trans. Faraday Soc.* **62**, 2785 (1966).
6 R. B. Cundall & D. G. Milne, *J. Am. Chem. Soc.* **83**, 3902 (1961).
7 R. B. Cundall & T. F. Palmer, *Trans. Faraday Soc.* **56**, 1211 (1960). R. J. Cventanovic, H. E. Gunning & E. W. R. Steacie, *J. Chem. Phys.* **31**, 573 (1959).
8 R. B. Cundall, 'Kinetics of *cis-trans* isomerisation'. *Progress in Reaction Kinetics*, ed. G. Porter (Pergamon, Oxford, 1963) vol. II, p. 165.
9 D. F. Evans, *J. Chem. Soc.* p. 1735 (1960).
10 A. Kuppermann & L. M. Raff, *Disc. Faraday Soc.* **35**, 30 (1963).
11 W. A. Noyes Jr. & I. Unger, *Advances in Photochemistry*, **4**.
12 R. B. Cundall, F. J. Fletcher & D. G. Milne, *J. Chem. Phys.* **39**, 3536 (1963); *Trans. Faraday Soc.* **60**, 1146 (1964); S. Sato, K. Kikuchi & M. Tanaka, *J. Chem. Phys.* **39**, 239 (1963).
13 R. B. Cundall & A. S. Davies, *Trans. Faraday Soc.* **62**, 1151 (1966).
14 W. A. Noyes Jr., W. A. Mulac & D. A. Harter, *J. Chem. Phys.* **44**, 2100 (1966).
15 J. Heicklen & W. A. Noyes Jr., *J. Am. Chem. Soc.* **81**, 3858 (1959).
16 G. Porter & P. West, *Proc. Roy. Soc.* (*London*) A, **279**, 302 (1964).
17 J. A. Poole, *J.Phys. Chem.* **69**, 1343 (1965).
18 P. Sigal, *J. Chem. Phys.* **42**, 1953 (1965).
19 G. B. Kistiakowsky & C. S. Parmenter, *J. Chem. Phys.* **42**, 2942 (1965).
20 I. Unger, *J. Phys. Chem.* **69**, 4284 (1965).
21 W. A. Noyes Jr., G. B. Porter & J. E. Jolley, *Chem. Rev.* **56**, 49 (1956).
22 J. R. Platt, *J. Chem. Phys.* **18**, 1168 (1950).
23 R. B. Cundall & A. S. Davies, *Proc. Roy. Soc.* (*London*) A, **290**, 563 (1966).
24 R. B. Cundall & A. S. Davies, *Trans. Faraday Soc.* **62**, 2793 (1966).
25 An extensive discussion is given by R. B. Cundall & A. S. Davies, *Progress in Reaction kinetics*, vol. IV (in the Press).

BIPHOTONIC PHOTOCHEMISTRY INVOLVING THE TRIPLET STATE: POLARISATION OF THE EFFECTIVE $T \to T$ TRANSITION AND SOLVENT EFFECTS

SEYMOUR SIEGEL AND HENRY S. JUDEIKIS

Abstract

One-photon photochemical processes involving the reactivity of photo-excited triplet state molecules toward abstraction or isomeration have been examined many times. However, the equally general biphotonic processes involving an absorption of a second photon by the excited triplet state molecule have only recently been receiving attention. Since the life-times of many photo-excited triplet state molecules dissolved in rigid matrices are quite long (i.e. 1–20 s), it is relatively easy to create systems with large concentrations of these excited molecules. (In our experiments, 70% depopulation of the ground state was routinely achieved.) The spin allowed $T \to T$ transitions have oscillator strengths which are as large, or even larger, than the corresponding $S \to S$ transitions. In several recent papers[1, 2, 3] it has been shown that the energy of a highly excited triplet state can be utilised either to ionise the solute molecule, to decompose the solute molecule, or to decompose a solvent molecule. This paper will be concerned with the sensitised solvent decomposition process.

For the case of naphthalene dissolved in ethanol, it has been shown that the $T \to T$ absorptions above 300 mμ are not effective in producing solvent radicals.[2] The effective absorption occurs at wavelengths ~ 260 mμ or less. The absorption gives the naphthalene molecule $\sim 7{\cdot}5$ eV of excitations for a time of $\sim 10^{-11}$–10^{-13} s. (Compare with the $\sim 8{\cdot}2$ eV ionisation potential of naphthalene in the vapor.) The details of the process by which this energy is transferred to the solvent and radicals are produced are open to considerable question. The observed result that the quantum yield for radical formation, based on the second photon, is approximately unity (for benzene) indicates that the transfer process must be fast in order to compete with radiationless cascade deactivation of the highly excited state. The two most obvious processes are (1) direct energy transfer to a repulsive dissociative state of the solvent, and (2) initial photo-ionisation of the solute, followed by an energy redistribution during the charge recombination step. It is difficult to distinguish kinetically between these two possibilities.

This paper will present the results of a study to obtain more information about the properties of the effective $T \to T$ transition(s) for naphthalene in ethanol and in 3-methyl-pentane glasses. In the study attention was focused on the effect of the solvent radical-aromatic solute interaction in the reduction of the steady state excited triplet state population. There is a direct proportionality between radical production and decay of the triplet state population. The technique of magneto-photoselection[4,5] was used to determine the polarisation of the $T \rightarrow T$ transition of interest. Analysis of the data for naphthalene-d_8 in ethanol at 77 °K indicates that the polarisation of the effective $T \rightarrow T$ transition lies predominantly (if not entirely) within the molecular plane. The effect of the wavelength of excitation on the biphotonic process was also studied. The latter data showed that the absorption corresponds to the $^3B_{2u} \rightarrow {}^3A_{1g}$ transition of naphthalene at 250–260 mμ, which is in agreement with the polarisation results.

Examination of the effect of solvent showed that while the *apparent* efficiency of the two-photon process was somewhat dependent on the nature of the solvent, the dependence upon wavelength was essentially independent of the solvent. Using a high-pressure mercury arc, it was found that the rate of radical production was a factor of two smaller in a 3-methyl-pentane glass than in an ethanol glass; however, the effective wavelengths for photolysis were $\sim$ 250–260 mμ in both glasses. The difference in efficiency is largely due to optical effects related to the optical properties of the two glasses; the quantum yield for radical production by the two-photon process is essentially constant in both systems. While the results do not lead to a completely unambiguous answer, the independence of the effective wavelengths and quantum yield on the identity of the solvent leads us to the conclusion that for naphthalene the direct energy transfer process (without the need of the intermediate solute ionisation step) is probably the most important process occurring.

References

1 B. Brocklehurst, W. A. Gibbons, F. T. Lang, G. Porter & M. I. Savadatti, *Trans. Faraday Soc.* **62**, 1793 (1966).
2 S. Siegel & K. Eisenthal, *J. Chem. Phys.* **42**, 2494 (1965).
3 Survey of Russian work by A. Terenin, *Recent Progress in Photobiology*, ed. Bowen (Blackwell, Oxford, 1964), p. 3.
4 M. A. El-Sayed & S. Siegel, *J. Chem. Phys.* **44**, 1416 (1966).
5 S. Siegel & H. S. Judeikis, *J. Phys. Chem.* **70**, 2205 (1966).

DIRECT AND SENSITISED PHOTO-OXIDATION OF AROMATIC HYDROCARBONS IN BORIC ACID GLASS

JACQUES JOUSSOT-DUBIEN
AND ROBERT LESCLAUX

Abstract

The biphotonic photo-oxidation of aromatic compounds in boric acid glass is now firmly established for low energy irradiation and monophotonic process at high energy. The determination of the energy at the limit of these two processes allows the evaluation of oxidation potentials for the compounds dissolved in the glass. From thermoluminescence data, we have obtained evidence for solvated electrons in the glass. The solvated electrons absorb in the near infrared. Sensitised photo-oxidation in the matrix can be carried out over distances longer than 50 Å. Several mechanisms are discussed.

We have been interested, during the past few years, in the photodetachment of electrons from aromatic hydrocarbons dissolved in boric acid glass, at room temperature and at liquid nitrogen temperature.

With low-energy ultraviolet irradiation, the photodetachment is a two-photon process, whereas with high-energy ultraviolet it is a one-photon process [1,2,3]. This has been verified by determining the rate of ion appearance as a function of the intensity of the incident radiations at different wavelengths. These experiments have enabled us to obtain oxidation potentials for aromatic compounds dissolved in boric acid glass. Oxidation potentials, defined as the threshold energy for which the relation between the rate of ion formation becomes linear with the light intensity, are given in table 1.

We have evidence for assuming that the second photon is absorbed by the triplet state of the molecule or a state X immediately derived from it.

The lifetime of the aromatic hydrocarbons in their triplet state is quite long in boric acid glass and we have found a relation between this lifetime and the rate of photo-ionisation.[1] This is not the only criterion for photodetachment since the compounds studied also have long triplet lifetime in matrices, such as sulfuric acid glasses at

77 °K, where apparently hardly any photo-oxidation occurs. However, the following experiment carried out with triphenylene which has a triplet lifetime of the order of 10 s and $T^* \leftarrow T$ absorption bands (27 800–28 800 cm^{-1}) distinct from $S^* \leftarrow S$ absorption bands seem to us quite conclusive.

Table 1. *Oxidation potentials of aromatic hydrocarbons*

I_{BA} determined in boric acid glass, Ig ionisation energy in gas phase.

compounds	I_{BA} (eV)	Ig (eV)	$I_{BA}-Ig$
Naphthalene	$>5{\cdot}4$	8·12[6]	$<2{\cdot}7$
Anthracene	5·05	7·38[6]	2·33
Tetracene	4·5	6·88[6]	2·38
Pyrene	5·35	7·55[7]	2·20
3,4-Benzopyrene	4·9	7·2[7]	2·30
Perylene	4·5	7·03[7]	2·53

The sample is irradiated with a high-pressure mercury arc through a filter made of triphenylene dissolved in alcohol. Under this condition, no photo-ionisation occurs. To excite the triphenylene molecules to their triplet state, the filter is mechanically removed a few tenths of a second every 8 s. If the quantities of photo-oxidised molecules obtained in this manner are compared with those obtained by irradiating the glass during the non-filtered intervals only, we obtain about twice as many ions when the molecules are irradiated in their triplet state than when the filter is replaced by an opaque screen.

Actually the model for photo-oxidation involving the triplet state as the intermediate step from which the second photon is absorbed, is losing ground. Johnson & Albrecht[4] at one time adopted it to explain the two-photon oxidation of tetramethyl paraphenylenediamine in 3 methyl-pentane, but abandoned it. One of the reasons was that the lifetime of the intermediate specie was longer than the lifetime of the triplet state. Studying the photo-conduction of triphenylamine in benzene and n-hexane, Pitts, Terry & Willets[5] found results supporting a biphotonic mechanism of carrier generation, but it was not obvious that the second photon was absorbed from the triplet state. Bagdasaryan *et al.*[6] and Kholmogorov *et al.*[7] however maintain that the production of free radical in rigid glasses involves a photon absorbed from the triplet state of the molecule.

Solvated electrons in boric acid glass

The overall photo-oxidation of aromatic hydrocarbons at room temperature in boric acid is a complex reaction. The photo-ejected electron reacts with boric acid hydrogen to give hydrogen atoms which diffuse through the matrix, recombine to give eventually H_2 molecules and aromatic free radicals.

To study the primary processes of photo-oxidation and recombination, we have carried out the irradiation at 77 °K and examined the light emission on warming the glass.

Each time we have been able to analyse the emitted radiation we have found that it corresponds to the $T \rightarrow S$ transition with no $S^* \rightarrow S$ component[8] in agreement with the findings of Linschitz *et al.*[9] The intensity of the light emission is much weaker than would be expected from the large depopulation of the ions on warming. The recombination is also a complex mechanism.

We have looked at the total light emission after irradiation at 77 °K and on warming. We have found that recombination occurs already at liquid nitrogen temperature. It is a very slow process that may last several hours. Upon warming we obtain the data plotted in fig. 1. If the sample has been only lightly irradiated, two humps appear very clearly. Irradiating the sample a short time favours the first peak, longer irradiation, the second peak. When the first peak appears alone, no photo-oxidation is detectable, except from the indirect evidence of light emission corresponding to recombination. When the second peak occurs, enough molecules have been oxidised to colour the glass. We believe that the first emission peak corresponds to the very efficient recombination reactions:

$$A^+ + e^-(\text{solvated}) \rightarrow {}^3A,$$

$${}^3A \rightarrow {}^1A, h\nu,$$

between the aromatic A^+ ions and solvated electrons formed at the beginning of the irradiation:

$$A \xrightarrow{h\nu} A^+e^-(\text{solvated}).$$

We have measured the height of the emission peak as a function of light intensity using neutral density filters and we have found a square law dependence. That which appears to us an elementary step is a biphotonic process.

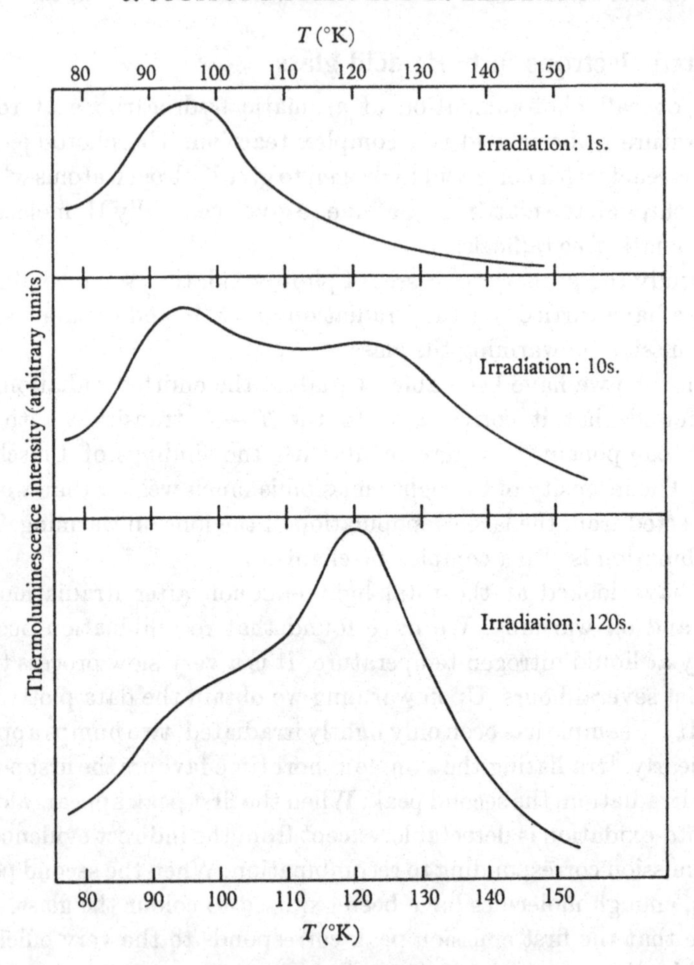

Fig. 1. Intensity of total light emission when a sample of boric acid glass containing an aromatic hydrocarbon irradiated by ultraviolet light at 77 °K is warmed. Heating rate 5° min.$^{-1}$.

We have found that near infrared light accelerates the recombination rate. This is not an effect due to heating, since the sample, a disk a few tenths of a millimeter thick, is immersed in liquid nitrogen. Using IR filters, we have shown that the wavelength of the bleaching light is between 7 000 and 8 500 Å.

This is the first time we have obtained evidence for solvated electrons in boric acid glass absorbing in the near infrared. Zaliouk-Gitter & Treinin[(10)] conclude, from their study of photo-oxidation of iodine in boric acid at room temperature, that the trapped electrons have their absorption centered at 2 500 Å. The compounds we have studied absorb so strongly in this region that we cannot check their conclusions.

From EPR data, the second thermoluminescence peak is assigned to H atoms untrapping, followed by complex recombination reactions.[8]

Under different experimental conditions, Walentynowicz[13] has obtained similar results using fluorescein.

Sensitised photo-oxidation

Perylene in boric acid glass is only very slowly photo-oxidized at room temperature or 77 °K. This is consistent with the fact that it is difficult to raise the molecule to its triplet state.

If a mixture of perylene and another aromatic hydrocarbon easily photo-oxidised such as phenanthrene is incorporated in the glass and irradiated then both the phenanthrene and perylene ions appear simultaneously even if the irradiation is carried through a filter containing perylene. Within experimental error, the spectra of the perylene ion obtained in the mixture and alone are identical (fig. 2). The phenanthrene can be replaced by triphenylene, naphthalene and biphenyl.

Mixed ions can be obtained, at about the same rate, at room temperature and 77 °K where we know that H atoms are trapped.

These experiments show clearly that on excitation of one type of molecule, all the molecules in the matrix initially excited or not participate in the ion formation.

Several tentative explanations, perhaps not mutually exclusive, can be put forward to interpret these results.

The perylene ion which appears during the photo-sensitised reaction could be the negative ion Pl^- formed according to the reactions:

$$S \xrightarrow{h\nu} S^+ + e^- \text{(solvated)},$$

$$e^-\text{(solvated)} + \text{perylene} \rightarrow Pl^-,$$

calling S the sensitiser and with perylene acting as traps. In this hypothesis, the spectra of the negative ion and positive ion should be identical. Furthermore, we have to assume that at 77 °K and room temperature photo-ejected electrons can travel distances longer than 50 Å without being trapped by hydrogen atoms. Evidence for such a mechanism should be given by recombination experiments, from

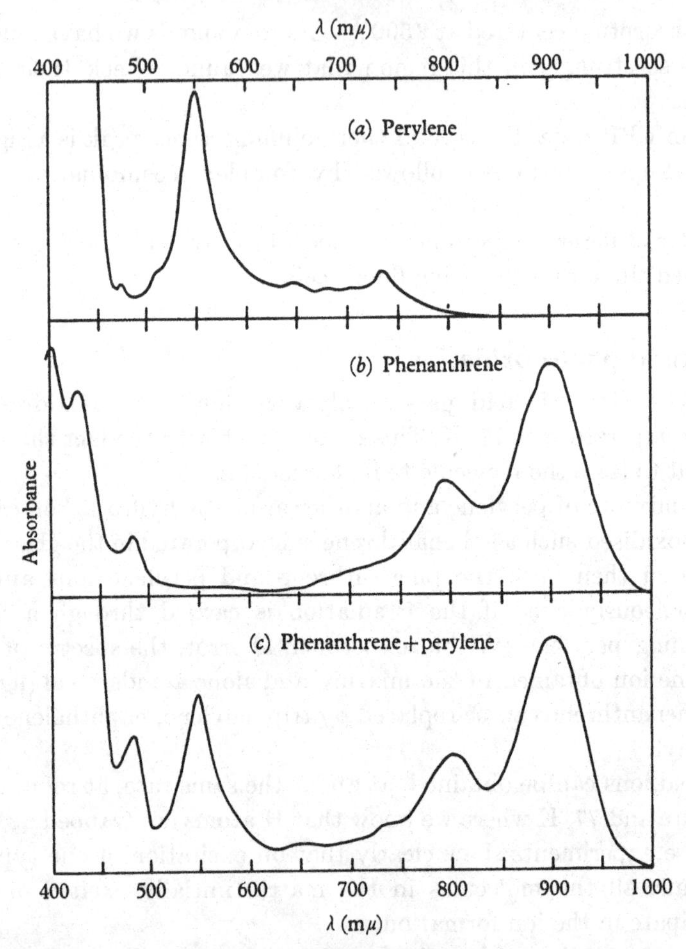

Fig. 2. Spectra: (*a*) of perylene ions obtained by irradiating perylene alone; (*b*) of phenanthrene ions alone; (*c*) obtained after irradiating a mixture of perylene and phenanthrene.

77 °K to room temperature, the solvated electrons decreasing the S^+ ion and increasing the Pl^- ion concentrations. That is what is observed at the beginning of the experiment in the temperature range corresponding to solvated electron mobility, but only to a very small extent. When the temperature reaches the range where H atoms become mobile, then a large proportion of S^+ and some perylene ions disappear simultaneously. It should be noted that irradiating the perylene ion does not lead to bleaching as would be expected from Pl^- ions (Van Voorst & Hoijtink[11]). Thus some Pl^- may be formed, but there is no clear evidence that it is the main species that is produced.

On the basis of the spectra and the data given above, however, it is quite reasonable to assume that the same ion Pl^+ is obtained by direct and by sensitized photo-oxidation. But, in this last case, either energy has to be transferred from the sensitiser to the acceptor or the photo-ejected electrons play a role in the sensitising process.

If an energy transfer is considered it can only be of the triplet-triplet type, since singlet excitation of perylene hardly leads to any ionisation:

$$^1S \xrightarrow{h\nu_1} {}^1S^* \rightarrow {}^3S,$$

$$^3S \rightarrow (X) \xrightarrow{h\nu_2} S^+ + e^-,$$

$$^3S + {}^1Pl \longrightarrow {}^1S + {}^3Pl,$$

$$^3Pl \rightarrow (Y) \xrightarrow{h\nu_3} Pl^+ + e^-,$$

where X and Y might be intermediate states in the second photon absorption step. This mechanism would again be consistent with the fact that triplet state is intimately involved in the photo-oxidation process. We have checked that all the substances that can sensitise perylene photo-oxidation have their lower triplet level above that of perylene. Furthermore, other donor-acceptor pairs, such as triphenylene naphthalene, verify this relation. However, perylene seems to be the best acceptor.

One objection to the T–T transfer is the distance over which it apparently proceeds. Sensitisation can be observed when the concentration of the sensitiser and sensitised molecules is 10^{-2}. Even at 10^{-3} the phenomenon can still be clearly observed. If one assumes a uniform distribution of molecules, this means triplet-triplet transfer over a distance of 25–50 Å. The glass becoming rigid above 100° C, no aggregation of solute molecules has ever been detected.

Terenin & Ermolaev[12] have reported triplet-triplet transfer from acetophenone to naphthalene at a distance of 15 Å in rigid glass. A similar experiment cannot be carried out on boric acid since in this medium the $n\pi$ absorption of acetophenone is not observed. The $T \rightarrow S$ emission of naphthalene and acetophenone are almost identical and have similar lifetimes, both being probably $\pi\pi^*$ transitions.

It would be unrealistic to suppose that electrons produced by near ultraviolet irradiations possess enough energy to detach another electron from a perylene molecule as in the case of electrons produced by ionising radiations. However, the presence of solvated electrons, H atoms and aromatic free radicals in the glass matrix might modify the spin-orbital coupling of the host molecules and enhance either direct $T \leftarrow S$ transitions or intersystem crossing in the case of perylene. This would increase indirectly the triplet state population from which, or from the X state that derives from it, a second photon may bring about the electron detachment.

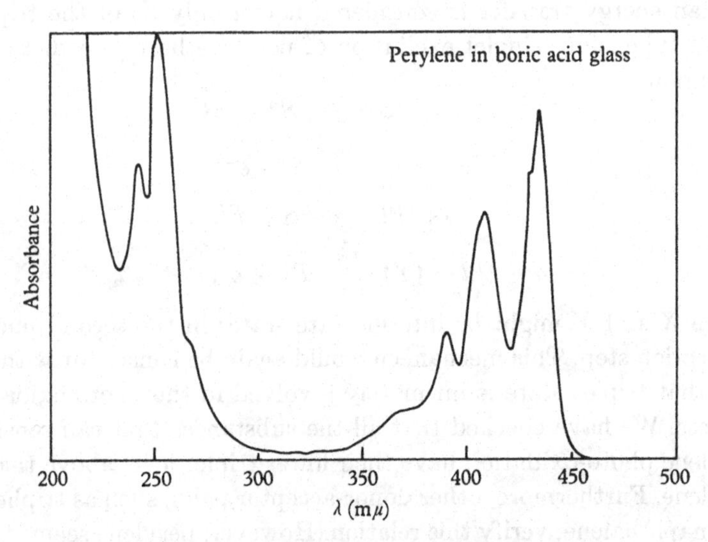

Fig. 3. Spectrum of perylene dissolved in boric acid.

One last point should be made. Among the acceptors we have used, perylene is unique, in that it has a large window in its spectrum in the near ultraviolet region as can be seen in fig. 3. This allows very efficient irradiation of sensitiser molecules and eventually perylene triplet or X state.

References

1 J. Joussot-Dubien & R. Lesclaux, *C.R. Acad. Sci. (Paris)*, **258**, 4260 (1964).
2 R. Lesclaux & J. Joussot-Dubien, *C.R. Acad. Sci. (Paris)*, **263**, C 1177 (1966).
3 J. P. Ray & T. D. S. Hamilton, *Nature*, **206**, 1040 (1965).
4 G. E. Johnson & A. C. Albrecht, *J. Chem. Phys.* **44**, 3162 (1966).
5 E. Pitts, G. C. Terry & F. W. Willets, *Trans. Faraday Soc.* **62**, 2851 (1966).
6 Kh. Bagdasaryan, Z. A. Sinitsyna & V. Muromtsev, *Dokl. Acad. Nauk SSSR*, **153**, 374 (1963).
7 V. Kholmogorov, V. Baranov & A. Terenin, *Dokl. Akad. Nauk SSSR*, **152**, 1399 (1963).

8 J. Joussot-Dubien & R. Lesclaux *Proceedings of the International Conference on Luminescence* (Budapest, 1966). To be published.
9 H. Linschitz, M. G. Berry & D. Schweitzer, *J. Am. Chem. Soc.* **76**, 5833 (1954).
10 A. Zaliouk-Gitter & A. Treinin, *J. Chem. Phys.* **42**, 2019 (1965).
11 J. D. W. Van Voorst & G. J. Hoijtink, *J. Chem. Phys.* **42**, 3995 (1965).
12 A. Terenin & V. Ermolaev, *Disc. Faraday Soc.* **20**, 471 (1956).
13 E. Walentynowicz, *Acta Physica Polonica,* **29**, 713 (1966).

DISCUSSION

KELLOGG

Wolf to Kellogg: Are your results conclusive in such a way that you are able to exclude other energy transfer mechanism than the Förster type transfer?

Kellogg to Wolf: In the acceptor concentration range of 10^{-3}–10^{-2} molar, the answer is yes. For completeness, this procedure should be repeated using a singlet donor, but that is a difficult experiment.

CUNDALL

Stevens to Cundall: Have you any estimate of the triplet state lifetime of the olefin? And if so is this sufficiently short to rule out the possibility of back transfer to regenerate the donor triplet state which would increase its effective lifetime?

Cundall to Stevens: The experimental evidence appears to rule out the possibility of back transfer in all cases where an olefine has been used as a triplet energy transfer agent.

The triplet state of a simple olefin must be very short-lived ($\leqslant 10^{-9}$ s). Attempts to find evidence for excitation of triplet states of lower energy than the olefin have all failed. It is certain that the ground state P.E. surface of the olefin triplet in the perpendicular form and the P.E. surface of the ground state when vibrationally activated are very close and probably overlap. Intersystem crossing will thus be rapid.

Parker to Cundall: I appreciate that Dr Cundall's determinations of triplet formation efficiencies are not likely to be affected by the presence of impurities, because the effects of the latter are swamped by the addition of a high concentration of acceptor. But is this true of the lifetime determinations?

Cundall to Parker: As Dr Parker remarks, on the basis of the scheme assumed, the triplet formation efficiencies should be independent (within the limits of error) of trace impurities. It is perhaps worth stating at this stage that vibrational de-activation of excited states might affect values for the yields. This should always be checked when the technique is used.

Impurities could affect the lifetimes measured, and the error would be greater the longer the lifetime of the triplet state, since the amount of impurity in the sensitiser relative to olefin acceptor would presumably be increased (assuming pure olefin). Data for benzene, for example, may be examined. The fall-off of isomerisation yield to about half the limit occurs at about 5×10^{-5} moles l.$^{-1}$ of butene-2. The same amount of impurity would be needed to explain the deviation from the 'theoretically acceptable' value. I cannot believe that our benzene contained nearly 10 % of impurity. The situation would be even worse for the fluoro-derivatives reported. The lifetime plots depend largely on values obtained when the olefin concentration is still high.

Robinson to Cundall: I believe there must be some unknown subtle effect that may cause your experimental triplet lifetimes to be much shorter than the real intrinsic lifetimes, since I can think of no good theoretical reason why the lifetimes should differ from those in the solid state by a factor of 10^7. At the moment, however, I can think of no direct criticism of your data nor your interpretation.

Windsor to Cundall: I have a comment which bears on Robinson's point about light-generated impurities quenching the triplet state in solution. The triplet-yield of naphthalene (Lamola & Hammond, *J. Chem. Phys.* **43**, 2129 (1965)) is $0{\cdot}40 \pm 0{\cdot}01$; the fluorescence yield is $0{\cdot}19 \pm 0{\cdot}01$ (M. W. Windsor & W. R. Dawson, *Mol. Crystals* (April 1967). The sum is appreciably less than unity. Non-radiative $S_1 \rightarrow S_0$ decay is therefore important. However, neither ϕT nor ϕF change significantly when perdeuterated naphthalene is used. This lack of isotope effect rules out an intramolecular $S_1 \rightarrow S_0$ radiationless transition and suggests rather that S_1 is quenched by a light-generated impurity which might be some relatively long-lived chemical modification of the parent compound. This explanation has been suggested by Robinson and Hammond.

Cundall to Windsor: Light-generated products are very probably formed. Isomerisation of aromatic sensitisers occurs under illumination, and intermediates may be formed although not identified (e.g the modifications of the benzene structure fulvene, benzvalene etc.). The external conversion $S_1 - S_0$ may proceed this way. Whether they are formed in sufficient quantity to be effective in the manner suggested by Dr Windsor is uncertain. We have no evidence on this so far but for reasons given in answer to Dr Parker, it is difficult to believe that they are formed in sufficient amount to be effective.

Birks to Cundall: Lipsky's observation that $\phi_F = 0$, and $\phi_T = 0$, for benzene vapour excited into a higher singlet state indicates that benzene behaves anomalously. Moreover, the total photochemical quantum yield, while finite, is much less than unity. Similar but less pronounced effects occur in alkyl benzene vapours.

Alkyl benzene liquids and strong solutions show a similar reduction in ϕ_F for excitation into higher singlet states. Birks & Conte (unpublished) have observed similar behaviour in alkyl naphthalenes and other hydrocarbons in the liquid phase and in concentrated solution. The effect is attributed to surface exciton quenching, and it is consistent with the theory of Birks (*Scintillation Counters*, Pergamon Press, 1953) of the response of organic scintillators to short-range ionising particles. The behaviour of benzene in the liquid phase is again anomalous, since the reduction of ϕ_F (the Lipsky β-factor) is *increased* when benzene is diluted, the reverse of the behaviour of all the other compounds studied.

Rice to Cundall: I wish to second Professor Robinson's disbelief in microsecond lifetimes for triplets of benzene, etc., in the gas phase.An examination of the high resolution spectrum of gaseous benzene shows the following interesting feature (pointed out to me by Mr A. V. Hazi). There is at least one sharp line, corresponding to a Rydberg transition, which overlaps a dense vibrational 'continuum' of other states. It is interesting that the line profile in this case looks as if antiresonance occurs, i.e. there appears to be configuration mixing of the Rydberg state and the vibrational states. If this deduction is correct, and if it also occurs for energies corresponding to the benzene triplet, it implies that there are enough vibrations in benzene to provide a quasi-continuum. A Robinson–Frosch type analysis is then possible. The non-stationary triplet 'passes' to a vibrationally hot ground singlet (actually the configurations are mixed). The prediction is that the triplet lifetime in the gas phase should be the same as the triplet lifetime in a glass, since intramolecular vibrations provide enough density of states that the presence or absence of an external medium is not rate limiting. Thus, if this interpretation is correct, the microsecond lifetimes of triplets in the gas phase, as currently reported, must be attributed to quenching of some sort.

Cundall to Rice: It is difficult to see what the quenching impurity may be. The lifetime is much shorter in the fluoro-derivatives and I do not think they are systematically of increasing impurity. The pressures of sensitiser are very much less in this case. The fact that three different experimental methods, that of Parmenter, Noyes and Ishikawa, and ourselves are fairly consistent is noteworthy. Perhaps, there may be some quenching process, but what is it?

Parmenter to Cundall: As Professor Birks has pointed out, the short liquid lifetimes may well be explained by triplet-triplet interaction. I feel quite sure, however, that the gas lifetimes are not established by such excited state interactions. The kinetics describing triplet gas decay in benzene, for example, are observed in our laboratory to be independent of exciting intensities. This situation could not be obtained if triplet interactions play a significant role in relaxation of the triplet state. We observe a gas lifetime that is very close to that quoted by Dr Cundall, about 25 μs when benzene pressures are 20 torr, and this is derived by a different type of experiment.

One may calculate that if such a lifetime is determined by impurities, their partial pressure must be at least about 0·01 torr. There are now three independent sets of data (those of Cundall, those of Ishikawa and Noyes, and those from our laboratories) which give completely consistent kinetics with respect to the triplet lifetime of benzene and energy transfer to acceptors. If impurities are significant in determining the triplet lifetime, the concentration of such must be the same in each of the different benzene sources used by these workers.

Van der Waals to Cundall and Parmenter: From the experiments of Kastler *et al.* we know that mercury atoms may change their magnetic state by collisions against the wall. Is it conceivable that the triplet state of benzene in the gas phase experiments is de-activated in a similar way?

Cundall to van der Waals: The pressure in our system is comparatively high (several torr). I do not think collisions with the wall affect our results. The phenomenon would be a very serious problem at lower pressures.

SIEGEL

Birks to Siegel: Can your data on naphthalene be analysed to give an independent estimate of T–T extinction coefficients? McClure and more recently Ramsay & Munro (Budapest Luminescence Conference, 1966) obtain values which are a factor of 20 less than those of Porter and Windsor and of Keller and Hadley. It is very desirable to have further improved data to resolve the discrepancy.

Silver to Siegel: The slow variation of the yield *vs* wavelength seems confusing in terms of a single final state. Can you explain this result?

Siegel to Silver: The apparent slow variation is due to two factors. The first factor is that some decay of the filters occur during the photolysis. The second factor is that the wavelength values given were 50% transmission values and the data must be analysed in terms of the complete transmission curves of the individual filters. Once these two factors are taken into account, the slow variation of yield with wavelength does not exist, and a peak at 250–260 mμ becomes quite apparent.

Stevens to Siegel: Is there any evidence for the recombination of solvent radicals leading to the population of an excited singlet state of the solute? This type of process could account for a biphotonic delayed fluorescence of aromatic solutes in the more dilute rigid solutions.

Siegel to Stevens: The answer is no. I do not see any evidence for this type of process. However, there has been some evidence published for a biphotonic thermal luminescence which has been attributed to change recombination by Porter and co-workers.

JOUSSOT-DUBIEN

Parker to Joussot-Dubien: The triplet formation efficiency of perylene is very small (about 0·01–0·02 in the fluid solutions) and it is to be expected therefore that photo-ionisation by the biphotonic process $S_0 \xrightarrow{h\nu} S_1 \rightarrow T_1 \xrightarrow{h\nu} P^+ + e$ will be inefficient. Furthermore, the triplet energy of perylene is very low, so that its triplet state could be populated by triplet-to-singlet energy transfer from most other compounds used as sensitisers. There is therefore at least some circumstantial evidence that the sensitised photo-ionisation proceeds via triplet-to-singlet energy transfer. Is it possible that the molecules of the solutes are concentrated in pockets in the basic acid matrix so that the local concentration is higher than 10^{-3}M?

Joussot-Dubien to Parker: It could indeed be important to show with certainty that there is no aggregation or higher concentration region. So far the experimental evidence we have makes us feel fairly confident that there is no aggregation.

Siegel to Joussot-Dubien: Would not a rotating sector method which would determine the lifetime of any intermediate, pinpoint the presence or absence of the perylene triplet state?

Joussot-Dubien to Siegel: That is one of the next experiments we intend to do, but I am not sure it will help to detect whether the triplet state is directly involved unless we get exactly the same lifetime.

Kommandeur to Joussot-Dubien: Could a bi-positive ion of perylene be involved in the two-photon process? The bi-positive ion appears to have the same spectrum as the mono-positive ion. Also, the second ionisation potential is lowered considerably, since the polarisation energy goes with the square of the charge in first order.

Joussot-Dubien to Kommandeur: This could be but we certainly have no evidence for this. Perhaps this could be obtained using ionising radiation.

4 RADIATIONLESS TRANSITIONS

RADIATIONLESS TRANSITION IN GASEOUS BENZENE

G. WILSE ROBINSON

Abstract

Earlier ideas of Frosch and Robinson about electronic relaxation in the solid phase are discussed in terms of gaseous molecules. Two limits are examined—the 'small molecule limit' where electronic relaxation cannot occur in the free molecule, and the 'big molecule limit', where electronic relaxation at a rate virtually identical with that in the solid can occur in the absence of any external perturbation. An 'intermediate case' is described where no relaxation can occur in the completely free molecule, but where only extremely minute perturbations are required to induce such a process. Intersystem crossing $^1B_{2u} \rightarrow {}^3B_{1u}$ in gaseous benzene may be an example of such an intermediate case. Theory therefore suggests that the absolute fluorescence yield may go to unity in the limit of zero pressure, while the implication from the experiments is that the yield becomes constant at some value lower than unity in the range $P < 0{\cdot}2$ torr. Other arguments that support the theoretical result, rather than the Kistiakowsky & Parmenter experimental result, are presented in the paper.

Introduction

Earlier papers by R. P. Frosch and the author[1, 2, 3] on subject matter entitled 'radiationless transitions' have discussed the mechan isms of what are commonly called, in the parlance of organic molecule spectroscopists, *intersystem crossing* and *internal conversion*. In each case the process is one where electronic energy, initially excited in a molecule in some condensed phase, is converted into other forms of electronic energy, molecular vibrations, phonons and eventually heat. Thus the process is of general occurrence and interest, going far beyond the rather limited applications to organic molecule spectroscopy.

The name *electronic relaxation*, in analogy with spin, rotational and vibrational relaxation, seems an appropriate generic term for these kinds of events. The close theoretical connection between electronic relaxation and electronic excitation transfer processes was emphasised in the last of the three papers mentioned above.

Electronic relaxation, while outwardly similar to other forms of energy dissipation, is physically distinct from them in many ways. One need only be familiar with the existing experimental observations regarding electronic relaxation of organic molecules in the solid or glassy state to recognise this fact. The most striking thing, for example, is that at low temperatures there is no temperature effect on the relaxation time; only at temperatures comparable to some characteristic temperature of the molecule or the solid does a mild temperature effect seem to set in.[4, 5] Furthermore, the rate of relaxation does not appear to go to zero at zero absolute temperature. These high energy relaxation processes apparently take place at absolute zero with a rate only slightly slower than that at room temperature. A factor of but two change in rate from 1·8 to 300 °K is typical.[1, 4] The basic reasons for this are that kT over the practical range of temperature is extremely small compared with the electronic energy difference between the two states undergoing the radiationless transition, and that molecular electronic and vibronic energies are only mildly affected by intermolecular interactions in the limit of weak van der Waals forces. In these respects electronic relaxation is much like nuclear decay processes. Thus perturbations by the surrounding heat bath are not effective in bringing about the transition, and any theoretical approach[6] through the correlation time of molecular motions of the surroundings is an inadequate way of solving these problems, just as it would be for nuclear decay. As stated in the earlier papers, the principal perturbation responsible for a high-frequency electronic radiationless transition resides in the free molecule itself and does not come from the solvent. The perturbation is at most only very slightly modulated by lattice motions when the molecule is immersed in its environment—and then at a frequency much too low to have much bearing on the high-frequency processes under discussion. The origin of the small temperature effect will be discussed in a forthcoming paper.[5]

Certain vibrational relaxation problems,[7] such as H_2 or N_2 in crystalline environments at low temperatures, also fit into the category of high-energy relaxation. Here, however, anharmonic coupling between the molecule and the lattice evidently takes the place of the intrinsic molecular perturbation in the electronic radiationless transition problem. But again, the anharmonic coupling is present in the crystal at 0 °K and is expected to be only slightly modified with increasing temperature up to, and perhaps even beyond, the melting-point of the crystal.

The above discussion does not mean to imply that the environment plays no part in electronic relaxation processes. Typically it acts as a heat bath that eventually soaks up the electronic energy originally localised in the molecule. Because the mixing of states is generally not affected by motions of the heat bath, the heat bath is characterless and little environmental effect is expected or observed in typical electronic relaxation events. One purpose of the present paper is to emphasise that even a gaseous environment at room temperature may behave similarly to a solid at 0 °K in furnishing this heat bath.

As emphasised earlier,[2] a solid environment merely acts, by reason of vibrational coupling to the relaxing molecule, to broaden levels that in the gas phase would otherwise be relatively sharp. It was assumed that this coupling, which in addition leads to vibrational relaxation with respect to the molecular levels, is the major cause of lifetime broadening in the solid or glassy states.† More specifically, any non-zero vibrational quantum state of the molecule was taken to have a width of order $\hbar/\Delta\tau_{\text{vib.}}$ determined by the vibrational relaxation time $\Delta\tau_{\text{vib.}}$ of the state. These broadened vibrational levels, in the case where the linewidths are large compared with their spacing, form a *quasi* continuum to which sharp molecular levels can couple. In the earlier papers we assumed that the *quasi* continuum is an effective one, that is, that the coupling between the sharp levels and the *quasi* continuum is sufficiently large that a large number of states in the *quasi* continuum are mixed with the sharp level. The picture of electronic relaxation that then emerges is one where discrete states, embedded in this *quasi* continuum, can undergo radiationless transitions into it. Irreversibility, a mathematical necessity for the applicability of the perturbation approach[2] is physically built into the process because of the preponderance of effective levels in the *quasi* continuum and the supposed high rate of vibrational relaxation compared with the rate of electronic relaxation. The process then is not unlike that of predissociation or the Auger effect, and identical theoretical machinery, modified by the vibrational intricacies and Franck–Condon factors[8] of the relaxing

† Spectral line broadening other than lifetime broadening occurs. For example, *inhomogeneous environmental broadening*, caused by the fact that energy levels are sensitive to environment and the environment of molecules is not often uniform in a condensed phase, is probably the greatest source of line broadening. Only when inhomogeneous broadening is absent are the observed spectral linewidths a measure of the lifetime of the state. Inhomogeneous broadening, just like Doppler broadening, is not related to any relaxation process.

molecule, may be used. In the present paper, we shall discuss not only this *quasi* continuum model but also cases where the *quasi* continuum becomes discrete.

In the case where electronic relaxation is fast compared with vibrational relaxation, the theoretical approach that emphasises the individualism of electronic states must be abandoned in favor of one that puts emphasis on the vibronic states of the system. This situation is expected to be met typically high up on the energy level diagram of most molecules where the electronic states are dense and electronic interactions are large. It is the lower, more coarsely spaced, electronic states that were emphasised in the earlier papers and the ones that will be emphasised in this one.

The big molecule and small molecule limits

One elaboration of the earlier ideas that we would like to make is to put more emphasis on the fact that a sufficiently complex molecule can act as its own environment. In other words, if the molecule has a large number of vibrational degrees of freedom whose normal frequencies are spread over a sufficiently large range, an environment may not be necessary to furnish the *quasi* continuum. After all, a crystal is nothing more than a giant molecule. The idea of treating the crystal as a 'super molecule' for the discussion electronic relaxation was advanced by the author.[1] Conversely, the idea of treating electronic relaxation in a complex molecule like that in a crystal has also been advanced before, particularly by Ross[9] and by Strickler.[10]

A complex molecule in the gaseous state in the absence of collisions may therefore provide the same ingredients for electronic relaxation as the molecule in a condensed state. We call this limit the 'big molecule limit'. In such a limit the rates of electronic relaxation in a collisionless environment and in the solid state are expected to be similar because the dominant factor is the intrinsic molecular perturbation and the presence, rather than the exact nature, of the broadening mechanism producing the *quasi* continuum. Naturally, as the pressure is raised, and collisions begin to take place, vibrational relaxation lifetime broadening sets in and the mechanism of electronic relaxation is much like that in the solid with energy being lost by the molecule into the kinetic continuum of the surroundings. However, the *rate* of electronic relaxation should be substantially the same with or without collisions in the big molecule limit even though the detailed mechanism of the eventual loss of energy may be quite different in the two cases.

In the 'small molecule limit', no relaxation can take place in the isolated molecule, say, a molecule in interstellar space. When the pertinent energy levels for the radiationless transition become so coarsely spaced compared with their widths, no *quasi* continuum exists. Each state of the system in the absence of collisions or other external perturbations is stationary except for the interaction with the radiation field. Therefore, the only kind of important level broadening† that can occur for a free molecule in this limit is radiative lifetime broadening, and such widths for excited electronic state levels are often substantially less than 0·01 cm^{-1}.

Thus, the small molecule limit can be defined as that where the inverse density of acceptable final states for a radiationless transition is much greater than the lifetime broadening, whatever its origin, and is also much greater than the interaction energy coupling 'initial' electronic levels to 'final' ones. In this limit, levels in the 'initial' electronic state are only rarely accidentally degenerate with levels with which they can mix in the 'final' electronic state. When an intramolecular perturbation mixing such levels is hypothetically 'turned on', the levels merely split apart, and, if the zero-order resonance were complete, the two resulting levels are 50-50 mixtures of the 'initial' and 'final' electronic states. Absorption of radiation from a light source giving uniform energy output over the two lines results in 50 % of the excited molecules being in one of these states and 50 % being in the other. It is important to note that the transition dipole associated with, say, a ground state transition to or from one of the two perturbed levels is now $2^{-\frac{1}{2}}(\mathbf{M}_1+\mathbf{M}_2)$. In the common case where $|M_1| \gg |M_2|$ or vice versa the absorption intensity to each of these levels is half what it would be if the levels were unsplit, and, what is more important, the lifetime of a molecule in one of these levels is twice as long as it would be if there were no perturbation. This result is associated with the well-known fact[11] that the oscillator strength f^{nm} for absorption is related to the Einstein A coefficient through a proportionality factor $G_n/(G_m \nu_{nm}^2)$, where ν_{nm} is the frequency of the transition, and G_n and G_m are spatial degeneracy factors for the upper and lower states, respectively. The result as it applies to the problem of electronic relaxation has been emphasised by Ross[12] and more recently by Douglas.[13]

Thus, in the small molecule limit one occasionally finds a region where a few levels of one electronic state are perturbed by those of another. Many such instances are known for diatomic molecules.

These 'regions of perturbation' cannot lead to electronic relaxation in the isolated molecule since the perturbed levels are discrete and stationary, except for the interaction with the radiation field. Such regions, however, are just the ones that would lead to relaxation among electronic states when level broadening is supplied by outside perturbations such as collisions or phonon interactions in the solid.

In the big molecule limit the degeneracy G_n is so large that the lifetime of even an isolated molecule placed in one of this multitude of nearly degenerate states is virtually infinite, and, since emission does not occur, one says that an electronic relaxation process has taken place from a discrete state into an adjoining region of the *quasi* continuum. In the real case where such a molecule is not in interstellar space but is in contact with an environment, further vibrational relaxation in the *quasi* continuum may lead again to a discrete level of the lower electronic state from which emission may occur. It is in this case that the process of electronic relaxation sometimes manifests itself in the laboratory.

The intermediate case

Perhaps the most interesting case of electronic relaxation to consider is the one where neither the big molecule limit nor the small molecule limit applies. In the former the continuum approximation may be used, while, in the latter, levels may be treated individually by using the results of high resolution spectroscopy. In the 'intermediate case' the *quasi* continuum in the final electronic state does not quite exist in the absence of external perturbations. That is, the radiative lifetime broadening and interaction energies are small compared with the level spacing. Yet the levels may be so densely spaced that their resolution cannot be affected because of the Doppler limitations ($\sim 0{\cdot}05\ \mathrm{cm}^{-1}$) of high resolution gas phase spectroscopy.

In the intermediate case not only are the details of the levels inaccessible to experimental study, but, because of the partial discreteness of the levels, the rate of electronic relaxation in the low pressure limit may be exceedingly sensitive to minute changes in excitation conditions or environment. For instance, the use of a highly monochromatic exciting light at two or more slightly different wavelengths, e.g. from a tunable laser, could reveal the dependence of the relaxation process on the fine details of the energy level spacings, providing no collisions or outside perturbations were present. In addition, exceedingly small perturbations might be expected to have large effects on the electronic relaxation process.

To illustrate this last point consider the energy of van der Waals interactions in two cases: C_6H_6–He and C_6H_6–C_6H_6. Note that we do not compute the collision broadening, which arises from a statistical distribution over all collisions, but simply energy perturbations caused by single collisions. Using force constants[14] $\epsilon/k = 10{\cdot}2$ °K and $\sigma = 2{\cdot}6$ Å for He–He interactions; and $\epsilon/k = 440$ °K and $\sigma = 5{\cdot}3$ Å for C_6H_6–C_6H_6 interactions; and by using the usual combining laws,[15] one can obtain the R^{-6} coefficient in the dispersion energy

$$V_{AB} = -4\epsilon_{AB}\sigma_{AB}R_{AB}^{-6}.$$

At 50 Å, the intermolecular energies are 4×10^{-5} and $1{\cdot}7\times10^{-3}$ cm^{-1} for C_6H_6–He and C_6H_6–C_6H_6 interactions, respectively. However, a *difference* of dispersion energy between two electronic states is the quantity required. It is easy to see, however, that, neglecting vibronic interactions, the difference, say, between ground and excited states, is about 10 % of the total ground state London dispersion energy.† Thus the van der Waals perturbation at 50 Å on the $C_6H_6\ {}^1B_{2u}$–${}^1A_{1g}$ electronic transition energy is around 4×10^{-6} cm^{-1} under He interactions and $1{\cdot}7\times10^{-4}$ cm^{-1} under C_6H_6 interactions. Such perturbations on the relative energies of ${}^1B_{2u}$ and ${}^3B_{1u}$ states are maybe smaller by perhaps another factor of two or three. Perturbations by molecules such as cyclohexane should be similar to those by C_6H_6. These are small energies but are, for instance, comparable to or larger than the $6{\cdot}6\times10^{-7}$ s lifetime broadening of the lowest singlet state (${}^1B_{2u}$) of benzene, which is around 10^{-5} cm^{-1}. They are also of course much, much larger than lifetime broadening in the triplet state.

One may now ask at what pressure does one 50 Å 'collision' with another benzene molecule take place on the average during the lifetime of the ${}^1B_{2u}$ state. Kinetic theory shows that, at 300 °K in order to avoid such collisions, pressures less than $1{\cdot}5\times10^{-3}$ torr would have to be used. This is lower than pressures normally used by kineticists for the study of electronic relaxation. For example, in the experiments by Kistiakowsky & Parmenter[16] on the effect of pressure on the fluo-

† The expression for London dispersion energy between molecules A and B contains an energy denominator equal approximately to $2I_A+2I_B$ where I represents ionisation potentials. In the excited state of C_6H_6, the energy denominator is lowered by roughly $\frac{1}{2}I_B$, if I_B is the ionisation potential of C_6H_6. Taking I (He) $\approx 2{\cdot}5I(C_6H_6)$, the increase in London energy between the ground and excited states of C_6H_6 is 7 % for He–C_6H_6 interactions and 13 % for C_6H_6–C_6H_6 interactions.

rescence yield of benzene, the lowest pressure used was $5{\cdot}5 \times 10^{-3}$ torr. Therefore in the case of benzene, even at very low pressures, collision broadening may be very important in the electronic relaxation process.

Benzene, an intermediate case?

For gaseous benzene to classify as a case intermediate between the big molecule limit and the small molecule limit for radiationless transitions, the inverse density of final states must be greater than radiative linewidths and interaction energies, yet smaller than Doppler widths. Let us examine this question for the $^1B_{2u} \rightsquigarrow {}^3B_{1u}$ radiationless transition in gaseous benzene, remembering that the radiative lifetime broadening is about 10^{-5} cm^{-1} for the $^1B_{2u}$ state and negligibly small for the $^3B_{1u}$ state because of the relatively low radiative transition probability of rotational, vibrational, and spin-forbidden electronic transitions.

Before we do this however we must also consider $^1B_{2u} \rightsquigarrow {}^3E_{1u}$ radiationless transitions, which under certain conditions may compete with $^1B_{2u} \rightsquigarrow {}^3B_{1u}$ transitions. It is now suspected[17] that the $^3E_{1u}$ state of C_6H_6 lies ~ 900 cm^{-1} below the $^1B_{2u}$ state. Because of this small energy difference, the Franck–Condon factors between the $^1B_{1u}$ state and the $^3E_{1u}$ state are expected to be relatively large. However, the vibrational levels in the $^3E_{1u}$ state are very coarsely spaced at 900 cm^{-1} above its own ground vibrational level, and it must be remembered that spin-orbit interactions in C_6H_6 are probably very small, ~ 1 cm^{-1} or less. These coarsely spaced levels therefore cannot mix strongly except when there are accidental resonances with the discrete $^1B_{1u}$ levels. They could mix vibronically with the *quasi* continuum of vibrational levels in the $^3B_{1u}$ state, and some broadening of the $^3E_{1u}$ levels is to be expected from this kind of mixing.† However, considering the $\sim 7\,670$ cm^{-1} electronic energy difference between the $^3E_{1u}$ state and the $^3B_{1u}$ state, vibronic broadening is not expected to exceed 140 cm^{-1}.‡ In any case the broadening reflects the high density of vibrational levels in the $^3B_{1u}$ state, not the $^3E_{1u}$ state, and in the case of a near

† Broadening, presumably of this type, seems to be prevalent among higher electronic states of complex molecules, for example, the $^1B_{1u}$ and $^1E_{1u}$ states of benzene, the second singlet state of azulene,[12] and most other higher excited states of large molecules.

‡ A vibronic coupling energy of 1 000 cm^{-1} is assumed. The observed broadening in the crystal[17] is about 150 cm^{-1}. While other broadening mechanisms have been suggested, it is likely that *vibronic broadening*, which was not mentioned in ref. (17), must give an important contribution to the 150 cm^{-1} breadth.

resonance between the $^1B_{2u}$ state and these broadened vibronic levels, the density of states appropriate for the discussion is still derived from the $^3B_{1u}$ state. We shall therefore ignore the $^3E_{1u}$ state in our discussion of low pressure intersystem crossing, keeping in mind that this state could either accidentally act like a bridge between the $^1B_{2u}$ and $^3B_{1u}$ states, or, at higher pressures and higher temperatures where harder collisions are more probable, may open up a new temperature and pressure-dependent path for the radiationless decay. Another possible role of the $^3E_{1u}$ state in the intersystem crossing process may arise under special excitation conditions, especially those involving higher vibrational levels of the $^1B_{2u}$ state, where adjoining vibronic levels belonging to the $^3E_{1u}$ state are themselves becoming rather dense.

We also ignore rotational levels, since in large molecules at high vibrational energies they are much more widely spaced than vibrational levels. In the absence of collisions, selection rules on J and K for radiationless transitions are rather strict. Under collisions these selection rules are relaxed, but in any case the presence of the rotational degrees of freedom can only increase the *effective* density of states by a small factor.

To estimate the density of vibrational levels in the $^3B_{1u}$ state adjacent to the zeroth vibrational level of the $^1B_{1u}$ state, a technique similar to that suggested in an earlier paper[(3)] is employed. Since almost nothing is known about the vibrational levels in the $^3B_{1u}$ state, ground state vibrations are used. Most vibrational frequencies in the excited state are somewhat smaller, but this should not affect our result substantially.

The thirty normal modes of benzene are divided up into four groups as shown in table 1: those having frequencies in the vicinity of 3 000 cm^{-1}, i.e. the C—H stretching vibrations; those having frequencies near 1 500 cm^{-1}; those having frequencies near 1 000 cm^{-1}; and those having frequencies near 500 cm^{-1}. The numbers of normal modes in each of these categories are 6, 6, 12 and 6, respectively. Therefore C_6H_6 can be approximated by a 30-dimensional oscillator, with four distinct fundamental frequencies having degeneracies of 6, 6, 12 and 6 each.

The degeneracy of the $(n-1)$-th overtone of a g-fold degenerate oscillator is given by,†

$$\frac{(n+g-1)!}{n!(g-1)!} \tag{1}$$

† This is the number of ways n indistinguishable objects can be placed in g boxes with no restrictions on the number per box.

Table 1. *Division of real benzene vibrations*† *for model benzene*

500 cm^{-1}	1000 cm^{-1}	1500 cm^{-1}	3000 cm^{-1}
$706b_{2g}$	$992a_{1g}$	$1328a_{2g}$	$3073a_{1g}$
$610e_{2g}$	$992b_{2g}$	$1599e_{2g}$	$3044e_{2g}$
$673a_{2u}$	$1180e_{2g}$	$1307b_{2u}$	$3068b_{1u}$
$404e_{2u}$	$851e_{1g}$	$1489e_{1u}$	$3079e_{1u}$
$g = 6$	$1012b_{1u}$	$g = 6$	$g = 6$
	$1148b_{2u}$		
	$984e_{2u}$		
	$1033e_{1u}$		
	$g = 12$		

† From the data of D. H. Whiffen, *Phil. Trans. Roy. Soc.* (*London*), A, **248**, 131 (1955); more modern values are somewhat different from these but for the model this set is adequate.

Note that the fundamental corresponds to the zeroth overtone where $n = 1$.

We now compute the total density of states, $\rho(E)$, of this multidimensional oscillator at energy intervals of 3000 cm^{-1} above the ground vibrational energy. This was done with the help of a computer program written by Mr Don Burland. The energy difference between the $^1B_{2u}$ state and the $^3B_{1u}$ state is about 8570 cm^{-1}. From the calculation, the density of states at 8570 cm^{-1} is about 10^5 states per cm^{-1}. Use of the correct excited state frequencies instead of the four artificial ground state frequencies would increase this density. However, not all states of the total density are by any means equally effective. A majority ($\sim 90\%$) have the wrong vibronic symmetry; B_{1u} or E_{2u} being required (i.e. a_{1g} or, less prominently, e_{1g} vibrations in the $^3B_{1u}$ electronic state), and many of the right symmetry have very small Franck–Condon factors. Taking into account these uncertainties, one can estimate within a factor of ten that there are about 3×10^3 states per cm^{-1} that can effectively form final states for the $^1B_{2u} \rightsquigarrow {}^3B_{1u}$ radiationless transition. The inverse density of effective states is therefore around 0·003 cm^{-1}.

This seems to qualify C_6H_6 as an intermediate case as far as the $^1B_{2u} \rightsquigarrow {}^3B_{1u}$ radiationless transition is concerned. However, one must still inquire about the magnitude of the spin-orbit coupling causing the radiationless transition between the $^1B_{2u}$ and $^3B_{1u}$ states. If the coupling were very much larger than the inverse density of states, the molecule would behave as if it lies in the big molecule limit, in spite of the discreteness of the levels, since a large number of levels in the $^3B_{1u}$ would then lie within the 'sphere of interaction' of the $^1B_{2u}$ level.

Magnitude of the coupling

To estimate the magnitude of the coupling energy between the discrete states of $^1B_{2u}$ and the $^3B_{1u}$ *quasi* continuum, we use solid phase data at low temperatures where the small temperature effect can be ignored and the line broadening is such that a continuum really exists. According to the continuum approximation,† the transition probability per unit time is

$$dW(t)/dt = W(t)/t = \frac{2\pi}{\hbar}\bar{\rho}(E)\bar{\beta}^2, \tag{2}$$

where $\bar{\rho}(E)$ is the density of 'effective molecular states', and $\bar{\beta}$ is an average interaction energy inclusive of Franck–Condon factors. The 'effective molecular states' of which we speak were called 'directly coupled final states' in the earlier papers. Our theoretical framework supposes that the environmental states affect neither $\bar{\beta}$ nor $\bar{\rho}(E)$, and thus $\bar{\rho}(E)$ is the density of states that we have just calculated. A knowledge of $W(t)/t$, together with the estimate of $\bar{\rho}(E) \approx 3\times 10^3$ states per cm^{-1} allows an empirical evaluation of $\bar{\beta}$.

Taking‡ $W(t)/t \equiv k_{\text{nonrad.}} = 6{\cdot}4\times 10^7\,s^{-1}$, one finds,

$$\bar{\beta} \approx 4{\cdot}2\times 10^{-5}\,cm^{-1}.$$

This small value places the $^1B_{2u} \rightsquigarrow {}^3B_{1u}$ radiationless transition in or near the 'intermediate molecule' range. For a really free benzene

† The connection between the results of ref. (2) and the continuum approximation can be illustrated by starting with equation (23) of ref. (2), which reads

$$W(t)/t = 2\pi/\alpha\hbar \sum_n \beta_n^2,$$

where $W(t)$ is the overall transition probability into the *quasi* continuum,

$$\alpha \gg (\sum_n \beta_n^2)^{\frac{1}{2}}$$

is the lattice coupling energy related to the vibrational relaxation time, and β_n is the matrix element mixing the initial state with the n-th final state of the molecular *quasi* continuum. Since lattice coupling broadens each molecular final state to an extent of order α, the range of interaction between the molecular levels is increased to the order of α in the solid. Taking as an approximation an average β_n, which we shall denote $\bar{\beta}$, there will therefore be terms of the order of $[\bar{\rho}(E)\times\alpha]$ in the above summation. Therefore the continuum approximation equation (2) follows. The point is, of course, that the larger α is, the more states of the molecular *quasi* continuum are brought into play, but the less of a role any single state plays. The magnitude of the coupling β_n to the n various final states is not expected to change over a range so small as α (or a range as small as kT for temperatures up to usual laboratory temperatures).

‡ The rate constant $k_{\text{nonrad.}}$ is estimated from the natural fluorescence lifetime of $6{\cdot}4\times 10^7\,s^{-1}$ corresponding to an oscillator strength $f = 0{\cdot}00147$, and a phosphorescence efficiency $\Phi_P \approx 0{\cdot}8$.

molecule, it is possible that a radiationless transition cannot occur, but a weak mixing with a lattice continuum or a kinetic energy continuum would bring the system into the 'big molecule limit'.

Résume of experimental results

Consider electronic relaxation in the gas phase between the $^1B_{2u}$ state of benzene and the lowest triplet state $^3B_{1u}$ of benzene. Experimentally, the situation is a bit foggy but is in the process of clearing. What seem to be presently accepted are the following results:

(1) At total pressures of benzene plus added inert gas greater than 10 torr (the 'high pressure region'), and for exciting wavelengths greater than about 2 500 Å (the 'long wavelength region') the absolute fluorescence efficiency Φ_f by direct measurement is around 0·2 and is pressure independent.[18, 19, 20]

(2) In the 'high pressure', 'long wavelength region' the intersystem crossing ($^1B_{2u} \rightsquigarrow {}^3B_{1u}$) efficiency Φ_T as measured by the sensitised isomerisation of butene-1 and of dideuteroethylene is around 0·63.[21, 22] This efficiency is pressure independent, and is also temperature independent between 20 and 250 °C. Collisions apparently do not affect fluorescence of intersystem crossing under these excitation conditions.

(3) The sum $\Phi_f + \Phi_T \approx 0{\cdot}83$ differs from unity either because of experimental errors in these difficult measurements, or because there is actually another process besides fluorescence and intersystem crossing that depletes the singlet state.

(4) In the 'high pressure' region, but for excitation wavelengths shorter than 2 500 Å, relative fluorescence yields decrease with decreasing wavelength down to 2 400 Å, where the yield becomes very small and perhaps zero.

(5) Triplet state yields as measured by chemical techniques[22] also appear to decrease with decreasing wavelength below 2 500 Å, but one is not altogether certain about the reliability of such measurements under a wide range of conditions.

(6) In the 'high pressure' region, vibrational relaxation, as evidenced from the fluorescence spectrum, appears complete.[16]

(7) One might therefore tentatively conclude that there is a process other than fluorescence and intersystem crossing that is fast enough to compete with vibrational relaxation under decreasingly short wavelength excitation.[22]

(8) In the pressure range below 10 torr, the relative fluorescence efficiency under 2 537 Å radiation increases to at least 1·6 times its 'high pressure' value.[16] In this same pressure range vibrational relaxation begins to become inefficient.

(9) Triplet state formation as measured by relative triplet yields using a modified Cundall technique[19] seems to be less efficient as the pressure is decreased below 10 torr, and the decrease is still apparent at the lowest total pressures used ($\sim 0{\cdot}07$ torr).[23]

(10) It has been claimed[16] that the fluorescence efficiency as a function of pressure in the pressure range around 0·05 torr abruptly levels off, becoming pressure independent, but the scatter of experimental points is large.

(11) Donovan & Duncan[24] have measured the lifetime of C_6H_6 over a pressure (of C_6H_6) range between 0·008 and 0·75 torr. The rate constant for decay of fluorescence in the wavelength range $\sim$2720–2 620 Å in this pressure range roughly fits the expression,

$$k(\mathrm{s}^{-1}) = [1{\cdot}69 + 20{\cdot}7P\,(\mathrm{torr})] \times 10^6.$$

(12) The radiative rate constant for fluorescence as calculated from the oscillator strength $f = 0{\cdot}00147$ of the ${}^1B_{2u} \leftarrow {}^1A_{1g}$ absorption transition,[25] and using an average fluorescence frequency of 40 100 cm^{-1}, is $1{\cdot}58 \times 10^6\,\mathrm{s}^{-1}$. This agrees, within the experimental uncertainties, with the zero-pressure value obtained by Donovan & Duncan.[24] Since the pressure range of these experiments does not include the 'high-pressure' region, the pressure dependence of lifetime may, however, be much different than that given by the above equation.

Discussion

The purpose of this section is to attempt to tie together experimental results and theoretical notions. The experimental results on benzene mentioned in the last section are fairly extensive. In addition, benzene could possibly lie in the intermediate range between the 'big molecule' and 'small molecule' limits as far as the ${}^1B_{2u} \rightsquigarrow {}^3B_{1u}$ radiationless transition goes (see above, pp. 220–224). For these reasons, the benzene problem is appropriate for discussion.

We shall concern ourselves only with the fluorescence efficiency Φ_f under $\lambda > 2\,500$ Å excitation conditions. It will be assumed for simplicity that $\Phi_f + \Phi_T = 1$. The deviation of the experimental results from this is not large compared with experimental uncertainties, and in any case a small yield, say $\Phi_x \approx 0{\cdot}17$, for some other process will not change the overall picture substantially anyway.

The basis of the discussion will be the experiments of Kistiakowsky & Parmenter[16] on the relative yield of fluorescence as a function of pressure. As mentioned in point 10 of the last section, these authors claim two pressure ranges over which Φ_f is constant, $p < 0{\cdot}05$ torr and $p > 10$ torr. The low pressure result is controversial for four reasons:

(1) Fluorescence yield experiments at such low total pressures are difficult to carry out.

(2) The experiments of Sigal[23] seem to show the triplet yield Φ_T to decrease towards zero as the pressure is decreased in the pressure range $0{\cdot}07 \leqslant P \leqslant 10$ torr, a result which might imply, but certainly does not prove, that Φ_f increases towards unity with decreasing pressure.

(3) The experimental lifetime[24] of the $^1B_{2u}$ state in the limit of zero pressure is equal, within experimental uncertainties, to that calculated from the absorption oscillator strength, a fact that strongly suggests not only a degeneracy factor $G_n = 1$ in the upper state but also $\Phi_f = 1$ at very low pressures.

(4) The theoretical estimate of vibrational state density in the $^3B_{1u}$ state shows the levels to be coarsely spaced relative to the spin-orbit interaction between the $^1B_{2u}$ and $^3B_{1u}$ states, and, therefore, there may be no radiationless process between the states in the limit of zero pressure.

In view of these arguments one would be tempted to discount the claim of Kistiakowsky & Parmenter[16] that the fluorescence yield becomes constant at pressures $P < 0{\cdot}05$ torr. A future paper, where vibrational relaxation among a number of fluorescing states is taken into account, will examine this question in more detail.

The constancy of Φ_f in the pressure range $P > 10$ torr seems to be well established by a number of different workers (see above, pp. 224–225), from relative yield measurements of both fluorescence and triplet formation. This behaviour, as well as a constant temperature behaviour in the $P > 10$ torr range would be expected from our earlier statement that it is the 'presence, rather than the exact nature, of the broadening mechanism producing the *quasi* continuum' that is important in electronic relaxation problems. Thus, intersystem crossing in gaseous benzene at a total (benzene or cyclohexane) pressure of $P > 10$ torr is very much like intersystem crossing for benzene in the solid.

References

1 G. W. Robinson, *J. Mol. Spectry* **6**, 58 (1961).
2 G. W. Robinson & R. P. Frosch, *J. Chem. Phys.* **37**, 1962 (1962).
3 G. W. Robinson & R. P. Frosch, *J. Chem. Phys.* **38**, 1187 (1963).
4 R. E. Kellogg & R. P. Schwenker, *J. Chem. Phys.* **41**, 2860 (1964).
5 R. P. Frosch & G. W. Robinson, manuscript in preparation.
6 C. P. Slichter, *Principles of Magnetic Resonance* (Harper and Row, New York, 1963).
7 D. S. Tinti & G. W. Robinson, unpublished results.
8 W. Siebrand, *J. Chem. Phys.* **44**, 4055 (1966); Symposium on the Triplet State, 100th Anniversary American University of Beirut, Beirut, 14–19 February 1967.
9 I. G. Ross, many stimulating discussions during the past five years.
10 See, for example, R. J. Watts & S. J. Strickler, *J. Chem. Phys.* **44**, 2423 (1966).
11 R. S. Mulliken, *J. Chem. Phys.* **7**, 14 (1939).
12 I. G. Ross, private discussions about the broadness of the absorption to the second singlet state of azulene.
13 A. E. Douglas, *J. Chem. Phys.* **45**, 1007 (1966).
14 J. O. Hirschfelder, C. F. Curtiss & R. B. Bird, *Molecular Theory of Gases and Liquids* (John Wiley, New York, 1954), pp. 1110–12.
15 *Ibid.* p. 168.
16 G. B. Kistiakowsky & C. S. Parmenter, *J. Chem. Phys.* **42**, 2942 (1965).
17 S. D. Colson & E. R. Bernstein, *J. Chem. Phys.* **43**, 2661 (1965).
18 H. Ishikawa & W. A. Noyes Jr., *J. Chem. Phys.* **37**, 583 (1962).
19 W. A. Noyes Jr., W. A. Mulac & D. A. Harter, *J. Chem. Phys.* **44**, 2100 (1966).
20 J. A. Poole, *J. Phys. Chem.* **69**, 1343 (1965).
21 R. B. Cundall, F. J. Fletcher & D. G. Milne, *J. Chem. Phys.* **39**, 3536 (1963); *Trans. Faraday Soc.* **60**, 1146 (1964).
22 W. A. Noyes Jr. & D. A. Harter, (to be published).
23 P. Sigal, *J. Chem. Phys.* **42**, 1953 (1965).
24 J. W. Donovan & A. B. F. Duncan, *J. Chem. Phys.* **35**, 1389 (1961).
25 D. S. McClure, *J. Chem. Phys.* **17**, 905 (1949).

LOW-LYING EXCITED TRIPLET STATES AND INTERSYSTEM CROSSING IN AROMATIC HYDROCARBONS

M. W. WINDSOR AND J. R. NOVAK

Introduction

The radiationless transition which carries molecules over from the first excited singlet state S_1 to the lowest triplet state T_1 is more than a million times faster than the radiationless transition which transfers T_1 molecules back to the ground state, S_0. Now it is true that the S_1T_1 energy gap is usually smaller than the T_1S_0 gap, but the difference is insufficient to account for such a large difference in rate. Also, for anthracene the two energy gaps are similar in magnitude, whereas the rate constants still differ by a factor of a million or more. Pariser[1] proposed that low-lying triplet levels between S_1 and T_1 might provide high probability pathways from S_1 into the triplet manifold. Experimental location of such levels would provide valuable data for calculation of rates of intersystem crossing. Most previous studies of T–T absorption in aromatic hydrocarbons have been restricted to the ultraviolet regions, whereas the postulated levels would give rise to absorption in the near infra-red. We have, therefore, carried out a search for triplet-triplet absorption in the near infra-red.

Experimental

The experiments were carried out at room temperature using flat rods of an epoxy plastic containing the aromatic in solution. The rods were placed in the sample chamber of a Cary 14 spectrophotometer and were irradiated by a 1000 W AH-6 mercury arc. An identical rod was placed in the comparison beam chamber but was left unirradiated. The steady-state triplet-triplet absorption was recorded from 0·4 to 1 μ, using double-beam detection and operating the instrument in its infra-red mode with the lead sulphide cell as detector. Under these conditions the double monochromator of the Cary effectively discriminates against scattered light from the excitation source. A logarithmic slide wire was employed to facilitate the comparison of new weak bands with known strong bands over as wide an optical density range as possible. Extinction coefficients were obtained by comparing the optical density with that of the strongly allowed T–T bands in the visible region. For lifetime measurements the signal beam was deflected by a mirror onto an Amperex infra-red photomultiplier and the decay of T–T absorption for each band maximum was displayed on an oscilloscope.

Results and discussion

New regions of T–T absorption have been observed for phenanthrene, anthracene, chrysene, picene and pyrene. The decay time of each new band was measured to confirm that it belonged to the compound under study and not to an impurity. The T–T absorption spectra for phenanthrene, anthracene and pyrene are shown in figs. 1 to 4.

Phenanthrene

The T–T absorption spectrum of phenanthrene from 0·4 to 1·0 μ is shown in fig. 1. In addition to previously reported bands between 400 and 500 nm,[5] three new bands are present in the red and near infrared at 653, 725 and 820 nm. The spectrum has been normalised to a value of 27 000 for the extinction coefficient at 495 nm.[2]

The new bands have extinction coefficients lying between 1 000 and 2 000 and probably correspond to $u-g$ allowed but $+/-$ forbidden transitions. The position of the 0–0 band locates an excited triplet level at 33 800 cm^{-1} which is 5500 cm^{-1} above the S_1 state. This triplet is probably not the lowest excited triplet of phenanthrene. According to Pariser,[1] in all alternant hydrocarbons the lowest triplet state T_1 is always a *plus* state. The ground state is always a *minus* state. The lowest excited singlet state can be either *plus* or *minus*. In the case that it is a *minus* state, which includes benzene, naphthalene and all non-linear polyacenes such as phenanthrene, chrysene and picene, there exists a triplet *minus* state which has the same energy as S_1. The S_1T_1 gap for phenanthrene is 6 700 cm^{-1} Thus one would expect a T–T transition at approximately 1·5 μ in the infra-red

Chrysene and Picene

A new band for chrysene was detected at 910 nm which locates an excited triplet level at 30 800 cm^{-1}. For picene new bands appeared at

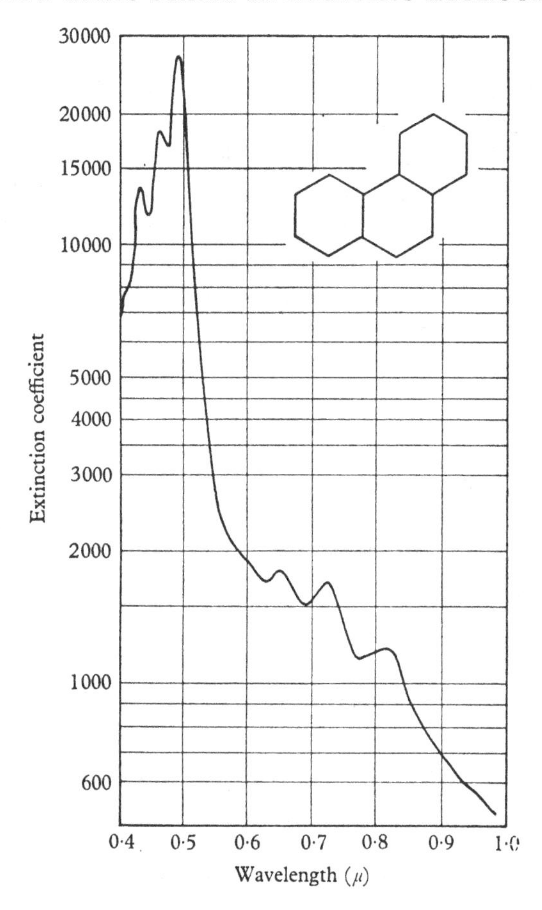

Fig. 1. T–T absorption spectrum of phenanthrene.

805 and 910 nm. For both chrysene and picene the lowest excited triplets found lie several thousand wave numbers above the lowest excited singlet state S_1. As in the case of phenanthrene, Pariser's theory predicts the existence of a triplet at about the same energy as S_1. The corresponding T–T absorptions would lie at 7 700 cm^{-1} (1·3 μ) for chrysene and about 6 500 cm^{-1} (1·5 μ) for picene. This conclusion, that the triplets so far observed are not the lowest excited triplets of these molecules, is also supported by measurements of triplet yields at low temperature in our laboratory. In EPA glass at 77° K the values of ϕ_T are 0·70 for chrysene and 0·36 for picene.[3] Values close to zero would be expected if intersystem crossing must proceed via the excited triplets we have found.

Anthracene

Our results for anthracene are shown in figs. 2 and 3. New bands are observed at 477, 517, 789, 805, 843 and 887 nm. The bands at 477 and 527 nm could also be detected photographically following flash photolysis of solutions of anthracene in mineral oil at room temperature. The bands in the region 750 and 950 nm show good agreement in wavelength and relative intensity with those reported by Kellogg.[4] In figs. 2 and 3, we have normalised the extinction coefficient scale to a value of 115,000 M^{-1} cm^{-1}, obtained in our laboratory for anthracene in EPA

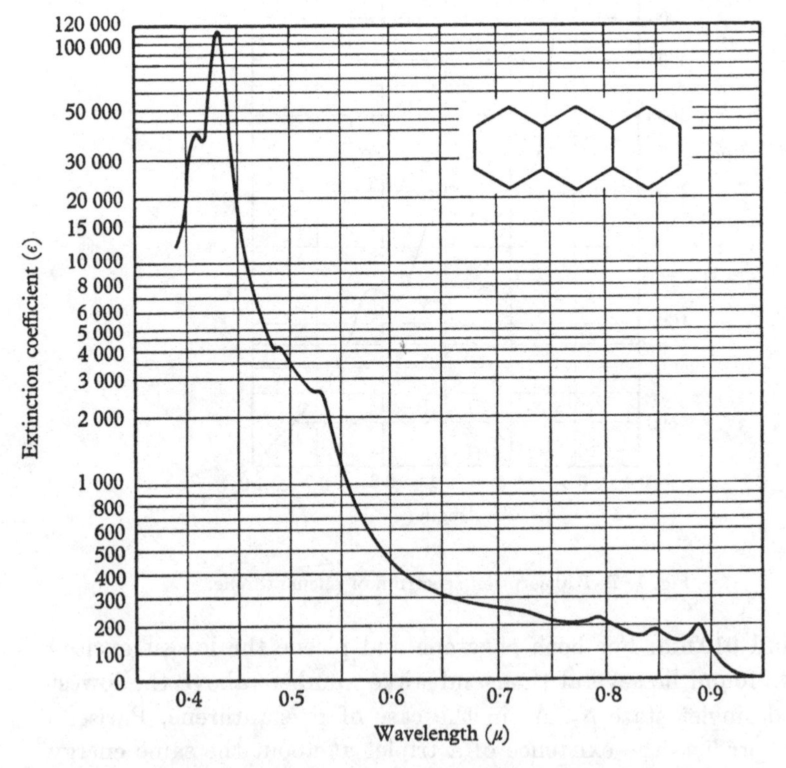

Fig. 2. *T–T* absorption spectrum of anthracene, 0·4 to 1·0 μ.

at 77 °K.[3] The true value of plastic is probably somewhat lower than this, since at room temperature in plastic the spectrum is broader than that obtained at 77 °K. However, the only available room temperature value is 71,500 ± 36,000 obtained in mineral oil by Porter & Windsor.[5] Using this value Kellogg obtained $\epsilon = 250$ for the 887 nm band. If we use the same value, we obtain $\epsilon = 130$.

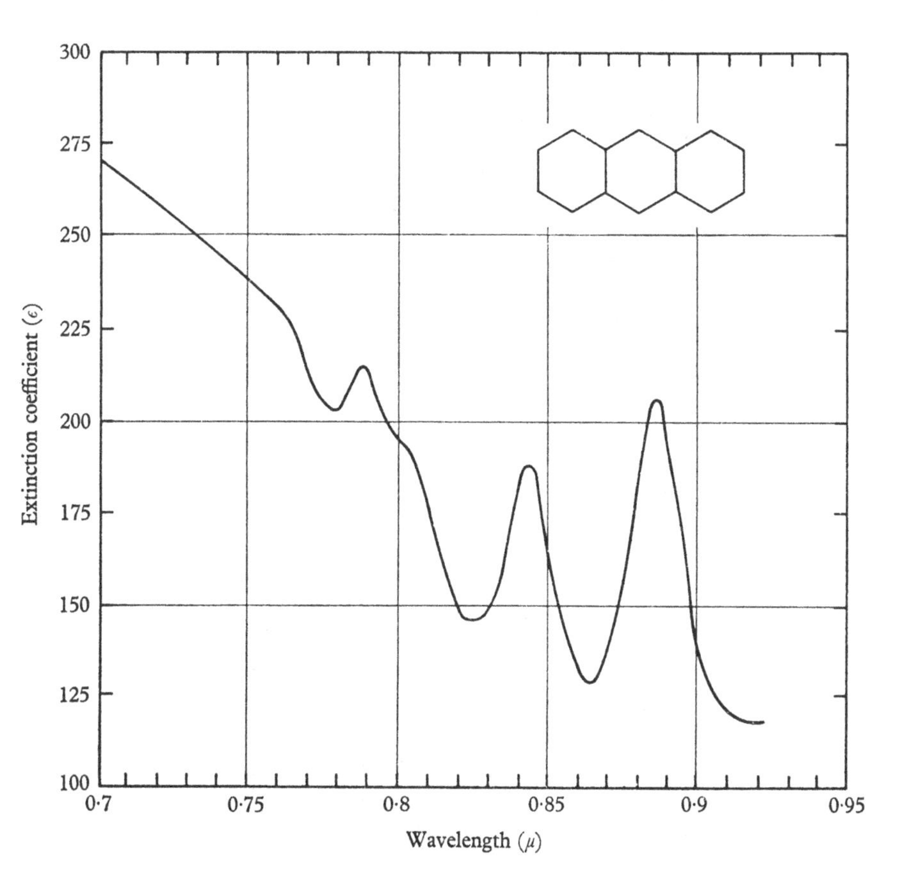

Fig. 3. *T–T* absorption spectrum of anthracene, 0·75 to 1·0 μ.

The 0–0 transition at 887 nm locates an excited triplet 11 200 cm^{-1} above the lowest triplet (14 850 cm^{-1}). This puts the excited triplet 3·25 eV above the ground state. Pariser predicts a $^3A^+_{1g}$ state at 3·98 eV and a $^3B^+_{1g}$ state at 2·84 eV. Transitions from the $^3B^+_{2u}$ lowest triplet would be $u-g$ allowed and *plus–plus* forbidden for both upper states, so either assignment would satisfy the experimental data. However, the new band at 527 nm is quite strong and defines an excited triplet state 4·24 eV above the ground state. The most likely assignment is $^3A^+_{1g}$ for the 527 nm band and $^3B^+_{1g}$ for the band at 887 nm.

Results for anthracene similar to ours but with better resolution of the bands have been obtained recently by Astier & Meyer from the flash photolysis of viscous solutions at low temperatures.[6]

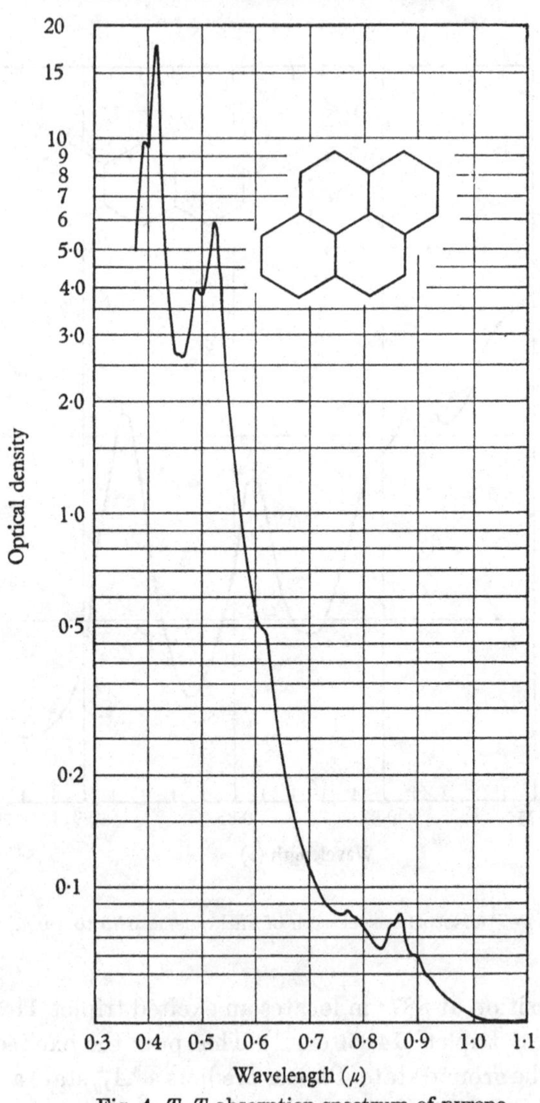

Fig. 4. *T–T* absorption spectrum of pyrene.

Pyrene

The *T–T* absorption of pyrene from 0·38 to 1·0 μ is shown in fig. 4. Two new transitions are detectable with 0–0 bands at 870 and 775 nm. These locate excited triplet levels at 28 300 and 29 700 cm^{-1} above the ground state. We assign these to $^3B^+_{1g}$ and $^3A^+_{1g}$ respectively which Evleth and Peticolas calculate to be 27 100 and 30 100 cm^{-1} above the ground state.[7]

With this assignment we do not expect there to be an excited triplet lower in energy than the one we observe at 28 300 cm^{-1}. Pyrene thus presents the interesting case of a compound whose lowest excited triplet lies appreciably (about 1 300 cm^{-1}) above its first excited singlet state. Intersystem crossing from S_1 to the triplet manifold via this excited triplet should thus require thermal activation and at 77 °K would be many orders of magnitude slower than fluorescence emission. This conclusion is supported by recent data of Stevens *et al.*[8] who infer from fluorescence lifetime studies, as a function of temperature, that the triplet yield of pyrene is zero at 77 °K.*

The above results demonstrate that plastic media can be successfully used to record T–T absorption spectra of aromatic hydrocarbons at room temperature to as far as 1·0 μ in the infra-red. Beyond 1·0 μ absorption by the plastic matrix begins to interfere. However, transmission windows are present and it may be possible eventually to obtain data out to 1·5 or even 2·0 μ.

References

1 R. Pariser, *J. Chem. Phys.* **24**, 250 (1956).
2 R. A. Keller & S. G. Hadley, *J. Chem. Phys.* **42**, 2382 (1965).
3 M. W. Windsor & W. R. Dawson, 'Quantum efficiencies of triplet formation in aromatic molecules.' *Molecular Crystals* (August 1967). (In the Press.)
4 R. E. Kellogg, *J. Chem. Phys.* **44**, 411 (1966).
5 G. Porter & M. W. Windsor, *Proc. Roy. Soc.* A, **245**, 238 (1958).
6 R. Astier & Y. H. Meyer, Proceedings of this Conference.
7 E. M. Evleth & W. L. Peticolas, *J. Chem. Phys.* **41**, 1400 (1964).
8 B. Stevens, M. F. Thomas & J. Jones, *J. Chem. Phys.* **46**, 405 (1967).

* [*Note added in proof*]:

We have recently studied the T–T absorption of perdeuterated pyrene in EPA at 77 °K. The triplet yield is definitely not zero. It might, however, be as low as a few per cent. It seems likely, therefore, that, in pyrene, T_1 can be populated both via the excited triplet and via the direct $S_1 \rightarrow T_1$ process. At room temperature the first route is dominant; at 77 °K thermal activation is much too slow and only the less effective (because of the large energy gap) $S_1 T_1$ process remains.

DE-EXCITATION RATES OF TRIPLET STATES IN CONDENSED MEDIA*

D. L. DEXTER AND W. B. FOWLER

Abstract

It is argued that under very general conditions, including those for the lowest triplet states of organic molecules in crystals, solutions, or glasses, the larger the vibrational relaxation time in the electronically excited state, the shorter will be its lifetime against both radiative and non-radiative processes. Numerical results of a model calculation for radiationless transitions will be presented, and an extension of the theory of Robinson and Frosch will be given for weak molecule-host interactions. The influence of even weak static solute-solvent interaction on the inter-system crossing probability will be demonstrated.

* Research supported in part by the U.S. Air Force Office of Scientific Research.

LIFETIMES OF THE TRIPLET STATE OF AROMATIC HYDROCARBONS IN THE VAPOUR PHASE

B. STEVENS, M. S. WALKER AND E. HUTTON

Abstract

The delayed fluorescence component exhibited by naphthalene, phenanthrene, anthracene, pyrene, perylene, 1,12-benzperylene, anthanthrene and fluoranthene in the vapour phase is found to decay exponentially under conditions of excitation such that its intensity varies as the second power of incident light intensity. This behaviour is identical with that observed for solutions of certain aromatic hydrocarbons A where the emitting state $^1A^*$ is believed to originate[1] in a process of triplet-triplet annihilation

$$^3A + {}^3A \rightarrow {}^1A^* + A, \tag{1}$$

which competes with unimolecular relaxation of the triplet state

$$^3A \rightarrow A \quad (\text{or } A + h\nu_p). \tag{2}$$

At relatively low excitation rates where

$$[^3A] \ll k_2/k_1,$$

this mechanism requires that the decay constant k_D characteristic of delayed fluorescence is given by

$$k_D = 2k_2, \tag{3}$$

which has been confirmed[2] for a number of liquid systems over a wide range of temperature.

If triplet-triplet annihilation is responsible for delayed fluorescence in aromatic vapours the relationship (3) provides a measure of the triplet state lifetime $^3\tau$ $(= 1/k_2 = 2/k_D)$ in the absence of observable phosphorescence emission and of the dominant effect of the annihilation process at the concentration of triplet state required to monitor its optical density by flash kinetic spectrophotometry.[3]

The triplet state lifetime $(2/k_D)$, which is of the order of 1 ms for the systems examined, is found to vary with the diameter (volume) of the cylindrical cell, with vapour pressure (controlled by the temperature of the condensed sample in a lower limit of the cell) and with the vapour temperature. Since there is no evidence for ground state quenching of the triplet state in solution,[1] these effects are largely attributable to quenching of the triplet state by an impurity in the solid sample which may or may not be completely vaporised at the prevailing sample temperature; this is consistent with an increase in $^3\tau$ with cell volume (lower impurity concentration).

The delayed fluorescence spectra of phenanthrene vapour and of fluoranthene vapour differ from the corresponding prompt fluorescence spectra and must originate in the sensitised delayed fluorescence of an impurity which in the former case has been identified as anthracene.[5] The triplet state relaxation equations are therefore examined for reversible and non-reversible quenching and for mixed and self-annihilation of the impurity triplet state in order to account for the types of pressure-dependence of k_D observed as follows:

(*a*) $^3\tau$ is independent of pressure at low pressures but is markedly reduced at higher pressures due to non-reversible quenching by an impurity of low volatility; this provides values for the triplet state lifetime in the absence of quenching and for the heat of vaporisation of the impurity in pyrene and perylene.

(*b*) The triplet state lifetime of anthracene vapour decreases with an increase in pressure to a minimum value at higher pressures which is independent of cell volume; this is described in terms of reversible quenching by a volatile (completely vaporised) impurity which affords an estimate of $^3\tau$, both for anthracene and the impurity, as well as the anthracene impurity content.

(*c*) $^3\tau$ *increases* rapidly with an increase in vapour pressure at very low pressures (anthanthrene, 1,12-benzperylene) due to diffusion of triplet states out of the observation zone; an approximate treatment[6] of the diffusional problem leads to diffusion coefficients of the magnitude expected for a molecule with a collision diameter of 9 Å.

The decrease in $^3\tau$ with increase in vapour temperature at constant pressure is qualitatively inconsistent with an increase in collisional quenching and is tentatively attributed to a dependence of the rate of radiationless relaxation of the triplet state on its vibrational energy content.

Introduction

A fluorescence component with a decay constant k_D of the order of $10^3 s^{-1}$, but spectroscopically indistinguishable from the normal fluorescence characterised by a decay constant in the region of $10^8 s^{-1}$, was first reported for aromatic hydrocarbons A in the vapour phase by Dikun.[7] The general features of delayed fluorescence in these systems were confirmed by Williams,[8] who found that the low temperature coefficient of the decay constant k_D is incompatible with a thermal repopulation of the fluorescent singlet from the long-lived triplet state, proposed to account for the α-phosphorescence exhibited by certain dyes in a rigid solution.[9] The linear variation in delayed fluorescence intensity, relative to the total fluorescence intensity, with the hydrocarbon vapour pressure is consistent with a bimolecular origin of the delayed component for which Williams suggested a reversible association of ground and excited singlet states A and A^* to produce a potentially long-lived species A_2^*. Stevens & McCartin[10] and Stevens & Hutton[11] subsequently interpreted the pressure-dependence of k_D in terms of a Lindemann mechanism for unimolecular excimer (A_2^*) dissociation which however required an excimer radiative lifetime of the order of 1 s.

More recently Parker & Hatchard [1] noted the dependence of delayed fluorescence intensity I_D on the square of the light intensity absorbed by similar molecules in solution, and suggested that of the unimolecular and bimolecular triplet relaxation processes the latter regenerates

$$^3A \rightarrow A, \quad (1)$$

$$^3A + {}^3A \rightarrow A^* + A, \quad (2)$$

the lowest excited singlet state A^* responsible for the delayed emission component. Under conditions of relatively low light intensity, such that

$$k_1 \gg k_2[^3A],$$

the decayed constant for delayed fluorescence is given by

$$k_D = \frac{-d\ln I_D}{dt} = -\frac{2d\ln[^3A]}{dt} = 2k_T, \quad (1)$$

where $k_T(=k_1)$ is the reciprocal of the triplet state lifetime under the conditions of observation; this relationship has been confirmed for a number of aromatic hydrocarbons in solution at different concentrations and over a wide temperature range.[1,12]

Owing to the negligible phosphorescence yield of aromatic hydrocarbons in the vapour phase, and the dominant second-order contribution to triplet-state relaxation at the concentrations required to follow this relaxation by kinetic absorption spectrophotometry,[3] the relationship I between k_D and k_T has yet to be established as proof of the triplet-triplet annihilation origin of delayed fluorescence in the vapour phase. However, the verification of the square-law intensity relationship under conditions for an exponential decay[13], leaves little doubt that this is the operative mechanism, in which case a method is available for the measurement of triplet state lifetimes in the vapour phase.

Experimental

Figure 1 shows schematic elevation of the apparatus used to measure decay constants of delayed fluorescence. Filtered radiation ($> 300\,\mathrm{m}\mu$) from the 125 W high pressure d.c. mercury arc (Mazda MBL/D) is focused by the lenses *L* through one of the wheels *D* of a Becquerel phosphoroscope onto the cell *C* containing the vapour. The delayed emission component isolated by the second rotating wheel

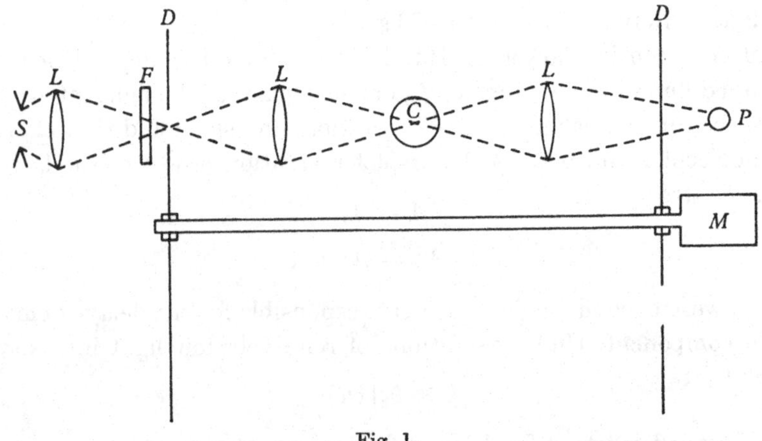

Fig. 1.

is focused onto the detector *P* (Mazda 27 M 3) the anode of which is earthed through a $500\,\mathrm{k}\Omega$, $\frac{1}{2}$ W resistor, and the potential across the resistor is fed to the *Y*-plates of a double beam oscilloscope (Solartron CD 71152) triggered internally by the emission signal; the decay curves displayed on the oscilloscope screen were photographed and projected for analysis.

The cylindrical cells of Pyrex or quartz were enclosed in an electrically heated aluminium block, fitted with quartz windows and aluminium sleeves to accommodate cells of different diameter. The vapour pressure was controlled by the temperature of the lower cell limb containing a solid or liquid hydrocarbon sample in a second, independently wound, aluminium block. Following initial purification by fractional microsublimation,[14] the hydrocarbons were sublimed into the cell before sealing off at a pressure of $\leqslant 10^{-6}$ mm Hg.

Calibrated wire-mesh screens were used to reduce the intensity of incident radiation, and the spectral distribution of delayed and normal fluorescence components were recorded photoelectrically on an Aminco-Keirs spectrophotophosphorimeter using a furnace specially designed to fit the Dewar compartment. The total emission spectrum was recorded without the phosphoroscope drum, at the same low resolution required to obtain an appreciable signal from the much weaker delayed component, and recordings were corrected for detector sensitivity as a function of emission wavelength.

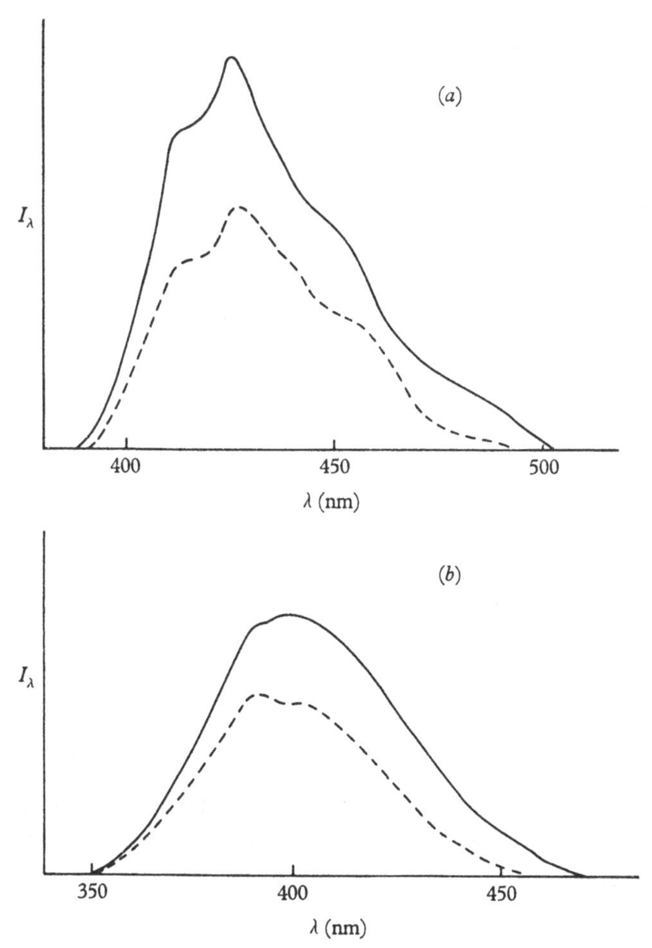

Fig. 2. Delayed (– – – –) and normal fluorescence spectra of (*a*) anthracene, and (*b*) pyrene vapour at same low resolution.

Results

Delayed fluorescence could not be detected for the vapours of naphthacene or of 9,10-diphenylanthracene but was observed for naphthalene, phenanthrene, anthracene, fluoranthene, pyrene, pery-

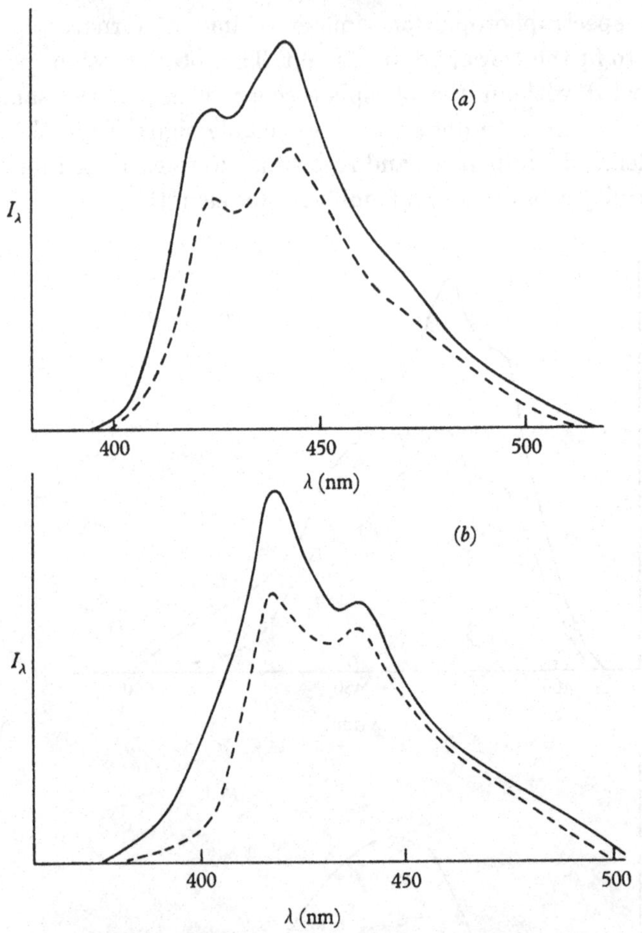

Fig. 3. Delayed (– – – –) and normal fluorescence spectra of (*a*) perylene and (*b*) 1,12-benzperylene vapour at same low resolution.

lene, 1,12-benzperylene and anthanthrene. Under the conditions described, the intensity of the delayed component invariably exhibited an exponential decay after excitation cut-off.

1. *Emission spectra*

Corrected photoelectric recordings of the delayed and normal fluorescence spectra are shown in figs. 2–4; except for fluoranthene and phenanthrene, which exhibited the delayed fluorescence spectrum of anthracene down to the lowest pressures,[5] the 'spectral' distributions of both components are identical within the limits of the low resolution employed.

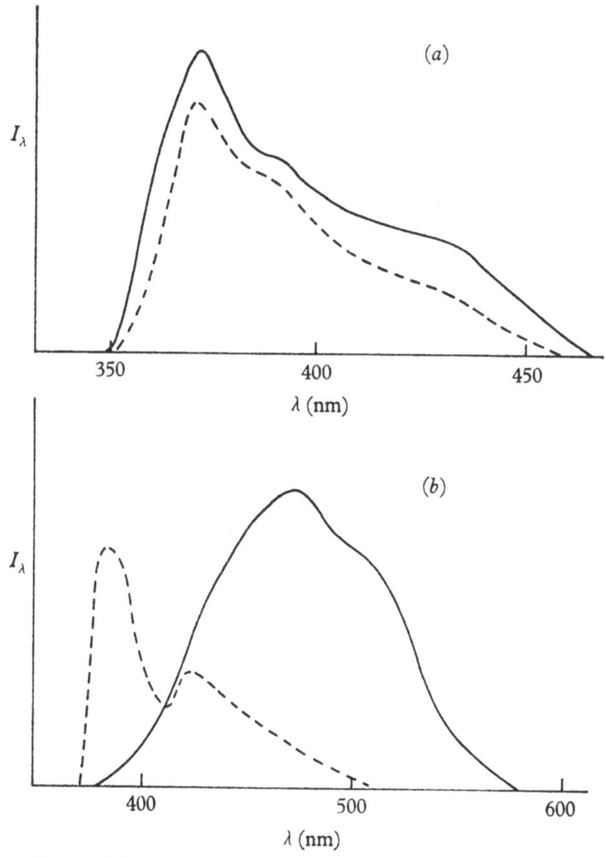

Fig. 4. Delayed (– – – –) and normal fluorescence spectra of (*a*) anthracene and (*b*) fluoranthene vapour at same low resolution.

2. *The dependence of delayed fluorescence intensity on incident light intensity*

For each compound examined the intensity I_D of delayed fluorescence was found to vary as the square of the incident light intensity I_S over the ranges of temperature and pressure employed. This is illustrated by fig. 5 in which $\log I_D$ is plotted as a function of the optical density, D, of the wire mesh screen, placed in the path of the excitation beam, which reduces the source intensity at the cell surface from I_S^0 to

$$I_S = I_S^0 10^{-D}.$$

The intensity of delayed fluorescence is correspondingly reduced to $I_D \propto I_S^2$ or

$$\log I_D = \text{constant} - 2D,$$

as shown by the slopes of the data lines in fig. 5.

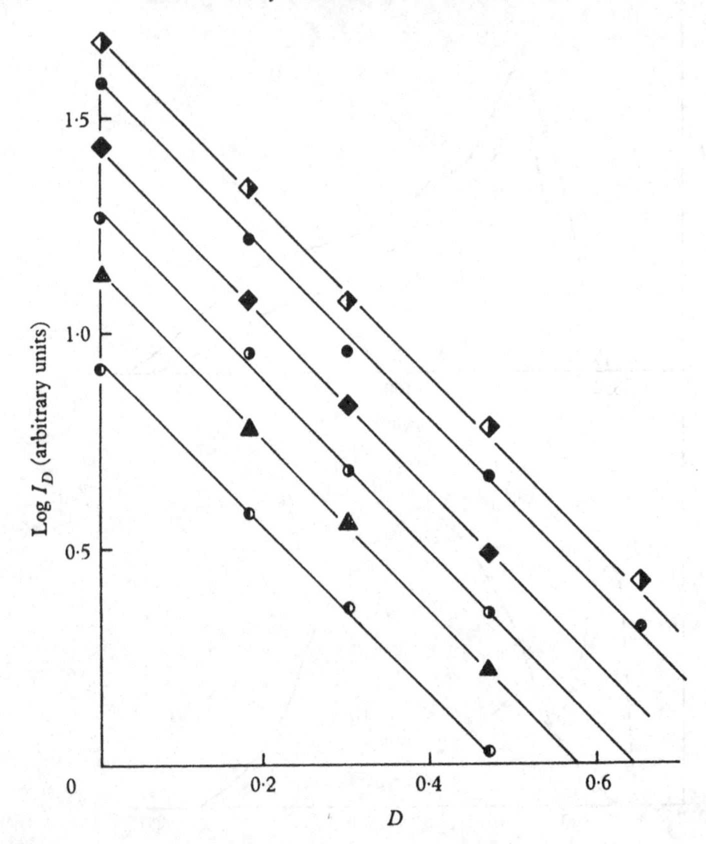

Fig. 5. Plot of logarithm of delayed fluorescence intensity against optical density D of wire mesh screen for: ◈, anthracene; ●, pyrene; ◆, anthranthrene; ◑, fluoranthene; ▲, 1,12-benzperylene; ◐, perylene in the vapour phase. Lines drawn with slope of $-2{\cdot}0$.

3. *The pressure-dependence of decay constant k_D of delayed fluorescence*

Preliminary investigations conducted with perylene vapour showed that k_D increases with vapour pressure (fig. 6, curve 1) as found by Williams.[8] However, as illustrated by curves 2 and 3 (fig. 6), the pressure-dependence is significantly reduced by prolonged periods of irradiation; this behaviour is attributed to quenching by oxygen occluded in the solid sample which is photochemically consumed, and is eliminated if the solid is vacuum-sublimed into the cell in three stages before sealing-off (curve 4), a procedure adopted in all subsequent work.

Fig. 7 illustrates the different variations in decay constant k_D

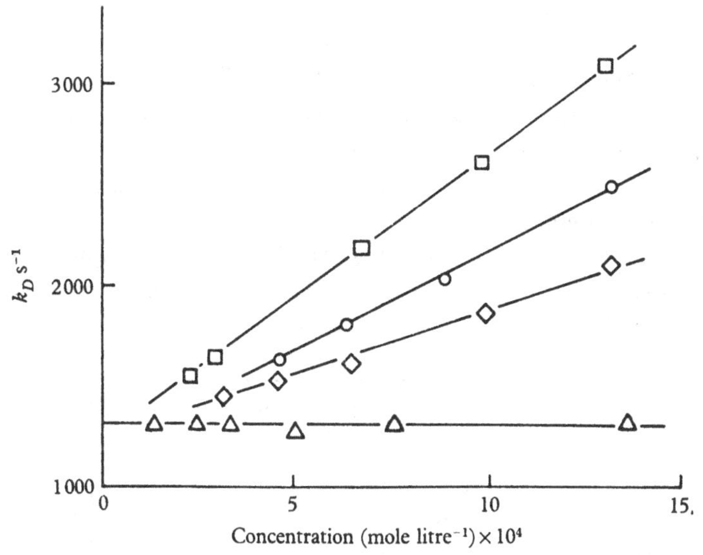

Fig. 6. Dependence of k_D concentration curve on history of perylene vapour sample: □, unsublimed; ○, after 5 h and ◇, after 24 h exposure; △, sublimed.

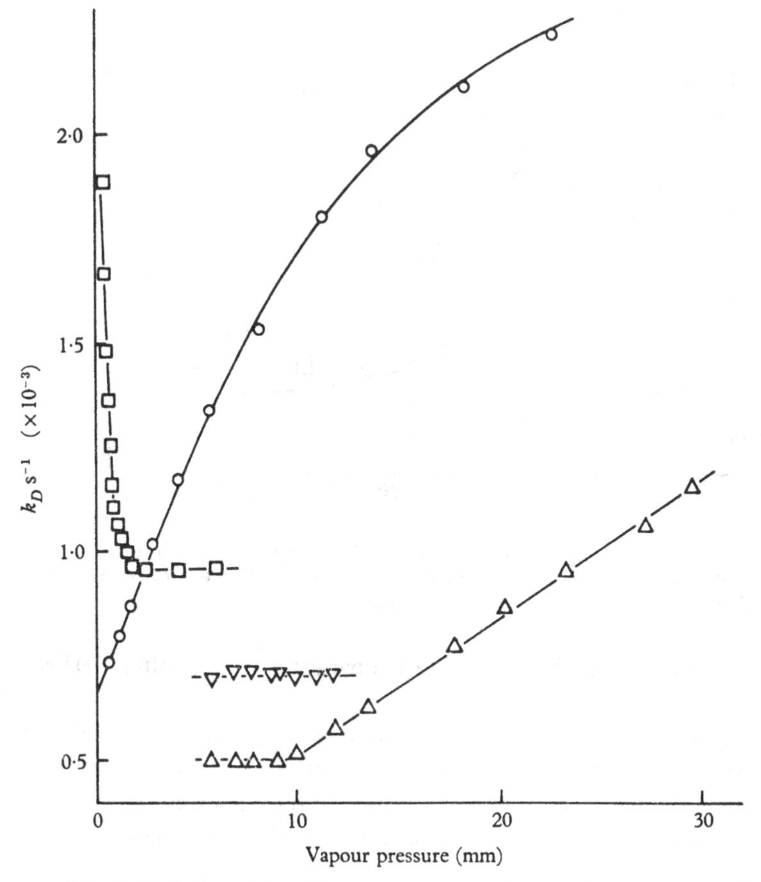

Fig. 7. Variation of k_D with vapour pressure for *o*-anthracene at 250 °C; △, pyrene at 250 °C; □, anthanthrene at 340 °C; ▽, naphthalene at 94 °C.

with hydrocarbon vapour pressure or temperature T_c of the condensed phase, which may be summarised as follows:

(i) k_D is independent of pressure over the whole pressure-range examined (naphthalene);

(ii) k_D remains constant at low pressures but increases rapidly at higher pressures (perylene, pyrene in the larger cells, phenanthrene and fluoranthene);

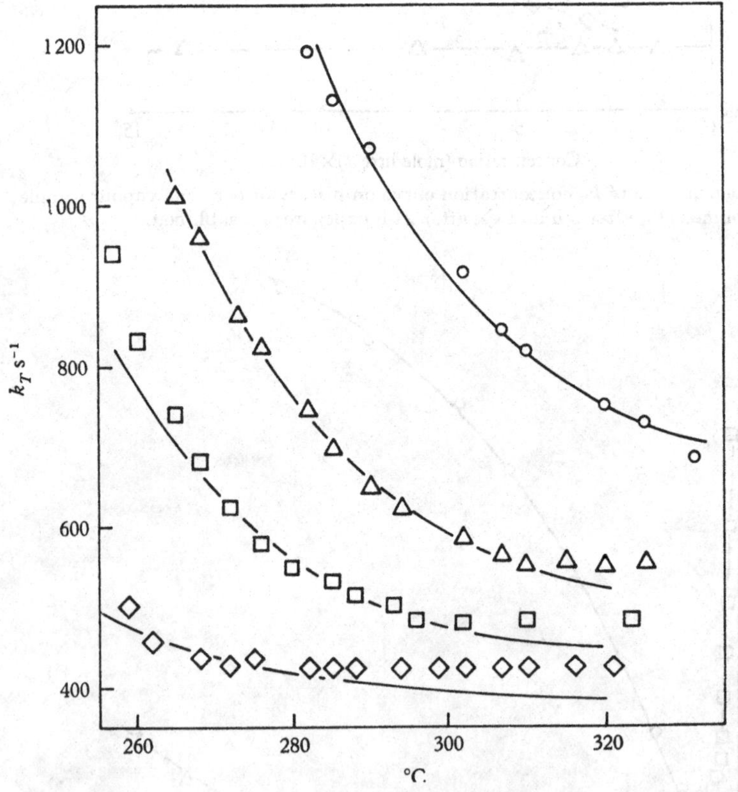

Fig. 8. Pressure-dependence of $\frac{1}{2}k_D$ for anthanthrene vapour at 340 °C in cells of diameter: 0·5 cm, ◇; 0·7 cm, ○; 1·1 cm, □, and 1·9 cm, △. Curves drawn in accordance with equation (9) and stated values of parameters.

(iii) k_D increases with pressure to a constant limiting value (anthracene);

(iv) k_D *decreases* rapidly with an increase in pressure at very low pressures to a constant minimum value (anthanthrene); the delayed fluorescence intensity exhibited by other compounds was too weak to permit reliable measurements at the same very low pressures.

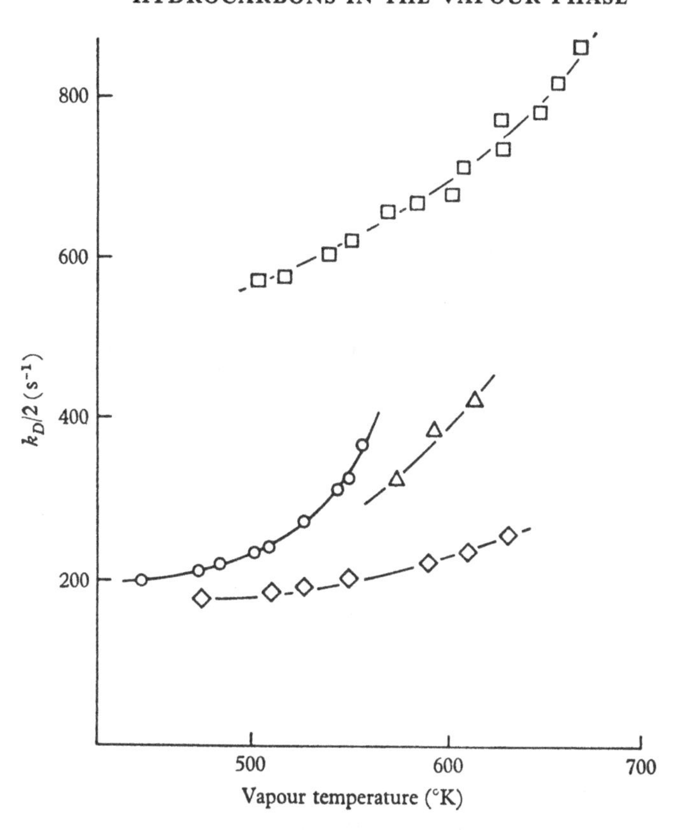

Fig. 9. Variation of k_D with vapour temperature at constant pressure for perylene (3 mm), □; phenanthrene (3 mm), ○; pyrene (5 mm), ◇; anthanthrene, △.

4. *The dependence of k_D on cell diameter*

If the *increase* in measured triplet state lifetime with anthanthrene vapour pressure at very low pressures (case (iv) above) is due to diffusion of triplet states out of the small observation zone (at the focal point of both source and detector assembly) under the influence of a concentration gradient maintained by quenching of the triplet state at the cell walls, the measured decay constant should decrease with an increase in cell diameter, which reduces this gradient; the effect of varying the cell diameter is shown in fig. 8.

5. *The temperature-dependence of k_D*

As noted by Williams[8] the temperature-coefficient of the decay constant k_D is less than would be required for a thermal (unimolecular) repopulation of the single fluorescent state, although by no means insignificant. The temperature-dependence of k_D in the pressure-independent region is shown for various compounds in fig. 9.

Discussion

Except for phenanthrene and fluoranthene, the spectral identity of delayed and normal fluorescence components establishes the lowest excited singlet state as the emitting species in each case. An interpretation of the experimental observations is presented below in terms of the following scheme:

$$^3A \rightarrow A, \quad (1)$$

$$^3A + {}^3A \rightarrow A^* + A, \quad (2)$$

$$^3A + Q \rightarrow A + {}^3Q \text{ (or } Q\text{)}, \quad (3)$$

$$^3Q + A \rightarrow Q + {}^3A, \quad (4)$$

$$^3A + {}^3Q \rightarrow A + Q^*, \quad (5)$$

$$^3A + {}^3Q \rightarrow A^* + Q, \quad (6)$$

$$^3Q + {}^3Q \rightarrow Q^* + Q, \quad (7)$$

$$^3Q \rightarrow Q, \quad (8)$$

$$A^* \rightarrow A + h\nu_A, \quad (9)$$

$$A^* \rightarrow {}^3A \text{ (or } A\text{)}, \quad (10)$$

$$Q^* \rightarrow Q + h\nu_Q, \quad (11)$$

$$Q^* \rightarrow {}^3Q \text{ (or } Q\text{)}, \quad (12)$$

where the asterisk denotes an electronically excited singlet state. The non-linear dependence of k_D on vapour pressure, exhibited by certain compounds, is attributed to collisional quenching of the triplet state ^3A by an impurity Q (process 3) present in the hydrocarbon sample; if the impurity is also an aromatic species this quenching process may involve the transfer of triplet excitation energy to produce the triplet state ^3Q of the impurity[15] in which case the reverse process 4 may be significant at the temperature employed.[12] Mixed annihilation (process 5) is undoubtedly responsible[16] for the sensitised delayed fluorescence of anthracene by phenanthrene[5] and of naphthacene by pyrene[17] in the vapour phase, and may also produce the original singlet state A^* (process 6). Finally self-annihilation of the impurity triplet state (process 7) is included to account for the sensitised delayed fluorescence of an impurity at shorter wavelengths than the normal fluorescence spectrum of fluoranthene.

The detailed treatment of delayed fluorescence parameters in terms of this scheme is considerably simplified for the condition of exponential decay, which requires that:

(i) mutual annihilation does not contribute significantly to overall relaxation of the triplet state,[1] i.e.

$$k_1 \gg k_2[{}^3A] + (k_5 + k_6)\,[{}^3Q]$$

$$k_8 \gg k_7[{}^3Q] + (k_5 + k_6)\,[{}^3A];$$

(ii) the states ${}^3A$ and ${}^3Q$ do not relax independently,[13] i.e.

$$d([{}^3Q]/[{}^3A])/dt = 0.$$

The relaxation equations are examined below for reversible and non-reversible quenching and the sensitised delayed fluorescence of an impurity, and the relevant data are collected in table 1.

Table 1. *Limiting values for triplet state decay constants in the vapour phase*

Compound	T °C	k_1 (s^{-1})
Naphthalene	93	$\leqslant 350$
Phenanthrene	230	$\leqslant 230$
Anthracene	220	$\leqslant 3000$
Fluoranthene	250	$\leqslant 270$
Pyrene	200	$\leqslant 180$
Perylene	230	$\leqslant 570$
1,12-Benzperylene	360	$\leqslant 940$
Anthracene	340	380 ± 20

1. *Non-reversible quenching of the triplet state*

If the quenching process 3 does not involve intermolecular energy transfer, or alternatively if $k_8 \gg k_4[A]$, the relaxation of ${}^3A$ is controlled by processes 1 and 3 only with

$$k_D/2 = k_T = k_1 + k_3[Q]. \tag{2}$$

$$= k_1 + k_3 q\omega/VM, \tag{3}$$

for a completely vaporised impurity in a sample of weight w, molecular weight M and impurity content q(mole/mole) in a cell of volume V. This behaviour explains the dependence of the high-pressure limiting value k_D for anthanthrene on the cell dimensions since $\frac{1}{2}k_D$ is a linear function of w/VM (fig. 10) and provides

$$\text{intercept} = k_1 = 380\,\text{s}^{-1}\,\text{at } 250\,°\text{C},$$

$$\text{slope} = k_3 q = 8{\cdot}35 \times 10^3\,\text{l. mole}^{-1}\,\text{s}^{-1}\,\text{at } 250\,°\text{C}.$$

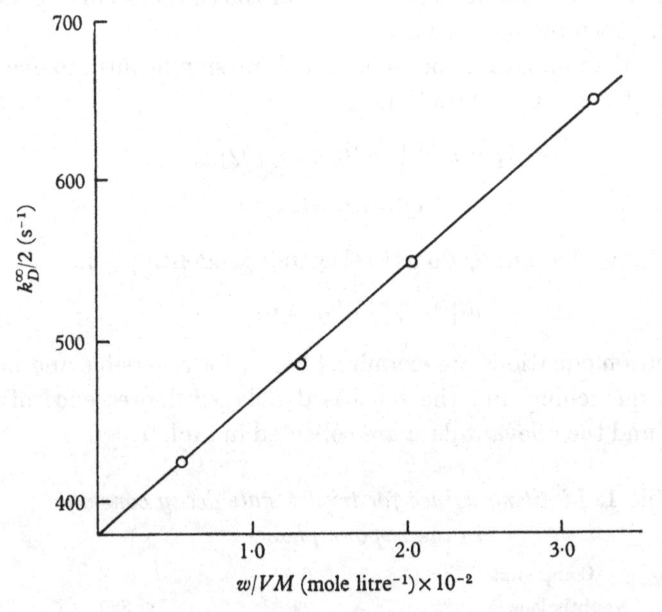

Fig. 10. Plot of k_D^∞ against w/VM for anthanthrene vapour at 340 °C in accordance with equation (3).

If it is assumed that process 3 has unit collisional efficiency for a collision diameter of 9 Å, the simple kinetic theory expression provides the estimated value

$$k_3 = 4{\cdot}33 \times 10^{11}\,\text{l. mole}^{-1}\,\text{s}^{-1}\,\text{at } 250\,°\text{C},$$

whence,
$$q \simeq 2 \times 10^{-8}\,\text{mole/mole}.$$

Incomplete vaporisation of the impurity requires that equation (1) be expressed as

$$\tfrac{1}{2}k_D = k_1 + k_3 B \exp\left(-\Delta H_Q/R_{Tc}\right), \tag{4}$$

where ΔH_Q is the heat of impurity vaporisation, T_c is the temperature of the condensed phase and B is a constant. Equation (4) is consistent with the observed variation of $\frac{1}{2}k_D$ with T_c for pyrene with

$$\Delta H_Q \simeq 51\,\text{kcal/mole}$$

corresponding to an aromatic-hydrocarbon impurity containing 32 carbon atoms.[18]

For naphthalene the experimental data are represented by

$$\tfrac{1}{2}k_D = 350\,\text{s}^{-1} \quad \text{at} \quad 93\,^\circ\text{C},$$

over the pressure range 0–12 mm Hg. Since it is unlikely that any quenching impurity (except O_2) is completely vaporised at 74 °C, the lowest condensed phase temperature, these data probably refer to the homogeneous unimolecular relaxation of the naphthalene triplet state; they certainly provide an upper limit to k_1, as in the case of perylene, where $k_D = 570\,\text{s}^{-1}$ at 230 °C, over the pressure range 0–5 mm Hg (fig. 6).

Reversible quenching of the triplet state

Processes 5, 7, 11 and 12 may be neglected in those systems which exhibit identical delayed and normal fluorescence spectra; the intensity $I_D(t)$ of the delayed component at time t after excitation cut-off is given by

$$I_D(t) = \gamma_A\{k_2[^3A]_t^2 + k_6[^3A]_t[^3Q]_t\},$$

where $\gamma_A = k_9/(k_9+k_{10})$ is the molecular fluorescence yield of A. With $d\,([^3A]/[^3Q])/dt = 0$ for an exponential decay, it is assumed that $k_3[Q] \gg k_1$ and $k_4[A] \gg k_8$, whence,

$$\frac{[^3Q]_t}{[^3A]_t} = \frac{[^3Q]_0}{[^3A]_0} = \frac{k_3[Q]}{k_4[A]}, \tag{5}$$

where $[\,]_0$ denotes the photostationary concentration, and

$$I_D(t) = \gamma_A[^3A]_t^2\{k_2 + k_3k_6[Q]/k_4[A]\},$$

or

$$k_D = -d\ln I_D/dt = -2d\ln[^3A]/dt = 2k_T. \tag{6}$$

The introduction of mixed triplet state annihilation (process 6) does not therefore affect the relationship (6) between k_D and k_T in so far as the decay curves remain exponential.

The overall rate of triplet state relaxation under these conditions is expressed by

$$-d[^3A]/dt - d[^3Q]/dt = k_1[^3A] + k_8[^3Q],$$

which rearranges with (5) to

$$\frac{-d[^3A]}{dt}\left\{1 + \frac{k_3[Q]}{k_4[A]}\right\} = [^3A]\left\{k_1 + \frac{k_3k_8[Q]}{k_4[A]}\right\}$$

with

$$\frac{k_D}{2} = -\frac{1}{[^3A]}\frac{d[^3A]}{dt} = \frac{k_1k_4[A] + k_3k_8[Q]}{[k_3[Q] + k_4[A]]}. \tag{7}$$

For a completely volatile impurity the limiting low- and high-pressure values are given by:

$$(k_D)_{[A]\to 0} = k_D^0 = 2k_8 \tag{7a}$$

$$(k_D)_{[A]\to \infty} = k_D^\infty = 2k_1 \tag{7b}$$

and since

$$\left\{\frac{d\,(k_D - k_D^0)}{d[A]}\right\}_{[A]\to 0} = \frac{2k_4(k_1 - k_8)}{k_3[Q]}, \tag{7c}$$

k_D should exhibit a positive or negative pressure-dependence according to the relative magnitudes of k_1 and k_8.

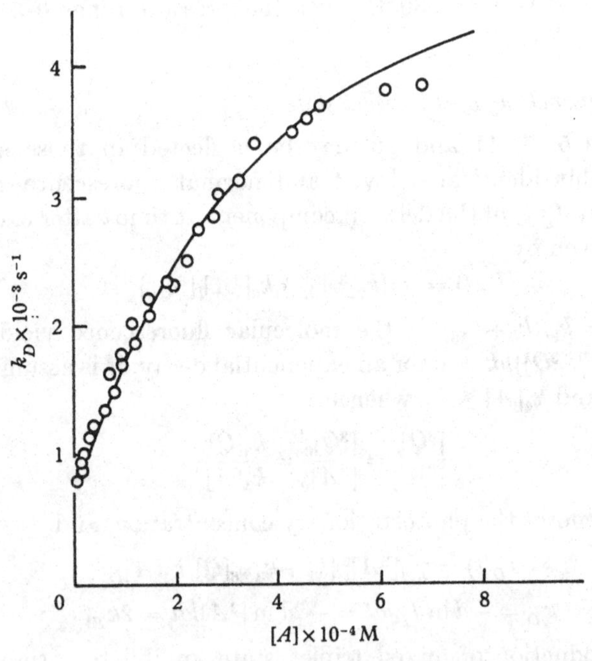

Fig. 11. Pressure-dependence of k_D for anthracene vapour at 250 °C according to equation (4).

Anthracene vapour provides a possible example of this behaviour, the solid curve in fig. 11 being drawn in accordance with equation (7) with the values

$$k_1 = 3000\,s^{-1},$$

$$k_8 = 400\,s^{-1},$$

$$k_3[Q]/k_4 = 4 \times 10^{-4}\,\text{mole l.}^{-1} \quad \text{at} \quad 220\,^\circ\text{C};$$

since $k_4 \not> k_3$ for exothermic transfer, 4×10^{-4} mole l.$^{-1}$ is an upper limit of the impurity concentration in this system.

3. *Sensitised delayed-fluorescence of the impurity*

For those vapours (phenanthrene and fluoranthene) in which the spectra of delayed and normal fluorescence are not identical, the delayed fluorescence emitter is identified as an impurity Q and the intensity of the delayed component is given by

$$I_{DQ}(t) = \gamma_{FQ}\{k_5[^3A]_t[^3Q]_t + k_7[^3Q]_t^2\}, \tag{8}$$

where $\gamma_{FQ} = k_{11}/(k_{11}+k_{12})$ is the molecular fluorescence yield of Q. If the sensitisation is non-reversible, i.e. $k_8 \gg k_4[A]$, the observation of an exponential decay requires that

$$[^3Q]_t/[^3A]_t = k_3[Q]/k_8,$$

whence
$$I_{DQ}(t) = \gamma_{FQ}[^3Q]_t^2\{k_7 + k_5k_8/k_3[Q]\}$$

and
$$-d\ln I_{DQ}/dt = k_D = -2d\ln[^3Q]/dt = 2k_{TQ}.$$

The overall triplet-state relaxation, expressed as

$$-d[^3A]/dt - d[^3Q]/dt = k_1[^3A] + k_8[^3Q],$$

becomes
$$-\frac{d[^3Q]}{dt} = \left\{1 + \frac{k_8}{k_3[Q]}\right\} = [^3Q]\left\{k_8 + \frac{k_1k_8}{k_3[Q]}\right\}$$

or
$$\tfrac{1}{2}k_D = k_{TQ} = \frac{k_8(k_1+k_3)[Q])}{k_8+k_3[Q]} \simeq k_1 + k_3[Q], \tag{9}$$

if $k_8 \gg k_3[Q]$. Equation (9), expressed in the form of equation (3) for an involatile impurity, accounts quantitatively for the variation in k_D recorded for phenanthrene and fluoranthene with the condensed phase temperature T_c (fig. 12), and provides the following data:

	$(k_1(s^{-1})$	ΔH_Q (kcal.mole^{-1})
phenanthrene	230 (230 °C)	14
fluoranthene	270 (250 °C)	29

The ΔH_Q for phenanthrene is consistent with the estimated value of 13·1 kcal mole^{-1} for the latent heat of vapourisation of anthracene,[19] known to be the impurity in this case.[20]

4. *Diffusion and surface quenching of the triplet state*

The decrease in k_D (increase in triplet state lifetime) with increase in pressure observed for anthanthrene (fig. 7) is qualitatively consistent with reversible quenching by a volatile impurity with a faster

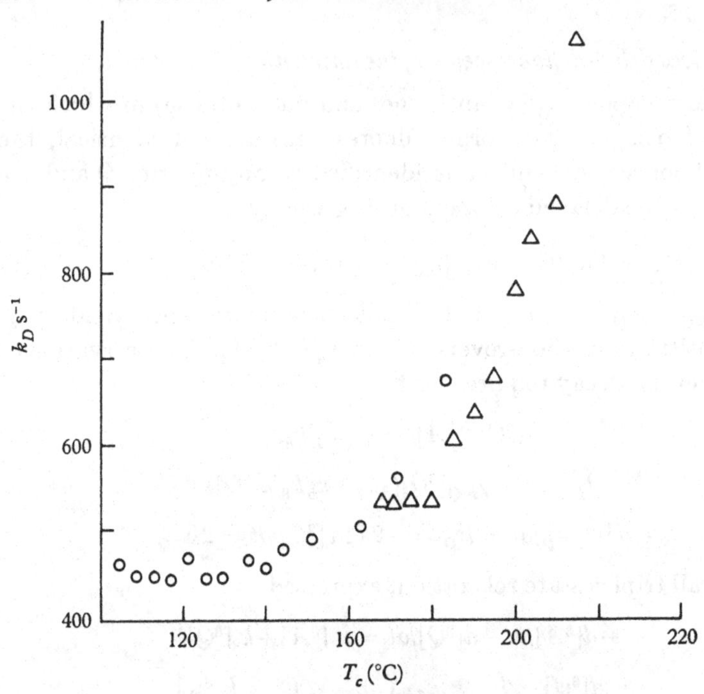

Fig. 12. Dependence of k_D with condensed phase temperature T_c for phenanthrene, ○, fluoranthene, Δ.

relaxation constant, i.e. $k_8 > k_1$ (equation 6). However, the limiting high-pressure decay constant given by $k_D^{\infty} = 2k_1$ should be independent of quencher concentration and hence of cell volume V (or radius R constant length) which is contrary to the experimental findings (fig. 8).

Since it was necessary to focus the incident radiation at the centre of the cell, and also to focus the fluorescence radiation from this small excitation zone onto the detector, in order to obtain an appreciable signal/noise ratio, diffusion of triplet states out of the observation zone under a concentration-gradient, maintained by surface quenching, may contribute to the measured decay constant k_D. The treatment of this diffusional process in three dimensions is complicated by the cylindrical symmetry of the system which imposes an angular dependence on the concentration gradient, and by homogeneous relaxation of the triplet state along the diffusion path. However a treatment[20] of the model system, in which the reacting species (^3A) diffuses from the principal axis of a cylinder of radius R to its curved walls where it is quenched, shows that diffusion from the axial observation region is characterised by the first-order rate constant

$$k_d = 2D/R^2 = \bar{c}/(\sqrt{2})\,\pi\sigma^2\mathbf{n}R^2,$$

where the diffusion-coefficient D is expressed in terms of the average molecular velocity $\bar{c}$, the collision diameter σ and molecular concentration n. If the concentration gradient is assumed to be independent of homogeneous triplet state relaxation, then the observed decay constants for delayed fluorescence are given by

$$\tfrac{1}{2}k_D = k_1 + k_3qw/VM + \bar{c}/(\sqrt{2})\,\pi\sigma\mathbf{n}R^2$$
$$= \tfrac{1}{2}k_D^{\infty} + \bar{c}/(\sqrt{2})\,\pi\sigma\mathbf{n}R^2, \qquad (10)$$

where the second term accounts for non-reversible quenching by a volatile impurity (equation (3) and subsequent discussion). The solid curves in fig. 8 are drawn [6] in accordance with equation (10) using the value of k_3q already cited and with $\sigma = 8{\cdot}7$ Å.

Temperature-dependence of k_D

Since the intensity of delayed fluorescence generally decreases with an increase in vapour temperature, while the lowest vapour temperature must exceed that of the condensed phase required to produce an appreciable pressure, the temperature range over which decay constants can be measured reliably is somewhat limited. However, for phenanthrene and pyrene a plot of $\log(\Delta k_D/2)$ against $1/T$ (where Δk_D is the increase in k_D over the lower temperature limiting value) leads to activation energies of 5 100 and 3 700 cm^{-1} respectively for first order relaxation processes of the corresponding triplet states: these are significantly lower than the energies required (6 700 and 10 200 cm^{-1}) for thermal repopulation of the fluorescent state.

Conclusions

The lifetime of the triplet state of an aromatic molecule in the gas phase is conveniently estimated from the decay constant of delayed fluorescence, under conditions such that the intensity of delayed fluorescence varies as the square of the incident light intensity and is an exponential function of the time following excitation cut-off.

However, the relatively long lifetime of the triplet state, and the high collision frequency of molecules in the gas phase promote impurity quenching effects which are already of significance in solution, with the result that the measured lifetime may reflect considerable quenching by a volatile impurity present at constant concentration throughout the whole pressure range. Moreover, under the experimental conditions necessary to observe the delayed fluorescence component, diffusion of the triplet state molecules out of the observation zone may contribute to the measured decay constant at low pressures.

Finally, the temperature coefficient of the delayed fluorescence decay constant, although small, is significant, and further investigations may well show that triplet state lifetimes in the gas phase, when corrected for impurity quenching effects, are of the same magnitude as those measured in the condensed phase at the same temperature.

This research was supported by the U.S. Department of Army through its European Research Office under contract number DA-91-591-EUC-2018.

References

1 C. A. Parker & C. G. Hatchard, *Proc. Chem. Soc.* p. 147 (1962); *Proc. Roy. Soc. (London)* A, **269**, 574 (1962).
2 B. Stevens & M. S. Walker, *Proc. Roy. Soc. (London)* A, **281**, 420 (1964).
3 G. Porter & P. West, *Proc. Roy. Soc. (London)* A, **279**, 302 (1964).
4 G. Porter & M. W. Windsor, *Disc. Faraday Soc.* **17**, 178 (1954).
5 B. Stevens, E. Hutton & G. Porter, *Nature*, **185**, 917 (1960).
6 B. Stevens & M. S. Walker, *Chem. Commun.* p. 18 (1965).
7 P. P. Dikun, *Zhur. Ekspti. et Teor. Fiz.* **20**, 193 (1950).
8 R. Williams, *J. Chem. Phys.* **28**, 577 (1958).
9 A. Jablonski, *Z. Physik. Lpz.* **94**, 38 (1935).
10 B. Stevens & P. J. McCartin, *Mol. Physics*, **3**, 425 (1960).
11 B. Stevens & E. Hutton, *Conference on Reversible Photochemical Processes*, Durham, North Carolina (April 1962).
12 B. Stevens & M. S. Walker, *Proc. Chem. Soc.* p. 181 (1963); *Proc. Roy. Soc. (London)* A, **281**, 420 (1964).
13 B. Stevens, M. S. Walker & E. Hutton, *Proc. Chem. Soc.* p. 62 (1963).
14 W. H. Melhuish, *Nature*, **184**, 1933 (1959).
15 G. Porter & F. Wilkinson, *Proc. Roy. Soc. (London)* A, **264**, 1 (1961).
16 C. A. Parker, *Proc. Roy. Soc. (London)* A, **276**, 125 (1963).
17 E. Hutton & B. Stevens, *Spectrochim. Acta*, **18**, 425 (1962).
18 H. Inokuchi, S. Shiba, T. Handa & H. Akamata, *Bull. Chem. Soc. Japan*, **25**, 299 (1952).
19 B. Stevens, *J. Chem. Soc.* p. 2973 (1953).
20 E. A. Moelwyn-Hughes, *Physical Chemistry*, p. 1169 (Pergamon Press, 1957),

DISCUSSION

ROBINSON

Rice to Robinson: I should like to make two comments which support Professor Robinson's interpretation of the lifetime of benzene: (1) One of my former students, Dr John Haebig, studied low-pressure quenching of the gas-phase fluorescence of anthracene and some other molecules. It happens that the emission from a Cd hollow cathode lamp falls on the 0–0 transition of the ground state to second electronic state of either the per hydro or per deutero anthracene (I cannot recall which at present). The emission spectra of the two compounds are very different. Excitation in the 0–0 transition leads to emission with many sharply-defined vibrational lines. Excitation above the 0–0 transition leads to emission which is broad and relatively structureless. These observations support the contention that importance should be attached to excitation into discrete vibronic levels, and that important differences of interpretation may result therefrom.

(2) I wish to repeat that there is at least one line in the benzene spectrum, a vibronic sublevel of a Rydberg transition, which lies over a vibrational *quasi* continuum and shows antiresonance in the line profile. This line profile suggests extensive state mixing, and I suspect such mixing occurs for other transitions (even at lower energy). If this interpretation is correct, Professor Robinson's proposal is supported, in a beautiful way, by the available data.

Robinson to Rice: I did not mention it in my talk, but there is other evidence of this kind in existence. Through conversations with Ian Ross, and from reading his papers, I can postulate a general rule (even though I do not like general rules) that: *Except for some Rydberg states, all but the lowest state of a given multiplicity in a complex molecule have broad structure.* This is true for the second singlet of azulene, the $^1B_{1u}$, $^1E_{1u}$, and $^3E_{1u}$ states of benzene, higher states of naphthalene, anthracene, etc. This broadening presumably comes from multiplicity-allowed vibronic interactions involving one sharp (to zero-order) state and the vibrational *quasi* continuum of another state (or states). This can happen in the gas phase in a complex molecule. The spectra in these examples do not look like the one mentioned by Rice since the Franck–Condon envelopes are fairly broad for transitions of this kind.

Rice to Robinson: I have a lot of questions but there will not be enough time to ask them. I have a short comment that I would like to make. I have done a little informal work on the coherent *vs.* hopping model for excitation transfer and more or less convinced myself that it is difficult, if not impossible, to tell the difference between the two experimentally without going to experimental time scales of 10^{-12} s or so. This is because the rate-determining step is not the transfer of excitation, which is very fast in both cases, but is the process (trapping, annihilation or whatever) that leads to the experimental detection of the transfer.

Birks to Robinson: May I congratulate the author on a masterly clarification of radiationless processes; he has explained *inter alia* why (i) the radiative lifetime of anthracene is the same in the crystal as in dilute solution, as shown by Birks and Dyson (1963)—this, by the way, favours the hopping crystal singlet exciton model, not the exciton band model, (ii) the existence of both temperature-independent and temperature-dependent intersystem crossing processes, and (iii) why the frequency factor of the latter process indicates that the probability of intersystem crossing can be of the order of unity. It is not time to abandon talking of intersystem crossing as having a spin-forbiddenness factor of 10^{-6}, as first introduced by Kasha sixteen years ago?

Windsor to Robinson: I do not understand Birk's comment that we can now ignore the spin-prohibition factor of 10^{-6}, first introduced by Kasha, in dealing with radiationless transitions. The rate constant for a radiationless transition according to Robinson's treatment is still made up of three factors, a Franck–Condon factor, a density of states factor and an electronic matrix element. Internal conversion between states of the same multiplicity, such as $S_2 \rightarrow S_1$, is predicted to be much faster than intersystem crossing, such as $S_1 \rightarrow T_1$, mainly because of the large difference in the size of the electronic matrix element connecting the initial and final states. The absence of radiation from upper singlet states leads to a rate constant $k(S_2 \rightarrow S_1)$ of at least $10^{11}\ s^{-1}$ which is 10^3 or more greater than the rates of intersystem crossing $S_1 \rightarrow T_1$.

Robinson to Windsor: Just like Dr Birks, I do not like strict rules. Because of the variation among all the factors that go into the radiationless transition rate, almost anything can happen. I would suspect that in general, though, multiplicity-forbidden rates are slower than multiplicity-allowed ones.

Birks to Robinson: The existence of pressure quenching of S_1 can be explained by excimer formation. The absence of pressure quenching of T_1 is consistent with a dissociated triplet excimer.

Robinson to Birks: Perhaps.

Silver to Robinson: Can you explain, according to the 'bed spring' model, why an excited singlet in anthracene vapour is quenched by collision with a ground singlet ($\simeq$ every collision) while the triplet is not (< 1 in every 10^5 collisions)?

Robinson to Silver: The only thing I can think of off hand is that for singlets you have the possibility of long-range transfer and subsequent trapping by adventitious impurities or by some unknown mechanism, while such mechanism would not be very efficient for triplets; but I really do not know enough about the experiments to comment intelligently.

Kommandeur to Robinson: Hutchison showed a contact interaction of neighbouring nuclear spins with the triplet electrons to be present. Could the deuteration effect of the solvent be due to a relaxation via the nuclear system? After all, you have to conserve angular momentum.

Robinson to Kommandeur: I would be willing to bet this is not the case, but certainly one should look into this possibility.

WINDSOR

Kellogg to Windsor: In corroboration of Dr Windsor's suggestion that T_2 probably lies lower still than the lowest excited triplets reported here, one can make a logarithmic plot of the rate of intersystem crossing against the energy gap for aromatic hydrocarbons where ΔE is certain, and get a relative smooth function varying from $K_{1s} \sim 10^{\circ}$ at large energy gaps and $K_{1s} \sim 10^1$ at small energy gaps. When T_1 is the 'highest known' triplet below S_1, and the log K_{1s} is plotted against $\Delta E(S_1 - T_1)$, the prints are significantly displaced from the above curve, indicating that T_2 states are generally below S_1 and that the genuine T_1–T_2 transitions lie well into the infrared ($\lambda \sim 2\mu$).

DEXTER

Robinson to Dexter: I do not understand where the big environmental perturbations on vibrational structure would come from, at least for the systems with which most persons here are familiar.

STEVENS

Robinson to Stevens: The triplet lifetimes in gaseous aromatics that you reported are still shorter than I would like to see, but they give me a much more comfortable feeling than the very short lifetimes being reported for benzene. Benzene may very well be some sort of a special case, but your talk has very elegantly brought up a number of experimental pitfalls that can occur in gas-phase measurements.

Birks and Kazzaz to Stevens: The intersystem crossing from the thermally excited triplet T_n, in perylene and pyrene, may readily occur into the excited singlet state S_1 and not into a high vibrational level of the ground singlet state S_0 as you appear to have assumed. If $T_n - S_1$ intersystem crossing occurs, it will lead to delayed fluorescence, and the reaction kinetics will need to be modified to include the process ${}^3A + H_{\text{est.}} - A^*$ (7). This corresponds to Jablonski's α-process or Parker's E-type delayed fluorescence. The activation energy W_7 of process 7 is 5300 cm^{-1} for perylene and 3600 cm^{-1} for pyrene. The magnitude of W_7 indicates that T_n is a higher triplet state, and not a high vibrational $\mathscr{H}$-level of T_1. Process 7 will occur provided that $T_1 + W_7 \geqslant S_1$ (8). Spectroscopic data in the literature (not available to us at the moment) will show whether (8) is satisfied for pyrene or perylene.

If condition (8) is satisfied by pyrene in solution, process 7 is relevant to the delayed fluorescence behaviour of this system. (See Birks and Moore, this Conference, and references therein.) In pyrene solutions, delayed monomer fluorescence (DMF) and delayed excimer fluorescence (DEF) are produced by ${}^3A - {}^3A$ association, probably via an excited state D^{**} of the excimer D^*, which then either (a) internally converts into D^* yielding DEF or (b) dissociates into A^* and A yielding DMF. k_{DT} and k_{MT}, the rate parameters of processes (a) and (b) respectively, are in the ratio $k_{DT}/k_{MT} = \alpha$, and their sum $k_{TT}(= k_{DT} + k_{MT})$ has been shown to be diffusion-controlled (Birks and Moore, *loc. cit.*). For pyrene in cyclohexane or ethanol, $\alpha = 2{\cdot}0$ at room temperature and above. Stevens (private communication) has suggested that the relative probabilities of ${}^3A$–${}^3A$ association yielding ${}^1D^{**}$, ${}^3D^{**}$ and ${}^5D^{**}$, the singlet, triplet and quintet states of D^{**}, may correspond to their multiplicities. The lowest triplet excimer state ${}^3D^*$ is dissociated (Birks and King, *Phys. Letters*, 1965), so that ${}^3D^{**}$ will also dissociate yielding A^* and A (process (b)). If

$^1D^{**}$ and $^5D^{**}$ are metastable (relative to internal conversion) they will internally convert into D^* (process (a)). Applying the Stevens hypothesis, $\alpha = (1+5)/3 = 2$, in agreement with experiment.

Below room temperature, α for pyrene in ethanol decreases below two (Moore and Munro, *Spectrochim. Acta*, in press). The $T_n \rightarrow S_1$ intersystem crossing leading to process 7 may be relevant to this behaviour since it shows an efficient mixing of the singlet and triplet manifolds in the monomer, and it suggests the possibility of similar mixing of the singlet, triplet and quintet manifolds in the excimer. Such mixing will be temperature dependent and it is to be expected that it will modify the simple 'multiplicity' value of $\alpha = 2$. A further experimental study of MF, EF, DMF, DEF and phosphorescence of pyrene solutions as a function of temperature is being made by the Manchester group to determine absolute values of k_{DT}, k_{MT} and the other rate parameters and their frequency factors and activation energies. It is hoped that these will enable the processes contributing to the temperature dependence of α to be identified, and at the same time provide a test of the Stevens multiplicity hypothesis.

Stevens to Birks and Kazzaz: If process 7 contributes significantly to the overall production of A^*, one might expect the intensity of delayed fluorescence to vary with the nth power of absorbed light intensity where $1 < n < 2$. It was precisely for this reason that the high temperature measurements were made, but experimentally it was found that $n = 2{\cdot}00 \pm 0{\cdot}05$ over the whole range; moreover, the observed activation energies are significantly lower than the energy difference between triplet and excited states.

An analysis of the experimental data for MF, EF, DMF, and DEF for pyrene in ethanol as a function of temperature was presented at the Organic Scintillator Symposium in Chicago in June 1966 (Stevens and Ban, *Molecular Crystals*, in press) in which it is shown that α does indeed fall below two at lower temperatures.

Kellogg to Stevens: A frequency factor of $10^7\,s^{-1}$ deactivation sounds very reasonable to me. What disturbs me is the high-frequency factor of 10^2 s at low temperature. Do you have any ideas on what this could be?

Stevens to Kellogg: The high-frequency factor $10^2\,s^{-1}$ at lower temperatures could reflect the rate of $T_1 \rightarrow S_0$ radiationless conversion from a vibrational level of T_1; this transition would be associated with the much lower temperature coefficient expected for population of the vibrational level.

5 TRIPLET EXCITONS

SOME COMMENTS ON THE PROPERTIES OF TRIPLET EXCITONS IN MOLECULAR CRYSTALS

STUART A. RICE

Introduction

The study of the triplet states of molecules and molecular crystals has been intensively pursued in recent years. The subject naturally divides into two parts: (*a*) the study of the triplet states of isolated molecules; (*b*) the study of triplet excitons in molecular crystals. In this paper I shall comment on both (*a*) and (*b*). It is not my intent to provide an extensive review of all that has been done, but rather to display, in abbreviated form, some of the studies made by me and my co-workers at the University of Chicago in the last three years. Extensive reports of these studies can be found elsewhere,[1] and herein I focus attention on only the major features of the analyses.

This report is concerned with three related topics: (*a*) The theory of triplet Rydberg states of molecules;[2] (*b*) The theory of triplet excitons in crystals of aromatic hydrocarbons;[3] (*c*) The theory of triplet excitons in crystalline I_2.[4] The work described in the second section (a pseudo-potential theory of triplet Rydberg states) was done in collaboration with Mr Andrew Hazi, that in the third section with Professors Joshua Jortner, Sang-il Choi and Robert Silbey, and that in the fourth section with Dr Ian Hillier.

Where do we stand in our understanding of the nature of triplet excitons? To answer this question we must pose many others, perhaps the most important of which is the following: do the triplet exciton states of the crystal bear a unique one-to-one correlation with the properties of the parent molecule, or must we also consider some states which are unique to the crystalline phase? At present no unequivocal answer to this question can be given, in part because the available wavefunctions for molecules are sufficiently inaccurate that quantitative arguments must be viewed with reserve. It was this situation, namely the absence of good molecular wave functions, that served as one of the stimuli to the development of the theory outlined on pages 266–270. Rice, Jortner and co-workers believe that even in the crystalline aromatic hydrocarbons, charge delocalisation is of some importance.[1] The rationale for, and the results of their analyses, are outlined on pages 280–300. The belief that intermediate excitons are appropriate to the description of many molecular solids stimulated the study, outlined on pages 300–307, of crystalline I_2, where some charge delocalisation is known to be a feature of the crystalline state. At present, it can be said that the available theory, including delocalisation effects, is able to account for the principal features of the experimental observations. There have even been some cases of striking quantitative agreement between theory and experiment, but such agreement must not be overemphasised because of the many approximations inherent in the several analyses employed. As will be clear from the following, there is ample room for more experimental work, more theoretical studies and, most important, more new and novel ideas.

A pseudo-potential theory of triplet Rydberg states

Hazi & Rice have developed a description of atomic and molecular Rydberg states using an effective Hamiltonian in the one-electron approximation. The advantage of this analysis lies in the possibilities of exploiting the physical concepts associated with an effective potential, i.e. the formalism suggests new approximations applicable to polyatomic molecules for which a treatment of all electrons is presently impractical. For this reason they introduce a pseudo-potential which represents the interaction between the outer electron and the core. This pseudo-potential has two parts: the Hartree–Fock potential of the core and the potential which prevents the collapse of a variational wavefunction for the outer electron.

The pseudo-potential formalism and Hartree–Fock formalism are related as follows. The Hartree–Fock wavefunction of the closed shell ground state of an atom or molecule is an antisymmetrised product of one-electron spin-orbitals. Each spin-orbital is an eigenfunction of the one-electron Hartree–Fock operator, F, corresponding to orbital energy ϵ_i:[5]

$$F\phi_i = \epsilon_i \phi_i, \tag{1}$$

$$\langle \phi_i | \phi_j \rangle = \delta_{ij}. \tag{2}$$

Consider some valence orbital, ϕ_v. Cohen & Heine,[6] and later Austin, Heine & Sham,[7] have shown that for every valence orbital ϕ_v there exists a set of pseudo-wavefunctions, χ_v, which satisfy the following equation:

$$(F+V_R)\chi_v = \epsilon_v \chi_v, \tag{3}$$

where

$$\epsilon_v = \langle \phi_v | F | \phi_v \rangle, \tag{4}$$

and the potential V_R is

$$V_R \psi = \sum_i^{\text{core}} \langle g_i | \psi \rangle \phi_i. \tag{5}$$

The functions g_i are arbitrary. In general V_R is a non-local potential and is positive for much of the range of interest. The important point to note is that the eigenvalues in equations (1) and (3) are the same.

The pseudo-wave function χ_v can be expanded in terms of the eigenfunctions of F, since these functions, ϕ_i, form a complete set:

$$\chi_v = \phi_v + \sum_i^{\text{core}} a_i \phi_i. \tag{6}$$

The summations in equations (5) and (6) are over the core eigenfunctions of F. The orthogonality condition (2) permits us to write

$$a_c = \langle \phi_c | \chi_v \rangle, \tag{7}$$

but it must be recognised that the expansion coefficients, a_c, are arbitrary to the same extent as the functions g_i in equation (5).

In the Hartree–Fock theory of open-shell configurations the core and valence orbitals are eigenfunctions of different one-electron operators.[8] The derivation of equation (5), however, explicitly uses the fact that for a closed shell the core and valence orbitals are eigenfunctions of the *same* one-electron operator. Thus, to use the pseudo-potential for the description of open-shell configurations the form of the pseudo-potential must be modified.

Let the core orbitals, ϕ_c, and the valence orbitals, ψ_v, be the eigenfunctions of operators F and G, respectively. Then

$$\left.\begin{aligned} F\phi_i &= \epsilon_i \phi_i, \quad \langle \phi_i | \phi_j \rangle = \delta_{ij}; \\ G\psi_i &= \eta_i \psi_i, \quad \langle \psi_i | \psi_j \rangle = \delta_{ij}. \end{aligned}\right\} \tag{8}$$

Further, let

$$\Delta\phi = (G-F)\phi, \tag{9}$$

where ϕ is an arbitrary function. The occupied valence and core orbitals are orthogonal to each other by assumption in both the closed-shell and open-shell Hartree–Fock theories,[8] i.e.

$$\langle \psi_v | \phi_c \rangle = 0. \tag{10}$$

We now seek a potential V_R with the property that

$$(G+V_R)\chi_v = \eta_v \chi_v, \tag{11}$$

where
$$\eta_v = \langle\psi_v|G|\psi_v\rangle / \langle\psi_v|\psi_v\rangle, \tag{12}$$

and
$$\chi_v = \psi_v + \sum_c \langle\phi_c|\chi_v\rangle \phi_c. \tag{13}$$

Hazi & Rice show that[(2)]

$$V_R\chi = \sum_c (\langle g_c|\chi\rangle \phi_c - \langle\phi_c|\chi\rangle \Delta\phi_c), \tag{14}$$

where, as before, the functions g_c are arbitrary. It is easily demonstrated that equation (14) has the correct form and that the eigenvalues of the pseudo-Hamiltonian $G+V_R$ are the same as the corresponding eigenvalues of the Hartree–Fock operator, G. The pseudopotential defined in (14) projects any χ onto the space spanned by the core eigenfunctions of the operator F.

We now consider the triplet Rydberg states of atoms and molecules which in the ground state have a closed-shell configuration, and for which the excitation involves a transition from a non-degenerate orbital to another non-degenerate orbital of the same symmetry species.

We assume that the $2N$ electron wavefunction of the Rydberg state may be represented as the single determinant

$$[(2N)!]^{-\frac{1}{2}} \det |\phi_1(1)\,\overline{\phi}_1(2) \ldots \overline{\phi}_N(2N-2)\,\phi_N(2N-1)\,\phi_{N+1}(2N)|, \tag{15}$$

where the superscript bar indicates β spin, and the absence of a bar α spin. For a triplet state, the electrons in orbitals ϕ_N and ϕ_{N+1} have parallel spins (either α or β) and form the open shell. All the other space orbitals are doubly occupied, and form the closed shell. We shall refer to ϕ_{N+1} as the Rydberg orbital, and $\phi_1 \ldots \phi_N$ as core orbitals.

In the Hartree–Fock approximation, the energy of the Rydberg state is given by

$$E_R = 2\sum_k^{N-1} H_k + H_N + H_{N+1} + \sum_{k,\,l}^{N-1} (2J_{kl} - K_{kl}) + \sum_k^{N-1} (2J_{kN} - K_{kN} + 2J_{kN+1}) + J_{NN+1} - K_{NN+1}, \tag{16}$$

where H_i, J_{ij} and K_{ij} are the usual one-electron, coulomb, and exchange integrals, respectively, involving the space orbitals ϕ_i and ϕ_j. The orbitals of the open shell, and in particular the Rydberg orbital ϕ_{N+1}, are determined from the open-shell Hartree–Fock equation

$$G\phi_i = \eta_i \phi_i, \tag{17}$$

$$G\psi = \left[H + \sum_k^{N-1} (2J_k - K_k) + J_N - K_N + J_{N+1} - K_{N+1}\right]\psi$$
$$+ \sum_k^{N-1} [\langle \phi_k | K_N + K_{N+1} | \psi \rangle \phi_k + \langle \phi_k | \psi \rangle (K_N + K_{N+1}) \phi_k], \tag{18}$$

where ψ is an arbitrary function, and H, J_i, K_i are the usual one-electron, coulomb and exchange operators, respectively, involving orbital ϕ_i. The last term displayed in equation (18) is the closed-shell exchange coupling operator defined by Roothaan.[8]

We now assume that the $2N-1$ electron wavefunction of the core is the wavefunction of the doublet ground state of the positive ion, i.e.

$$[(2N-1)!]^{-\frac{1}{2}} \det |\phi_1(1)\bar{\phi}_1(2) \ldots \bar{\phi}_{N-1}(2N-2)\phi_N(2N-1)|. \tag{19}$$

The corresponding Hartree–Fock energy is

$$E_{\text{core}}^{\text{ion}} = 2\sum_k^{N-1} H_k + H_N + \sum_{k,l}^{N-1} (2J_{kl} - K_{kl}) + \sum_k^{N-1} (2J_{kN} - K_{kN}). \tag{20}$$

The core orbitals ϕ_i are eigenfunctions of the open-shell Hartree–Fock operator F, with eigenvalues ϵ_i:

$$F\psi = [H + \sum_k^{N-1} (2J_k - K_k) + J_N - \tfrac{3}{2}K_N]\psi$$
$$+ \sum_k^{N-1} [\langle \phi_k | K_N | \psi \rangle \phi_k + \langle \phi_k | \psi \rangle K_N \phi_k]$$
$$+ \tfrac{1}{2}[\langle \phi_N | K_N | \psi \rangle \phi_N + \langle \phi_N | \psi \rangle K_N \phi_N], \tag{21}$$

$$F\phi_i = \epsilon_i \phi_i \quad (i = 1, \ldots, N). \tag{22}$$

The last two terms displayed on the right-hand side of equation (21) are the closed- and open-shell exchange operators for the positive ion. Taking the expectation value of operator G with respect to the Rydberg orbital, ϕ_{N+1}, we obtain

$$\eta_{N+1} = H_{N+1} + \sum_k^{N-1} (2J_{kN+1} - K_{kN+1}) + J_{NN+1} - K_{NN+1}. \tag{23}$$

By comparison of equations (16), (20) and (23) we find that

$$E_R = E_{\text{core}}^{\text{ion}} + \eta_{N+1}. \tag{24}$$

If the core orbitals $\phi_1 \dots \phi_N$ are known, then in principle equation (17) can be solved, the resulting eigenvalue η_{N+1} being the binding energy of the Rydberg electron. Of course, the eigenvalue problem must be solved by using an iterative procedure, since the Rydberg orbital ϕ_{N+1} enters G through the operators J_{N+1} and K_{N+1}.

The usual procedure for solving equation (17) involves the expansion of ϕ_{N+1} in a finite basis set, each function of which is orthogonal to the core orbitals. However, η_{N+1} may be obtained from a pseudo-wavefunction χ, not orthogonal to the core orbitals, and which is an eigenfunction of the pseudo-Hamiltonian, $G+V_R$, with V_R given by equation (14). All effects of orthogonality are included in the non-local potential.

An explicit expression for the pseudo-potential follows once the operator Δ has been determined. Using the definition of Δ and subtracting equation (21) from equation (17) we obtain

$$\begin{aligned}\Delta\psi = {} & (J_{N+1}-K_{N+1}+\tfrac{1}{2}K_N)\psi \\ & + \sum_k^{N-1} [\langle\phi_k|\,K_{N+1}\,|\psi\rangle\phi_k + \langle\phi_k|\psi\rangle K_{N+1}\phi_k] \\ & - \tfrac{1}{2}[\langle\phi_N|\,K_N\,|\psi\rangle + \langle\phi_N|\psi\rangle K_N\phi_N], \qquad (25)\end{aligned}$$

where ψ is an arbitrary function. Operating with Δ on the core functions $\phi_1 \dots \phi_N$, and using the fact that the core orbitals are mutually orthogonal, we find that

$$\begin{aligned}\Delta\phi_k = {} & (J_{N+1}-K_{N+1})\phi_k + (\tfrac{1}{2}K_N + K_{N+1})\phi_k + \sum_l^{N-1} \langle\phi_l|K_{N+1}|\phi_k\rangle\phi_l \\ & - \tfrac{1}{2}\langle\phi_N|\,K_N\,|\phi_k\rangle\phi_N \quad (k = 1, \dots, N-1), \qquad (26)\end{aligned}$$

$$\Delta\phi_N = (J_{N+1}-K_{N+1})\phi_N + \sum_l^{N-1} \langle\phi_l|\,K_{N+1}\,|\phi_N\rangle\phi_l - \tfrac{1}{2}\langle\phi_N|\,K_N\,|\phi_N\rangle\phi_N. \qquad (27)$$

To proceed further it is necessary to select that set of functions, g_i, which enter into the expression defining V_R. Since the functions g_i are arbitrary, physical criteria suitable to different situations may be used to select different forms for the g_i. For example, Cohen & Heine suggested that the g_i be chosen so that the pseudo-wavefunction be as smooth as possible, or so that the cancellation between the kinetic energy and the potential be most complete, etc. The only important point to bear in mind is that as soon as a finite basis set for the pseudo-wavefunction is selected the exact form of the pseudo-potential is determined. For, once a finite function space is selected for the pseudo-wave function, the projection of this space onto the space spanned by the core orbitals is also determined, and the pseudo-potential is essentially the operator which projects any wave-function onto the space spanned by the core functions. Therefore, in general it is impossible to choose independently the pseudo-potential and a finite basis set for the pseudo-wave function.

There is one form of the pseudo-potential, the form originally proposed by Phillips & Kleinman, which allows the choice of an arbitrary finite basis set.[9] The valence function ϕ_{N+1} and the (first N) core eigenfunctions of the operator F (see equation (22)) are all eigenfunctions of the Phillips–Kleinman pseudo-Hamiltonian, with identical eigenvalues. In other words, for this choice of pseudo-potential the core orbitals are degenerate (with eigenvalue η_{N+1}) with respect to $(G+V_R)$. The following equation

$$(G+V_R)\phi_k = (F+\Delta)\phi_k + \sum_l^N \langle\phi_l|\phi_k\rangle[(\eta_{N+1}-\epsilon_l)\phi_l - \Delta\phi_l], \quad (28)$$

together with equation (22) and the orthonormality condition,

$$\langle\phi_k|\phi_l\rangle = \delta_{kl},$$

lead to

$$(G+V_R)\phi_k = \eta_{N+1}\phi_k \quad (k \leqslant N). \quad (29)$$

The degeneracy of the core orbitals in this case implies that the expansion of the pseudo-wave function in terms of the valence and core orbitals (see equation (6)) is still arbitrary, even though the functions g_i defining the pseudo-potential have been selected. The remaining arbitrariness in the expansion of χ can be removed by expanding the pseudo-wave function in a suitable finite basis.

Using equations (18), (26) and (27), we obtain the following explicit expression for the pseudo-Hamiltonian:

$$\begin{aligned}(G+V_R)\chi_i = {} & \left[H + \sum_k^{N-1}(2J_k - K_k) + J_N - K_N + J_{N+1} - K_{N+1}\right]\chi_i \\ & + \sum_k^{N-1}[\langle\phi_k|\,K_N + K_{N+1}\,|\chi_i\rangle\phi_k + \tfrac{1}{2}\langle\phi_k|\chi_i\rangle K_N\phi_k] \\ & + \sum_k^{N}\langle\phi_k|\chi_i\rangle(\eta_i - \epsilon_k)\phi_k - \sum_k^{N}\langle\phi_k|\chi_i\rangle[(J_{N+1} - K_{N+1})\phi_k \\ & + \sum_l^{N-1}\langle\phi_l|\,K_{N+1}\,|\phi_k\rangle\phi_l - \tfrac{1}{2}\langle\phi_N|\,K_N\,|\phi_k\rangle\phi_N].\end{aligned} \quad (30)$$

Let the normalised Rydberg orbital be written in the form

$$\phi_{N+1} = \frac{1}{N}(\chi_i - \sum_k^N \langle\phi_k|\chi_i\rangle\phi_k), \tag{31}$$

$$N = [1 - \sum_k^N |\langle\phi_k|\chi_i\rangle|^2]^{\frac{1}{2}}. \tag{32}$$

Then using equation (31) and the identity

$$(J_{N+1} - K_{N+1})\phi_{N+1} = 0, \tag{33}$$

we can reduce equation (32) to the form

$$(G+V_R)\chi_i = \left[H + \sum_k^{N-1}(2J_k - K_k) + J_N - K_N\right]\chi_i + \sum_k^{N-1}\{[\langle\phi_k|K_N|\chi_i\rangle + N\langle\phi_k|K_{N+1}|\phi_{N+1}\rangle]\phi_k + \tfrac{1}{2}\langle\phi_k|\chi_i\rangle K_N\phi_k\} + \sum_k^N\langle\phi_k|\chi_i\rangle(\eta_i - \epsilon_k)\phi_k + \tfrac{1}{2}\sum_k^N\langle\phi_k|\chi_i\rangle\langle\phi_N|K_N|\phi_k\rangle\phi_N. \tag{34}$$

Note that equation (34) must be solved interatively, since both the Rydberg orbital, ϕ_{N+1}, and the energy, η_{N+1}, enter into the pseudo-Hamiltonian $G+V_R$. Of course, one can always use an approximate form of equation (34) to obtain an approximate solution for η_{N+1} and χ. Indeed, our numerical calculations indicate that the only term in equation (34), which explicitly depends on the Rydberg function, i.e. $\langle\phi_k|K_{N+1}|\phi_{N+1}\rangle$, can be neglected for all practical purposes.

Using the formalism outlined, Hazi & Rice have studied the ^3S states of He and Be. For He,

$$(G+V_R)\psi = (H + J_{1s} - K_{1s})\psi + \langle\phi_{1s}|\psi\rangle \times [\eta_R - \epsilon_{1s} + \tfrac{1}{2}\langle\phi_{1s}\phi_{1s}|\phi_{1s}\phi_{1s}\rangle]\phi_{1s}. \tag{35}$$

Equation (35) was solved by iteration using as the basis set for the Rydberg orbital an orthogonalised set of Slater type orbitals with fixed orbital exponents. The expansion coefficients were obtained from a variational calculation of the energy. The results obtained are shown in tables 1 and 2.

Table 1. *Basis sets of Slater type orbitals for He ^3S states*

Core orbital: 1s (2·0)†

Rydberg basis no. 1:				
2s (0·575),	3s (0·33333),	4s (0·25),	5s (0·2),	6s (0·16667)
Rydberg basis no. 2:				
2s (0·575),	3s (0·33333),	4s (0·25),	5s (0·2),	6s (0·16667)
1s (0·575),	2s (0·33333),	3s (0·25),	4s (0·2),	5s (0·16667)

† Notation for Slater type orbitals: $ns(\xi) = [(2n)!]^{-\frac{1}{2}}(2\xi)^{n+\frac{1}{2}}r^{n-1}e^{-\xi r} \times Y_0^0(\theta,\phi)$; $Y_0^0(\theta,\phi) = (4\pi)^{-\frac{1}{2}}$.

Table 2. *Comparison of calculated and experimental term values for He ^3S states*

	1s, 2s	1s, 3s	1s, 4s	1s, 5s	1s, 6s
Exp. term values†	38 455	15 074	8 013	4 964	3 375
Basis set no. 1	37 848	14 756	7 773	4 725	2 872
	607	318	240	239	503
Basis set no. 2	38 242	15 031	7 997	4 956	3 370
	213	43	16	8	5
Davidson‡	38 233	15 021	7 991	—	—
	222	53	22	—	—

† Term values are given in units of cm^{-1}. ‡ Reference (10).

For the Be atom, the analogue of equation (35) is

$$\begin{aligned}(G+V_R)\psi = {} & [H+2J_{1s}-K_{1s}+J_{2s}-K_{2s}]\psi+\tfrac{1}{2}\langle\phi_{1s}|\psi\rangle K_{2s}\phi_{1s}\\ & +[\langle\phi_{1s}|K_{2s}|\psi\rangle+N\langle\phi_{1s}\phi_R|\phi_R\phi_R\rangle]\phi_{1s}\\ & +\langle\phi_{1s}|\psi\rangle(\eta-\epsilon_{1s})\phi_{1s}+\langle\phi_{2s}|\psi\rangle(\eta-\epsilon_{2s})\phi_{2s}\\ & +\tfrac{1}{2}[\langle\phi_{1s}|\psi\rangle\langle\phi_{1s}\phi_{2s}|\phi_{2s}\phi_{2s}\rangle+\langle\phi_{2s}|\psi\rangle\langle\phi_{2s}\phi_{2s}|\phi_{2s}\phi_{2s}\rangle]\phi_{2s}\\ N = {} & [1-|\langle\phi_{1s}|\psi\rangle|^2-|\langle\phi_{2s}|\psi\rangle|^2]^{\frac{1}{2}}.\end{aligned} \quad (36)$$

Equation (36) was solved by iteration for each Rydberg orbital.

Hazi & Rice also carried out one calculation (using basis set no. 4) with an approximate form of equation (36). The term which explicitly depends on the Rydberg orbital, i.e. $\langle\phi_{1s}\phi_R|\phi_R\phi_R\rangle$, was neglected. The binding energies obtained differed by less than 1 cm^{-1} from the energies obtained using the full pseudo-Hamiltonian. The results of the calculations are displayed in tables 3 and 4.

It is important to remark that in the numerical calculations cited no attempt was made to optimise the orbital exponents of the basis orbitals, or to include correlation effects (apart from what correlation is inherent in calculating term values instead of absolute energies). Thus, a comparison of the calculations of Hazi & Rice with experimental data does not test fully the adequacy of the theory. Nevertheless, despite the lack of optimisation, the comparison between theory and experiment displayed in tables 2 and 4 clearly shows that accurate calculations of the energies of Rydberg states can be made.

A more revealing comparison is that between the theory discussed herein and other calculations of similar complexity. There have been many calculations of the ^3S (1s, 2s) state of the He atom. Recently,

Table 3. *Basis sets of Slater type orbitals for Be ^3S states*

Core basis no. 1†
$1s$ (6·5),‡ $1s$ (3·4), $1s$ (1·2), $2s$ (6·5), $2s$ (3·4), $2s$ (1·2)

Core basis no. 2§
$1s$ (4·0), $3s$ (4·549), $3s$ (3·451), $3s$ (2·0103), $3s$ (1·3309)

Rydberg basis no. 1
$3s$ (0·38333), $4s$ (0·25), $5s$ (0·2), $6s$ (0·1667)

Rydberg basis no. 2
$3s$ (0·43333), $4s$ (0·29), $5s$ (0·23), $6s$ (0·18667)

Rydberg basis no. 3
$3s$ (0·38333), $4s$ (0·25), $5s$ (0·2), $6s$ (0·16667)
$2s$ (0·38333), $3s$ (0·25), $4s$ (0·2), $5s$ (0·16667)

Rydberg basis no. 4
$3s$ (0·43333), $4s$ (0·29), $5s$ (0·23), $6s$ (0·18667)
$2s$ (0·575), $3s$ (0·29), $4s$ (0·23), $5s$ (0·18667)

† Taken from reference (11).
‡ Notation for Slater type orbitals same as in table 1.
§ Taken from reference (12).

Table 4. *Comparison of calculated and experimental term values for Be ^3S states*

	$1s^22s\,3s$	$1s^22s\,4s$	$1s^22s\,5s$	$1s^22s\,6s$
Experimental term value†	23 110	10 685	6 183	4 031
Basis set no. 1	21 746	10 179	5 921	3 856
	1 364	506	262	175
Basis set no. 2	21 980	10 344	6 021	3 914
	1 130	341	162	117
Basis set no. 3	22 010	10 369	6 046	3 960
	1 100	316	137	71
Basis set no. 4	22 008	10 367	6 046	3 961
	1 102	318	137	70

† Term values are given in units of cm^{-1}.

Sharma & Coulson[13] obtained an energy differing by only 234 cm^{-1} from the experimental energy with a calculation retaining the three leading terms in the expansion of the energy in inverse powers of the nuclear charge. Cook & Murrell,[14] who used a single fully optimised Slater type orbital for the Rydberg orbital and also varied the exponent of the $1s$ core orbital, obtained an energy within 615 cm^{-1} of the observed energy. The energy of the $2s$ state as calculated by Hazi & Rice differs from the experimental energy by only 213 cm^{-1} and thus compares favorably with those obtained in the other calculations cited. It should be remarked that to reduce the difference between the calculated and observed energies to 4 cm^{-1}, Weiss[15] found it necessary to use a 19-term configuration interaction wavefunction in a calculation of much greater complexity than that reported herein. Recently, Davidson has reported the numerical results of a single configuration calculation on the three lowest ^3S states of He.[10] The Hazi–Rice binding energies are slightly better than Davidson's, but the overall agreement between the two sets of computed energies is very encouraging. Davidson's results are included in table 2 for comparison.

In developing the formalism we have assumed that it was sufficiently accurate to represent the core wavefunction of the neutral excited atom (molecule) by the ground state wavefunction of the corresponding ion. This appears to be a good approximation for the ^3S Rydberg states of both He and Be. However, we should note that the assumption made is one of convenience, not necessity. For example, one can use the wavefunction obtained from the ground state wavefunction of the neutral atom (molecule) by the removal of one electron as the core function for the Rydberg states.

It is worth while, before considering extension to molecules, to examine in physical terms the role played by the pseudo-potential. To do this it is convenient to consider the binding energies of the valence electrons in the alkali atoms. Consider, as a first approximation, that the valence orbitals are $1s$ orbitals of some effective Hamiltonian, say of the form $-\frac{1}{2}\nabla^2 - Z_c/r$. The binding energy is $-\frac{1}{2}Z_c^2$. We have argued in this section that the exclusion of the valence electron by the core is of prime importance. Suppose, as the simplest next approximation, that the valence electron sees no potential inside the core, which begins at some distance r_c from the nucleus. This implies that $V_R = 0$, $r < r_c$, to use our previous notation. We represent this pseudo-potential as $\Theta(r < r_c)(1/r)$ where $\Theta(r < r_c)$ is the unit step function, which has the values unity for $r \geqslant r_c$, and zero for $r < r_c$. If a $1s$ hydrogenic orbital is integrated over this potential, Z_c is obtained as a function of r_c. The function so obtained has a minimum at $Z_c^{-1} = 1{\cdot}236 r_c$, which we use as a crude estimate of Z_c for each atom. But the kinetic energy is also determined by Z_c. Assuming that r_c is proportional to the size of the atom, one is led to the simple functional relationship

$$E = -\frac{A}{r_{\text{atom}}} + \frac{B}{r_{\text{atom}}^2}.$$

If A and B are now determined from the ionisation energies of Li and Cs, it is found that the above relationship reproduces the ionisation energies of Na, K and Rb with an error of only a few per cent. Thus, the introduction of the simplest possible pseudo-potential serves to accurately describe these atomic states.

There are two deductions which can be drawn from the elementary argument of the preceding paragraph:

(*a*) Despite its complex appearance, the role of the pseudo-potential can be simply and accurately described as altering the potential near the ion cores so as to simultaneously account for the deep coulomb well, the increased kinetic energy of the electron in that well, and the requirements imposed by the Pauli principle.

(*b*) In the absence of suitable functions, or for the construction of simple analyses, it is often sufficiently accurate to make guesses at the form of the pseudo-potential which will reproduce the major features of the complete non-local potential.

Condition (*a*) is, of course, just a restatement of the philosophy to the introduction of the leading pseudo-potential analysis. However, in so far as it leads to (*b*), the analysis may be used in new ways.

Consider, for example, the Rydberg states of complex molecules. Now it seems unlikely that useful SCF functions for benzene or larger molecules will become available in the near future. But we might ask whether or not a guess at the effective potential of the core might not provide a pseudo-Hamiltonian of sufficient accuracy that the Rydberg states of complex molecules could be consistently classified and ordered. One might assume, for example, that the effective potential of the benzene positive ion core is adequately approximated by the sum of the pseudo-potentials of the C (and H?) atoms in their appropriate valence state (*s*). The pseudo-Hamiltonian then has the same symmetry as the multiple field and simple calculations are possible.

To examine the hypothesis that the molecular pseudo-potential is adequately approximated as a sum of atomic pseudo-potentials, Hazi & Rice[16] have studied some of the Rydberg states of H_2, and N_2 and benzene. We here consider only the studies of H_2. The problem posed can be rephrased in the following terms: Is it possible to find a simple 'model potential', V_M, such that, to an adequate approximation, the operator $T+V_M$ will have the same eigenvalues as the operator $T+V_R$. Recently, Abarenkov & Heine have investigated several model potentials with adjustable parameters for the case of atoms, and found that these potentials could adequately represent the potential of ions with closed-shell electron configurations. They obtained the best results with a potential of the form

$$V_M(r) = \sum_l V_l(r) P_l(\theta, \phi), \tag{37}$$

$$V_l(r) = A_l, \quad (r < R_M), \tag{38}$$

$$V_l(r) = -Z/r, \quad (r \geqslant R_M), \tag{39}$$

where $P_l(\theta, \phi)$ is the projection operator on the subspace of spherical harmonics with the same value of l, and Z is the net charge on the ion. The two adjustable parameters in V_M are the constant A_l and the range R_M. R_M is usually set equal to the radius of the ion.

What general criteria can be used to define a model potential for a molecule?

(*a*) The model potential must have simple mathematical properties so that the solution of the eigenvalue problem, once the potential is specified, is not too difficult.

(*b*) The model potential must approach the potential of a monopole, $-Z/r$, in the limit as $r \to \infty$.

(*c*) In molecules the quantum defect is determined by the symmetry of the charge distribution in the core, penetration of the Rydberg electron into the core, and by restrictions imposed on the wavefunction of the Rydberg electron by the Pauli principle. Accordingly, V_M must have the proper symmetry and should properly account for the penetration and exclusion effects.

Consider the case of H_2. We adopt the model potential

$$\left.\begin{aligned} V_M^{(1)} &= \frac{1}{2}\left(-\frac{1}{r_A}-\frac{1}{r_B}\right) \quad (r \geqslant R_0), \\ V_M^{(1)} &= A \quad (r < R_0). \end{aligned}\right\} \tag{40}$$

Clearly (40) has the correct symmetry and satisfies the asymptotic condition as $r_A \to \infty$, $r_B \to \infty$, $R_{AB} =$ constant, where r_A and r_B are the distances of the electron from nucleus A and nucleus B, and R_{AB} the distance between nuclei. The form of the potential for $r < R_0$ is suggested by the electron density map of H_2^+, which shows that for $R_{AB} = 2{\cdot}0$ a.u. and $r \geqslant 2$ a.u., the constant density surfaces are nearly spherical. A is selected as follows: for a given R_0 the binding energy computed for some one state is adjusted to agree with experiment. If V_M is an adequate potential, the energies of all other states of the Rydberg series must be correctly predicted.

Calculations have been carried out for the $^3\pi_u$ states of H_2 by both a variational method and a perturbation method. The results are shown in table 5. We note that:

(*a*) The variational calculation and perturbation calculation agree in all states except the lowest, where second-order corrections are needed.

(*b*) The computed binding energies are not very sensitive to the basis sets used as long as the electron density outside the core is well represented.

(*c*) There is good agreement between theory and experiment.

Table 5. $^3\pi_u$ *states of* H_2*—model potential no.* 1: $R_0 = 2{\cdot}0\,a.u.$, $A = -0{\cdot}85\,a.u.$, $R_{AB} = 2{\cdot}0\,a.u.$

	Experimental energies (eV)	Variational calculation				Perturbation calculation
		Basis 1	Basis 2	Basis 3	Basis 4	
$2p\pi$	−3·689	−3·589	−3·589	−3·590	−3·597	−3·555†
$3p\pi$	−1·588	−1·573	−1·573	−1·574	−1·575	−1·566
$4p\pi$	−0·885	−0·877	−0·877	−0·877	−0·878	−0·874
$5p\pi$	−0·565	−0·558	−0·558	−0·558	−0·559	−0·556
$6p\pi$	−0·388	−0·385	−0·386	−0·378	−0·368	−0·385
$4f\pi$	−0·875‡	−0·852	−0·852	−0·852	−0·852	−0·852

† First-order energies, see text. ‡ Experimental data is uncertain.

Basis 1: $2p$ (0·65), $3p$ (0·38333), $4p$ (0·29), $5p$ (0·23), $6p$ (0·19)
$2p$ (0·38333), $3p$ (0·29), $4p$ (0·23), $5p$ (0·19), $4f$ (0·25), $5f$ (0·2), $6f$(0·1667)

Basis 2: $2p$ (0·725), $3p$ (0·43333), $4p$ (0·325), $5p$ (0·26), $6p$ (0·21)
$2p$ (0·43333), $3p$ (0·325), $4p$ (0·26), $5p$ (0·21), $4f$ (0·25), $5f$ (0·2), $6f$ (0·16667)

Basis 3: $2p$ (0·8), $3p$ (0·48333), $4p$ (0·36), $5p$ (0·29), $6p$ (0·24)
$2p$ (0·48333), $3p$ (0·36), $4p$ (0·29), $5p$ (0·24), $4f$ (0·25), $5f$ (0·2), $6f$ (0·16667)

Basis 4: $2p$ (0·875), $3p$ (0·53333), $4p$ (0·4), $5p$ (0·32), $6p$ (0·27)
$2p$ (0·53333), $3p$ (0·4), $4p$ (0·32), $5p$ (0·27), $4f$ (0·25), $5f$ (0·2), $6f$ (0·16667)

To check that the results obtained are not fortuitously good because of the nearly spherical electron density surfaces of H_2^+, whereas the electron density in other molecules is likely to be very far from spherical in symmetry, we have also examined the model potential

$$\left.\begin{aligned} V_M^{(2)} &= V_A + V_B, \\ V_A(r_A) &= -\tfrac{1}{2}r_A \quad (r_A \geqslant R_0), \\ V_A(r_A) &= C \quad (r_A < R_0). \end{aligned}\right\} \tag{41}$$

Note that (41) has a more complicated, non-spherical, form for $r < R_0$. This is a very important feature of the potential—$V_M^{(2)}$ is constant only in a complex subset of the core region. Indeed, the core consists of the three subregions

$$\left.\begin{aligned} r_A < R_0,\quad r_B < R_0;\quad V_M^{(2)} &= 2C,\\ r_A < R_0,\quad r_B \geqslant R_0;\quad V_M^{(2)} &= C - \tfrac{1}{2}r_B,\\ r_A \geqslant R_0,\quad r_B < R_0;\quad V_M^{(2)} &= C - \tfrac{1}{2}r_A.\end{aligned}\right\} \tag{42}$$

The form (41) may be thought of as a sum of atomic pseudo-potentials. By an atomic potential we mean one centered on one of the constituent atoms, and which depends on the nature of the atom, but is relatively insensitive to the rest of the molecule. Of course, (41) also satisfies the general criteria cited for the choice of a model potential.

A study of the $^3\pi_u$ states of H_2 using $V_M^{(2)}$ as defined by (41) leads to the results displayed in table 6. As can be seen, there is very good agreement between the computed and observed binding energies.

Table 6. $^3\pi_u$ *states of* H_2—*model potential no.* 2: $R_0 = 1{\cdot}5\,a.u.$, $C = -0{\cdot}85\,a.u.$, $R_{AB} = 2{\cdot}0\,a.u.$

	Experimental binding energies (eV)	Variational calculation		Perturbation calculation
		Basis 1†	Basis 2	
$2p\pi$	−3·689	−3·566	−3·566	−3·520‡
$3p\pi$	−1·588	−1·571	−1·572	−1·558
$4p\pi$	−0·885	−0·877	−0·877	−0·871
$5p\pi$	−0·565	−0·558	−0·558	−0·555
$6p\pi$	−0·388	−0·387	−0·378	−0·384
$4f\pi$	−0·875§	−0·852	−0·852	−0·852

† Basis set compositions are shown in table 5.
‡ First-order energies, see text. § Experimental data is uncertain.

Results of comparable accuracy have been obtained using model potential 2 (and its obvious extension to polyatomic molecules) to describe the Rydberg states of nitrogen and benzene.

We conclude that, at least for triplet Rydberg states, it is possible to obtain accurate binding energies from simple analytic pseudo-potentials. Thus, the pseudo-potential transformation is useful in providing a simple and accurate description of the field of the positively charged core in which the Rydberg electron moves. The extension of these techniques to the valence states and to the non-Rydberg triplet states is currently under investigation.

Triplet excitons in crystals of aromatic compounds

The lowest excited state of all aromatic molecules is a triplet, to which direct optical transition from the ground state is extremely weak because of spin selection rules. There are, however, other ways of populating the triplet level, and these have been employed to make studies of energy transfer, energy trapping, exciton-exciton interaction, and several other phenomena. In addition, despite the difficulties created by small oscillator strength in the $S_0 \rightarrow T_1$ transition, recent experiments to measure directly the singlet-triplet absorption spectrum have succeeded. Much of this information has been analysed elsewhere, but it seems appropriate to repeat, for the record of this Symposium on the Triplet State, what I believe to be the current state of our understanding of triplet excitons in crystals of aromatic compounds.

A. Stationary state properties of triplet excitons

We consider π^* type triplet states of aromatic molecules. In crystals of these compounds the intermolecular overlap is small ($\sim 10^{-3}$) in both the ground state and the excited states, so that the interactions may be expected to be weak. In the approximation that the interaction between the molecules of the crystal is small, we take the Hamiltonian of the crystal to be

$$H = \sum_n \{h_n + \sum_{m \neq n} V_{mn}\}, \tag{43}$$

where h is the Hamiltonian of the free molecule, and V_{mn} is the interaction between the molecules m and n. The molecular wavefunctions are the eigenfunctions of the molecular Hamiltonian:

$$h_n \phi_n^f = \epsilon^f \phi_n^f. \tag{44}$$

For the case of crystals of aromatic compounds it is convenient to take for the wavefunctions of the excited states of the crystal the excitation wave representation. We remind the reader that the excitation waves have the form[18]

$$E(\boldsymbol{\beta}, \mathbf{K}) = (\sigma N_c)^{-\frac{1}{2}} \sum_{n=1}^{\sigma N_c} a_n(\mathbf{K}) \,|\mathbf{n} + \tfrac{1}{2}\boldsymbol{\beta}, \mathbf{n} - \tfrac{1}{2}\boldsymbol{\beta}\rangle, \tag{45}$$

where the state $|m, p\rangle$ represents a hole in the molecule at site m and an extra electron on the molecule at site p, the momentum of the hole electron pair is given by $\hbar\mathbf{K}$, N_c is the number of unit cells in the crystal, and σ is the number of molecules per unit cell.

Let us consider the case in which $\boldsymbol{\beta}$ is zero, i.e. the Frenkel exciton. In this case, the states may be represented in the form

$$\Phi_m^f \equiv |m, m\rangle = \phi_m^f \prod_{n \neq m} \phi_n^0, \tag{46}$$

where ϕ_m^0 represents a ground state molecule at site m in the crystal and ϕ_m^f represents the f-th excited state of the molecule at site m. The ground state, in this approximation, may be represented in the form

$$\Phi_G = \prod_n \phi_n^0. \tag{47}$$

We may now find the matrix elements of the Hamiltonian in the set of states $E(\mathbf{0}, \mathbf{K})$. When this secular matrix is diagonalised, the energies of the several states are determined.

The coefficients $a_n(\mathbf{K})$ of equation (45) may be found using group theoretical arguments. The crystal has a defined translational symmetry, as well as other symmetry elements (mirror planes, rotations, and the inversion). Since these operations form a group, the functions $E(\boldsymbol{\beta}, \mathbf{K})$ must transform under the cited operations in a manner prescribed by group theory. Indeed, the functions $E(\boldsymbol{\beta}, \mathbf{K})$ must transform as do the irreducible representations of the appropriate space group. Bouckaert, Smoluchowski & Wigner[(19)] have shown how to find the irreducible representations of the symmorphic space groups, and Herring[(20)] has extended this to the non-symmorphic space groups. The relevant tables, and a good discussion of these methods, may be found in Koster's article[(21)] in *Solid State Physics*, volume 5. Now, the Hamiltonian is periodic, i.e. there is a Bravais lattice, so that Bloch's theorem[(22)] requires that the $a_n(\mathbf{K})$ be given by

$$a_n(\mathbf{K}) = a_n e^{-i\mathbf{K}.\mathbf{x}_n}, \tag{48}$$

where $\mathbf{x}_n$ is a vector representing the position of the n-th molecule in the crystal. It has been shown that the a_n are completely determined by the group of the wave vector $\mathbf{K}$. This group consists of all operations in the group under which the wave vector is invariant. The operations of the space group are performed in real space, but because of the structure of the wavefunctions, these operations may be transformed into operations in reciprocal space, and those that leave the wave vector invariant form the group of the wave vector. For example, for the wave vector parallel to the direction in reciprocal space that corresponds to the monoclinic axis in a monoclinic crystal, the group of the wave vector is C_2. If $\mathbf{K}$ is exactly zero, the group of the wave vector becomes the factor group of the space group with respect to the translation group of the Bravais lattice.

For the crystals we shall consider, the factor groups are isomorphic to Abelian groups, and hence have only one-dimensional irreducible representations. Thus, the a_n are simply the characters of the various factor group operations multiplied by phase factors due to the operation itself. The site group[23] represents those operations that are common to the factor group and to the molecular point group. These operations are those that leave the molecule at any site invariant except for a phase factor. The other operations in the factor group change a molecule of one site to one of another site in the unit cell. In all cases dealt with in this review, the site group is C_i, so that the identity and the inversion are the operations that transform a molecule into itself.

To include in the wavefunction the types of excitation waves in which $\boldsymbol{\beta}$ is not zero, it is necessary to find the coefficients, $C(\boldsymbol{\beta})$. These are found by diagonalisation of the Hamiltonian in the basis $E(\boldsymbol{\beta}, \mathbf{K})$. For example, the crystals naphthalene and anthracene fall into the space group $C_{2h}^5(P_{2_1}/a)$. The factor group is then isomorphic to the point group C_{2h}. In crystalline naphthalene and anthracene there are two molecules per unit cell. The second molecule is generated from the first molecule by motion along a glide plane perpendicular to the b (monoclinic axis), or along a screw axis parallel to the same axis. We choose to use the first definition of the transformation. The site group is C_i, and since only states that are excited by light are to be considered, attention is restricted to those states that are antisymmetric with respect to the inversion (the ground state is symmetric with respect to the inversion). Since the molecular point group (D_{2h}) and the site group both contain the inversion, all transitions in the molecule that are antisymmetric with respect to inversion will be allowed in the crystal. In fact, since no other point group operations except the identity and the inversion are common to the factor group, it is possible to say that every antisymmetric state in the free molecule is allowed in the crystal. By elementary considerations it is possible to show that, for the naphthalene and anthracene crystals,

$$E^f_{A_u}(\mathbf{0},\mathbf{0}) = (2N)^{-\frac{1}{2}} \sum_{n=1}^{N} (\Phi^f_{1n} - \Phi^f_{2n}), \tag{49}$$

$$E^f_{B_u}(\mathbf{0},\mathbf{0}) = (2N)^{-\frac{1}{2}} \sum_{n=1}^{N} (\Phi^f_{1n} + \Phi^f_{2n}). \tag{50}$$

For the excited states in which an electron has been transferred from one site to another site, it is necessary to start with 'two site' excitons (ion pair excitons)

$$\Phi(\boldsymbol{\beta}_0, \mathbf{0}) = N^{-\frac{1}{2}} \sum_{n=1}^{N} |\mathbf{n}, \mathbf{n}-\boldsymbol{\beta}_0\rangle, \tag{51}$$

where the notation is the same as in equation (45). We consider only those states in which the separation between charges is the distance between near neighbors. Let us examine again the case in which the crystal may be represented by the C_{2h}^5 space group, and has two molecules per unit cell (naphthalene and anthracene). Given that the distance between near neighbors in these crystals is $\boldsymbol{\tau} = \frac{1}{2}(\mathbf{a}+\mathbf{b})$, the only states we will consider are the following:

$$\left.\begin{array}{ll} |\mathbf{R}_m, \mathbf{R}_m+\boldsymbol{\tau}\rangle, & |\mathbf{R}_m+\boldsymbol{\tau}, \mathbf{R}_m\rangle, \\ |\mathbf{R}_m+\mathbf{c}, \mathbf{R}_m\rangle, & |\mathbf{R}_m, \mathbf{R}_m+\mathbf{c}\rangle, \\ |\mathbf{R}_m+\mathbf{b}, \mathbf{R}_m\rangle, & |\mathbf{R}_m, \mathbf{R}_m+\mathbf{b}\rangle, \\ |\mathbf{R}_m+\mathbf{c}+\boldsymbol{\tau}, \mathbf{R}_m\rangle, & |\mathbf{R}_m, \mathbf{R}_m+\mathbf{c}+\boldsymbol{\tau}\rangle, \end{array}\right\} \tag{52}$$

in the notation of equation (45). By use of the Wigner projection operator the following symmetrised states are obtained at $\mathbf{K} = \mathbf{0}$, i.e. for the factor group:

$$|\mathbf{0}, \boldsymbol{\tau}; \mp\rangle = (4N)^{-\frac{1}{2}} \sum_{m=1}^{N} \{|\mathbf{R}_m, \mathbf{R}_m+\boldsymbol{\tau}\rangle \mp |\mathbf{R}_m-\mathbf{b}+\boldsymbol{\tau}, \mathbf{R}_m\rangle \mp |\mathbf{R}_m-\mathbf{a}+\boldsymbol{\tau}, \mathbf{R}_m\rangle + |\mathbf{R}_m, \mathbf{R}_m-\mathbf{a}-\mathbf{b}+\boldsymbol{\tau}\rangle\} \tag{53}$$

$$|\boldsymbol{\tau}, \mathbf{0}; \mp\rangle = (4N)^{-\frac{1}{2}} \sum_{m=1}^{N} \{|\mathbf{R}_m+\boldsymbol{\tau}, \mathbf{R}_m\rangle \mp |\mathbf{R}_m, \mathbf{R}_m-\mathbf{b}+\boldsymbol{\tau}\rangle \mp |\mathbf{R}_m, \mathbf{R}_m-\mathbf{a}+\boldsymbol{\tau}\rangle + |\mathbf{R}_m, \mathbf{R}_m-\mathbf{a}-\mathbf{b}+\boldsymbol{\tau}\rangle\}, \tag{54}$$

$$|\mathbf{0}, \mathbf{c}+\boldsymbol{\tau}; \mp\rangle = (4N)^{-\frac{1}{2}} \sum_{m=1}^{N} \{|\mathbf{R}_m, \mathbf{R}_m+\mathbf{c}+\boldsymbol{\tau}\rangle \mp |\mathbf{R}_m-\mathbf{b}+\boldsymbol{\tau}, \mathbf{R}_m-\mathbf{c}\rangle \mp |\mathbf{R}_m-\mathbf{a}+\boldsymbol{\tau}, \mathbf{R}_m+\mathbf{c}\rangle + |\mathbf{R}_m, \mathbf{R}_m-\mathbf{c}+\boldsymbol{\tau}\rangle\}, \tag{55}$$

$$|\mathbf{c}+\boldsymbol{\tau}, \mathbf{0}; \mp\rangle = (4N)^{-\frac{1}{2}} \sum_{m=1}^{N} \{|\mathbf{R}_m+\mathbf{c}+\boldsymbol{\tau}, \mathbf{R}_m\rangle \mp |\mathbf{R}_m-\mathbf{c}, \mathbf{R}_m-\mathbf{b}+\boldsymbol{\tau}\rangle \mp |\mathbf{R}_m+\mathbf{c}, \mathbf{R}_m-\mathbf{a}+\boldsymbol{\tau}\rangle + |\mathbf{R}_m-\mathbf{c}+\boldsymbol{\tau}, \mathbf{R}_m\rangle. \tag{56}$$

Note that only those functions in which both types of molecule participate in the ion pair state will have unequal mixing coefficients into any neutral state for the two different symmetries. This corresponds to neglecting terms in which three molecules are excited, for example $\langle \mathbf{R}_m^f \mathbf{R}_n^0 | H | \mathbf{R}_n^+ \mathbf{R}_1^- \rangle$. The neglect of the overlap in the normalisation factor will be shown to be a good approximation.

Consider now the neutral exciton states (Frenkel states), to describe which it is necessary to solve the Schrodinger equation using the Hamiltonian of equation (43). If the intermolecular potentials were absent, the functions displayed in equation (49) for naphthalene and anthracene would be eigenfunctions of the crystal Hamiltonian. Moreover, the observed spectra of these crystals resemble the respective spectra of the isolated molecules. Hence, it is natural to assume that we may treat the interactions between molecules as a perturbation. The energy levels in the crystal will be given to first order by

$$W_i^f = \langle E_i^f(\mathbf{0},\mathbf{0})|\, H_0 \,|E_i^f(\mathbf{0},\mathbf{0})\rangle + \langle E_i^f(\mathbf{0},\mathbf{0})|\, V \,|E_i^f(\mathbf{0},\mathbf{0})\rangle,$$

where f refers to the excited state, i to the symmetry of the zero-order crystal function, H_0 is the sum of single molecule Hamiltonians, and V is the sum of intermolecular potentials. Group theoretical arguments show that states of different symmetry do not mix, since the matrix element $\langle E_i|\, H \,|E_j\rangle$ is identically zero. To find the excitation energy of the crystal it is necessary to subtract the energy of the ground state. Starting with

$$W^0 = \sum_{n=1}^{N} \epsilon_n^0 + \sum_{p>q}^{\sigma N} \langle \phi_p^0 \phi_q^0|\, V_{pq} \,|\phi_p^0 \phi_q^0\rangle, \tag{57}$$

the energy of the excited state is given by

$$\begin{aligned} W_i^f = (\sigma N)^{-1} \sum_{n=1}^{\sigma N} \sum_{l=1}^{\sigma N} a_n^{i*} a_l^i \exp\{-i\mathbf{K}.(\mathbf{R}_l - \mathbf{R}_n)\} \\ \times \{ \sum_{p=1}^{\sigma N} \langle \phi_n^f \Pi \phi_m^0|\, h_p \,|\phi_l^f \Pi \phi_q^0\rangle + \sum_{r>s}^{\sigma N} \langle \phi_n^f \Pi \phi_m^0|\, V_{rs} \,|\phi_l^f \Pi \phi_q^0\rangle\} \\ = (\sigma N)^{-1} \sum_{n=1}^{\sigma N} \sum_{l=1}^{\sigma N} a_n^{i*} a_l^i \exp\{i\mathbf{K}.(\mathbf{R}_n - \mathbf{R}_l)\} \langle \phi_n^f \phi_l^0|\, V_{nl} \,|\phi_n^0 \phi_l^f\rangle \\ + (\sigma N)^{-1} \sum_{n=1}^{\sigma N} |a_n^i|^2 \{\epsilon^f + (\sigma N - 1)\epsilon^0 + \sum_{r>s \neq n}^{\sigma N} \langle \phi_r^0 \phi_s^0|\, V_{rs} \,|\phi_r^0 \phi_s^0\rangle \\ + \sum_{q \neq n}^{\sigma N} \langle \phi_n^f \phi_q^0|\, V_{nq} \,|\phi_n^f \phi_q^0\rangle\}. \end{aligned} \tag{58}$$

The coefficients of translationally equivalent molecules are equal and in all the cases dealt with here, the square of any one coefficient is unity; hence it is found that

$$\begin{aligned} W_i^f = \epsilon^f + (\sigma N - 1)\epsilon^0 + \sum_{r<s \neq n}^{\sigma N} \langle \phi_r^0 \phi_s^0|\, V_{rs} \,|\phi_r^0 \phi_s^0\rangle + \sum_{q \neq n}^{\sigma N} \langle \phi_n^f \phi_q^0|\, V_{nq} \,|\phi_n^f \phi_q^0\rangle \\ + \sum_{l=1}^{\sigma N} a_n^{i*} a_l^i \exp\{i\mathbf{K}.(\mathbf{R}_n - \mathbf{R}_l)\} \langle \phi_n^f \phi_l^0|\, V_{nl} \,|\phi_n^0 \phi_l^f\rangle, \end{aligned} \tag{59}$$

from which

$$\Delta W^f = (\epsilon^f - \epsilon^0) + \sum_{q \neq n}^{\sigma N} \{\langle \phi_n^f \phi_q^0 | V_{nq} | \phi_n^f \phi_q^0 \rangle - \langle \phi_n^0 \phi_q^0 | V_{nq} | \phi_n^0 \phi_q^0 \rangle\}$$

$$+ \sum_{n \neq l}^{\sigma N} a_n^{i*} a_l^i \exp\{-i\mathbf{K}.(\mathbf{R}_l - \mathbf{R}_n)\} \langle \phi_n^f \phi_l^0 | V_{nl} | \phi_n^0 \phi_l^f \rangle, \quad (60)$$

or

$$\Delta W = \Delta w^f + D + \epsilon_i^f(\mathbf{K}). \quad (61)$$

In equation (61), Δw^f is the excitation energy of an isolated molecule, D is the environmental energy shift, and $\epsilon_i^f(\mathbf{K})$ is that part of the energy that depends on the wave vector and the symmetry of the wavefunctions. The differences in energy between the different factor group symmetry species for the same excited state are called the Davydov splittings. For the space group C_{2h}^5 at $\mathbf{K} = \mathbf{0}$, all the coefficients for the B_u states are unity, and those for the A_u states are 1 for the equivalent set of molecules, and -1 for the other set. Hence the Davydov splitting for the f-th excited state, to first order, is given by

$$\epsilon_{B_u}^f - \epsilon_{A_u}^f = 2 \sum_{l=1}^{\sigma N} \langle \phi_n^f \phi_l^0 | V_{nl} | \phi_n^0 \phi_l^f \rangle, \quad (62)$$

where the sum is taken over inequivalent sites.

Now, if more than one excited state is considered, then it is necessary to include matrix elements of the Hamiltonian between any two excited states of the same symmetry (in the factor group).

$$I_i^{fg} = (2\sigma N)^{-1} \sum_{n \neq r}^{\sigma N} a_n^{i*} a_l^i \exp\{-i\mathbf{K}.(\mathbf{R}_l - \mathbf{R}_n)\}$$

$$\times \{\langle \phi_n^f \phi_l^0 | V_{nl} | \phi_n^0 \phi_l^g \rangle + \langle \phi_l^f \phi_n^0 | V_{nl} | \phi_n^g \phi_l^0 \rangle\} + \sum_{r=1}^{\sigma N} \langle \phi_n^f \phi_r^0 | V_{nr} | \phi_n^g \phi_r^0 \rangle. \quad (63)$$

Thus at $\mathbf{K} = \mathbf{0}$,

$$I_i^{fg} = \sum_n a_n^{i*} a_l^i J_{nl}^{fg} + \sum_n K_{nl}^{fg}. \quad (64)$$

Note that the functions I_i^{fg} will be a function of the symmetry of the representation of the factor group.

The interaction between molecule m and molecule n, V_{mn}, may be written as the sum of the electron-electron potentials and the sum of all the electron-nuclear potentials. Following other workers, for the case of aromatic molecules it is assumed that the pi electrons are those responsible for all the low-lying excited states of the molecules under consideration. Hence, we write

$$V_{nm} = \sum_{i \in n} V_{mi} + \sum_{j \in m} V_{nj} + \sum_{i > j} \frac{e^2}{r_{ij}}. \quad (65)$$

where V_{mi} is the interaction of the i-th pi electron on n with the nuclei and sigma electrons of m, hereafter referred to as the sigma frame of molecule m. The second term is similar to the first with the roles of molecule m and n reversed, while the last term represents the pi-pi electron-electron repulsion.

It is now necessary to evaluate the integrals introduced in the above equations. As long as spin-orbit coupling is neglected the excitation exchange matrix element vanishes (spin orthogonality). An estimate of the integral J can be obtained by considering the spin-orbit coupling between the triplet state of interest and the perturbing singlet states. Application of perturbation theory leads to

$$(\phi_m^f)_1 = (\phi_m^f)_0 + \sum_s \frac{(H_{\mathrm{s.o.}})_{fs}}{E_f - E_s} \phi_m^s, \tag{66}$$

where $(\phi_m^f)_1$ is the perturbed triplet wavefunction, ϕ_m^s an unperturbed excited singlet state, E_f and E_s the triplet and singlet energies, and $(H_{\mathrm{s.o.}})_{fs}$ the spin-orbit coupling matrix element combining the two states. We then find for the coulomb contribution to the excitation exchange interaction

$$J_{mn}^f = \left\langle \sum_s \frac{(H_{\mathrm{s.o.}})_{fs}}{E_f - E_s} \phi_n^s \phi_m^0 \right| V_{nm} \left| \phi_n^0 \sum_{s'} \frac{(H_{\mathrm{s.o.}})_{fs'}}{E_f - E_{s'}} \phi_m^{s'} \right\rangle. \tag{67}$$

In this expression the interaction of the ground state wavefunction with excited triplet states by spin-orbit coupling is not taken into account. For approximate numerical estimates, this omission is not serious.

Spin-orbit coupling in aromatic hydrocarbons has been studied by several investigators.[25, 26, 27] McClure[25] has shown that when only coupling between $\pi \to \pi^*$ transitions is considered, one- and two-center contributions to the spin-orbit coupling matrix elements vanish, and three-center contributions have to be included. Mizushima & Koide[26] considered the possibility of spin-orbit coupling between a triplet $\pi \to \pi^*$ state and a singlet $\sigma \to \pi^*$ state, where one-center contributions do not vanish. In any case, the spin-orbit coupling is extremely weak in these systems, leading to long radiative (phosphorescence) lifetimes. Following McClure,[25] we assume that $(H_{\mathrm{s.o.}})_{fs} \approx 1\ \mathrm{cm}^{-1}$, and to make a rough estimate of the excitation exchange integral we consider just one perturbing singlet state. Then taking $E_s - E_f = 20\,000\ \mathrm{cm}^{-1}$, the triplet state coulomb integral, J^f, is related to the perturbing singlet state coulomb integral, J^s, by

$$J_{mn}^f \approx \frac{|(H_{\mathrm{s.o.}})_{fs}|^2}{(E_f - E_s)^2} J_{mn}^s. \tag{68}$$

The contribution of these coulomb type integrals to the Davydov splitting can be estimated by summing over all translationally inequivalent molecules. For strongly allowed singlet-singlet transitions,

$$\sum_m J^s_{mn} \approx 10^3\text{–}10^5\,\mathrm{cm}^{-1},$$

so that the contribution of the coulomb integrals to the Davydov splitting of the triplet state is of the order of $10^{-3}\,\mathrm{cm}^{-1}$, and is therefore negligibly small. We conclude that, within the tight binding approximation, the triplet exciton band width arises from intermolecular electron exchange interactions.

In the case of the molecular crystals with which we deal, the overlap between molecular orbitals on different molecules is small, and has been neglected to this point. This is equivalent to the assumption that the crystal wavefunctions used are orthonormal. Now, this is not strictly true, and the deviation from orthogonality in some cases may be extremely important. We adopt the Löwdin[28] symmetric orthogonalisation technique, since for the cases considered all overlap integrals are small. The reader will recall that, using the Löwdin method, from a set of functions $\{\Phi_i\}$ an orthonormal set is formed by setting

$$\Phi_i' = \Phi_i - \tfrac{1}{2}\sum_{i \neq j} \Phi_j \langle \Phi_i | \Phi_j \rangle + \tfrac{3}{8} \sum_{\substack{j \neq i \\ k \neq j}} \Phi_k \langle \Phi_k | \Phi_j \rangle \langle \Phi_j | \Phi_i \rangle + \dots \tag{69}$$

The matrix elements of interest are the off-diagonal elements between the $\{\Phi_i'\}$:

$$\langle \Phi_i' | H | \Phi_k' \rangle = \langle \Phi_i | H | \Phi_k \rangle - \tfrac{1}{2} \sum_{j \neq i} \langle \Phi_i | \Phi_j \rangle \langle \Phi_j | H | \Phi_k \rangle - \tfrac{1}{2} \sum_{j \neq k} \langle \Phi_j | \Phi_k \rangle \langle \Phi_i | H | \Phi_j \rangle + \mathcal{O}(S^2), \tag{70}$$

where we have set aside and will neglect all terms of order S^2. Equation (70) then assumes the form

$$H_{ik}' = H_{ik} - \tfrac{1}{2} \sum_{j \neq i} S_{ij} H_{jk} - \tfrac{1}{2} \sum_{j \neq k} S_{jk} H_{ij}. \tag{71}$$

For the cases considered in this section, the off-diagonal elements of H are extremely small relative to the diagonal elements; hence, we write

$$H_{ik}' = H_{ik} - S_{ik}[H_{kk} - \tfrac{1}{2}(H_{kk} - H_{ii})]. \tag{72}$$

Again, for the cases considered, $H_{kk} - H_{ii} \ll H_{kk}$, so that

$$H_{ik}' = H_{ik} - S_{ik} H_{kk}. \tag{73}$$

The mixing of neutral exciton states with neutral exciton states will be given by the above, but S_{ik} will be of order S^2 where S is the overlap between molecular orbitals, while S_{ik} for the mixing between neutral and ion pair exciton states will be of order S. In all that follows, we consider terms of order S. Thus, we will need equation (73) only for neutral state-ion-pair state mixing.

One more general topic must be discussed. In the discussion thus far, it has been assumed that the molecules are non-vibrating. That, of course, is an unjustified approximation. To illustrate the nature of the effects of nuclear motion consider the properties of a model system, a dimer.[29] The Hamiltonian of the system contains the following terms:

(*a*) an electronic part which contains intramolecular electron-electron and electron-nuclear interactions;

(*b*) a potential due to intramolecular nuclear-nuclear potentials;

(*c*) a potential due to intermolecular electron-electron and electron-nuclear interactions.

There are two limiting cases to be distinguished: (1) the intermolecular interaction is small compared to that part of the intramolecular potential that may be called the electronic-vibrational interaction, and (2) the opposite case, in which the intermolecular interaction is large compared to the electronic vibrational interaction. In case (1) the vibrational electronic interaction is diagonalised before the intermolecular interaction terms (and after the other intramolecular terms, of course). This is known as the weak coupling case since

$$\Delta E_{\text{intermolecular}} \ll \Delta E_{\text{vibronic}}. \tag{74}$$

In this case the basis functions for diagonalisation for the intermolecular interactions are vibronic functions, i.e. functions whose vibrational states depend on the electronic state. In case (2), on the other hand, the intermolecular interactions are diagonalised prior to the vibrational electronic interaction, hence the basis states for the intermolecular interactions are states with vibrational parts independent of the electronic part. These will be taken as harmonic oscillator states in both cases (1) and (2), but in case (1) the position of the minimum in potential energy will depend on the electronic state, while in case (2) it will be the same for all states. The difference between these limits is extremely important. In the strong coupling case, the exciton theory predicts for the dimer that there will be two well-separated progressions (which arise from the diagonalisation of vibronic interactions after the intermolecular forces). For the weak coupling case, the vibronic interaction is larger than the intermolecular interaction and the splittings between corresponding lines (e.g. the two 0–0 transitions) will be the total electronic splitting multiplied by a vibrational overlap squared. Actually, of course, one must diagonalise the intermolecular interaction in a manifold of vibronic states and each matrix element is the product of the full electronic terms multiplied by two vibrational overlaps.

The extension to a crystal of N molecules is non-trivial,[30, 31] but it is clear that the same qualitative effects are expected. That is, for the crystal it is possible to adopt the same basis states as above for the two limiting cases. The relevant vibrational overlaps may be obtained in two ways: (*a*) from vapor phase spectra, the ratio of the intensity of one vibrational state to the entire vibrational band is proportional to the vibrational overlap squared to that state; or (*b*) the overlap may be calculated by the method of McCoy & Ross.[32] Thus, in the weak coupling limit the molecular wavefunctions are constructed from vibronic functions, i.e.

$$\phi_n^{f(i)} = \phi_n^f \chi_n^{f(i)}, \tag{75}$$

where $\chi_n^{f(i)}$ is a vibrational wavefunction for the n-th excited molecule in the i-th vibrational state. When the functions defined by equation (75) are used as the basis for calculating the matrix elements of the intermolecular potential, the integrals are all modified as follows:

$$\langle \phi_n^{f(i)} \phi_m^0 | V_{nm} | \phi_n^0 \phi_m^{f(i)} \rangle = J_{nm} \langle \chi_n^{0(0)} | \chi_n^{f(i)} \rangle \langle \chi_m^{0(0)} | \chi_m^{f(i)} \rangle. \tag{76}$$

Therefore, in the weak coupling limit the total electronic matrix element is modified by vibrational overlap factors. It should be noted that, in this case, off-diagonal matrix elements between vibronic states corresponding to the same electronic transition must be included. The vibrational sum rule,

$$\sum_i |\langle \chi^{0(0)} | \chi^{f(i)} \rangle|^2 = 1$$

implies that the total electronic contribution to the Davydov splitting can be obtained by summation over all the experimental splittings corresponding to the individual vibronic components.

We now examine the magnitude of the exchange interaction. Three types of molecular wavefunctions were used for the calculation of the required exchange integrals: antisymmetrised products of Hückel orbitals, antisymmetrised products of π electron SCF orbitals calculated by Hoyland & Goodman,[33] and π electron configuration interaction wavefunctions computed by Pariser.[34] The matrix elements calculated from all three wavefunctions are nearly equal to one another. These interactions lead to a total electronic contribution of 44 cm^{-1} for the Davydov splitting in the first triplet exciton state of the anthracene crystal. In the weak coupling scheme, which is obviously appropriate for this case, the contribution of electron exchange interactions to the splitting in the 0–0 vibronic band of the anthracene $^3B_{2u}$ state is about 11 cm^{-1}.

We consider, now, the configuration interaction between the neutral and charge-transfer triplet exciton states, and its effect on the Davydov splitting. As long as the mixing matrix elements are small compared with the separation between the two exciton states, the energy is simply computed from perturbation theory. For the case of the triplet exciton, the energy separation cited is of order 1·5 eV in naphthalene and anthracene. It is, therefore, not necessary to diagonalise the energy matrix. In the calculations of Jortner, Rice, Choi & Katz,[3] second-order perturbation theory was applied to a non-orthogonal set of unperturbed basis functions. It was shown by these investigators that the overlap is sufficiently small that the effects of non-orthogonality are small. Of course, some error is introduced into the second-order energy because of the use of only a limited number of ion-pair exciton wavefunctions. This error was also shown to be negligibly small.

The actual computation of the contribution of state mixing to the Davydov splitting requires the evaluation of matrix elements of the Hamiltonian. Details of this calculation and of the treatment of the relevant vibrational overlap factors can be found elsewhere.[1] Here we merely quote the final results of the calculations for naphthalene and anthracene, which lead to the following contribution by the ion-pair states to the Davydov splittings of the first triplet state:

$$\text{naphthalene} = 37/\Delta\,\text{eV cm}^{-1}, \quad \text{anthracene} = 24/\Delta\,\text{eV cm}^{-1},$$

where Δ is the energy difference between the charge transfer and the neutral exciton states. Setting $\Delta = -1{\cdot}7$ eV for naphthalene and $\Delta = -1{\cdot}5$ eV for anthracene, the contributions of the ion-pair states to the first triplet exciton state are finally obtained. The ion-pair states-neutral state mixing increases the triplet band splitting by about 50 %. The numerical values quoted in this section of the review differ slightly from those originally published. The new figures result from a correction to the computer program used to obtain the matrix elements, an error having been discovered after the original publications appeared. The reader should note that no conclusions have been altered by the changes, which are small.

The data available to test the theory of triplet exciton spectra is very scanty, and most of the evidence we shall produce is indirect and comes from studies of the diffusion length and triplet-triplet annihilation rates in anthracene. Recently, however, Hanson and Robinson[35] have succeeded in measuring the Davydov splitting of the transition to the lowest triplet state in naphthalene.[35] The observed and calculated splittings are in good agreement, as seen in table 7.

Table 7. *Triplet Davydov splitting in naphthalene and anthracene*

	Expt	Theory	Exchange contribution	CT contribution
Anthracene	-48 cm^{-1}	-60 cm^{-1}	-44 cm^{-1}	-16 cm^{-1}
Naphthalene	-40 cm^{-1}	-67 cm^{-1}	-45 cm^{-1}	-22 cm^{-1}

Only a few other substances have been examined (i.e. *p*-dichloro, *p*-dibromo and *p*-diodobenzene).[36] Since our calculations do not refer to these substances, it is pertinent only to remark that in crystalline *p*-diodobenzene, the triplet manifold shows strong coupling behavior. In view of the enhanced spin-orbit coupling, and the possibility of I–I, I–C, and I–H interactions in this crystal, it is not surprising that the total triplet manifold is split by ~ 900 cm^{-1}. Clearly, when hetero-atoms are present in the molecule, interactions between carbon atoms represent only a part of the total intermolecular interaction, and this must be borne in mind when examining any particular substance.

B. Exciton dynamics

As mentioned, many of the studies of triplet excitons have been concerned with nonstationary phenomena, e.g. diffusion, triplet-triplet interaction, etc. It is out of place to attempt anything approaching a complete analysis of exciton dynamics in this review. For such an analysis we refer the reader elsewhere,[1] and herein concentrate on only a few points of major interest. These are encompassed in the following questions:

(*a*) Is the crystal momentum a good quantum number for describing exciton motion?

(*b*) How are excitons transferred in a molecular crystal?

(*c*) How important is exciton-lattice phonon scattering, and what are its manifestations?

(*d*) What trapping mechanisms, if any, exist?

(*e*) How do electronic and vibrational relaxation influence exciton behavior?

Consider first the interaction of excitons with the *intra*molecular vibrations. It was hinted in subsection A that the use of the Born–Oppenheimer approximation is only valid under restricted conditions. Indeed, it is now well established the Born–Oppenheimer approximation is not, in general, valid for molecular exciton states. There exists, in fact, an interaction between Frenkel exciton states and intramolecular vibrations arising from the difference in the molecular potential energy surfaces in different electronic states. If b_σ^+ and b_σ are intramolecular phonon creation and annihilation operators, σ the phonon wave number and ω the ground state vibrational frequency, whilst $a_\mathbf{k}^+$, $a_\mathbf{k}$ are exciton creation and annihilation operators, it may be shown that the effective Hamiltonian describing the interaction between the excitons and the intramolecular phonons is

$$H = \omega \sum_\sigma b_\sigma^+ b_\sigma + \sum_\mathbf{k} W(\mathbf{k}) a_\mathbf{k}^+ a_\mathbf{k} + A\left(\frac{m\omega}{2}\right)^{\frac{1}{2}} \frac{w}{\sqrt{N_c}} \sum_\mathbf{k} \sum_\sigma a_{\mathbf{k}+\sigma}^+ a_\mathbf{k} (b_\sigma + b_{-\sigma}^+). \tag{77}$$

In equation (77), A denotes the change in the internal vibrational coordinate in the upper state, and $W(\mathbf{k})$ is the exciton energy. It is, in general, very difficult to find the eigenfunctions of H. For the case that the unperturbed exciton band has zero (very small) width it may be shown that:[37]

(*a*) The main effect of exciton-intramolecular phonon coupling is the reduction of the bandwidth relative to the free exciton value. The narrowing is asymmetric, with the upper half of the band being more compressed than the lower half of the band. This asymmetry results from the relatively larger interaction of the vacuum state with high-energy phonon states than with low-energy phonon states. Note that the predicted asymmetric narrowing of the exciton band is at variance with the results of the simple weak coupling vibronic overlap argument. In the latter case the exciton bandwidth is reduced uniformly by the vibrational overlap factor.

(*b*) The effective mass of the exciton increases with increasing exciton-phonon coupling.

(*c*) The wider the free exciton band, the less effective is the exciton-phonon coupling.

(*d*) If $\langle n_{ph} \rangle$ is the average number of phonons in interaction with (clothing) an exciton, then the Franck–Condon factor for the electronic transition from the ground state becomes

$$F_0^2 = \exp(-\langle n_{ph} \rangle) \tag{78}$$

and the intensity distribution in the crystal is thereby altered.

Of course, in addition to intramolecular vibrations there are *inter*-molecular vibrations in the crystal. The interaction of an exciton with these latter vibrations leads to a broadening and a shift of the exciton states and a relaxation of the crystal momentum selection rules. In terms of the intermolecular phonon creation and annihilation operators, $\alpha_{\sigma f}^+$, $\alpha_{\sigma f}$, the effective Hamiltonian describing the exciton-phonon coupling is (to terms linear in the phonon operators)[38, 39]

$$H = \sum_{\mathbf{K}} W_\nu(\mathbf{K}) a_{\nu\mathbf{K}}^+ a_{\nu\mathbf{K}} + \sum_{\sigma f} \hbar\omega_{\sigma f} \alpha_{\sigma f} \alpha_{\sigma f}^+$$

$$+ \sum_{\mathbf{K}} \sum_{\mathbf{K}'} \sum_{\sigma f} \delta(\mathbf{K}' - \mathbf{K} + \boldsymbol{\sigma})\, iG_{\sigma f}(\nu\mathbf{K}, \nu'\mathbf{K}')\, a_{\nu\mathbf{K}}^+ a_{\nu'\mathbf{K}'} (\alpha_{\sigma f} - \alpha_{\sigma f}^+). \tag{79}$$

The first term in (79) represents the energy of pure exciton states corresponding to the ν-th exciton band, and with wave vector $\mathbf{K}$, while the second term represents the energy of the phonon states. The third term in (79), corresponding to the exciton-phonon coupling, has two contributions: (*a*) the term involving $a_{\mathbf{K}}^+ \alpha_{\mathbf{K}'} \alpha_{\sigma f}$ destroys a phonon with wave vector $\boldsymbol{\sigma} = \mathbf{K} - \mathbf{K}'$ and scatters an exciton from a state with wave vector $\mathbf{K}$ to a state with wave vector $\mathbf{K}'$. (*b*) The term $a_{\mathbf{K}}^+ a_{\mathbf{K}'} \alpha_{\sigma f}^+$ creates a phonon of wave vector $\boldsymbol{\sigma} = \mathbf{K}' - \mathbf{K}$ and scatters an exciton from a state with wave vector $\mathbf{K}'$ to a state with wave vector $\mathbf{K}$. The coupling constant, $G_{\sigma f}$, is determined by the nature of the phonon field and by the nature of the exciton wavefunction.

The exciton-phonon coupling constant for singlet Frenkel excitons has been evaluated in the dipole approximation.[40] Because the interaction potential cannot be satisfactorily represented as having only dipole character, it is not surprising that the calculated and observed diffusion lengths of singlet excitons in crystalline anthracene do not agree. In the case of Wannier excitons, intraband scattering by long wavelength polar modes is less effective than is scattering by acoustical modes. The polar modes are diminished in effectiveness by the tendency towards cancellation of the interactions of the electron and hole with the polarisation field.

The study of triplet exciton-phonon scattering phenomena is even less advanced than are the studies for the case of singlet excitons. Instead of attempting to find the proper effective Hamiltonian, and some of its eigenfunctions, attention has been focused on crude model calculations. These models are all based on the assumption that the exciton states created by photon absorption are very rapidly thermalised. The time scale for thermalisation is usually asserted to be $\sim 10^{-12}$ s, and is determined by phonon and imperfection scattering.

To describe the behavior of thermalised excitons, the band structure of the crystal must be known. We have constructed the triplet exciton states in several aromatic crystals in terms of states which involve the symmetric and antisymmetric combinations of molecular wavefunctions corresponding to the two molecules in the unit cell.[3] The corresponding **k**-dependent eigenvalues give the two energy bands (only **k**-dependent terms are displayed):

$$\begin{aligned} W'_{\pm}(\mathbf{k}) = 2K_2 \cos(\mathbf{k}.\mathbf{c}) + 2K_3 \cos(\mathbf{k}.\mathbf{b}) + 2K_4[\cos\mathbf{k}.(\mathbf{b}+\mathbf{c}) \\ + \cos\mathbf{k}.(\mathbf{b}-\mathbf{c})] + 2K_5 \cos(\mathbf{k}.\mathbf{a}) + 2K_6 \cos\mathbf{k}.(\mathbf{c}+\mathbf{a}) \\ + 2K_7[\cos\mathbf{k}.(\mathbf{a}+\mathbf{b}) + \cos\mathbf{k}.(\mathbf{a}-\mathbf{b})] + 2K_8[\cos\mathbf{k}.(\mathbf{a}+\mathbf{b}+\mathbf{c}) \\ + \cos\mathbf{k}.(\mathbf{a}-\mathbf{b}+\mathbf{c})] \pm 2K_9[\cos\mathbf{k}.(\tfrac{1}{2}[\mathbf{a}+\mathbf{b}]) + \cos\mathbf{k}.(\tfrac{1}{2}[\mathbf{a}-\mathbf{b}])] \\ \pm 2K_{10}[\cos\mathbf{k}.(\tfrac{1}{2}[\mathbf{a}+\mathbf{b}+2\mathbf{c}]) + \cos\mathbf{k}.(\tfrac{1}{2}[\mathbf{a}-\mathbf{b}+2\mathbf{c}])], \end{aligned} \quad (80)$$

and the ten K_i are the electron exchange integrals between the reference molecule 1 and i. The band can be readily visualised if we consider now the special cases when **k** is parallel to a reciprocal lattice vector, $\mathbf{a}^{-1}$, $\mathbf{b}^{-1}$, or $\mathbf{c}^{-1}$:

$$W'_{\pm}(\mathbf{k}\|\mathbf{a}^{-1}) = A + B\cos(\mathbf{k}.\mathbf{a}) \pm C\cos(\mathbf{k}.\tfrac{1}{2}\mathbf{a}) \quad (81)$$

$$W'_{\pm}(\mathbf{k}\|\mathbf{b}^{-1}) = D + E\cos(\mathbf{k}.\mathbf{b}) \pm F\cos(\mathbf{k}.\tfrac{1}{2}\mathbf{b}) \pm E_{14}\cos(\mathbf{k}.\tfrac{3}{2}\mathbf{b}) \quad (82)$$

$$W'_{+}(\mathbf{k}\|\mathbf{c}^{-1}) = G + H\cos(\mathbf{k}.\mathbf{c}); \quad W'_{-}(\mathbf{k}\|\mathbf{c}^{-1}) = I + L\cos(\mathbf{k}.\mathbf{c}), \quad (83)$$

where we have defined A through L by the relations

$$\left.\begin{aligned} A &= 2(K_2+K_3+2K_4), & F &= 4(K_9+K_{10}), \\ B &= 2(K_5+K_6+2K_7+2K_8), & G &= 2(K_3+K_5+2K_7+2K_9), \\ C &= 4(K_9+K_{10}), & H &= 2(K_2+2K_4+K_6+2K_8+2K_{10}), \\ D &= 2(K_2+K_5+K_6), & I &= 2(K_3+K_5+2K_7-2K_9), \\ E &= 2(K_3+2K_4+2K_8), & L &= 2(K_2+2K_4+K_6+2K_8-2K_{10}). \end{aligned}\right\} \quad (84)$$

For the cases when **k** is parallel to the reciprocal lattice vectors $\mathbf{a}^{-1}$ and $\mathbf{b}^{-1}$, the two bands are degenerate at $\pi\mathbf{a}^{-1}$ and at $\pi\mathbf{b}^{-1}$ and intersect at these points with equal and opposite slopes. This conclusion is a general result of the effect of time reversal symmetry on the energy bands of crystals. The energy bands are expected to stick together in pairs on the boundary plane of the Brillouin zone that is perpendicular to a twofold screw axis of the reciprocal lattice.

The first exciton band structure of anthracene, naphthalene and biphenyl shows no especially interesting features and will not be displayed herein. The interested reader is referred to ref. (3) for further details.

We now proceed to discuss the fate of the thermalised exciton. When the exciton mean free path is considerably larger than the lattice spacing, a coherent motion band model may be used to describe exciton migration. The group velocity of the exciton packet can be expressed in terms of the exciton band structure $W(\mathbf{k})$:

$$\mathbf{V}(\mathbf{k}) = \frac{1}{\hbar}\nabla_{\mathbf{k}} W(\mathbf{k}). \tag{85}$$

We now introduce an exciton diffusion tensor $\mathbf{D}$ with components $D_{\mu\nu}$. Adopting a phenomenological approach, we assume that the exciton scattering can be described in the relaxation time approximation, so that

$$D_{\mu\nu} = \langle\langle \tau_s(\mathbf{k})\, V_\mu(\mathbf{k})\, V_\nu(\mathbf{k})\rangle\rangle. \tag{86}$$

$\tau_s(\mathbf{k})$ is a relaxation time, V_μ is the μ-th component of the group velocity $\mathbf{V}(\mathbf{k})$ and the angular brackets represent the average over the Boltzmann distribution of the excitons in the band. Alternatively, we can define an isotropic mean free path $\Lambda(\mathbf{k})$ by the relation

$$\Lambda(\mathbf{k}) = \tau_s(\mathbf{k})\,\mathbf{V}(\mathbf{k}), \tag{87}$$

so that

$$D_{\mu\nu} = \left\langle\left\langle \Lambda(\mathbf{k}) \frac{V_\mu V_\nu}{|\mathbf{V}(\mathbf{k})|}\right\rangle\right\rangle. \tag{88}$$

The exciton mean free path is determined by the various scattering mechanisms. The scattering cross-section $\sigma_0(\nu\mathbf{k}, \nu'\mathbf{k}')$ for the process $|\nu\mathbf{k}\rangle \rightarrow |\nu'\mathbf{k}'\rangle$, induced by the perturbation H_p, can be represented within the framework of first-order time-dependent perturbation theory

$$\sigma_0(\nu\mathbf{k}, \nu'\mathbf{k}') = \frac{2\pi\rho(\nu', \mathbf{k}')}{\hbar I_0(\nu, \mathbf{k})} \langle\langle \nu'\mathbf{k}' | H_p | \nu\mathbf{k}\rangle\rangle, \tag{89}$$

where $\rho(\nu', \mathbf{k}')$ is the density of final states and $I_0(\nu, \mathbf{k})$ is the exciton flux. The total scattering cross-section is obtained by summing over all the possible scattering channels. For triplets the intermolecular pair interactions are so much smaller than the intramolecular vibration energies, that the weak vibronic coupling scheme is applicable. For a given exciton state we consider a set of vibronic sublevels separated by the intramolecular vibrational quantum (which is of the order of 1 000 cm^{-1}). As already noted, in this extreme limit, the coupling can be considered to be a stationary effect, modifying the electronic interaction matrix elements by multiplying them by a Franck–Condon vibrational overlap factor. Exciton scattering occurs by interaction with the intermolecular vibrations, and in particular with acoustic modes. The off-diagonal matrix elements of the crystal Hamiltonian are small even when compared with the frequency of the intermolecular lattice vibrations. Under these conditions the rate of excitation transfer will be small relative to the lattice relaxation time. The coupling of the exciton state with the lattice phonons is then expected to be strong enough to cause elastic scattering at every lattice site, and thus to destroy the crystal momentum as a good quantum number. Under these extreme conditions the mean free path is of the order of magnitude of the intermolecular separation in the crystal, and the motion of the excitation may be described as a random walk. The small but finite off-diagonal matrix elements result in a random hopping from one localised state to another. We shall thus consider resonance transfer of the excitation between localised states. The resonance transfer probability is a sensitive function of the vibrational state of the lattice, in view of the exponential dependence of the transfer integrals on the intermolecular spacing. A detailed theory of acoustic mode scattering within the framework of the tight binding approximation must include these effects.[41] For the study of the random walk model we shall limit ourselves to a discrete probability distribution based on the intermolecular interaction integrals at the equilibrium configuration of the crystal. These intermolecular interaction integrals must be related to the probability of exciton transfer. Application of first-order time-dependent perturbation theory leads to the conventional expression for the transition probability for energy transfer from the reference molecule 1 in the vibrational state ξ to molecule i in the vibrational state ζ:

$$W(1\xi|i\zeta) = \frac{2\pi}{\hbar}|J(1\xi|i\zeta)|^2 \rho(E_f), \tag{90}$$

where $J(1\xi|i\zeta)$ is the transfer integral determined by the intermolecular interactions and $\rho(E_f)$ is the density of final states. The spectral weight function representing the local density of states is approximated by assuming a uniform distribution of states through the exciton band, characterised by a total width ΔE, so that $\rho(E_f) \approx (\Delta E)^{-1}$. Using these approximations the mean lifetime, τ, of an excitation on the reference molecule is given by

$$\frac{1}{\tau} = \sum_i W(1\xi|i\zeta) \tag{91}$$

and the components of the diffusion tensor **D** are given by

$$D_{\mu\nu} = \sum_i \left[\frac{\pi}{\hbar\Delta E}|J(1\xi|i\zeta)|^2(\mathbf{R}_1-\mathbf{R}_i)_\mu(\mathbf{R}-\mathbf{R}_i)_\nu\right]. \tag{92}$$

In the general expression derived for the mobility tensor, the individual jump frequencies are calculated from the contributions of molecular pair interactions. Each jump frequency is weighted according to the square of the displacement vector.

The application of first-order perturbation theory might appear questionable, because the effects of intermediate virtual states have been neglected, and because the final density of states is not well defined within the framework of the localised state model. However, an analogous result can be derived for a similar case, that of excess electron mobility in aromatic solids,[42] on the basis of Kubo's formalism of linear response theory.

Triplet exciton states in crystals of aromatic molecules are particularly suitable for experimental study of exciton migration in view of the long lifetime of the triplet state, which is of the order of 10 ms in crystalline anthracene and naphthalene. Utilising the effects of triplet-triplet annihilation in crystalline anthracene, the space intermittency method when applied to the study of the diffusion length of triplet excitons in anthracene leads to

$$L = 15 \pm 2\mu \quad \text{and} \quad D = 2(\pm 0{\cdot}5) \times 10^{-4}\,\text{cm}^2/\text{s}.$$

On the other hand, experimental data[45] on the effect of crystal thickness on the lifetime of triplet excitons generated by a Q switched ruby laser in crystalline anthracene lead to a diffusion coefficient of the order of $D = 2 \times 10^{-2}\,\text{cm}^2/\text{s}$. Because the surface-quenching mechanism is rather obscure, and because alternative mechanisms (such as singlet exciton ionisation) may interfere with the measurements, we believe that the results based on the space intermittency method are more reliable. A theoretical study of the diffusion coefficient of triplet excitons in molecular organic crystals has been performed using the random walk strong scattering model.[3, 44] The calculated value of the diffusion coefficient for crystalline anthracene ($D\,(\text{cal}) = 5 \times 10^{-4}\,\text{cm}^2/\text{s}$)

is in reasonable agreement with the experimental data. This theoretical treatment also provides an adequate interpretation of the anisotropy of the triplet exciton diffusion tensor in the *ab* crystal plane (though not for the *c* crystal direction). The reader is referred to the original work,[3, 44] for further details.

Further evidence for the validity of the strong scattering random walk model comes from the study of triplet-triplet annihilation in crystalline anthracene. This annihilation reaction leads to the formation of a singlet exciton state.[46–49] The direct population of the first $^3B_{2u}$ state of crystalline anthracene is possible using intense light sources. When anthracene is illuminated with a ruby laser, delayed fluorescence on the time scale of milliseconds is observed. The intensity of the delayed fluorescence depends on the square of the light intensity and follows bimolecular kinetics. These results may be interpreted in terms of triplet-triplet annihilation, whereupon the collision of two triplet excitons gives rise to a singlet-exciton state which decays radiatively. The rate constant for triplet-triplet annihilation is found to be $\gamma_t = 2\times 10^{-11}\,\text{cm}^3\,\text{s}^{-1}$. In the strong scattering random walk model, we suppose that the rate-determining step in this process involves the production of two adjacent excited molecules, i.e. the bimolecular annihilation reaction is diffusion-controlled. The rate constant is then given by the rate for the encounter of two triplet excitons,

$$\gamma_D = 8\pi D\langle R\rangle,$$

and a statistical factor, $\eta = \frac{1}{9}$, representing the probability for the formation of a singlet state from a pair of triplet excitons. Thus

$$\gamma_t = \eta\gamma_D,$$

and setting $\langle R\rangle = 6A$, $D = 2\times 10^{-4}\,\text{cm}^2\,\text{s}^{-1}$ we get $\gamma_t = 2\times 10^{-11}\,\text{cm}^3\,\text{s}^{-1}$ which is in good agreement with the experimental observations. These results imply that the mechanism recently proposed by Singh *et al.*[50] for the decomposition of a singlet exciton to give a pair of triplet excitons, which is just the reverse of the triplet-triplet annihilation reaction, is not amenable to experimental observation, because the rate of recombination of triplet excitons is large on the time scale of triplet-exciton migration.

As a last topic in this section we consider the probability for the emission of a photon $\hbar\omega$ by radiative decay of a pair of localised excited atoms or molecules. This is given by the conventional expression

$$T_{f0\to 00} = \frac{4\omega^3}{\hbar c^3}K^2, \tag{93}$$

where K is the radiative transition probability matrix element connecting the initial state $|fg\rangle$ with the final state $|00\rangle$.

To estimate the rate of the radiative annihilation of a double exciton state in molecular crystals, we again restrict our discussion to triplet-exciton bands with small width (of the order 1–10 cm^{-1}), so that the strong scattering limit can be used. Within the framework of the random walk model, a simple kinetic scheme leads to the following formula for the reaction rate for radiative annihilation of the double exciton state:

$$RR = \frac{\gamma_{\text{coll.}}\, n^2}{1+\gamma_{\text{trans.}}/T_{fg\to 00}}, \tag{94}$$

where n is the exciton concentration, $\gamma_{\text{coll.}}$ the rate of collision of two excitons, and $\gamma_{\text{trans.}}$ the excitation transfer rate in the crystal. When the rate of dissociation of the exciton pair (where two excitations are located on neighbour molecules) is large compared with the radiative annihilation rate (i.e. $\gamma_{\text{trans.}} \gg T$) the annihilation rate is determined by the radiative transition rate to the final state

$$RR = \frac{\gamma_{\text{coll.}}\, T_{fg\to 00}}{\gamma_{\text{trans.}}} n^2. \tag{95}$$

On the other hand, when the transition to the final state is fast on the time scale of exciton migration (i.e. $\gamma_{\text{trans.}} \ll T$), the radiative annihilation rate is diffusion-controlled, being determined by the rate of collision of the excitons

$$RR = \gamma_{\text{coll.}}\, n^2. \tag{96}$$

Consider, now, the radiative annihilation of a double triplet-exciton state in anthracene. We use the numerical values:

$$\gamma_{\text{coll.}} = 10^{-10}\,\text{cm}^{+3}\,\text{s}^{-1} \quad \text{and} \quad \gamma_{\text{trans.}} = 10^{10}\,\text{s}^{-1}.$$

For a pair of triplet excitations located on adjacent molecules the radiative annihilation rate is, approximately, $T_{fg\to 00} \approx \tau_a (J/W)^2$, where τ_a is the radiative lifetime of singlet anthracene, J is an electron exchange matrix element combining the double exciton state and the intermediate singlet state, and W is the energy difference between the double triplet-exciton state and the singlet-exciton state. We shall set $W = 0{\cdot}5\,\text{eV}$, which is the energy difference between the double triplet-exciton state (populated by a ruby laser) and the 0–0 band of the $^1B_{2u}$ state in anthracene, and also set $\tau_a = 3\times 10^8\,\text{s}^{-1}$. The exchange matrix element is about $J = 10\,\text{cm}^{-1}$. Using the quoted values, we obtain $T_{fg\to 00} = 10^4\,\text{s}^{-1}$. Thus it follows immediately that for radiative triplet-triplet annihilation in the case of crystalline anthracene, $\gamma_{\text{coll.}} \gg T_{fg\to 00}$, and the reaction rate is dominated by the transition probability to the final state, the annihilation rate being $RR = 10^{-18}\,\text{n}^2\,\text{cm}^{-3}\,\text{s}^{-1}$. This latter rate is seven orders of magnitude smaller than that in the other diffusion-controlled channel, which involves direct transition to the singlet state.

Intermediate triplet excitons in crystalline I_2[51]

Experimental and theoretical studies of the electronic states of molecular crystals have largely been confined to the examination of the properties of aromatic hydrocarbons. In these compounds the lowest electronic transitions are of $\pi \to \pi^*$ character, and the intermolecular overlap is small in both the ground state and the excited states. In addition, the spin-orbit coupling in the molecule is very weak. It is of obvious interest to extend the range of studies to include solids in which the intermolecular overlap is significant, and in which spin-orbit coupling effects are important.

Consider, for example, crystalline Cl_2, Br_2 and I_2. The ratio of the heat of sublimation to the heat of dissociation varies from 0·11 in Cl_2 to 0·41 in I_2. It might be guessed, from just the figures cited, that the Cl_2 molecule in the Cl_2 crystal is relatively less altered from the Cl_2 molecule in the free state than is the I_2 molecule in the I_2 crystal from the I_2 molecule in the free state. This expectation implies, in turn, that the excited states of the Cl_2 crystal are nearer the Frenkel limit than are the excited states of crystalline I_2. Indeed, in the cases of crystalline Br_2 and I_2, the increased intermolecular overlap suggests that the excited states in these solids will be in the intermediate exciton domain. Some evidence for electron delocalisation has already been discussed in the literature. In a previous paper we have shown that electron delocalisation, represented in terms of the mixing of charge-transfer exciton states and neutral exciton states, makes an important contribution to the ground state binding energy of the halogen crystals. For the case of Cl_2 it is estimated that this contribution amounts to 25 % of the cohesive energy of the crystal.[4] By extension of the argument, we also expect the energies of the excited states to be altered because of the overlap. In the case of crystalline I_2 more direct evidence comes from nuclear quadrupole resonance studies, which have shown that the molecular bond in I_2 is weakened to $\sim 82\%$ of its value in the free molecule.[52]

Apart from the increased intermolecular overlap, the halogens show a further difference when compared with crystals of the aromatic hydrocarbons. We have already remarked that spin-orbit coupling in the aromatic hydrocarbons is very weak. In the halogens, the spin-orbit coupling is large, and the various orbital transitions are split. As a consequence of this splitting, the spin quantisation in the crystal must be treated carefully.

We now briefly outline an analysis of the excited states of crystalline I_2, the crystal spectrum of which has been reported.[53] Because of the large intermolecular overlap and large spin-orbit coupling in I_2, effects to be expected in all of the crystalline halogens will be relatively pronounced in solid I_2.

The treatment used follows so closely the methods introduced in the study of crystals of aromatic hydrocarbons, that only a few points need be emphasised.

The lowest orbital transitions of molecular I_2 correspond to the elevation of an electron from the highest filled orbital of π_g symmetry to the lowest unfilled orbital of σ_u symmetry. The lowest excited triplet state, ${}^3\pi_u(\pi_g \rightarrow \sigma_u)$ is split by spin-orbit interaction into the states 0_u^+, 0_u^-, 1_u, 2_u. The two lowest energy transitions of iodine correspond to transitions from the ground state ${}^1\Sigma_g^+$ to the states 0_u^+ and 1_u. The first transition, occurring at 2·37 eV, is polarised along the molecular axis. The second, at 1·75 eV, is polarised perpendicular to this axis.[54] In discussing the excited states of molecular crystals, it is customary to construct zero-order crystal states from the free molecule excited states. This, of course, assumes that the intermolecular interactions are sufficiently weak that the free molecule state is only slightly distorted by the crystal field. If there is large intermolecular overlap, then this condition is not satisfied, and the crystal states will be far from the Frenkel limit (zero overlap). In the case of large overlap in solids with a large dielectric constant, it is usual to describe the exciton states in the quasi-hydrogenic Wannier limit. All real solids lie between these two limits and we expect the halogens to be described in the Frenkel picture with increasing importance of charge-delocalisation as the series is ascended.

We now take the molecular excited states 0_u^+, 0_u^-, 1_u, 2_u as starting-points for the description of the crystal excited states. These may be written in terms of the orbital transitions $X(\pi_g^x \rightarrow \sigma_u)$ and $Y(\pi_g^y \rightarrow \sigma_u)$

with the spin assignments T_x, T_y, and T_z as shown in table 8, and with

$$\left.\begin{aligned} T_x &= \frac{1}{\sqrt{2}}(\alpha_1\alpha_2+\beta_1\beta_2), \\ T_y &= \frac{1}{\sqrt{2}}(\alpha_1\alpha_2-\beta_1\beta_2), \\ T_z &= \frac{1}{\sqrt{2}}(\alpha_1\beta_1+\alpha_2\beta_2). \end{aligned}\right\} \quad (97)$$

In table 8 we show how these I_2 triplet excited states combine to form symmetry adapted functions for $\mathbf{k} = \mathbf{0}$. (I_2 crystallises in the space group D_{2h} with four molecules per unit cell.) Consider the mixing between the charge-transfer states and the neutral excited states. For the various neutral triplet excited states, the spin is quantised according to the molecular axes. In the construction of the triplet charge-transfer states, the spin will be quantised along some other axes, say the crystal axes. In the evaluation of the neutral exciton charge-transfer matrix elements, it is convenient to preserve the orthogonality of the spin functions α and β. For this reason we transform the neutral exciton functions such that the spin is quantised along the crystal axes. Using the well-known transformation properties of T_x, T_y, and T_z it may readily be shown that

$$\left.\begin{aligned} T_{x1,3} &\to -\cos\vartheta T_x - i\sin\vartheta T_z, \\ T_{x2,4} &\to \cos\vartheta T_x - i\sin\vartheta T_z, \\ T_{z1,3} &\to \cos\vartheta T_z + i\sin\vartheta T_x, \\ T_{z2,4} &\to \cos\vartheta T_z - i\sin\vartheta T_x, \end{aligned}\right\} \quad (98)$$

where ϑ is the angle the I_2 molecule makes with the c axis in the bc plane and T_x, T_z refer to the crystal axes. $T_{y,i}$ will remain unaltered in this new coordinate system. In table 8 we display the neutral exciton states for which this transformation has been carried out.

The calculation is now carried out in the same fashion as the analogous analysis of the states of crystalline aromatic hydrocarbons (see Hillier & Rice for details).[51] Only nearest neighbour ion-pair exciton states were considered, and a modified Goeppert–Mayer–Sklar procedure was used to evaluate the relevant matrix elements. The principal approximations made are the following:

(*a*) We have not specifically calculated the effect of spin-orbit

Table 8

Molecular symmetry ($D_{\infty h}$)	Molecular spin-state function	Crystal function	Spin-transformed crystal functions	Symmetry
$\left.\begin{matrix}\Sigma_u^+\\ 0_u^+\end{matrix}\right\}$	$XT_y - YT_x$	$(X_1T_{y_1} - Y_1T_{x_1}) \substack{+\\-\\-\\+} (X_2T_{y_2} - Y_2T_{x_2}) \substack{-\\+\\-\\+} (X_3T_{y_3} - Y_3T_{x_3}) \substack{-\\-\\+\\+} (X_4T_{y_4} - Y_4T_{x_4})$	$\left(X_1 \substack{+\\-\\-\\+} X_2 \substack{-\\+\\-\\+} X_3 \substack{-\\-\\+\\+} X_4\right) T_y + \cos\vartheta \left(Y_1 \substack{-\\+\\+\\-} Y_2 \substack{-\\+\\-\\+} Y_3 \substack{+\\+\\-\\-} Y_4\right) T_x + i\sin\vartheta \left(Y_1 \substack{+\\-\\-\\+} Y_2 \substack{-\\+\\-\\+} Y_3 \substack{-\\-\\+\\+} Y_4\right) T_z$	$\begin{cases} A_u \\ B_{1u} \\ B_{2u} \\ B_{3u} \end{cases}$
$\left.\begin{matrix}\Sigma_u^-\\ 0_u^-\end{matrix}\right\}$	$XT_x + YT_y$	$(X_1T_{x_1} + Y_1T_{y_1}) \substack{+\\-\\-\\+} (X_2T_{x_2} + Y_2T_{y_2}) \substack{+\\-\\+\\-} (X_3T_{x_3} + Y_3T_{x_3}) \substack{+\\+\\-\\-} (X_4T_{x_4} + Y_4T_{y_4})$	$\left(Y_1 \substack{+\\-\\-\\+} Y_2 \substack{+\\-\\+\\-} Y_3 \substack{+\\+\\-\\-} Y_4\right) T_y - \cos\vartheta \left(X_1 \substack{-\\+\\+\\-} X_2 \substack{+\\-\\+\\-} X_3 \substack{-\\-\\+\\+} X_4\right) T_x - i\sin\vartheta \left(X_1 \substack{+\\-\\-\\+} X_2 \substack{+\\-\\+\\-} X_3 \substack{+\\+\\-\\-} X_4\right) T_z$	$\begin{cases} A_u \\ B_{1u} \\ B_{2u} \\ B_{3u} \end{cases}$
π_u	XT_z	$X_1T_{z_1} \substack{+\\-\\-\\+} X_2T_{z_2} \substack{+\\-\\+\\-} X_3T_{z_3} \substack{+\\+\\-\\-} X_4T_{z_4}$	$\cos\vartheta \left(X_1 \substack{+\\-\\-\\+} X_2 \substack{+\\-\\+\\-} X_3 \substack{+\\+\\-\\-} X_4\right) T_z + i\sin\vartheta \left(X_1 \substack{-\\+\\+\\-} X_2 \substack{+\\-\\+\\-} X_3 \substack{-\\-\\+\\+} X_4\right) T_x$	$\begin{cases} A_u \\ B_{1u} \\ B_{2u} \\ B_{3u} \end{cases}$
1_u	YT_z	$Y_1T_{z_1} \substack{+\\-\\-\\+} Y_2T_{z_2} \substack{-\\+\\-\\+} Y_3T_{z_3} \substack{-\\-\\+\\+} X_4T_{z_4}$	$\cos\vartheta \left(Y_1 \substack{+\\-\\-\\+} Y_2 \substack{-\\+\\-\\+} Y_3 \substack{-\\-\\+\\+} Y_4\right) T_z + i \sin\vartheta \left(Y_1 \substack{-\\+\\+\\-} Y_2 \substack{-\\+\\-\\+} Y_3 \substack{+\\+\\-\\-} Y_4\right) T_x$	$\begin{cases} A_u \\ B_{1u} \\ B_{2u} \\ B_{3u} \end{cases}$
Δ_u	$XT_x - YT_y$	$(X_1T_{x_1} - Y_1T_{y_1}) \substack{+\\-\\-\\+} (X_2T_{x_2} - Y_2T_{y_2}) \substack{+\\-\\+\\-} (X_3T_{x_3} - Y_3T_{y_3}) \substack{+\\+\\-\\-} (X_4T_{x_4} - Y_4T_{y_4})$	$-\left(Y_1 \substack{+\\-\\-\\+} Y_2 \substack{+\\-\\+\\-} Y_3 \substack{+\\+\\-\\-} Y_4\right) T_y - \cos\vartheta \left(X_1 \substack{-\\+\\+\\-} X_2 \substack{+\\-\\+\\-} X_3 \substack{-\\-\\+\\+} X_4\right) T_x - i\sin\vartheta \left(X_1 \substack{+\\-\\-\\+} X_2 \substack{+\\-\\+\\-} X_3 \substack{+\\+\\-\\-} X_4\right) T_z$	$\begin{cases} A_u \\ B_{1u} \\ B_{2u} \\ B_{3u} \end{cases}$
2_u	$XT_y + YT_x$	$(X_1T_{y_1} + Y_1T_{x_1}) \substack{+\\-\\-\\+} (X_2T_{y_2} + Y_2T_{x_2}) \substack{-\\+\\-\\+} (X_3T_{y_3} + Y_3T_{x_3}) \substack{+\\+\\-\\-} (X_4T^5_{y_4} + Y_4T_{x_4})$	$\left(X_1 \substack{+\\-\\-\\+} X_2 \substack{-\\+\\-\\+} X_3 \substack{-\\-\\+\\+} X_4\right) T_y - \cos\vartheta \left(Y_1 \substack{-\\+\\+\\-} Y_2 \substack{-\\+\\-\\+} Y_3 \substack{+\\+\\-\\-} Y_4\right) T_x + i\sin\vartheta \left(Y_1 \substack{+\\-\\-\\+} Y_2 \substack{-\\+\\-\\+} Y_3 \substack{-\\-\\+\\+} Y_4\right) T_z$	$\begin{cases} A_u \\ B_{1u} \\ B_{2u} \\ B_{3u} \end{cases}$

Table 9. *Iodine interaction integrals using double zeta set*†

Overlap integrals

Distance	$\langle\pi/\pi\rangle$	$\langle\sigma/\sigma\rangle$
6	0·6174E–01	0·2534E–00
8	0·9256E–02	0·6762E–01
10	0·1001E–02	0·1081E–01
12	0·8466E–04	0·1222E–02

Coulomb integrals $(\langle + - | - + \rangle = \langle +_a^*(1) -_a(1) | 1/r_{12} | -_b^*(2) +_b(2) \rangle)$

Distance	$\langle + - \| - + \rangle$	$\langle + + \| + + \rangle$	$\langle 0 + \| + 0 \rangle$	$\langle 0\,0 \| + + \rangle$	$\langle 0\,0 \| 0\,0 \rangle$
6	0·1637E–02	0·1685E–00	0·2914E–02	0·1937E–00	0·2106E–00
8·5	0·3028E–03	0·1232E–00	0·6032E–03	0·1344E–00	0·1392E–00
11·0	0·8351E–04	0·9678E–01	0·1660E–03	0·1021E–00	0·1035E–00

Hybrid integrals $(\langle - + | + - \rangle = \langle -_a^*(1) +_a(1) | 1/r_{12} | +_a^*(2) -_b(2) \rangle)$

Distance	$\langle - + \| + - \rangle$	$\langle + + \| + + \rangle$	$\langle + 0 \| 0 + \rangle$	$\langle + 0 \| - 0 \rangle$	$\langle 0\,0 \| + + \rangle$	$\langle + + \| 0\,0 \rangle$	$\langle 0\,0 \| 0\,0 \rangle$
6·0	0·1060E–02	0·1614E–01	0·2195E–02	−0·3003E–02	0·1727E–01	0·7146E–01	0·8440E–01
8·5	0·4850E–04	0·1156E–02	0·1469E–03	−0·2514E–03	0·1255E–02	0·9971E–02	0·1159E–01
11·0	0·1420E–05	0·5246E–04	0·5572E–05	−0·1082E–04	0·5661E–04	0·6910E–03	0·7813E–03

Exchange integrals $(\langle + - | - + \rangle = \langle +_a^*(1) -_b(1) | 1/r_{12} | -_a^*(2) +_b(2) \rangle)$

Distance	$\langle + - \| - + \rangle$	$\langle + + \| + + \rangle$	$\langle 0 + \| 0 - \rangle$	$\langle 0 + \| + 0 \rangle$	$\langle 0\,0 \| + + \rangle$	$\langle 0\,0 \| 0\,0 \rangle$
6·0	0·1479E–03	0·1211E–02	−0·1448E–02	0·1251E–02	0·5696E–02	0·3054E–01
8·5	0·1006E–05	0·8710E–05	−0·1775E–04	0·1594E–04	0·7978E–04	0·8160E–03
11·0	0·2398E–08	0·2329E–07	−0·6988E–07	0·6381E–07	0·3295E–06	0·5110E–05

Goeppert–Mayer–Sklar integrals

Distance	$\langle\pi \| V_I \| \pi\rangle$	$\langle\sigma \| V_I \| \sigma\rangle$
6·0	−0·9675E–02	−0·4354E–01
8·5	−0·7031E–03	−0·6209E–02
11·0	−0·3169E–04	−0·4247E–03

† All units are atomic units. The symbols +, −, 0 give the m value associated with the $5p$ atomic orbital, i.e. +, $m = +1$; −, $m = -1$; 0, $m = 0$.

The basis set used was: $\phi_{5p} = -0{\cdot}06454\chi_{21}(24{\cdot}15) + 0{\cdot}0099\chi_{21}(47{\cdot}25) - 0{\cdot}0019\chi_{31}(16{\cdot}52)$

$$+0{\cdot}1563\chi_{31}(12{\cdot}39) - 0{\cdot}6299\chi_{41}(5{\cdot}50) + 0{\cdot}0609\chi_{41}(8{\cdot}34) + 0{\cdot}8824\chi_{51}(1{\cdot}95) + 0{\cdot}4864\chi_{51}(4{\cdot}62).$$

Here χ_{nl} represents a Slater orbital with quantum numbers n and l, and orbital exponent ξ.

coupling on the crystal excited states, rather assuming that it is taken into account when we take the free molecule wavefunctions to be eigenfunctions of the monomer Hamiltonian, with eigenvalues related to the experimental transition energies. It is commonly assumed that the spin-orbit coupling is predominantly determined by one-center terms, and we have taken this to be so by ignoring any contribution from spin-orbit terms in specifying the interaction part of the crystal Hamiltonian.

(*b*) We have not considered the effect of singlet-triplet mixing on the off-diagonal elements of the energy matrix, but the dominant effect of such mixing in the neutral diagonal elements is implicitly allowed for by the use of the experimental transition energies.

(*c*) In the specification of the energy of the charge-transfer states we have used simple classical electrostatic considerations, and so have not considered these states to be split by spin-orbit coupling. In view of the uncertainty in the values of the molecular electron affinity, and the crystal polarisation energy, this approximation is probably not too serious. The calculated polarisation energy ($-0{\cdot}35$ eV) is unexpectedly small because of a partial cancellation between the ion-dipole, and dipole-dipole contributions. The approximations inherent in the use of a simple electrostatic model, particularly the neglect of any dielectric screening effects, render the polarisation energy an adjustable parameter in the calculation reported. It is to be particularly noted that when the polarisation energy is taken to be $-0{\cdot}35$ eV, the lowest charge-transfer state is located at 3·7 eV. The free molecule state, 0_u^+, is located at 2·37 eV, so that any increase in the polarisation energy will tend to decrease the separation of these states and increase the perturbation of the neutral state by the charge-transfer states.

(*d*) A further approximation in this work is the use throughout of an SCF $5p$ iodine wavefunction in the calculation of all the required matrix elements. Although it is known that such atomic wavefunctions are good starting-points for SCF molecular orbital calculations, the change in this orbital on bond formation is difficult to assess in the absence of such SCF MO calculations. However, it is to be expected that the use of atomic orbitals will tend to overestimate the intermolecular overlap.

(*e*) We have not considered the mixing of higher charge-transfer states, originating from the transitions $\pi_u \rightarrow \sigma_u$, $\sigma_g \rightarrow \sigma_u$. The effect of these states will be smaller than that of those arising from the transitions $\pi_g \rightarrow \sigma_u$, both because of their higher energy, and the smaller intermolecular overlap of the more contracted, tightly bound π_u, σ_g orbitals. Our neglect of these charge-transfer states will tend to compensate for the overestimation of the influence of the $\pi_g \rightarrow \sigma_u$ charge-transfer states resulting from the use of atomic orbitals.

Table 10. *Neutral exciton splitting of iodine triplet states*

	A_u	B_{1u}	B_{2u}	B_{3u}
Σ_u^+, Δ_u	50	−50	−40	42
Σ_u^-, Δ_u	40	−90	−48	−98
1_u	74	−96	−68	90
	135	−81	−111	57

Table 11. *Exciton states of crystalline iodine*

Free molecule state	Free molecule transition energy. Polarisation energy = −0·35 eV	A_u	$B_{1u}(z', c)$	$B_{2u}(y', a)$	$B_{3u}(x', b)$
0_u^+	2·37	2·07	2·20	2·24	2·20
1_u	1·75	1·72	1·69	1·71	1·67
		1·66	1·56	1·67	1·58
	Polarisation energy = −1·0 eV				
0_u^+	2·37	1·96	2·14	2·19	2·15
1_u	1·75	1·72	1·68	1·70	1·66
		1·64	1·51	1·66	1·54

Table 12. *Summary of iodine crystal absorption bands*

(eV)	Polarisation
3·65	*c* (strongly pol.)
3·01	*a* (strongly pol.)
2·50	*c* (weakly pol.)
2·30	*c* (strongly pol.)
2·06	*a* (strongly pol.)

We turn now to a comparison of our calculations with the experimental iodine crystal absorption bands reported by Schnepp *et al.*[53] These investigators have observed five bands in the region 2·06–3·65 eV, two lying below the free molecule transition and three above it. Our calculations predict the 0_u^+ state to be depressed by 0·13 to 0·3 eV, or 0·18–0·41 eV, depending upon the value of the polarisation energy used. It is thus probable that the two states observed at 2·06 and 2·30 eV are of 0_u^+ parentage, but our calculation does not allow the assignment of the symmetry of the crystal states, as the four states are calculated to be close together, and the accuracy of our calculation is certainly not greater than the calculated separation. With regard to the three transitions observed by Schnepp above that of the 0_u^+ state, it is probable that these are transitions to charge-transfer states. The separation of the charge-transfer states we consider here is ~ 1 eV, their exact position depending upon the value of the polarisation energy taken. If the polarisation energy were taken to be 1·5 eV, the lowest state would be at 2·5 eV and the highest at 3·5 eV, which would correlate well with the experimental data. The intensity of these states would arise from the strong mixing with the nearby neutral exciton states. There are no data with which we may compare our predictions of the positions of the states of 1_u parentage. Our calculations indicate that these states are depressed by about 0·1–0·2 eV from the free molecule values.

Acknowledgements

The research briefly summarised in this review has been supported by the Directorate of Chemical Sciences, Air Force Office of Scientific Research, The United States Public Health Service, and indirectly by the Advanced Research Projects Agency through a grant for materials science research at the University of Chicago.

References

1 For a review see S. A. Rice & J. Jortner, *Physics and Chemistry of the Organic Solid State*, Vol. III (in the Press).
2 A. U. Hazi & S. A. Rice, *J. Chem. Phys.* **45**, 3004 (1966).
3 J. Jortner, S. A. Rice, S. I. Choi & J. L. Katz, *J. Chem. Phys.* (in the Press); J. L. Katz, S. A. Rice, S. I. Choi & J. Jortner, *J. Chem. Phys.* **39**, 1683 (1963); S. I. Choi, J. Jortner, S. A. Rice & R. Silbey, *J. Chem. Phys.* **41**, 3294 (1964).
4 I. H. Hillier & S. A. Rice, *J. Chem. Phys.* (in the Press).
5 C. C. J. Roothaan, *Rev. Mod. Phys.* **23**, 69 (1951).
6 M. H. Cohen & V. Heine, *Phys. Rev.* **122**, 1821 (1961).
7 B. J. Austin, V. Heine & L. J. Sham, *Phys. Rev.* **127**, 276 (1962).
8 C. C. J. Roothaan, *Rev. Mod. Phys.* **32**, 179 (1960).
9 J. C. Phillips & L. Kleinman, *Phys. Rev.* **116**, 287 (1959).
10 E. R. Davidson, *J. Chem. Phys.* **42**, 4199 (1965).
11 C. C. J. Roothaan, L. M. Sachs & A. W. Weiss, *Rev. Mod. Phys.* **32**, 186 (1960).

12 P. S. Kelly, *Proc. Phys. Soc.* A, **83**, 533 (1964).
13 C. S. Sharma & C. A. Coulson, *Proc. Phys. Soc.* A, **80**, 81 (1962).
14 D. B. Cook & J. N. Murrell, *Mol. Physics*, **9**, 417 (1965).
15 A. W. Weiss, *Phys. Rev.* **122**, 1826 (1961).
16 A. U. Hazi & S. A. Rice (to be published).
17 V. Heine & I. Abarenkov, *Phil. Mag.* **9**, 451–65 (1964).
18 J. C. Slater & W. Shockley, *Phys. Rev.* **50**, 705 (1936).
19 L. P. Bouckaert, R. Smoluchowski & E. Wigner, *Phys. Rev.* **50**, 58 (1936).
20 C. Herring, *J. Franklin Inst.* **233**, 525 (1942).
21 G. Koster, *Solid State Physics*, **5**, 174 (1957).
22 J. C. Slater, *Quantum Theory of Molecules and Solids*, Vol. II (McGraw-Hill, New York, 1965).
23 H. Winston, *J. Chem. Phys.* **19**, 156 (1951).
24 A. S. Davydov, *Theory of Molecular Excitons* (McGraw-Hill, New York, 1962).
25 D. S. McClure, *J. Chem. Phys.* **20**, 682 (1952).
26 M. Mizushima & S. Koide, *J. Chem. Phys.* **20**, 765 (1952).
27 H. F. Hameka & L. J. Oosterhoff, *Mol. Physics*, **1**, 358 (1958).
28 P. O. Löwdin, *J. Chem. Phys.* **18**, 365 (1950).
29 A. Witkowski & W. Moffitt, *J. Chem. Phys.* **33**, 872 (1960).
30 E. G. McRae, *Aust. J. Chem.* **14**, 329 (1961).
31 W. Siebrand, *J. Chem. Phys.* **40**, 2231 (1964).
32 E. F. McCoy & I. G. Ross, *Aust. J. Chem.* **15**, 573 (1962).
33 J. R. Hoyland & L. Goodman, *J. Chem. Phys.* **36**, 12 (1962).
34 R. Pariser, *J. Chem. Phys.* **24**, 250 (1956).
35 D. M. Hanson & G. W. Robinson, *J. Chem. Phys.* **43**, 4174 (1965).
36 G. Castro & R. M. Hochstrasser, *J. Chem. Phys.* **44**, 412 (1966).
37 R. E. Merrifield, *J. Chem. Phys.* **40**, 445 (1964).
38 A. I. Ansel'm & I. U. A. Firsov, *Soviet Phys.—JETP* **1**, 139 (1955); *JETP* **3**, 564 (1956).
39 Y. Toyoazawa, *Progr. Theoret. Phys.* **20**, 53 (1958); **27**, 89 (1962).
40 V. M. Agranovich & I. U. V. Konobeev, *Opt. Spectr.* **6**, 155 (1959).
41 L. Friedman, *Phys. Rev.* **140**, A 1649 (1965).
42 P. Goar & S. I. Choi, *J. Chem. Phys.* (in the Press).
43 P. Avakian & R. E. Merrifield, *Phys. Rev. Letters*, **13**, 541 (1964).
44 M. Levine, J. Jortner & A. Szoke, *J. Chem. Phys.* (in the Press).
45 R. G. Kepler & A. C. Switendick, *Phys. Rev. Letters*, **15**, 56 (1965).
46 R. G. Kepler, J. C. Caris, P. Avakian & E. Abramson, *Phys. Rev. Letters*, **10**, 400 (1963).
47 J. L. Hall, D. A. Jennings & R. M. McClintock, *Phys. Rev. Letters*, **11**, 364 (1963).
48 S. Z. Weisz, A. B. Zahlan, M. Silver & R. C. Jarnigan, *Phys. Rev. Letters*, **12**, 71 (1964).
49 P. Avakian, E. Abramson, R. G. Kepler & J. C. Caris, *J. Chem. Phys.* **39**, 1127 (1963).

50 S. Singh, W. J. Jones, W. Siebrand, B. P. Stoicheff & W. G. Schneider, *J. Chem. Phys.* **42**, 330 (1965).
51 I. H. Hillier & S. A. Rice (to be published).
52 C. H. Townes & B. P. Dailey, *J. Chem. Phys.* **20**, 35 (1952).
53 O. Schnepp, J. L. Rosenberg & M. Gouterman, *J. Chem. Phys.* **43**, 2767 (1965).
54 R. S. Mulliken, *Phys. Rev.* **46**, 549 (1934); **57**, 500 (1940).
L. Mathieson & A. L. G. Rees, *J. Chem. Phys.* **25**, 753 (1956).

EXACT TREATMENT OF COHERENT AND INCOHERENT TRIPLET EXCITON MIGRATION

H. HAKEN AND G. STROBL

Abstract

We set up an exactly solvable model by assuming that the excitation energy of a localised exciton at a lattice site, as well as the transition matrix elements, are random functions of the time (due to lattice vibrations). For the explicit example for triplet exciton motion in a linear chain we show that there is an abrupt change from coherent to incoherent motion depending on the fluctuating energies. If the corresponding condition is fulfilled, after a sufficiently high time the motion becomes a diffusion process.

Both coherent and incoherent migration of excitons have been treated theoretically in past literature. In our present analysis we want to shed light on the question of which parameters determine whether coherent or incoherent exciton migration takes place, and if there is a continuous or discontinuous transition from one kind of migration to the other. To this end we choose a model which allows for an exact solution and which contains the relevant parameters. In this paper we present the mathematical skeleton and refer the reader for the detailed discussion of the implications and, particularly for connection with previous theoretical and experimental work on this subject, to the detailed paper of the authors.

We assume that both the energy of an exciton at the lattice site n as well as its transition matrix element from a lattice site n to m is a random function of time t caused by lattice vibrations. We assume that one may neglect the reaction of the exciton on the lattice vibrations so that the lattice acts as a thermal reservoir in the thermo-dynamic sense. This is a reasonable assumption for not too low temperatures. We assume that the single fluctuations of the transition matrix elements and of the excitation energy of the atoms at the lattice sites are due to many independent lattice modes with a broad spectrum. (Mathematically speaking: the fluctuating coefficients are supposed to be gaussian and markoffian.) By means of this Hamiltonian we find equations for the averaged density matrix which can be solved by the method of Green's function exactly. We use second quantisation denoting the electronic ground state by Φ_0. The generation and annihilation respectively of exciton n is described by the operators b_n^+ resp. b_n. The Hamiltonian then reads

$$H = \hbar \left\{ \Sigma h_{nn}(t) b_n^+ b_n + \sum_{n \neq n} (H_{n-n'} + h_{nn'}(t)) b_n^+ b_{n'} \right\}, \qquad (1)$$

where the first sum describes the energy of an exciton at lattice sites n, whereas the second term describes the transition of the exciton from lattice site n' to n. $H_{n-n'}$ is the time averaged part and $h_{n,n'}$ describes the residual fluctuations. The markoffian property is represented by

$$\langle h_{nn'}(t) \,.\, h_{n''n'''}(t) \rangle = [\delta_{nn''}\,\delta_{n'n'''} + \delta_{nn'''}\,\delta_{n'n''}(1-\delta_{nn'})]\,\gamma_{n'-n}\, 2\delta(\tau). \qquad (2)$$

We may further assume $\qquad \langle h_{nn'}(t) \rangle = 0 \qquad (3)$

and $\qquad \gamma_m = \gamma_{-m}. \qquad (4)$

The density matrix is determined by

$$\dot{\rho} = -\frac{i}{\hbar}[H_1 \rho] = -\frac{i}{\hbar}[\bar{H}_1 \rho] - [h(t)_1 \rho]. \qquad (5)$$

In the following, we are interested in the density matrix averaged over the fluctuations. A detailed analysis shows that the averaged density matrix obeys the equation

$$\langle \dot{\rho} \rangle_{nn'} = -i[\bar{H}_1 \langle \rho \rangle]_{nn'} - 2\Gamma \langle \rho \rangle_{nn'} + 2\delta_{nn'} \sum_{\nu} \gamma_{n-\nu} \langle \rho \rangle_{\nu\nu} + 2(1-\delta_{nn'})\gamma_{n-n'} \langle \rho \rangle_{n'n}, \qquad (6)$$

where Γ is given by $\sum_m \gamma_m$. The first term on the right-hand side of equation (6) describes the completely coherent motion. The solution of this equation would be the usual Frenkel exciton. The remaining part alone would represent a master equation describing a random walk problem. The crux of the problem therefore consists of determining the interplay between these two parts and of deciding which one is dominant. An obvious way to treat such a problem would be to start with the assumption that one part is dominant over the other so that the equation is first solved for this part alone, and that the other part is then taken into account as a perturbation. We have done this but we do not represent the results here because they do not give an answer about the transition region. We rather want to solve equation

(6) exactly. For this end we first eliminate the time dependence by $\rho_{nn'}(t) = \rho_{nn'} e^{Rt}$ and use the internal translational symmetry of the averaged equation (6) by the relation

$$\rho_{nn'} = \exp(iKn) . N^{-\frac{1}{2}} \rho^K_{n-n'}. \tag{7}$$

With $n - n' = m$ we find for the new variables $\rho^K_{n-n'}$ the following set of equations

$$R^{(K)}\rho^K_m = -i \sum_{m'} H_{m-m'}(\exp[-iK(m-m')]-1)\rho^K_{m'}$$

$$-2\Gamma\rho^K_m + 2(1-\delta_{m0})\gamma_m \exp(-iK_m)\rho^K_{-m} + 2\delta_{m0}\gamma_K\rho^K_0, \tag{8}$$

where
$$\gamma_K = \sum_\lambda \gamma_\lambda \exp(iK\lambda). \tag{9}$$

In the following we drop the index K where it is not necessary and write

$$(\rho^K_m) = \boldsymbol{\rho}.$$

Equation (8) takes the form of a matrix equation:

$$r\boldsymbol{\rho} = D_1\boldsymbol{\rho} + D_2\boldsymbol{\rho}, \tag{10}$$

where

$$r = R + 2\Gamma; \quad D_{1_1 m_1 m'} = -iH_{m-m'}\exp(-iK(m-m')-1)$$

and $$D_{2_1 m_1 m'} = \delta_{mm'} 2\delta_{m0}\gamma_K + \delta_{m_1-m'} 2(1-\delta_{m0})\gamma_m \exp(-iKm).$$

Equation (10) is solved by the Green's function method where the Green's function is defined by

$$G_r = (D_1 - r)^{-1}. \tag{11}$$

Thus (10) is reduced to a new eigenvalue problem

$$\boldsymbol{\rho} = -G_r D_2 \boldsymbol{\rho}. \tag{12}$$

As is known from the problem of determining wavefunctions of impurities in solids, the Green's function formulation as expressed by (12) has the advantage that due to the short-range character of D_2 equation (12) is of a very low rank. We have solved equation (12) explicitly for triplet excitons in a linear chain assuming

$$H_m = H_1(\delta_{m,1} + \delta_{m_1-1}),$$

$$\gamma_m = \gamma_1(\delta_{m,1} + \delta_{m,-1}) + \gamma_0\delta_{m0} \quad (\gamma_0 \gg \gamma_1);$$

where δ is the usual Kronecker symbol. If we assume

$$\gamma_1 \ll \gamma_0 \approx \gamma_K \tag{13}$$

it turns out that D_2 contains only one matrix element so that the secular equation reduces to

$$1+2\gamma_K G_{00,r} = 0. \tag{14}$$

The Green's function for the problem (11) is given by

$$G_{00} = -\frac{\cot \frac{1}{2}Nz}{ic\cos(z+\frac{1}{2}Ka)}, \tag{15}$$

where $$ic\sin(z+\tfrac{1}{2}Ka) = r \quad (\operatorname{Im} z \geqslant 0);$$

has to hold.

Equation (14) is thus an equation for c and therefore an equation for the decay constant r. A detailed discussion of the equation (14) yields the following result: if the fluctuation constant γ_0 for the excitonic excitation energy is bigger than two times the transition matrix element H_1, that is

$$\gamma_K \approx \gamma_0 > 2H_1, \tag{16}$$

then after a sufficiently long time the excitonic motion is described by a diffusion process. If (16) does not hold, the diffusion character does not show up. The characteristic time after which the exciton motion becomes a diffusion process is given by

$$t \gg \{(2\gamma_0)^2-(4H_1)^2\}^{-\frac{1}{2}}. \tag{17}$$

If (16) and (17) hold, the mean square deviation of the exciton obeys the equation

$$\langle n^2(t)\rangle \equiv \Sigma n^2 \rho_{nn}(t) \equiv 2tD+\langle n^2(0)\rangle. \tag{18}$$

As a second example we have treated energy transfer within a two-atomic molecule assuming both a fluctuating energy for the transition as well as for the excitation energy which represents an extension of Sewell's calculation.[1]

Reference

1 G. Sewell, *Phys. Rev.* **129**, 597 (1963).

MAGNETIC SUSCEPTIBILITY OF A SYSTEM OF TRIPLET EXCITONS:† WÜRSTER'S BLUE PERCHLORATE

J. KOMMANDEUR AND G. T. POTT

Abstract

The anomalous magnetic susceptibility of the low-temperature phase of Würster's Blue Perchlorate can be explained on the model of the inequivalent ion pair. Quantitative agreement between the calculated and experimental susceptibilities, between the calculated and experimental heat of transition as well as between the calculated and experimental integrated difference in enthalpy content of the high- and low-temperature phase has been obtained. The statistics to be used depend on an assumption about the way in which triplets can be formed in this system. In general a system, in which any pair of ions can form a triplet, is favored.

Introduction

A number of systems containing triplet excitons have been reported. A good review can be found in a paper by Nordio, Soos & McConnell.[1] There are usually a number of difficulties concerned with the magnetic susceptibility of these systems. Normally, the triplet state in these systems is a very low-lying excited state and for the magnetic susceptibility one then expects a form of the type

$$= C[T(3+e^{J/kT})]^{-1},$$

where J is the excitation energy and C is a constant.

There are, however, only a few cases which obey this formalism and the discrepancies have stimulated a number of authors to discuss it in detail.[2, 3]

A case in hand is the magnetic susceptibility of Würster's Blue Perchlorate ($WBClO_4$), the perchlorate salt of the N-N-N'-N'-tetramethyl-p-phenylene-diamine ($TMPD^+$) ion. This substance has a transition point at 190° K, below which triplet excitons occur. In the temperature range between 20 and 77 °K McConnell *et al.*[4] have clearly shown these excitons to be present. At these low temperatures they

† The Research reported in this paper has been supported in part by *Air Force Cambridge Research Laboratories* under Contract AF 61(052)-769 with the European Office of Aerospace Research, United States Air Force.

found an exponential dependence of the triplet population on temperature with $J = 246 \pm 20\,\text{cm}^{-1}$. At higher temperatures, that is to say, at higher exciton densities, the susceptibility is considerably greater as given by the simple triplet excitation law, as was shown by measurements of Duffy.[5]

This paper reports on a new measurement of the susceptibility, a measurement of the heat of transition and a theoretical model, which can explain the observed discrepancies. It is based on earlier work[6, 7] showing that the low temperature phase of $WBClO_4$ is a lattice in which the $TMPD^+$ ions have disproportionated into TMPD and $TMPD^{++}$ species. The observed triplet state is then of a $TMPD^+$–$TMPD^+$ charge distribution, which is excited with respect to the disproportionated form. Zero-field splittings[4] confirm this charge distribution. The disproportionation is maintained by an inequivalence potential arising from the asymmetrically oriented ClO_4^- ions, and from the disproportionated TMPD ions. Our theory is based on the consideration that this inequivalence potential will be reduced as the triplet density increases, thus giving rise to more magnetic species than required by the simplet triplet excitation function.

After a short discussion of the experimental work, the theoretical work and the results of both will be discussed.

Experimental

Susceptibility measurements as a function of temperature were carried out on a Faraday balance, as well as on a Varian V-4502 ESR spectrometer. The Faraday balance consisted essentially of a Beaudouin electromagnet with 'constant force' polepieces and a Mettler semi-microbalance, type H 16/14. The sample in a small quartz sample holder was suspended from the balance's arm into a Dewar vessel, the temperature of which could be adjusted to within 0·1 °K between 90 and 400 °K. The temperature gradient over a sample of 3 mm height was negligible. The volume inside the Dewar vessel was filled with nitrogen gas. The apparatus was calibrated with a sample of Gd_2O_3 measuring the susceptibility as function of temperature. Our values for χ and θ (Curie temperature) were compared with known values from the literature[8] and agreed within 1 %. The complete apparatus will be described elsewhere. For absolute measurements great care was taken that the sample was positioned in the properly constant part of the $H(dH/dx)$ field. For measurements at low temperatures the paramagnetism is too slight to allow using the small-sized samples commonly desired for this work. Then absolute susceptibility measurements are not reliable, but the relative accuracy of the measurements is not at all affected. Therefore, absolute calibrations were performed at room temperature, while for other temperatures all measurements were referred to the room temperature value. To obtain the highest possible density, samples were powdered for this experiment. A separate experiment was carried out determining that no change in susceptibility through formation of defects occurred upon dispersion of the sample.

The overall accuracy for absolute measurements of the total susceptibility was 2 %, while for relative measurements 0·6 % at high temperatures and 3 % at the lowest temperatures was obtained for the paramagnetic susceptibility alone.

The ESR measurements were performed with the Varian-4502 apparatus equipped with a variable temperature rig. The detection was replaced by an integrating circuit, which allowed digital registration of the intensity of the signal. The normal modulation was replaced by a 22 c/s triangular sweep, which is applied for a known number of times. The total integral of the absorption curve can then be read off an electronic counter. A complete description of this technique will be given elsewhere. The relative ESR measurements were referred to the absolute value of the room temperature susceptibility obtained with the Faraday balance. The ESR measurements were performed on single crystals.

The differential thermal analysis experiments were performed with the aid of a sensitive 200 μV/cm Moseley X–Y recorder. The junctions of very thin thermocouple wire, together with a small crystal (about 1 mg in weight) on one side and some CCl_4 on the other side, were inserted into thin-walled glass tubes which in their turn were suspended in an air space in an aluminium block, which was cooled by placing it in liquid nitrogen. The X-axis of the recorder gave the temperature, while the temperature difference was registered on the Y-axis. The heat of the transition of the Würster's Blue Perchlorate was obtained from comparison with the well-known heats of melting and transition of CCl_4.[9]

The Würster's Blue Perchlorate used in these measurements was prepared as reported earlier.[6]

Results and Discussion

A. *The magnetic susceptibility*

The measured values of the Curie constant C and the Curie temperature θ, together with the measured values of Duffy, are tabulated in table 1. The last column of the table gives the Curie constant calculated for a 100 % pure substance of Würster's Blue Perchlorate from

$$C = \frac{Ng^2s(s+1)\beta^2}{3k},$$

where N is the number of $TMPD^+$ ions per gram, g is the Landé splitting factor, in this case $g = 2{\cdot}00$, s is the magnetic spin quantum number, for $TMPD^+$ ions $s = \frac{1}{2}$, β is the Bohr magneton and k the Boltzmann constant. Comparing the theoretical and experimental Curie constants, the Faraday balance measurements show 97·5 % free radical, ESR measurements 104 % and Duffy's measurements 94 % free radical.

Table 1

	Faraday balance	ESR	Duffy	Calculated
$C \times 10^3$ (e.m.u.) °K	$1{\cdot}386 \pm 0{\cdot}015$	$1{\cdot}471 \pm 0{\cdot}02$	1·33	1·421
θ (°K)	$-14{\cdot}30 \pm 3$	$-34{\cdot}0 \pm 2$	-36 ± 5	—

The results of the ESR and Faraday balance measurements are given in fig. 1. They are compared with Duffy's results. The discrepancy between the data can probably be explained on the difference in purity of the samples. Particularly the difference in the value of the transition temperature is remarkable. It may be, that a higher impurity content has some influence on this value, the transition being 'pinned' on impurity sites.

The experimentally found susceptibility is compared in fig. 1 with the susceptibility expected from a simple triplet excitation with an activation energy of $246 \pm 20\,\text{cm}^{-1}$ as found by McConnell.[3] at low temperatures. It will be clear, that although the experimental curve has some of the characteristics of the triplet susceptibility curve, there occurs a considerable excess susceptibility, which on the simple model cannot be accounted for. The equivalence of the ESR results, the Faraday balance measurements and—to some extent—Duffy's results assures that the excess susceptibility is not an experimental artifact.

In a rather involved paper[10] Soos extended previous treatments of triplet exciton theories[11, 12] to higher exciton densities. Although a modification of the simple triplet theory is obtained the effect on the singlet triplet excitation energy seems too small to account for the decrease from about 300 to 100 cm^{-1} required for $TMPD^+ ClO_4^-$. Also,

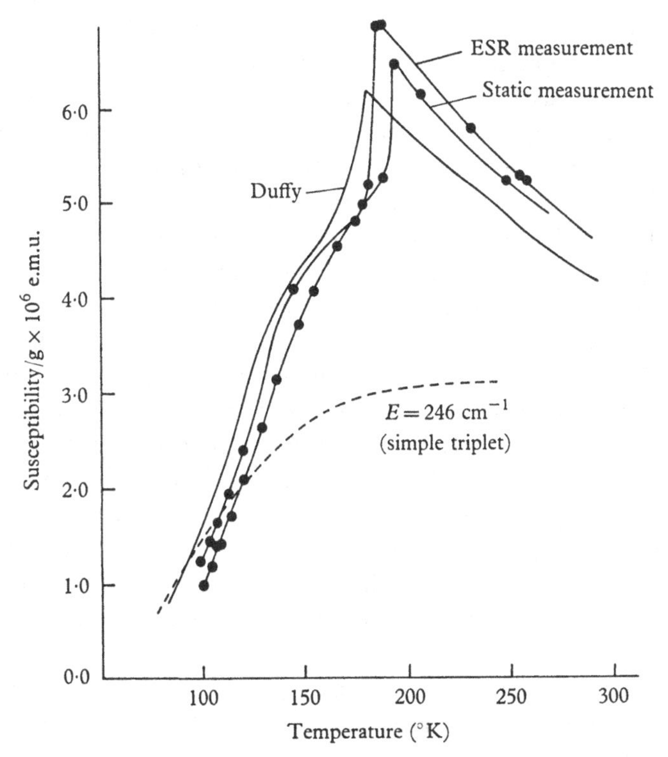

Fig. 1. Relative paramagnetic susceptibility as a function of temperature for Würster's Blue Perchlorate as measured by Duffy[5] and by static measurements and ESR as reported in this paper. The lower curve gives the susceptibility expected from a simple triplet state activation.

the susceptibility curve cannot be fitted with the simple temperature dependent exchange energy $E = E_0 - CT$, where C is a constant, as was done by Kepler[13] and later explained by Soos[10] for the TCNQ-radical ion salts. It appeared therefore useful to consider some other mechanism to account for the excess susceptibility.

An X-ray measurement† performed at 180 °K showed that the nuclear structure of the solid in the region of the excess susceptibility is

† We are indebted to Mr van Bolhuis of the Department of Structural Chemistry of this University for performing these measurements.

still the same as at 100 °K, therefore the transition appears to be first order in the nuclei. This is also consistent with the differential thermal analysis results and the conductivity experiments reported in a previous paper. One must assume then, that the excess susceptibility stems from the thermally accessible triplet states. The theory of the inequivalent ion pair as set forth in our earlier work[7] provides a very convenient explanation. It was assumed there, that the triplet state arose through an inequivalence of the crystal lattice sites, as suggested by the X-ray analysis.[4] In the theory this effect was allowed for by introducing the 'inequivalence parameter' Δ, which is a crystal potential present at the site of one of the constituting species and caused by the disproportionation of the lattice.

No such inequivalence occurs in the high-temperature phase; there, to a high degree of approximation, the singlet and the triplet state can be regarded as being degenerate. Therefore, the triplet singlet separation should be taken as zero when the inequivalence is removed. On our model the charge distributions of singlet and triplet state differ significantly. In first approximation the ion pair in the singlet state can be regarded as $TMPD/TMPD^{++}$, while in the triplet charge-transfer state it is best approximated by $TMPD^{+}/TMPD^{+}$. Therefore, if two triplet excitons are close to each other, the crystal sites in between will lose their inequivalence, the electron will proceed towards the neutral molecule and a row of 'free' spins will be established. This will give rise to a susceptibility higher than expected on the simple triplet excitation model.

A quantitative treatment of this model proceeds as follows. If we assume n triplet dimers to be present in a system of N monomers, the free energy and the entropy are given by the thermodynamic relations:

$$G = H - TS, \tag{1}$$

$$S = K \ln W + S_{sp.}, \tag{2}$$

where W is the number of ways in which n triplets can be distributed over N molecules and $S_{sp.}$ is the entropy due to the magnetic moments of the triplets. The ground state of the system ($n = 0$) is taken as the zero of energy.

The calculation of W now involves an assumption how the triplet states can be generated. On the model of the disproportionated lattice with the TMPD cores remaining equally spaced, a triplet state can be generated on any neighboring pair. On a dimer model, however, the triplet state should be restricted to pairs having the closest approach. We will treat both cases in this order.

If the triplet state can be found by any two neighboring molecules, the calculation of W proceeds as follows: since there are $(N-2n)$ molecules not involved in forming the triplet state, and since there are $(n+1)$ interstices in a linear chain of molecules containing n triplets, our problem reduces to finding the number of ways in which $(N-2n)$ molecules can be distributed over $(n+1)$ interstices. This problem is formally identical to that of the Bose–Einstein distribution and therefore

$$W = \frac{(N-n)!}{(N-2n)!\,n!}. \tag{3}$$

Adding the spin entropy term $nk\ln 3$ the problem can now be solved. However, if interstices of one and two sites also carry a spin, as demanded by our model, we should allow for this in our spin entropy. The chance of finding one site in an interstice is

$$p = \frac{(N-n-2)!}{(N-2n-1)!\,(n-1)!}\cdot\frac{(N-2n)!\,n!}{(N-n)!} \approx \frac{(N-2n)\,n}{(N-n)^2}. \tag{4}$$

Then the average number of single sites in $(n+1)$ interstices becomes

$$m_1 = p(n+1) \approx \frac{(N-2n)\,n^2}{(N-n)^2} \tag{5}$$

and with the mole fractions $x_1 = m_1/N$ and $x = n/N$

$$x_1 = \frac{(1-2x)\,x^2}{(1-x)^2}. \tag{6}$$

Similarly, we find for the mole fraction of two sites per interstice

$$x_2 = \frac{(1-2x)^2 x^2}{(1-x)^3}. \tag{7}$$

The spin entropy thus becomes

$$\begin{aligned} S_{\text{sp.}} &= Nk[(x-2x_1-2x_2)\ln 3 + x_1 \ln 32 + x_2 \ln 64] \\ &= Nk[x\ln 3 + x_1 \ln \tfrac{32}{9} + x_2 \ln \tfrac{64}{9}], \end{aligned} \tag{8}$$

where the triplet entropy has been reduced appropriately and the last two terms stand for the magnetic entropy of the row of 5 resp. 6 spins formed. Then equation (1) becomes

$$\begin{aligned} G = NxE - NkT[&(1-x)\ln(1-x) - (1-2x)\ln(1-2x) \\ &- x\ln x + x\ln 3 + x_1 \ln \tfrac{32}{9} + x_2 \ln \tfrac{64}{9}], \end{aligned} \tag{9}$$

where E is the singlet triplet excitation energy. The equilibrium condition $(\partial G/\partial x)_{P,N} = 0$ then yields

$$\frac{E}{kT} = \ln\frac{3(1-2x)^2}{x(1-x)} + \frac{2x(1-3x+x^2)}{(1-x)^3}\ln\tfrac{32}{9} + \frac{x(2-11x+16x^2-4x^3)}{(1-x)^4}\ln\tfrac{64}{9}. \tag{10}$$

Assuming a value for the triplet-singlet separation, (10) can be solved numerically. Table 2 gives the results for $E = 316\ \text{cm}^{-1}$.

Table 2. *Susceptibility and mole fractions of triplets and 5- and 6-spin rows*

°K	$\chi \times 10^6$	x	x_1	x_2
100	1·24	0·035	0·001	0·001
120	2·46	0·082	0·007	0·006
140	3·71	0·141	0·019	0·016
160	4·35	0·188	0·033	0·026
180	4·57	0·221	0·045	0·032
186	4·56	0·227	0·047	0·033

The susceptibility can now be calculated from

$$\chi = N\frac{g^2\beta^2}{3kT}[(x-2x_1-2x_2)2 + 5x_1\cdot\tfrac{3}{4} + 6x_2\cdot\tfrac{3}{4}],$$

where the triplets and rows of 5 and 6 spins respectively have been summed separately.

The statistics are somewhat different when it is assumed that only pairs of closest approach—as in the dimer lattice—can form a triplet state. Treated in an analogous way W is now the number of ways in which n triplets can be distributed over $\frac{1}{2}N$ dimers. Then

$$W = \frac{(\frac{1}{2}N)!}{(\frac{1}{2}N-n)!\,n!}, \tag{12}$$

which, without including the extra spins, would lead to the triplet excitation function hitherto used.[1] If now, however, interstices containing two molecules lose their inequivalence and become paramagnetic as well, we should again add those to the spin entropy. For the mole fraction of these interstices we find similarly as above

$$x_2 = m_2/N = 2(1-2x)x^2,$$

where it should be remembered that interstices containing only one molecule are excluded by this model. Then

$$\begin{aligned} S_{\text{sp.}} &= Nk\{(x-2x_2)\ln 3 + x_2\ln 64\} \\ &= Nk\{x\ln 3 + x_2\ln\tfrac{64}{9}\}. \end{aligned} \tag{13}$$

Proceeding as pointed out above, minimising the free energy:

$$G = NxE - NkT\{\tfrac{1}{2}(1-2x)\ln(1-2x) + x\ln 2x + x\ln 3 + x_2 \ln \tfrac{64}{9}\} \tag{14}$$

yields
$$\frac{E}{kT} = \ln\frac{3(1-2x)}{2x} + 4x(1-3x)\ln\tfrac{64}{9}. \tag{15}$$

In the first term of the right-hand side of equation (15) we recognise the triplet excitation function normally used (with $x' = 2x$, the mole fraction being taken with respect to the dimeric lattice containing N dimers); the second term is due to the extra entropy of the rows of 6 spins occurring at higher concentrations. The susceptibility can then be calculated from

$$\chi = \frac{Ng^2\beta^2}{3kT}\{(x-2x_1-2x_2).2+6x_2.\tfrac{3}{4}\}, \tag{16}$$

where again the triplets and the 6-spin rows have been summed separately. The equation can be solved numerically and for $E = 309\,\text{cm}^{-1}$ the results are given in table 3.

Table 3. *Susceptibility and mole fractions of triplets and 6-spin rows*

°K	$\chi \times 10^6$	x	x_2
100	1·20	0·035	0·001
120	2·21	0·077	0·005
140	3·24	0·132	0·015
160	4·02	0·188	0·029
180	4·41	0·234	0·042
186	4·42	0·244	0·045

The two models can now be compared with experiment by plotting both equations (11) and (16) as a function of temperature, the singlet-triplet separation chosen so as to give the best fit with the experimental data. Fig. 2 gives such a plot, showing that both models give very good agreement, there being a slight preference for the model in which any pair can form a triplet state. The activation energies of $316\,\text{cm}^{-1}$ or $309\,\text{cm}^{-1}$ are in some disagreement with $E = 246 \pm 20\,\text{cm}^{-1}$ found by McConnell *et al.*[4] at lower temperatures, where no complications occur. This disagreement, however, would persist on any model. Even the simple triplet excitation function yields a susceptibility, which compared with our experimental data is too high by 20% at 100 °K if $E = 246\,\text{cm}^{-1}$, but would, at that temperature only, fit in with Duffy's data. The disagreement does not seem serious, the main point being the resulting excess susceptibility.

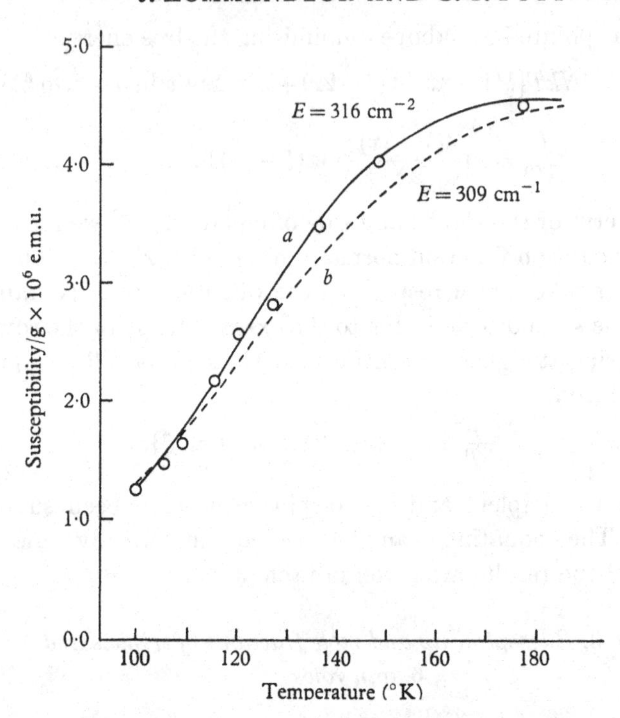

Fig. 2. Calculated and experimental susceptibility of the low temperature phase of $TMPD^+ClO_4^-$: (*a*) for the model in which any two molecules can participate in the triplet state; (*b*) for the model in which only dimers can participate. Circles represent the experimental data.

B. *The heat of transition*

Further information on the nature of the initial and final stages of the transition can be obtained from the heat of transition. To obtain its value a differential thermal analysis experiment was carried out. Because the transition shows hysteresis, and since the amounts of heat evolved are small, it proved necessary to carry out the experiments on single crystals. Because of the inherent inaccuracy of the experiment a droplet of CCl_4 was used as an internal calibrator. Fig. 3 shows the result of a typical experiment. At 250 °K a peak occurs due to the solidification of CCl_4 and at 222 °K there is a peak for the phase transformation of solid CCl_4. The peak at 186 °K is due to the phase transformation of $TMPD^+ClO_4^-$. The latter heat of transition was determined by measuring the areas under the peaks and comparing them with each other. Because the heats of solidification and transition of CCl_4 are well known, the heat of transformation of Würster's Blue

Perchlorate can be calculated. It was determined to be $0 \cdot 22 \pm 0 \cdot 02$ kcal mole^{-1}. This value is smaller by a factor of two than the one reported by Chihara *et al.*[14] obtained from molar heat capacity measurements. A difference between these numbers could, however, be expected, since they pertain to different quantities. The D.T.A. measurement as performed in our case with a high rate of heating or

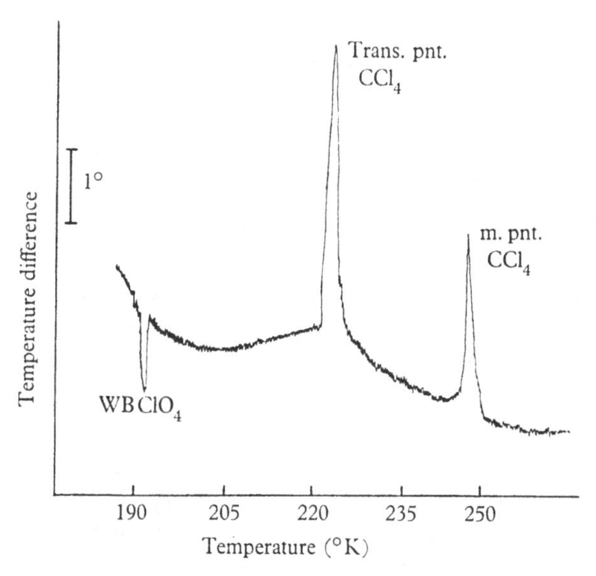

Fig. 3. Differential thermal analysis results for CCl_4 and Würster's Blue Perchlorate.

cooling and using CCl_4 as an internal calibration does not measure the difference in heat capacity of the two systems, but the heat evolved or taken up in the transition: Q_0. This is the difference in enthalpy between the two phases at the transition point. Chihara *et al.*, however, measure essentially

$$Q = \int_0^{186} (C_p^l - C_p^h)\, dT$$

where C_p^l and C_p^h stand for the molar heat capacities of the low- and high-temperature phase, respectively. We will return to this point later. On the basis of the two models for the excess susceptibility a calculation of the heat of transition can be made. As was pointed out in another paper on this subject[6] the disproportionation of the WB lattice is removed by the rotation of the perchlorate ions destroying largely the inequivalence potential. It is sufficient to assume that the ClO_4^- ions will then be distributed over two positions. Together with the spin entropy this yields for the entropy content per mole of high temperature lattice:

$$S = S_0(T) + R \ln 2 + R \ln 2, \tag{17}$$

where $S_0(T)$ is all the other entropy due to lattice vibrations, etc. As pointed out in the previous section, the entropy content per mole of monomeric lattice of the low-temperature lattice is, from equation (9),

$$S = S_0'(T) + R[(1-x)\ln(1-x) - (1-2x)\ln(1-2x) - x\ln x + x\ln 3 + x_1 \ln \tfrac{32}{9} + x_2 \ln \tfrac{64}{9}] \tag{18}$$

for the model, where any pair of TMPD's can form a triplet state and from equation (13)

$$S = S_0''(T) + R[\tfrac{1}{2}(1-2x)\ln(1-2x) + x\ln 2x + x\ln 3 + x_2 \ln \tfrac{64}{9}] \tag{19}$$

for the model, where only dimeric pairs can form a triplet state. Since the high- and low-temperature phase are almost identical, apart from a doubling of the unit cell it is probably very reasonable to assume $S_0(T) = S_0'(T) = S_0''(T)$. Then the heat of transition can be calculated from: $\Delta H = T\Delta S$, where T stands for the transition temperature (186 °K) and ΔS for the entropy difference between the high- and low-temperature phase.

Taking the values of x, x_1 and x_2 as found from the solution of equation (10) we find $\Delta H = 200$ cal in almost too good agreement with the experimental value of 220 ± 20 cal. Taking the values of x and x_2 from the solution of equation (14) we find $\Delta H = 260$ cal, which is somewhat too high and again a slight preference for the non-dimeric model is established. One can also endeavour to calculate

$$\int_0^{486} (C_p^l - C_p^h)\, dT$$

as measured by Chihara *et al.*[14] We will split this quantity into two parts: a heat of transition at the transition point ($Q_0 = 200$ cal mole^{-1}) and a molal heat capacity difference, which will be almost solely due to the enthalpy content of the triplet system in the low-temperature phase. In both our models it requires 310 cm^{-1} = 900 cal to 'produce' a mole of triplets. At the transition point with the mole fraction $x = 0{\cdot}25$ the enthalpy content of the system is therefore 225 cal per mole of monomer lattice. For the quantity

$$\int_0^{T} (C_p^l - C_p^h)\, dT$$

we therefore calculate $Q_0 + 225 = 425$ cal mole^{-1}, which compares very well with the 440 cal mole^{-1} as reported by Chihara *et al.*

Conclusions

In conclusion it can be said that the excess susceptibility and the heat of transition of $TMPD^+ ClO_4^-$ can be adequately accounted for by the mechanism of the inequivalence potential, together with the realisation that this inequivalence is removed by the excitation of triplet states. Unfortunately, the experiments do not allow a distinction to be made between a dimeric model in which only TMPD's in close approach can form triplets and one in which any pair of TMPD's can do so, although a slight preference for the latter model is established.

It is probably of interest to point out that our models do not allow a population of triplets engaging significantly more than half of the lattice, since in that case all inequivalence will be removed and a phase transformation should occur. For the model in which any pair forms a triplet the fraction of magnetic molecules becomes

$$f = 2(x-2x_1-2x_2)+5x_1+6x_2 = 2x+x_1+2x_2,$$

which at 180 °K gives $f = 0{\cdot}537$, indeed half of the lattice at the temperature at which the transition becomes co-operative!

Finally, it should be noted that consideration of other susceptibility versus temperature curves in the light of the statistics to be used could be valuable since there appear to be a number of other cases in which the susceptibility does not follow the simple singlet-triplet excitation law. This is in particular the case for the TCNQ radical ion salts, which in other aspects also appear to have properties similar to those of Würster's Blue Perchlorate. The extension of our interpretation to these salts is presently being investigated in our laboratory.

References

1 P. L. Nordio, Z. G. Soos & H. M. McConnell, *Ann. Rev. Phys. Chem.* **17**, 237 (1966).

2 D. B. Chesnut, *J. Chem. Phys.* **40**, 405 (1964).

3 H. M. McConnell, *Molecular Biophysics*, B. Pullmann, ed. (Academic Press Inc., New York, 1965.)

4 D. D. Thomas, H. Keller & H. M. McConnell, *J. Chem. Phys.* **39**, 2321 (1963).

5 W. Duffy Jr., *J. Chem. Phys.* **36**, 490 (1962).

6 G. T. Pott & J. Kommandeur, *J. Chem. Phys.* (to be published).

7 H. J. Monkhorst, G. T. Pott & J. Kommandeur, *J. Chem. Phys.* (to be published).

8 S. Arajs & R. V. Colvin, *J. Appl. Phys.* **33**, 2517 (1962).

9 L. Börnstein, *Zahlenwerte und Funktionen*. 6 Auflage, II Band, 4 Teil, 294.
10 Z. G. Soos, *J. Chem. Phys.* **43**, 1121 (1965).
11 R. M. Lynden-Bell & H. M. McConnell, *J. Chem. Phys.* **37**, 794 (1962).
12 Z. G. Soos & H. M. McConnell, *J. Chem. Phys.* **43**, 3780 (1965).
13 R. G. Kepler, *J. Chem. Phys.* **39**, 3528 (1963).
14 H. Chihara, M. Nakamura & S. Seki, *Bull. Chem. Soc. Japan*, **38**, 1776 (1965).

A STUDY OF TRIPLET EXCITONS IN ANTHRACENE CRYSTALS UNDER LASER EXCITATION

T. A. KING AND H.-G. SEIFERT

Abstract

Carefully purified and grown anthracene crystals are excited into their first triplet state by absorption of high intensity radiation from a ruby laser. The time dependence of the resulting delayed fluorescence in the time region from 2 to 2000 μs can be described simply by the mechanism of triplet exciton diffusion and triplet-triplet interaction to form excited singlet states. The observed intensity-time functions are analysed in three ways and agreement with the triplet-triplet interaction mechanism and internal consistency of the analyses is found. The triplet-triplet interaction rate constant (X) is obtained and found to have a value of $(3 \pm 1{\cdot}5) \times 10^{-11}$ $cm^3\ s^{-1}$, when measured under vacuum conditions. This value is reduced on exposure to air.

Introduction

The knowledge of the first excited triplet state in anthracene in solution and crystalline phases has, over the last few years, increased considerably. This has been due partly to the fact that the triplet state can undergo a triplet-triplet process resulting in observable delayed fluorescence from the singlet state. This stems from the original proposal by Czarnecki(1) to account for the experimental observation in solution(2) for the delayed fluorescence of naphthalene, and on pure and mixed organic crystals.(3, 4) Electron spin resonance measurements(5) demonstrated that the excited triplet state in molecular crystals is mobile. The near-coincidence of the first triplet level in anthracene with the ruby laser frequency promoted the study of anthracene crystals using laser excitation. This offered time-resolved measurements with high triplet density. A number of theoretical discussions(6, 7) of the properties of the triplet exciton have established that it is a Frenkel-type exciton; however, the nature of its mobility is still a matter of uncertainty. We considered that it was possible to study the properties of the triplet exciton by studying the effect of triplet density on the triplet-triplet annihilation. In this type of experiment we were to neglect any prompt fluorescence arising from double-photon processes during the excitation time.

Experimental details

The anthracene used as starting material in these experiments was either of puriss or scintillation grade.† This material was passed about 50 times through an automatic zone-refining apparatus with

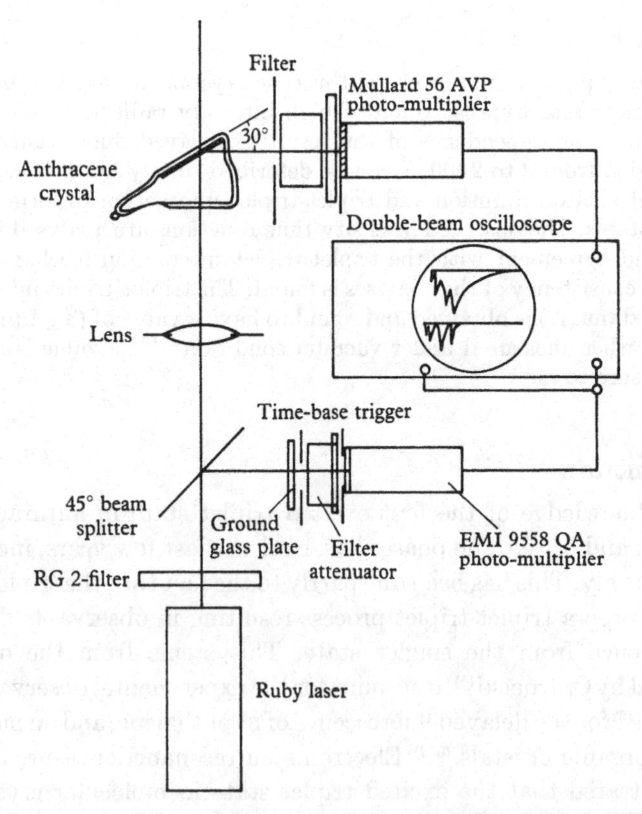

Fig. 1. Apparatus for observation of time dependence of delayed fluorescence excited by a ruby laser.

the inside of the zone-refining tube in a nitrogen atmosphere. The crystals of anthracene were grown under vacuum in the cell used in the fluorescence experiments. The cell had two 'Spectrosil' optical windows inclined at an angle of 30° to each other as shown in fig. 1. The cell was evacuated to a pressure of 8×10^{-5} torr and carefully heated

† Obtained from Koch-Light Ltd and Nuclear Enterprises Ltd.

up to 35 °C, to accelerate the release of absorbed oxygen and to sublime off the surface layer of anthracene, until the pressure reached a constant value. The sealed cells were then wrapped in aluminium foil except for the area over which the crystal was to be deposited. The cell and sample were heated to about 230 °C in a heavy oven and allowed to cool slowly over a period of about 24 h to prevent cracking of the crystalline sample. The crystals prepared in this way formed clear and flat plates of about 1 cm^2 with their *ab*-plane parallel to the inclined pyrex window with a thickness between 0·5 and 1 mm.

The exciting light source was a ruby laser† emitting radiation at 6943 Å (1·79 eV). The ruby crystal was 6 in. in length and 3/8 of an in. in diameter and was suspended with the flash lamp in an elliptical mirror arrangement. The output aperture of the crystal had a dielectric mirror coating with about 80 % reflectivity at 6943 Å while the other reflector forming the laser cavity was a Porro prism. The laser was operated slightly above threshold, when it was possible to obtain reasonably regularly spaced light pulses (spikes) of about 0·5 to 2 μs half-intensity width and at 3 to 10 μs intervals. The total number of spikes per complete laser pulse varied between 2 and 10 for these experiments. For the measurements on the build-up process (see below, p. 333) this number was increased and the regularity of the spiking was lost.

The laser light passed through a Schott RG 2 red filter to eliminate the spurious light shorter than $\lambda = 610$ nm originating from the laser pumping flash lamp. A 45° beam-splitter reflected about 10 % of the light on to a ground glass screen. This light passed through a copper sulphate liquid attenuator and a second RG 2 filter and onto the monitoring photo-multiplier. The working resistor of this photo-multiplier was an emitter follower and matched the high input impedance of the display oscilloscope and also acted as the common trigger for the two oscilloscope beams.

The fluorescence was observed in reflection at the vacuum-crystal interface at 90° to the laser beam and at 30° to the *ab*-plane. Also, the time response of the detector photo-multiplier was checked to be less than 0·1 μs using a pulsed light source. All measurements were made with the sample at an ambient temperature of (23 ± 2) °C. No stray light could be detected with this experimental arrangement.

† Made by G. and E. Bradley Ltd.

Theoretical equations, assumptions and solutions

We make the general assumption that the triplet exciton is mobile in the crystal lattice, and that when two of the excitons have approached each other to within a certain distance an excited singlet state is formed from this collision.

Under monochromatic light excitation (e.g. from the laser source) the singlet-triplet rate equations have the following general form:

$$\left.\begin{aligned} -\frac{\partial T(r,t)}{\partial t} &= \beta T + XT^2 - \alpha I - D\nabla^2 T - kS, \\ -\frac{\partial S}{\partial t} &= \alpha S - g\frac{X}{2}T^2 - \alpha_s I, \end{aligned}\right\} \quad (1)$$

with boundary conditions for $S(t=0)$, $T(t=0)$.

T, S: Triplet, first singlet excited state concentrations.
β^{-1}, α^{-1}: Triplet, singlet decay times.
X: Triplet-triplet annihilation rate constant.
g: Statistical term for triplet-triplet annihilation leading to formation of S.
D: Triplet diffusion constant.
k: $S \to T$ intersystem-crossing rate constant.
α_T, α_s: Triplet, singlet absorption coefficients.

If the volume excited by the light is large, the contribution of the diffusional term, $D\nabla^2 T$, is negligible. If observations are made at times which are long compared to the prompt fluorescence decay time, and outside the duration of the exciting light pulse, such that the absorption terms can be included into the boundary conditions, then the triplet differential equation reduces to

$$-\frac{dT}{dt} = \beta T + \left(X - \frac{kXg}{2\alpha}\right) T^2 \quad T(t=0) = T_0, \quad (2)$$

which can, by experimental evidence, be further reduced to

$$-\frac{dT}{dt} = \beta T + XT^2.$$

This equation can be solved to give:

$$T = T(t) = T_0 \frac{\beta e^{-\beta t}}{XT_0 + \beta - XT_0 e^{-\beta t}}.$$

There are two different approximations that can be made:

(*a*) $\beta \gg XT_0$ which results in $T \simeq T_0 e^{-\beta t}$ for all times $t > \Delta t$, where Δt represents the larger of α^{-1} or the duration of the exciting light pulse.

(*b*) $\beta \ll XT_0$ (i) this also reduces for $t > \beta^{-1}$ to an exponential decay law: $T \propto e^{-\beta t}$. (ii) Making the additional assumption to $\beta \ll XT_0$ that $\Delta t < t < \beta^{-1}$ then,

$$T = T_0 \frac{1}{1+XT_0 t}. \tag{3}$$

Analysis from $F^{-\frac{1}{2}}$ *vs* t curves

This method in its simple form is only valid for $XT_0 \gg \beta$ and for times smaller than the triplet decay time. In such a case the delayed fluorescence intensity is

$$I_d = g\frac{X}{2}\frac{T_0^2}{(1+XT_0 t)^2},$$

which can be rewritten into

$$I_d^{-\frac{1}{2}} = \left(\frac{2}{gXT_0^2}\right)^{\frac{1}{2}} + \left(\frac{2X}{g}\right)^{\frac{1}{2}} t \quad \text{and} \quad F^{-\frac{1}{2}} = \left(\frac{2c}{gXT_0^2}\right)^{\frac{1}{2}} + \left(\frac{2cX}{g}\right)^{\frac{1}{2}} t, \tag{4}$$

where g is a conversion factor which takes into account processes leading from triplet-triplet annihilation to products other than the first excited singlet. These processes include allowance for spin conservation, wave-vector conservation and reversibility.

$F(= I_d/c)$ is the measured signal deflection on the oscilloscope (in mm); c the conversion factor depending on the apparatus.

Plotting the delayed fluorescence values of the function $F = F(t)$ in the form of equation (4) results therefore in straight lines as shown in fig. 2 and the ratio of slope to intercept gives (XT_0) without the knowledge of c. To find X (and hence T_0) one has to know c. This was measured with natural light of two different bands covering the transmission of the filter combination before the multiplier. These filters are matched to the emission spectrum of crystalline anthracene. The conversion factor was determined against a calibrated thermopile. The bandwidths of the filters used were taken into account by a weighting factor in connection with the response of the photo-multiplier. The fluorescence emission was assumed to be isotropic and the corresponding solid angle calculated from the geometry of the set-up.

Analysis from build-up process

Let the density of triplet excitons at time t in a certain test volume be $T(t)$, and let the dimensions of the test volume be large compared

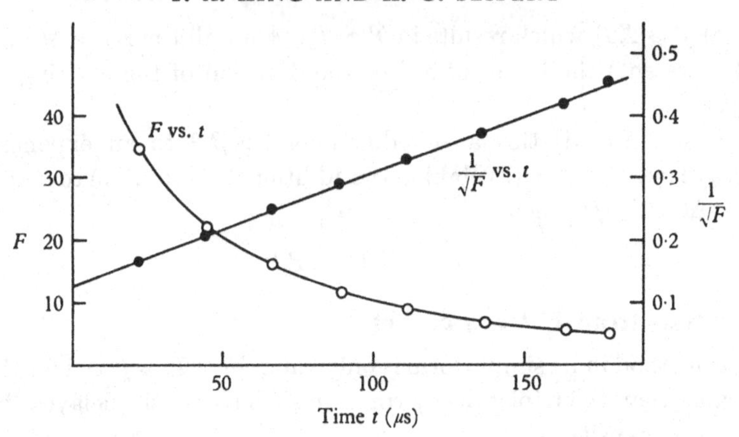

Fig. 2. Delayed fluorescence intensity (F) as a function of time-yielding in this case $XT_0 = 1{\cdot}4 \times 10^4\ \text{s}^{-1}$.

with the diffusion length of the triplet excitons ($\sim 5.10^{-4}$ cm), so that diffusional processes can be neglected, and, furthermore, let $XT_0 \gg \beta$ and $t < \beta^{-1}$, then the triplet density at a later time t' is

$$T(t') = \frac{T(t)}{1+XT(t)\,(t'-t)} + T_n \delta_{t',t_n}. \tag{5}$$

Here t' is measured and lies within the range $t \leqslant t' \leqslant t_n$, where t_n = time at the n-th laser spike.

The last term is a source term adding T_n at times $t'_n = t_n$ when a laser spike occurs. The total number of red light photons per unit sample area from a spike can be presented as $I_{\text{max.}}\Delta t$ where Δt is a characteristic width and $I_{\text{max.}}$ the maximum light intensity of the exciting pulse, $I_{\text{max.}} = Q.\,M$; M being the measured deflection of signal and Q a conversion factor depending on the apparatus. If α_T represents the absorption coefficient, then the triplet density formed by $I_{\text{max.}}$ will be:

$$T = \alpha_T I_{\text{max.}} \Delta t = \alpha QM\Delta t = \alpha QW.$$

The quantity W (measured spike area) is the quantity experimentally measured; this is substituted in equation (1), so that

$$W(t') = \frac{W(t)}{1+\alpha_T QXW(t)\,(t'-t)} + W_n \alpha_T Q \delta_{t',t_n}. \tag{6}$$

Since, under the assumptions above, the density of triplets is an additive quantity, one can calculate from equation (2) the variation of triplet density with time, knowing $W_n(t_n)$ and $p = \alpha_T QX$. The conversion factor Q was measured for the apparatus using a vacuum thermopile of known sensitivity and natural light from a tungsten lamp operated at constant power. The absorption coefficient of

$$\alpha_T = 2{\cdot}1 \times 10^{-5}\,\text{cm}^{-1}\ (6943\,\text{Å})$$

measured by Avakian *et al.*[10] was used. In the measured experimental curves of delayed fluorescence F, against t, the intensity ($\sim F$) is proportional to W^2

$$F = \frac{I_d}{c} = \frac{gXT^2}{2c} = \frac{gX\alpha_T^2 Q^2 W^2}{2c}. \tag{7}$$

Therefore the time variation of W^2 was computed for various values of p from the laser input data $W_n(t_n)$, and the right p value found by trying out the best fit between the measured build up and decay curves and the ones computed from the monitor signal.

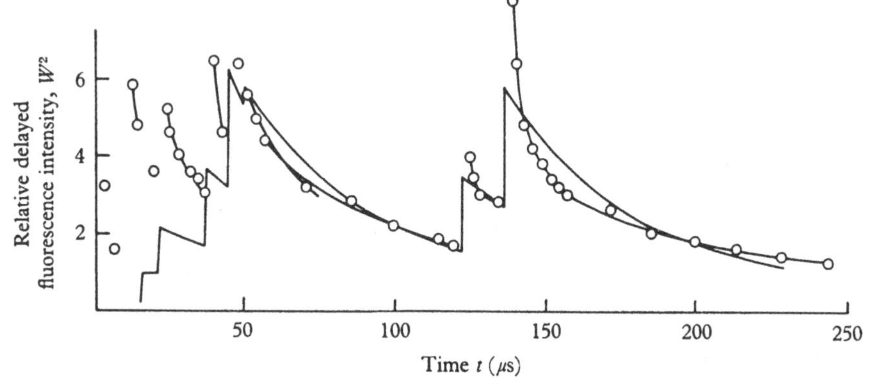

Fig. 3. Build-up and decay of delayed fluorescence during irregular laser spike excitation. Full lines: calculated time dependence from laser excitation function for $p = 4 \times 10^{-4}\,\mu\text{s}^{-2}\,\text{mm}^{-1}$. Circles: measured intensity; anthracence crystal in air.

Since α and Q are known, $p(= \alpha QX)$ gives X directly. Figs. 3 and 4 give examples of these fittings.

Some of the build-up processes reach a plateau for which the following analysis can be made.

Defining that the first laser spike of the plateau region is at $t_1 = t_{n-1}$ and T increases to $T(t_{n-1})$, at the next laser spike, $t = t_n$:

$$T(t = t_n) = T(t = t_{n-1}) = T_p = \frac{T(t_{n-1})}{1 + XT(t_{n-1})\,\Delta t} + T_n\,\delta_{t,t_n}$$

$$T_p = \frac{T_p}{1 + T_p X \Delta t} + T_n, \tag{8}$$

then
$$\frac{1}{XT_p} = \Delta t \left(\frac{T_p}{T_n} - 1\right). \tag{9}$$

The time position of the maximum concentration is strongly dependent on the p value. The start of the build-up is normally a bad fit, since in the experiment the first laser spike triggering the trace could not be displayed with the oscilloscope that was used. This value was therefore only known to be higher than the trigger level, which allowed for considerable variation.

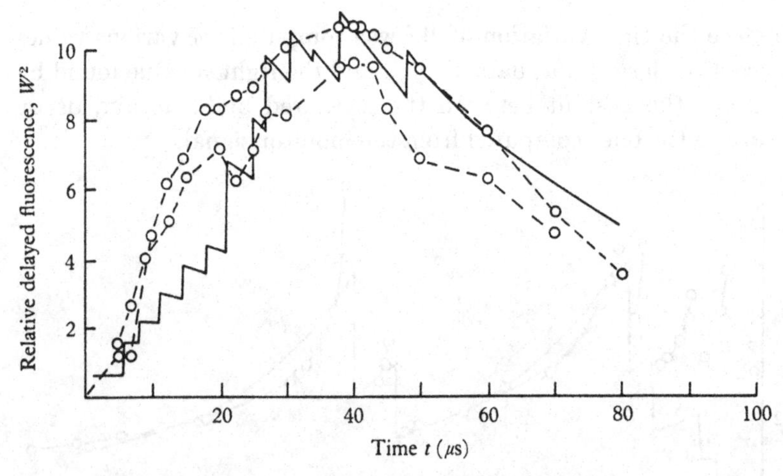

Fig. 4. Build-up and decay of delayed fluorescence during irregular laser spike excitation. Full lines: calculated time dependence from laser excitation function for $p = 8 \times 10^{-5}\ \mu s^{-2}\ mm^{-1}$. Circles: measured intensity, commercial anthracene crystal in prevacuum showing possible spread in experimental values.

Analysis from restricted decay time measurements

The experimentally measured decay curves (F *vs* t) were plotted as $\log F$ *vs* t and the extrapolated value of F_0 at the time of the last large laser spike (sometimes one or two very small spikes were neglected) noted as parameter.

$$F = \frac{I_d}{c}$$

where F is measured signal deflection (mm), I_d is intensity of delayed fluorescence (phot. $cm^{-2}\,s^{-1}$), c is instrumental conversion factor,

and
$$F_0 = \frac{gXT_0^2}{2c} = \frac{I_{0d}}{c}.$$

The curve was then approximated in the vicinity of $t = R$ by an exponential decay law:

$$I_R(t) = I_{0R} \exp(-t/\tau_{\text{ex.}} R)$$

with $$I_d(t = R) = I_R(t) \quad \text{and} \quad \left(\frac{dI_d}{dt}\right)_R = \left(\frac{dI_R}{dt}\right)_R.$$

For $$I_d = \frac{gX}{2} \frac{T_0^2}{(1+XT_0 t)^2},$$

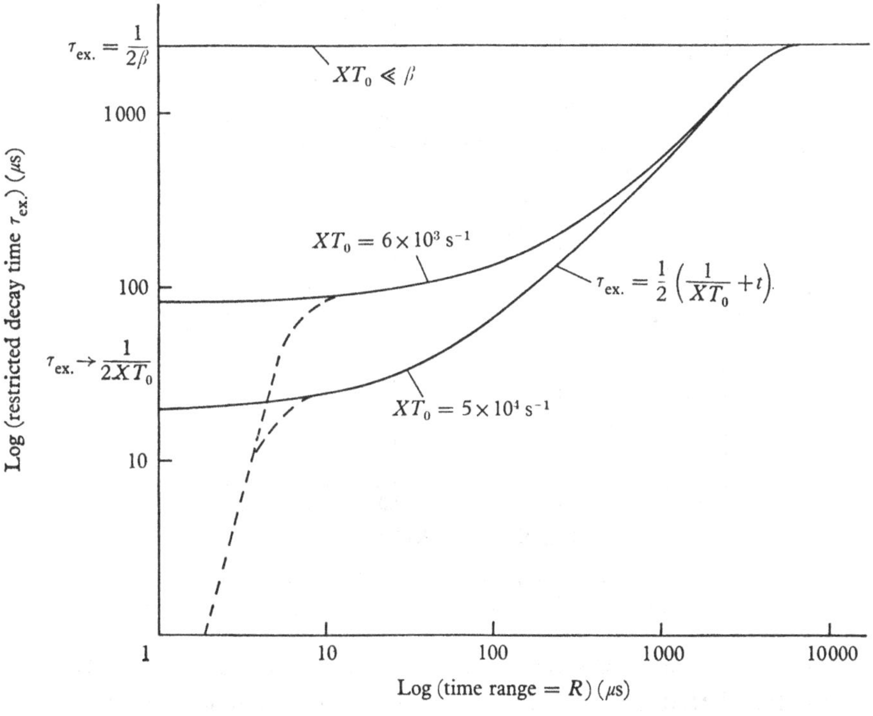

Fig. 5. Restricted time $\tau_{\text{ex.}}$ as a function of time range calculated from theoretical equations.

every curve in the $\log \tau_{\text{ex.}}$ *vs* $\log R$ plot, $(R = t)$, corresponds to one value of XT_0.

$$\tau_{\text{ex.}} = \frac{1}{2}\left(\frac{1}{XT_0} + t\right). \tag{10}$$

This function is shown in fig. 5. Then, curves having the same F_0 value should give

$$F_{01} = F_{02} = \frac{X_1}{2c} T_{01}^2 = \frac{X_2}{2c} T_{03}^2.$$

Let $X_2 = aX_1 (a > 0)$ then, $F_{01} = \dfrac{aX_1}{2c} T_{02}^2$, $T_{01}^2 = aT_{02}^2$.

If $a > 1$, curves having the same F_0 but different X-values will lie on characteristic curves with $X_2 T_{02} = a^{\frac{1}{2}} X_1 T_{01}$.

Therefore finding from the build-up that $X_{air} < X_{vac.}$, as shown in the results, should produce a difference between the air and vacuum curves on this analysis.

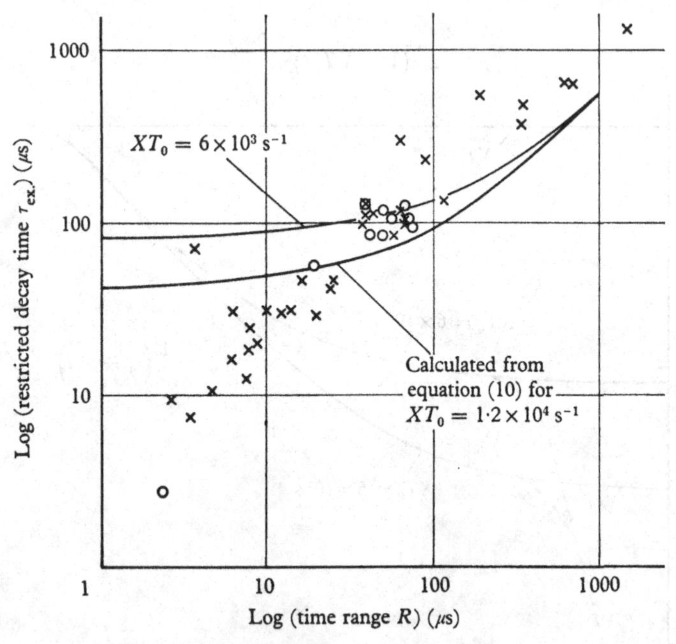

Fig. 6. Experimentally determined restricted decay time $\tau_{ex.}$ as a function of time range R. ×, Crystal in vacuum F_0: 4–20; ○, crystal in air.

In fact, because the restricted decay time analysis is insensitive, it is found that the air curves lie at the same value as the vacuum curves with perhaps a tendency to be lower (see figs. 6–8). It should be noted that the tailing off at short times in the log $\tau_{ex.}$ *vs* log R plot shows a reversed oxygen effect from that at longer times. There is the possibility that this corresponds to the build-up analysis, but this does not show why, as will be seen later, the decay analysis of ($F^{-\frac{1}{2}}$ *vs* t) at longer time ranges agrees with the build-up.

Results

A series of preliminary experiments established that all decay curves, in the time range of 2–2000 μs were non-exponential, indicating (together with the anticipated values of XT_0) that the bitriplet

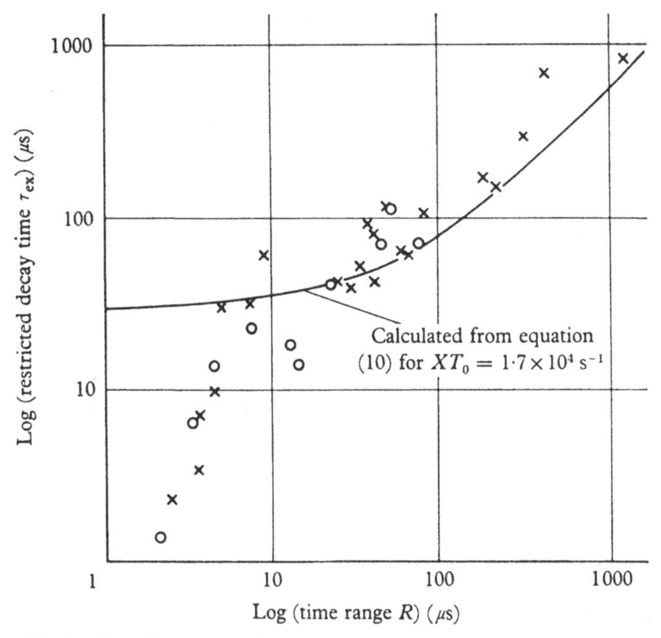

Fig. 7. Experimentally determined restricted decay time $\tau_{ex.}$ as a function of time range R. ×, Crystal in vacuum F_0 = 20–50; ○, crystal in air.

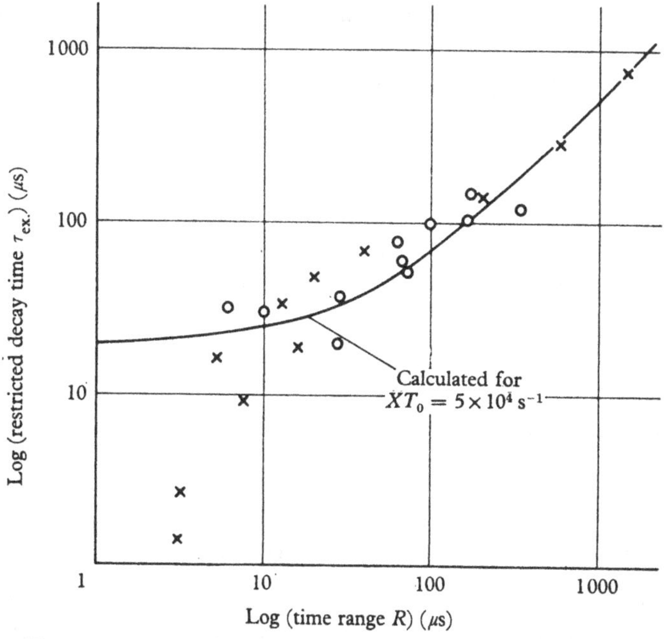

Fig. 8. Experimentally determined restricted decay time $\tau_{ex.}$ as a function of time range R. ×, Crystal in vacuum F_0 > 50; ○, crystal in air.

annihilation effect was not negligible. The experimental results were first analysed using the *restricted decay time method* as shown in figs. 6–8. This type of analysis is not very accurate because it is possible to approximate over quite wide time intervals with one value of the restricted time, $\tau_{ex.}$, but it gives at once a normalised graph. The experimental results were divided, according to their F_0 values, into three large groups and showed, within the expected accuracy of this method, behaviour describable by triplet-triplet annihilation. A comparison of the results for samples in vacuum and in air indicated that there was no difference in these conditions. The measurements for this method were made mainly with laser excitation of only a few spikes and which showed little build-up. In the short time region up to 8 μs. the experimental decay times showed a marked departure from the theoretical form. If a gaussian shape were assumed, with a width characteristic of the longest laser spike, one could only just explain the deviation from the triplet-triplet behaviour. This is shown as a dotted line in fig. 5 and is to be compared with the points in figs. 6–8. This is still not completely satisfactorily solved.

The $F^{-\frac{1}{2}}$ *vs t analysis* gave reasonably straight lines. However, this method is sensitive to any systematic errors in the decay curve baseline and the time position of the maximum triplet concentration. The calibration of the fluorescence detection enables a value of X/g to be obtained from this analysis. The following values have been obtained from the slope $= (2cX/g)^{\frac{1}{2}}$.

Crystal in vacuum: $X/g = (4 \pm 4)\, 10^{-11}\, \text{cm}^3\, \text{s}^{-1}$.

Crystal in air: $X/g = (0{\cdot}7 \pm 0{\cdot}7)\, 10^{-11}\, \text{cm}^3\, \text{s}^{-1}$.

These values apply to the same crystal. Because there is a large degree of uncertainty as to the value of g we are not able to calculate X from this analysis.

The *build-up analysis* is complementary to the above analysis and uses a common theory. The laser excitation function was determined independently and, by applying the build-up theory, allowed a value of X to be estimated from the values of α_T, Q and $p(= \alpha_T QX)$:

Crystal in vacuum: $X = (3{\cdot}0 \pm 1{\cdot}5)\, 10^{-11}\, \text{cm}^3\, \text{s}^{-1}$,

Crystal in air: $X = (1{\cdot}5 \pm 0{\cdot}7)\, 10^{-11}\, \text{cm}^3\, \text{s}^{-1}$.

These values are mean values over many crystals. Results for commercial samples of anthracene, unpurified by us and maintained under different excitation geometry and measured only for a few cases, gave values of $X \simeq 0{\cdot}4 \times 10^{-11}\,\text{cm}^3\,\text{s}^{-1}$. This type of analysis also showed that it is possible to describe the growth and decay during the laser excitation time by the triplet-triplet assumption. The accuracy of the build-up process is limited by the prompt fluorescence occurring at each spike. The actual value of p strongly affects the time position of the maximum triplet concentration and plateau. For individual crystals, the values of XT_p obtained from $F^{-\frac{1}{2}}$ *vs* t at long time range and from the plateau were in reasonable agreement and gave a check on the internal consistency of the analysis: for example,

$$\text{from } F^{-\frac{1}{2}}\, vs\, t\colon XT_p > 6 \times 10^4\,\text{s}^{-1},$$

$$\text{from the plateau effect: } XT_p = 7{\cdot}6 \times 10^4\,\text{s}^{-1}.$$

Both the build-up and $F^{-\frac{1}{2}}$ analyses show a difference between the derived values for the vacuum and air conditions.

The inaccuracy in the values obtained for X/g and X are mainly contributed to the errors in the calibration quantities c and Q. Further uncertainty in these results is introduced because the crystals were not oriented in any special way with the crystallographic axes.

Conclusion

In general, delayed fluorescence decay curves excited by laser radiation creating $T_0 > 10^{12}\,\text{cm}^{-3}$ are non-exponential; we find this behaviour applies generally for the anthracene crystal over the time range 2–2000 μs and indicates the presence of bimolecular processes occurring in the crystal. It is likely that crystal purity can affect the kinetics of triplet-triplet interactions and it is found from the two more sensitive analyses that exposure to air almost certainly reduces the effective triplet-triplet interaction rate constant and therefore the effective diffusion coefficient. However, neither of these effects appear large in the time region considered. The role of defects[(8)] other than chemical impurities may well be important in this type of work.

The derived values of X are well within the range of previous measurements of this quantity.[(6, 7, 9–12)]

In the short time region it is found that the triplet-triplet interaction model can only be justified on this scheme if the laser spike is assumed to have a large time width. It is possible that in the time range up to 10 μs effects other than those allowed for here (e.g. volume diffusion) are present.

In view of the large theoretically predicted anisotropy of D in the *ab*-plane compared to the *c*-direction,[6] and known anisotropy of the crystal structure,[13] it is possible that the crystal orientation conditions of these experiments may affect the results and future experiments with controlled changes of orientation are under way. However, recent experiments would suggest that in fact the anisotropy if present is small.[7, 14]

It is interesting to note that in this type of experiment it is not possible to obtain a value for the diffusion coefficient (D) and the diffusion length (l), since there is no spatially dependent term in equation (2). The measured values of D reported in the literature fall into two sets with a numerical discrepancy of about two orders of magnitude.[6, 7, 11, 12, 14–16] This variation could be observed in this type of experiment by traversing the sample along the optic axis and through the focal plane of the lens and thus achieving a change in the triplet exciton density.

References

1 S. Czarnecki, *Bull. Acad. Polon. Sci. Ser. Math. Astr. Phys.* **9**, 561 (1961).

2 C. A. Parker & C. G. Hatchard, *Proc. Roy. Soc.* (*London*), A, **269**, 574 (1962).

3 G. C. Nieman & G. W. Robinson, *J. Chem. Phys.* **37**, 2150 (1962).

4 M. Kinoshita, T. N. Misra & S. P. McGlynn, *J. Chem. Phys.* **45**, 817 (1966).

5 (*a*) R. W. Brandon, R. E. Gerkin & C. A. Hutchison Jr., *J. Chem. Phys.* **37**, 447 (1962).
(*b*) N. Hirota & C. A. Hutchison Jr., *J. Chem. Phys.* **42**, 2869 (1965).

6 J. Jortner, S. A. Rice, J. L. Katz & S. Choi, *J. Chem. Phys.* **42**, 309 (1965).

7 M. Levine, J. Jortner & A. Szöke, *J. Chem. Phys.* **45**, 1591 (1966).

8 W. Helfrich & F. R. Lipsett, *J. Chem. Phys.* **43**, 4368 (1965).

9 R. G. Kepler, J. C. Caris, P. Avakian & E. Abramson, *Phys. Rev. Letters*, **10**, 400 (1963).

10 P. Avakian, E. Abramson, R. G. Kepler & J. C. Caris, *J. Chem. Phys.* **39**, 1127 (1963).

11 S. Singh, W. J. Jones, W. Siebrand, B. P. Stoicheff & W. G. Schneider, *J. Chem. Phys.* **42**, 330 (1965).

12 G. F. Moore, *Nature*, **211**, 1170 (1966); G. F. Moore & I. H. Munro, *Nature*, **208**, 772 (1965).

13 I. Nakada, *J. Phys. Soc. Japan*, **17**, 113 (1962).

14 R. G. Kepler & A. C. Switendick, *Phys. Rev. Letters*, **15**, 56. (1965).

15 V. Ern, P. Avakian & R. E. Merrifield, *Phys. Rev.* **148**, 862 (1966).

16 D. F. Williams, J. Adolph & W. G. Schneider, *J. Chem. Phys.* **45**, 575 (1966).

THE ELECTRONIC STATES IN CRYSTALLINE ANTHRACENE

M. SILVER AND J. HERNANDEZ

Abstract

In light of recent recombination results,[1] the experiments of Castro & Hornig[2] and Pope[3] are analysed in terms of band-to-band transitions.

While the recombination coefficient is very large, this process does not limit carrier production under the experimental conditions used by Castro & Hornig. Thus, their quantum yield represents the results of the primary excitation process. However, because of recombination, no net carrier generation is obtained from excitation directly to the narrow tight binding band.

Carriers are generated by excitation to an almost free electron band. The quantum yield is shown to be α/k, where α is the absorption coefficient due only to the band-to-band transition, and k is the total absorption coefficient.

Attempts have been made to calculate α. Because of symmetry, perturbation theory predicts a cross-section which tends to zero in the dipole approximation. Higher multipole terms give an energy dependent matrix element. Recent experiments by Pope[3] and others suggest no energy dependence. If vibronic coupling is included to lower the symmetry of the ground state, a non-energy dependent transition probability may be obtained. In any event, a small cross-section is predicted in agreement with experiment.

On the basis of these theoretical considerations, photo-ionisation of the first excited singlet should be more efficient than the band-to-band transition.

References

1 M. Silver, *Bull. Am. Phys. Soc.* **11**, 269 (1966).
2 G. Castro & J. F. Hornig, *J. Chem. Phys.* **42**, 1459 (1965).
3 M. Pope, *J. Chem. Phys.* (to be published).

DISCUSSION

RICE

Wolf to Rice: Linewidth of exciton emission can answer the question of coherence or incoherence of exciton motion.

Rice to Wolf: I agree that the use of the linewidth to determine the dynamical properties of the exciton offers exciting possibilities. However, the available theory does not yet permit interpretation of the experimental data. At present even the crudest estimates of the interaction strength predict that vibration of the molecular centers leads to energy fluctuations larger than the triplet band-width. I believe, therefore, in agreement with the strong scattering model, that the motion of the triplet exciton in anthracene will be found to be incoherent when we finally know how to do the analysis properly.

Silver to Rice: Concerning the mixing of charge-transfer states into excited neutral exciton states, Pope sees no direct transitions to the CT state but that these can only be populated by a very inefficient indirect path (photo-conduction followed by recombination); does not this mean that the mixing is very small?

Rice to Silver: Our calculations show that the mixing of charge transfer exciton states into the neutral singlet exciton state of anthracene is very small.

The extent of mixing is predicted to be somewhat larger in the singlet exciton states of benzene and naphthalene. In all of the cases we have studied the charge-transfer state is sufficiently far from the neutral triplet exciton state that the mixing is predicted to be small. I believe our results are, therefore, in agreement with the results of Pope's experiments.

Siebrand to Rice: I feel that the motion of a triplet exciton in anthracene, say, is neither coherent nor incoherent, but semi-coherent. You assume one-phonon scattering at every lattice site, which leads to a mean free path approximately equal to the lattice spacing, in agreement with experiment. In the limit of coherent motion the exciton passes a large number of lattice sites before being scattered. In the limit of incoherent motion the exciton undergoes *multiple* scattering at every lattice site. Multiple scattering leads to a complete loss of coherence and thus to a random-walk model, whereas a single scattering event only leads to a partial loss of coherence. These definitions of coherent and incoherent transport agree with the definitions of band-type motion and hopping given by Holstein in his analysis of the small-polaron model.

From this point of view one might be tempted to label your choice of scattering parameters as arbitrary. However, experimentally a situation corresponding to a mean free path equal to the lattice spacing seems to occur in a large number of molecular crystals. I do not understand why this is so. A mathematical analysis of existing polaron models indicates that this situation is rather improbable and unstable, since a small change in one of the energy parameters (band-width and exciton-phonon coupling) would lead to a mean free path very large or very small compared to the lattice spacing (using the language of the band model). Thus there appears to be a discrepancy between theory and experiment or at least a gap in our knowledge of how they are connected.

Rice to Siebrand: I agree that our introduction of the strong scattering model is not mathematically satisfactory. However, I am not surprised that one lattice spacing is found as a 'universal' mean free path, since this is the relevant mesh on which we define the motion of the exciton. Thus, in a proper theory connecting coherent and incoherent motion, the strong scattering limiting step length must come out one lattice spacing. Incidentally, our treatment does not assume one phonon scattering of the exciton, but allows for multiphonon scattering, in principle, in the transfer integral $J(l\xi/i\zeta)$. Until a satisfactory and complete analysis of the influence of lattice scattering on the motion of the exciton has been completed, our model must be regarded as heuristic in nature—although *no* adjustable parameters are involved.

Tanaka to Rice: In connection with the question of Professor Silbey, I would like to comment that the oscillator strength of the charge-transfer absorption band was actually observed in case of 9,10-dichloroanthracene crystal although it is difficult to discern in anthracene crystal. The position of the *CT* band is in good agreement with the calculated value, but I would like to emphasise that the oscillator strength is coming through the interaction with the $2p\pi \rightarrow ns$, or continuum states. The spectrum will be shown later.

KOMMANDEUR

Tanaka to Kommandeur: As regards the nature of the phase transition occurring at low temperature, I would not agree with the mechanism proposed by Professor Kommandeur, since the argument based on the behavior of the longest band might not be reliable since it is much influenced by the crystal structure. The origin of this band was assigned as an inter-cation charge transfer by Hausser and Murrell, and the wavefunction for the dimer state will be written in an abbreviated form as follows:

$$\Psi_G = a\Psi(W^+ \dots W^+) + b\Psi(W^{++} \dots W^0),$$

$$\Psi_E = b\Psi(W^+ \dots W^+) - a\Psi(\overline{W}^{++} \dots W^0)$$

$$[a^2 + b^2 = 1. \quad \int \Psi(W^+ \dots W^+)\Psi(W^{++} \dots W^0)\,dv = 0].$$

In order to have an appreciable magnitude of oscillator strength for the transition between those two states, both a and b must have comparable value, and the values of a and b will be determined by the exchange integrals between cation pairs. Therefore at room temperature it might be that $a \gg b$ since the overlap is not large, and at the lower temperature, in which the phase transition takes place, the cations will be close together, the overlap will be increased, and it might be that $a \gtrsim b$. This reasoning is based on the spectroscopic and crystallographic studies of other Würster radicals in our laboratory which are now in progress.

KING

Siebrand to King: I wish to mention that D. F. Williams and J. Adolph of the National Research Council of Canada, who reported a high value of the diffusion coefficient of triplet excitons in anthracene crystals, in agreement with the results of Kepler and Switendick, feel now that this result is erroneous. It appears that the crystal surface is damaged by cutting off small slices with a microtome in this method. Their new results, which are in the course of publication, agree with the low value of the diffusion coefficient first reported by Avakian and Merrifield.

SILVER

Hochstrasser to Silver: Could you clarify to what extent your considerations are dependent on the Beer–Lambert assumption for the distribution of absorbed radiation in the crystal? Would it not be that for those collective-like excited states Beer's law would be an invalid assumption?

Silver to Hochstrasser: It is, of course, simpler to assume the Beer–Lambert Law, but the conclusions are not strongly dependent on the variation of intensity with distance in the crystal. It only matters that the absorption be strong and dependent on wavelength. The stronger the decay the smaller is the volume of the crystal in which there is a light available for band to band transitions. Thus one should, under these conditions, always get some sort of inverse dependence upon absorption coefficient. The total number of carriers produced is of course proportional to this volume and it is the total number one measures rather than the density in a flash experiment.

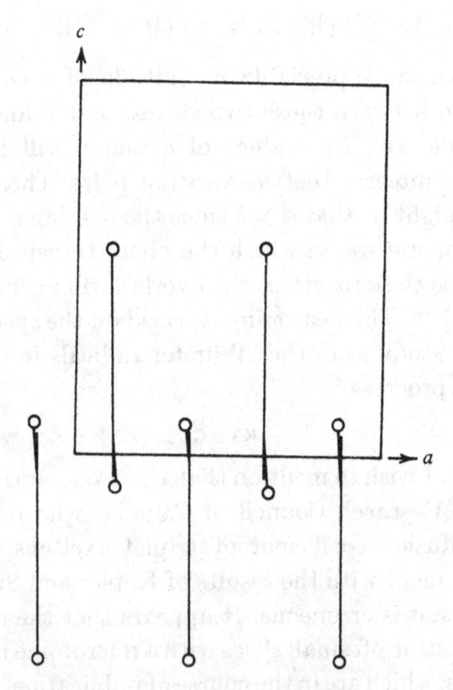

Fig. 1. Projection of 9, 10-dichloroanthracene molecules onto the (0|0) plane according to the crystal data of Trotter (*Acta Cryst.*).

Tanaka to Silver: In favour of the chairman's kindness, I would like to show the spectra of 9,10-dichloroanthracene, with a view to clarifying the nature of the conduction band in the anthracene-type crystal. The crystal shows a cleavage plane as shown in fig. 1, in which both the ${}^1B_{2u}$ and ${}^1B_{3u}$ excited states are expected to appear only along the c axis. The spectra shown in fig. 2 have prominent peaks for a axis polarised direction. The peak around 27000 cm^{-1} is reasonably regarded as an intermolecular charge-transfer band as discussed by Professor Rice; the band starting from 34000 cm^{-1} with gradual increase of intensity to the higher energy region is regarded as a band to band transition. However, the important point is its polarisation character: it has an out-of-plane component, so that the

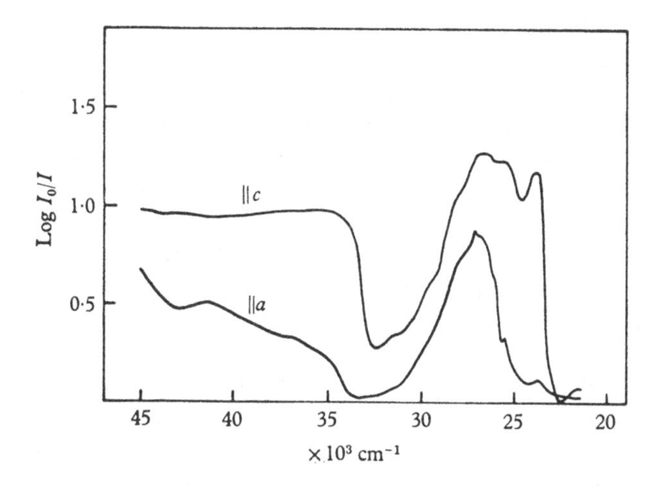

Fig. 2. Crystal absorption spectra of 9,10-dichloroanthracene taken with (0|0) plane with the light polarised parallel to a and c axes (unpubished work of Tanaka and Shibata).

assignment should be as $2p\pi \rightarrow 3s, 4s, \ldots, ns$ or ionisation continuum state but not to the $2p\pi$–$2p\pi$ band to band transition. The latter band should be rather narrow in energy as was calculated by LeBlanc, and it might have little contribution to the free carrier formation. Recent experiments by Castro and Hornig, and Kearns should be connected with the broader band of Rydberg nature, or with the true ionisation continuum, since those states are found in this energy region by the present spectroscopic measurement. The details of this experiment will be published in *Bull. Chem. Soc. Japan* (Tanaka and Shibata).

Rice to Silver: The Rydberg levels of polyatomic molecules often 'break off' before the extrapolated ionisation limit. R. S. Berry has interpreted this phenomenon as an internal mixed Auger effect, i.e. the molecular core 'pumps' the electron up in energy by itself dropping down in energy, say by a quantum of vibrational energy. I suspect that the photo-conduction threshold behaviour you describe will involve similar electron-nuclear motion coupling and will *not* be determined by the purely electronic part of the transition probability to the conduction band.

Silver to Rice: I agree that this effect may be at least as large as any purely electronic part of the transition probability. Nevertheless, by the use of a coulomb wave, we should be able to estimate to some modest accuracy the contribution due to the pure electronic part. This may be useful in discussing the difference between the cross-section for band to band transition from the ground state and from the excited singlet state. Also, this effect should be temperature-dependent and a temperature dependence has not been observed.

Joussot-Dubien to Silver: Why do you use anthracene? In a glass, anthracene is one of the most difficult aromatic hydrocarbons to photo-ionise. Other compounds would be easier to study.

Silver to Joussot-Dubien: The question of how hard or easy it is to ionise anthracene in a glass, relative to other aromatics, is a difficult one and involves not only the cross-section but secondary processes such as recombination and trapping. Further, since my theoretical discussion involved crystal data, anthracene is a suitable crystal because it is principally here that one has reliable data to compare with theory.

Some results have been obtained on pyrene by Kearns and tetracene by Pope and are, in my opinion, in general agreement with the ideas presented on the possible importance of band to band transitions in 'intrinsic' photo-conductivity.

Birks to Silver: May I support Joussot-Dubien's comment that anthracene is a most unsuitable material to study. It has a very large self-absorption which increases the fluorescence decay time by a factor of 3·5 and 'removes' the 0–0 fluorescence band. It is photochemically active, forming photodimers and photo-oxides. Its surface deteriorates on exposure to the atmosphere and it is sensitive to chemical impurities and physical defects. Work at Dupont Central Research Laboratories at Wilmington has shown that a pure anthracene crystal has a fluorescence quantum yield $\phi_F = 1{\cdot}0$ at room temperature, so that the triplet quantum yield $\phi_T = 0$. Far too much research effort has been expended on this difficult material. If a similar effort had been put on a simpler material like *para*-terphenyl we would be much further forward in our physical understanding of organic crystal phenomena.

Silver to Birks: I do not believe that this comment, as made by Dr Birks at the time of the Symposium, is relevant to a theoretical discussion and interpretation of well-established results on band to band transitions by at least six different laboratories and by at least three different techniques. Perhaps Dr Birks was referring to the well-known 'extrinsic' surface-generated photo-conduction which plays no role here since it is obviously due to impurity effect. These impurity effects account for the photo-conduction below photon energies of 4 eV.

We and the N.R.C. have measured in the bulk, and away from the surface, ϕ_{ST} and it is about 5 %. Thus the comment on zero yield of triplets is not correct. We have seen triplets in very pure anthracene from intersystem crossing (some of these crystals were kindly donated by DuPont).

6 DELAYED FLUORESCENCE AND PHOSPHORESCENCE

DELAYED FLUORESCENCE OF SOLUTIONS

C. A. PARKER

Abstract

This paper reviews the historical development of the phosphorescence and delayed fluorescence of organic compounds in fluid and rigid solution. The kinetics of E-type, P-type, recombination and sensitised delayed fluorescence are discussed. Some applications of the measurement of long-lived photoluminescence to photochemical and analytical problems are presented.

With solutions of complex molecules having an even number of electrons it is generally necessary to consider only two kinds of radiative transition, namely that from the first excited singlet state to the ground state, and that from the lowest triplet state to the ground state. Each of the radiative transitions gives rise to a characteristic emission spectrum which serves to identify the excited state concerned and gives a basis for classifying the luminescence without reference to the process by which the excited state was formed. We shall employ the general name 'fluorescence' for the $S_1 \rightarrow S_0$ transition and reserve the name 'phosphorescence' for the $T_1 \rightarrow S_0$ transition. The former is spin-allowed and the natural lifetime of the molecule in the excited singlet state is thus short (approximately 10^{-8} s). The second is spin-forbidden and in the absence of radiationless processes the lifetime of the molecule in the lowest triplet state is very large. If there were no interconversion between T_1 and S_1, the lifetime of fluorescence would always be short. In fact, there are a variety of processes by which the molecule in the triplet state can pass over to the excited singlet state and these processes result in the emission of luminescence having a spectral distribution characteristic of the $S_1 \rightarrow S_0$ transition. but a lifetime much longer than that characteristic of the S_1 state. To distinguish the long-lived emission arising from the process $T_1 \rightarrow S_1 \rightarrow S_0$ from the short-lived emission from molecules that have not previously passed over to the triplet state, we shall call the former 'delayed fluorescence' and the latter 'prompt fluorescence'. We then further subclassify delayed fluorescence into '*E*-type' and '*P*-type' delayed fluorescence according to the mechanism by which the process $T_1 \rightarrow S_1$ occurs.[1]

There are, in addition, other forms of long-lived emission for which the long lifetime does not depend directly on the long lifetime of molecules in the triplet state. These emissions are observed when irradiation in rigid medium leads, directly or indirectly, to the decomposition of the solute into fragments that recombine slowly with the formation of an excited singlet or a triplet state. We classify these emissions as 'recombination delayed fluorescence' and 'recombination phosphorescence'. In this paper we shall be concerned mainly with *E*-type and *P*-type delayed fluorescence.

The characteristics of the various kinds of photoluminescence are summarised in table 1. It should be noted that this method of classifying long-lived luminescence according to its spectral distribution has

Table 1. *Mechanisms of photoluminescence*

Type	Mechanism (see fig. 1 and table 2)	Spectrum	Form of decay
Prompt fluorescence	1, 2	$S_1 \rightarrow S_0$	Exponential (τ_f)
Phosphorescence	1, 4, 5	$T_1 \rightarrow S_0$	Same as that of T_1 and exponential at low $I_a(\tau_p)$
E-type delayed fluorescence	1, 4, 8, 2	$S_1 \rightarrow S_0$	Equals τ_p
P-type delayed fluorescence	1, 4, 9*a*, 2	$S_1 \rightarrow S_0$	Equals $\frac{1}{2}\tau_p$ at low I_a—non-exponential at high I_a
Sensitised *P*-type delayed fluorescence	10, 9*a*, 2	$S_1 \rightarrow S_0$	Equals $\frac{1}{2}\tau_p$ at low I_a
Sensitised *P*-type delayed fluorescence	10, 9*b*, 2	$S_1 \rightarrow S_0$	Equals $\tau_p\tau_{p'}/(\tau_p+\tau_{p'})$ at low I_a
Recombination delayed fluorescence	11 or 12, followed by return to S_1	$S_1 \rightarrow S_0$	Often non-exponential
Recombination phosphorescence	11 or 12, followed by return to T_1	$T_1 \rightarrow S_0$	Often non-exponential if rate of return to T_1 is small

not been universally employed. Many workers have used the term 'phosphorescence' for all emission having a lifetime long enough to be resolved with a mechanical phosphoroscope. Some of this alternative terminology will be indicated in discussing the earlier work.

In most of what follows we shall be concerned with the rates of transitions between the states S_0, S_1 and T_1. Most of the relevant transitions are shown in fig. 1, and their rate constants are defined in table 2. We shall usually deal with systems in which the triplet life-

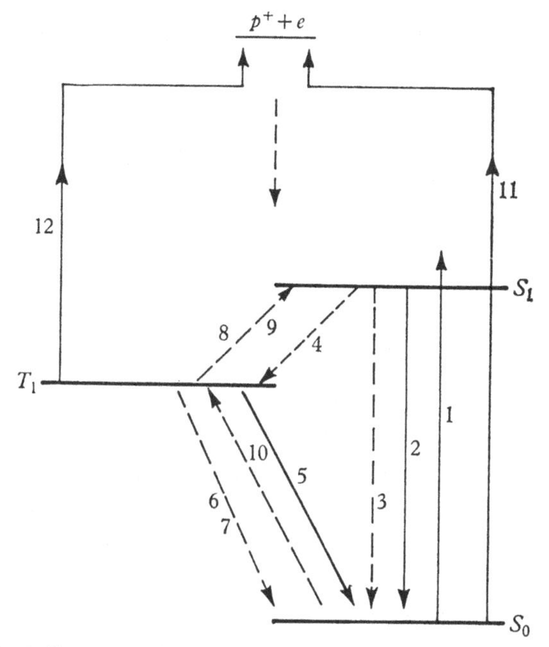

Fig. 1. Processes affecting the emission of photoluminescence.

times are greater than 10^{-4} s, and the steady-state concentrations of S_1 and T_1 can thus be treated separately. For example, we can then write the following general expressions for three fundamental quantum efficiencies (for light absorption by S_0) in terms of the appropriate rate constants shown in table 2:

Quantum efficiency of prompt fluorescence,

$$\phi_f = k_f/(k_f + k_n + k_g). \tag{1}$$

Quantum efficiency of triplet formation,

$$\phi_t = k_g/(k_f + k_n + k_g). \tag{2}$$

Quantum efficiency of phosphorescence,

$$\phi_p = k_p \phi_t/(k_p + k_m + k_e + k_q[q] + k_i[T_1]). \tag{3}$$

Table 2. *Rate constants of processes shown in fig.* 1

No.	Process	Rate
1	Light absorption and internal conversion to S_1, without dissociation or chemical reaction	I_a
2	Emission of prompt fluorescence	$k_f[S_1]$
3	$S_1 \rightarrow S_0$ radiationless transition	$k_n[S_1]$
4	$S_1 \rightarrow T_1$ intersystem crossing	$k_g[S_1]$
5	Emission of phosphorescence	$k_p[T_1]$
6	$T_1 \rightarrow S_0$ intersystem crossing	$k_m[T_1]$
7	Triplet quenching by solute q	$k_q[T_1][q]$
8	Thermal activation of triplet	$k_e[T_1]$
9*a*	Triplet to triplet energy transfer (like triplets)	$pk_c[T_1]^2$
9*b*	Triplet to triplet energy transfer (unlike triplets)	$p'k_c[T_1][T_1']$
10	Triplet to singlet energy transfer	$p_e k_c[S_0][T_1']$
11	Absorption by ground state to produce ionisation, etc.	$\propto I_0[S_0]$
12	Absorption by triplet state to produce ionisation, etc.	$\propto I_0^2[S_0]$ (at low I_0)

Notes. (1) For simplicity quenching of S_1 by added solute and chemical reaction of S_1 are assumed to be negligible. (2) I_a is the rate of light absorption by S_0 in Einstein litre^{-1}. (3) k_c is the diffusion-controlled rate constant, normally taken to be $8RT/3000\eta$, and p and p_e are probability factors. p_e is normally taken as unity. (4) The *total* rate of disappearance of $[T_1]$ by bimolecular reaction is defined as $k_i[T_1]^2$.

There is some evidence that processes analogous to no. 4 in table 2 can proceed from upper excited singlet states. Such processes are not considered in the present paper. If they do occur, they could cause a variation of the quantum efficiencies (equations (1), (2), (3)), with wavelength of exciting light.

E-type delayed fluorescence in rigid media

It has long been known that photo-excitation of rigid solutions of certain dyes produced a long-lived emission, or 'after glow'.[2, 3, 4] In 1929, J. & F. Perrin[5] suggested that by analogy with excited atoms, the excited molecules could pass over to a metastable state, α, of lower energy from which spontaneous emission was impossible, and that, if the decrease in energy was not too great, the molecules could return to the fluorescent state and ultimately give out light. Such substances would thus give rise to two kinds of photoluminescence, a short-lived fluorescence ('fluorescence directe') and a long-lived or 'delayed' fluorescence ('fluorescence retardée par le passage dans l'état α').

The Perrins had assumed that thermal activation was the only mechanism by which a molecule in the metastable state could return to the ground state. In 1933, Jablonski[6] proposed a scheme that was essentially an extension of that just described, but included the possibility of a radiative transition of low probability direct from the metastable state to the ground state. He thus explained the observation of two emission bands, one appearing at high temperatures and having the spectral distribution of fluorescence ('phosphorescence at room temperature') and the other appearing at low temperature with a spectrum displaced towards the red ('phosphorescence at low temperatures'). Jablonski realised that the phosphorescence would be 'quenched' more strongly than the prompt fluorescence, i.e. that non-radiative transitions direct to the ground state would occur. He also stated that a weak absorption band corresponding to a transition direct to the metastable state must also be present, and that under the influence of disturbing fields the probability of this transition would be increased. Thus, in this and his subsequent paper[7] Jablonski laid the foundation on which later studies of photoluminescence are based, and his scheme provides the essential features of the diagram shown in fig. 1.

In 1936, Levshin & Vinokurov[8] showed that the rates of decay of the two emission bands from fluorescein in boric acid glass at room temperature were identical, and thus provided direct experimental evidence that the same metastable state was returning to the ground state by two independent processes, each giving rise to a different emission. The fluorescein–boric acid system was investigated in more detail by Lewis, Lipkin & Magel.[9] They called the two emission bands 'alpha phosphorescence' (corresponding to our E-type delayed fluorescence) and 'beta phosphorescence' (corresponding to our phosphorescence). They also succeeded in making direct measurements of the absorption spectrum of the metastable state. They found that the decay of both emission bands was exponential at all temperatures at which they were observable, and from the measured lifetime (τ) they attempted to determine the specific rate of the alpha process (our k_e in table 2) by deducting from the observed values of $1/\tau$ at high temperatures the values of the beta process extrapolated from the observed values of $1/\tau$ at low temperatures, where the beta process alone was assumed to occur. The logarithm of the resulting rate plotted against $1/T$ gave a straight line from which the derived activation energy was equal to the spectroscopic energy difference between the two emission bands. Lewis and co-workers brought this forward as evidence that the alpha phosphorescence was indeed produced by thermal activation of the metastable state. It should be noted that their calculation depends on the assumption that there are no other processes contributing to the decay of the metastable state at high temperatures. Although this seemed to be the case with fluorescein in boric acid glass, it is by no means true of all systems (e.g. in fluid solution) and the occurrence of other decay processes in some systems may have been partly responsible for the earlier criticisms[4, 10] of the simple Jablonski scheme. As we shall see later, the rate of thermal activation can be determined precisely, whatever the rates of the other decay processes, by measuring the ratio of the *intensities* of *both* emission bands as a function of temperature, or by measuring the intensities of delayed and prompt fluorescence, and the lifetime of the delayed fluorescence, all as a function of temperature.

In a later publication, Lewis & Kasha[11] identified the metastable state with the triplet state and this has been amply confirmed by subsequent work. Lewis & Kasha measured the triplet-singlet emission spectra of a variety of organic compounds in rigid solutions at 77 °K, and indicated that such measurements could provide useful information about the properties of both the triplet state and ground state, and could also be used for analytical purposes. This technique has since been employed by many other workers.

E-type delayed fluorescence and phosphorescence in fluid solution

In fluid solution the rates of $T_1 \rightarrow S_0$ intersystem crossing (k_m) and/or impurity quenching ($k_q[q]$) of the triplet state are generally so great compared with the rate of the radiative process (k_p) that ϕ_p is nearly always small (see equation (3)) and, in fact, phosphorescence from fluid solutions had, until recently, been observed only rarely.

The first quantitative measurements of long-lived photoluminescence in fluid solution were made as long ago as 1930 by Mlle Boudin[12] working in the laboratory of J. Perrin. Mlle Boudin used a visual phosphoroscope and investigated the effect of potassium iodide on the lifetime of the long-lived emission from eosin in glycerol at room temperature. With 10^{-4} M eosin, in the absence of iodide, she observed a lifetime of 1·1 ms at room temperature and estimated that the intensity was about $\frac{1}{400}$th of that of the prompt fluorescence. These values are somewhat lower than those found in later work[13] with vacuum-deoxygenated solutions, but they are nevertheless remarkable in view of the difficulties inherent in visual measurements of weak luminescence. A few years later, Kautsky[14] made qualitative observations with deoxygenated solutions of a variety of dyestuffs at room temperature and observed long-lived emissions which were probably due to what we now call *E*-type delayed fluorescence. The subject then received little attention until the advent of the technique of flash photolysis which was used by Bäckström & Sandros[15] to investigate the quenching of the triplet state of biacetyl, a substance long known for its phosphorescence.

The question of long-lived luminescence in fluid solution was taken up again by Parker & Hatchard[13, 16] using a spectrophosphorimeter[1] of high sensitivity and unusual design that could be used to measure not only the spectrum and lifetime of the long-lived emission, but also its precise intensity relative to that of the prompt fluorescence. The spectra observed with deoxygenated solutions of eosin in glycerol and ethanol (fig. 2) show two long-lived emission bands, the relative intensities of which are strongly temperature-dependent. The visible band at $\sim 1{\cdot}8\,\mu m^{-1}$ had a contour identical with that of the prompt fluorescence and it was therefore classified as delayed fluorescence. It no doubt corresponds to the long-lived luminescence observed by Boudin.[12] The far-red band at $\sim 1{\cdot}4\,\mu m^{-1}$ was attributed to triplet-singlet phosphorescence.

The facts that (*a*) the lifetime of the delayed fluorescence was at all temperatures identical with that of the far-red band, (*b*) the intensity of the delayed fluorescence relative to that of the far-red band varied exponentially with $-1/T$, and (*c*) the intensities of both bands were proportional to the first power of the intensity of the exciting light, left no doubt that the emissions were analogous to the two bands discussed by Jablonski,[6, 7] i.e. they were completely analogous to the 'α and β phosphorescence' observed by Lewis, Lipkin & Magel[9] with fluorescein in rigid medium. Because this type of delayed fluorescence was first clearly demonstrated by Boudin[12] with eosin, Parker & Hatchard[17] proposed that it should be called 'eosin-type' or '*E*-type' delayed fluorescence, and they reserved the term 'phosphorescence' for the red band corresponding to the transition from the triplet state direct to the ground state. Similar emissions were subsequently observed from fluid solutions of proflavine[16] and acridine orange[18] and they are to be expected with fluid solutions of all fluorescent dyestuffs that have an appreciable triplet formation efficiency and a triplet level lying close to the first excited singlet

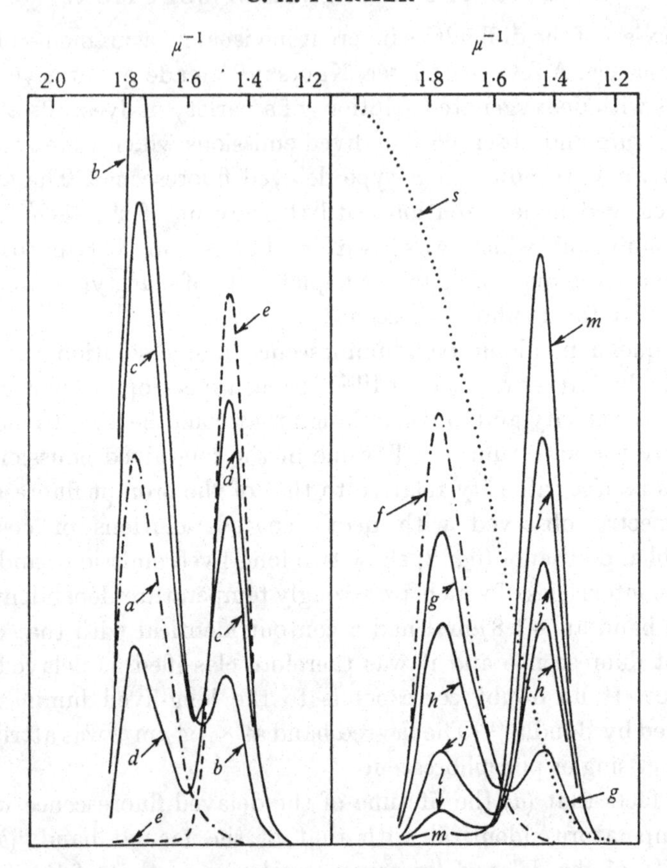

Fig. 2. (1) Eosin in glycerol (7×10^{-5} M). (2) Eosin in ethanol ($1{\cdot}5 \times 10^{-5}$ M). (*a*) Fluorescence emission spectrum at +30 °C; (*b*) delayed emission spectrum (DES) at +69 °C; (*c*) DES at +48 °C; (*d*) DES at +18 °C; (*e*) DES at −40 °C (delayed emission spectra at a sensitivity 600 times greater than that for the fluorescence emission spectrum); (*f*) fluorescence emission spectrum at +22 °C; (*g*) delayed emission spectrum (DES) at +71 °C; (*h*) DES at +43 °C; (*j*) DES at 22 °C; (*l*) DES at −7 °C; (*m*) DES at −58 °C. (*s*) Sensitivity of 9558 photomultiplier with quartz monochromator (units of quanta and frequency). Delayed emission spectra at a sensitivity 3000 times greater than that for the fluorescence emission spectrum. Sensitivity values not corrected for the phosphorimeter factor. (From Parker and Hatchard[13].)

level. As will be seen later, such dyestuffs can also give rise to another type of delayed fluorescence under appropriate conditions.

Relationships between the rate constants in table 1, and the quantum efficiencies and lifetimes of E-type delayed fluorescence and triplet-singlet phosphorescence can be derived as follows. Since all molecules raised from T_1 to S_1 (see fig. 1) by thermal activation are expected to fluoresce with normal efficiency (ϕ_f), the relationship between the efficiency of E-type delayed fluorescence and the rate constant k_e (with $k_i[T_1]$ assumed to be zero) is given by

$$\phi_e = (k_e \phi_t \phi_f)/(k_p + k_m + k_e + k_q[q]), \tag{4}$$

in which ϕ_t is given by equation (2). Combining this with equation (3) gives

$$\phi_e/\phi_p = (k_e \phi_f)/k_p. \tag{5}$$

Thus the ratio of the intensities of the two bands should be completely independent of ϕ_t and of all triplet quenching processes.[13] Since k_e represents a thermally activated process:

$$\phi_e/\phi_p = \tau_R \phi_f A\, e^{-\Delta E/RT}, \tag{6}$$

where A is a frequency factor, ΔE is the activation energy and $\tau_R\,(= 1/k_p)$ is the radiative lifetime of the triplet state. Parker & Hatchard found that plots of $\ln(\phi_e/\phi_p)$ for eosin against $1/T$ were linear and the derived value of ΔE (10 kcal) agreed within experimental error with the energy difference between the maxima of the two emission bands. Similar results were obtained with proflavine ($\Delta E = 8$ kcal).

Equation (3) can be rewritten in the form

$$\phi_p/\tau = \phi_t/\tau_R, \tag{7}$$

where τ is the observed lifetime of phosphorescence or E-type delayed fluorescence at any temperature. Since the radiative lifetime τ_R is independent of temperature, measurements of ϕ_p and τ as a function of temperature can be used to determine how ϕ_t varies with temperature.[13] With a sensitive spectrophosphorimeter, such measurements can be made in fluid solution, i.e. under conditions of most interest to the photochemist. Using this technique, the triplet formation efficiencies of eosin and proflavine were found to be practically independent of temperature.[1]

The observation of E-type delayed fluorescence can also be used to determine the triplet energy of a substance for which the phosphorescence cannot be observed directly. From equations (6) and (7)

$$\phi_e/(\phi_f \tau) = \phi_t A\, e^{-\Delta E/RT}. \tag{8}$$

Thus if ϕ_t does not vary with temperature, a plot of ln $(\phi_e/\phi_f \tau)$ against $1/T$ will give a straight line from which ΔE, and hence the triplet energy, can be determined. Parker & Joyce[19, 20] have used this method to determine the triplet energies of chlorophylls a and b.

The radiative lifetimes ($1/k_p$) of many aromatic hydrocarbon triplets are greater than 1 s and hence in fluid solution at room temperature, where triplets are strongly quenched by non-radiative processes, the intensity of phosphorescence is small. The weak triplet-singlet phosphorescence from an aromatic hydrocarbon in fluid solution was first observed by Parker & Hatchard[16] with solutions of phenanthrene. These observations were of particular importance, not because they showed the expected weak phosphorescence,

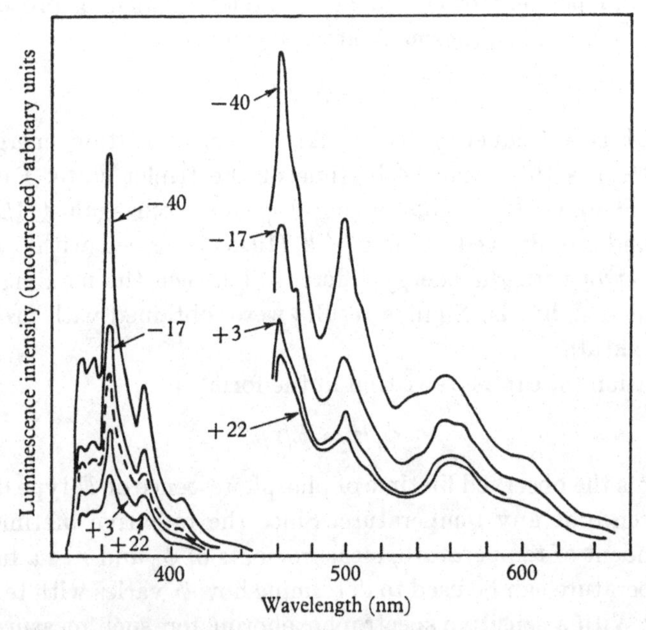

Fig. 3. Triplet-singlet phosphorescence (right-hand curves) and delayed fluorescence (left-hand curves) of phenanthrene in ethanol at the temperatures indicated. The rate of light absorption was approximately 10^{-5} Einstein litre^{-1} s^{-1} at 313 nm. The sensitivity for the right-hand curves was 8500 times, and for the left-hand curves 108 times, that for the measurement of prompt fluorescence (as indicated by the pecked curve). (From Parker[75].)

but because of the additional appearance of relatively intense delayed fluorescence (see fig. 3). This could not be *E*-type delayed fluorescence because the triplet level of phenanthrene is situated at a much lower energy than the excited singlet state, and the degree of thermal activation—and hence the intensity of *E*-type delayed fluorescence—must be exceedingly small, even at room temperature. That the delayed fluorescence was not produced by the *E*-type mechanism was confirmed by the fact that it did not follow the

$e^{-\Delta E/RT}$ law and indeed over the higher temperature range the value of ϕ_{DF}/ϕ_f *decreased* with rise in temperature. This was an example of a new kind of delayed fluorescence to be described in the following section.

P-type delayed fluorescence

The early work on *P*-type delayed fluorescence was confused by two factors. First, the materials used for some of the experiments were impure, and secondly, several workers incorrectly attributed the long lifetime of the emissions to the formation of long-lived excited dimers.

In 1950, Dikun[21] reported that the emission from phenanthrene vapour contained two components, one having the expected lifetime of prompt fluorescence and the other having a much longer lifetime of about 10^{-3} s. In 1958, Williams[22] observed long-lived emissions from the vapours of anthracene, perylene, pyrene and phenanthrene. The spectra of the emissions from the first two compounds were identical with those of the prompt fluorescence. With phenanthrene the ratio of the intensity of the long-lived emission to that of the total emission *increased* as the concentration of phenanthrene increased. On the basis of the last observation, Williams suggested that the long-lived emissions from all four compounds were due to long-lived excited dimers formed by reaction of the respective excited singlet monomers with a second molecule in the ground state. He supposed that the excited dimer dissociated by thermal activation and regenerated a singlet excited monomer, and that the emission from the latter gave rise to the delayed fluorescence. Williams did not report on the spectrum of the long-lived emission from his phenanthrene, but Stevens and co-workers[23] repeated the work of Dikun and showed that the long-lived emission from phenanthrene vapour had a spectrum characteristic of *anthracene* fluorescence. They suggested that the anthracene was present as a trace impurity in the phenanthrene and that its delayed emission was sensitised by collision with the 'long-lived excited dimer' of phenanthrene already proposed by Williams.

It should be noted that several years earlier, Förster & Kasper[24] had published their now celebrated work on the prompt fluorescence of concentrated pyrene solutions in which they had demonstrated conclusively that singlet excited pyrene does indeed react with a second molecule of pyrene in the ground state to form an excited dimer, and that the radiative transition of the latter to the ground state is responsible for the broad structureless band that appears in concentrated solutions of pyrene to the long wavelength side of the monomer emission. They did not determine the lifetime of the excited dimer, but in 1960 Stevens & Hutton[25] observed a long-lived emission from concentrated pyrene solutions and reported that it had a spectral distribution identical with the excited dimer emission of Förster & Kasper (they did not observe any long-lived emission from the monomer, although Parker & Hatchard[17] later showed that this was present also). Stevens & Hutton interpreted their results by assuming that the pyrene dimer was long-lived, i.e. they adopted the mechanism of Williams. They proposed the term 'excimer' for such 'long-lived' excited dimers and subsequently explained the sensitised delayed fluorescence from the vapours of mixtures of other compounds by a similar mechanism.[23, 26]

At this stage, the experimental evidence for long-lived excited dimers, or 'excimers' seemed to be strong. However, in 1961, Parker & Hatchard[27] showed that the greater part of the dimer emission from pyrene solutions must have a lifetime less than 5×10^{-5} s. Later, Birks & Munro[28] obtained a value of 4×10^{-8} s by direct measurement. It was then difficult to see how the long-lived component observed by Stevens & Hutton[25] could be accounted for by the Williams mechanism. The problem was first resolved in 1962 by the work of Parker & Hatchard with solutions of anthracene and phenanthrene[29, 30] and later with pyrene[17] and other compounds. They observed that the intensity of delayed fluorescence was proportional to the *square* of the rate of absorption of exciting light, and showed that the delayed fluorescence was produced by the transfer of energy on collision between two triplet molecules, whereby one of the molecules was ultimately raised to the excited singlet state from which it then emitted luminescence having a spectral distribution characteristic of the excited singlet state, but a rate of decay equal to twice that of the triplet state. They suggested[29, 30] that the same mechanism might explain the results of Williams[22] and Stevens and co-workers[23, 26] with aromatic hydrocarbons in the vapour phase, and this was subsequently found to be the case by Stevens, Walker & Hutton.[31] This process of *triplet to triplet energy transfer* has since come to be known as 'triplet-triplet annihilation'. Thus the postulation of *long-lived* excited dimers, or 'excimers', was not necessary to explain any of the observed phenomena, whether in the gas phase or in solution. Nevertheless, the term 'excimer' has been adopted by many workers for *short-lived* excited dimers of the Förster–Kasper type.

Before describing the detailed mechanisms of P-type delayed fluorescence in fluid solution, we must note that the process of triplet to triplet energy transfer was proposed independently in papers from two other laboratories to explain the emission of delayed fluorescence from aromatic hydrocarbons in *rigid media*. The first was that of Czarnecki[32] who reported the observation of delayed fluorescence (he called it 'phosphorescence') from solutions of naphthalene in rigid methyl methacrylate polymer. He could not observe it with phenanthrene and did not investigate the effect of varying the intensity of the exciting light. The second was that of Muel[33] who reported the observation of delayed fluorescence from frozen solutions of benz(a)-pyrene at 77 °K. He showed that the intensity was proportional to the square of the intensity of the exciting light and proposed that triplet to triplet energy transfer took place. We shall discuss later the more complicated effects that can occur in rigid solutions.

The mechanisms of triplet to triplet energy transfer in fluid and rigid solution are in some respects analogous to the annihilation of triplet excitons in the crystalline phase (see, for example, references (34) and (35)). However, we shall not deal with the crystalline state in this review.

Mechanisms of P-type delayed fluorescence in fluid solution

To explain the observation that the intensity of delayed fluorescence from solutions of anthracene and phenanthrene was proportional to the square of the rate of absorption of exciting light, Parker & Hatchard[29, 30] proposed the following mechanism involving the intermediate formation of a short-lived excited state X as a product of the interaction between two molecules in the triplet state:

$$^3A + {^3A} \rightarrow X(+{^1A}), \qquad \text{rate} = k_i[^3A]^2; \tag{9}$$

$$X \rightarrow {^1A}(+{^1A}), \qquad \text{rate} = k_\beta[X]; \tag{10}$$

$$X \rightarrow {^1A^*}(+{^1A}), \qquad \text{rate} = k_\alpha[X]. \tag{11}$$

At low rates of light absorption the rate of (9) is small compared with the sum of the rate constants of first-order triplet decay

$$(1/\tau = k_p + k_m + k_e + k_q[q], \quad \text{see table 1})$$

and hence

$$\theta/\phi_f = \tfrac{1}{2}k_i I_a(\phi_t\tau)^2 k_\alpha/(k_\alpha + k_\beta), \tag{12}$$

where τ is the actual lifetime of the triplet under the conditions concerned. Clearly with this scheme, the *efficiency* of delayed fluorescence is proportional to the rate of light absorption, as is actually observed. Furthermore, the mechanism does not require the postulation of any new long-lived species: the lifetime of the delayed fluorescence is accounted for in terms of the lifetime of the triplet. Thus when the exciting light is shut off the triplet concentration will decay according to

$$[^3A] = [^3A]^0 e^{-t/\tau}. \tag{13}$$

If the lifetime of X is short compared with that of the triplet, the rate of production of delayed fluorescence will be proportional to $[^3A]^2$ and its intensity when the exciting light is shut off will be proportional to $e^{-2t/\tau}$. Thus the lifetime of delayed fluorescence will be one-half of that of the triplet-singlet phosphorescence. This law was found to be obeyed for phenanthrene in ethanol[30] and was later confirmed by the more precise measurements[36, 37] with solutions of four other aromatic hydrocarbons in liquid paraffin at various temperatures.

Parker & Hatchard originally suggested[30] that the species X might be a short-lived dimer, or perhaps a single molecule in the quintuplet state

$$^3A + {}^3A \rightarrow {}^5A + {}^1A. \tag{14}$$

This question was at least partly resolved by their subsequent work with solutions of pyrene.[17] With this compound they observed that the spectra of prompt fluorescence (fig. 4) were similar to those observed by Förster & Kasper,[24] i.e. the relative intensity of the dimer emission decreased as the concentration of pyrene decreased, and tended to zero at low concentrations. In the spectra of delayed fluorescence (fig. 5) the relative intensity of the dimer band was greater than that in the corresponding spectrum of prompt fluorescence, and at infinite dilution the intensity of the dimer emission tended to a constant finite value. They therefore suggested that the following mechanisms operated. The sequence of reactions giving rise to singlet excited monomer and excited dimer as observed by *prompt* fluorescence is

$$^1P \xrightarrow{h\nu} {}^1P^* \xrightarrow{+P} {}^1P_2^*. \tag{15}$$

When the exciting light is shut off the concentrations of both singlet excited states decay rapidly to much lower values that are maintained by triplet-triplet interaction

$$^3P + {}^3P \longrightarrow {}^1P_2^* \longrightarrow {}^1P^* + {}^1P. \tag{16}$$

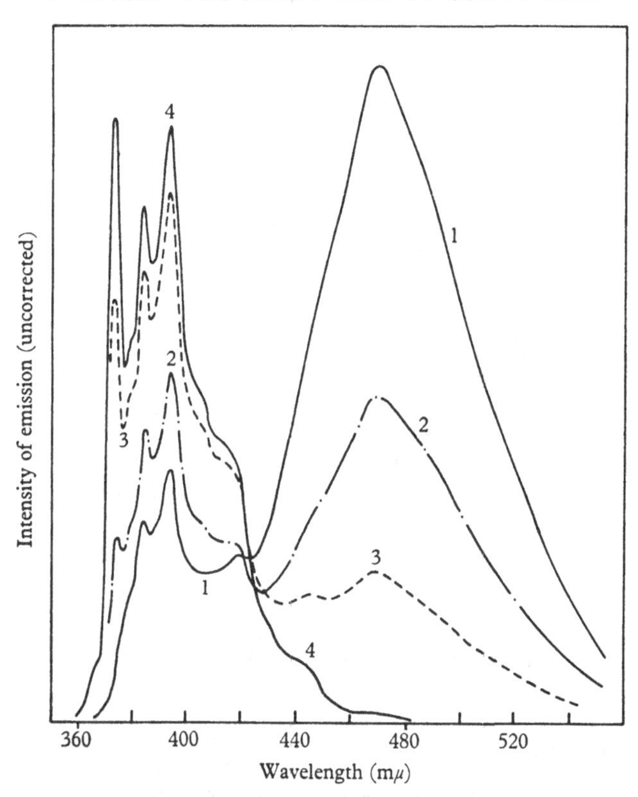

Fig. 4. Normal fluorescence of pyrene in ethanol. (1) 3×10^{-3} M, (2) 10^{-3} M, (3) 3×10^{-4} M, (4) 2×10^{-6} M. The instrumental sensitivity for curves (1) and (4) were approximately 0·6 and 3·7 times that for curves (2) and (3). The short wavelength ends of the spectra in the more concentrated solutions are distorted by self-absorption. (From Parker & Hatchard[17].)

Thus in prompt fluorescence it is the singlet excited *monomer* that is first formed, but in delayed fluorescence it is the excited dimer that is first formed. Clearly if $^1P^*$ and $^1P_2^*$ fluoresce before equilibrium between them is established the proportion of dimer observed in delayed fluorescence will always be greater than that observed in prompt fluorescence and as the concentration of pyrene is decreased the relative intensity of the delayed fluorescence of the dimer will decrease to a constant finite value. This outline scheme is shown in fig. 6 and quantitative spectral measurements[17] with pyrene solutions at room temperature were in agreement with it.

Parker[38] later obtained qualitatively similar results with solutions of naphthalene at lower temperatures, following the observations of Döller & Förster[39] of the prompt fluorescence of the excited dimer

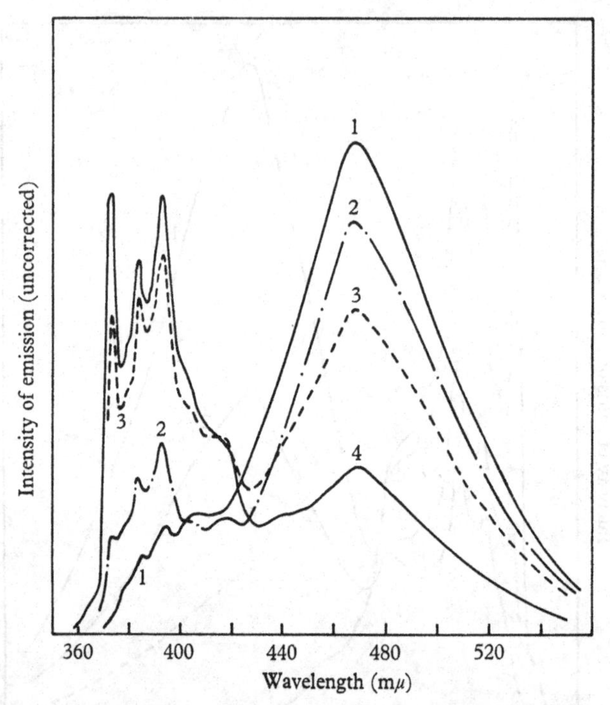

Fig. 5. Delayed fluorescence of pyrene in ethanol. (1) 3×10^{-3} M, (2) 10^{-3} M, (3) 3×10^{-4} M, (4) 2×10^{-6} M. The instrumental sensitivity settings were approximately 1000 times greater than those for the corresponding curves in fig. 1. The short wavelength ends of the spectra in the more concentrated solutions are distorted by self-absorption. (From Parker & Hatchard[17].)

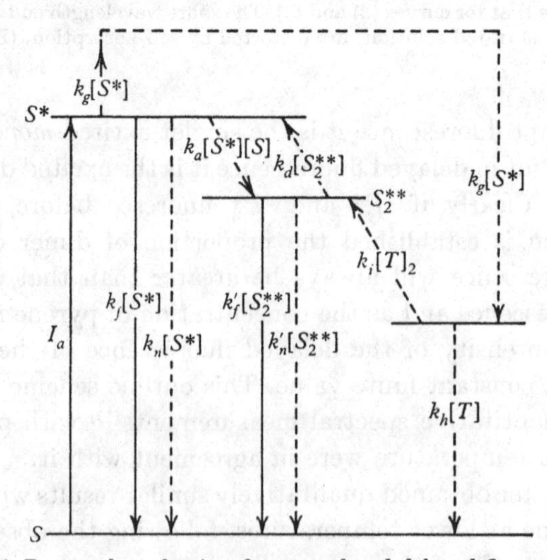

Fig. 6. Proposed mechanism for normal and delayed fluorescence.

in this system. He suggested that the mechanism for the emission of delayed fluorescence from solutions of anthracene and phenanthrene was the same as that for pyrene and naphthalene, and that the absence of dimer emission was simply due to the rapid dissociation of the excited dimer formed by triplet-triplet interaction, i.e. he identified the species X in equation (9) with the excited dimer. However, it soon became apparent that matters were not as simple as this. Stevens and co-workers[40] measured the relative intensities of the dimer band (θ_D) and the monomer band (θ_M) in the spectra of delayed fluorescence of relatively concentrated solutions of pyrene in ethanol. They found that below a certain temperature the value of θ_D/θ_M *decreased*. This is the reverse of what would be expected if all delayed fluorescence of the monomer were produced by dissociation of the

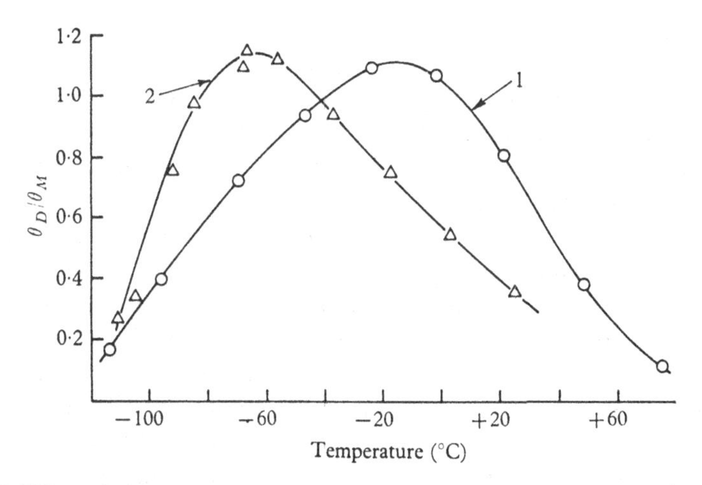

Fig. 7. Effect of temperature on delayed emission of dimer. Ethanolic solutions of (1) 2×10^{-5} M pyrene; (2) 10^{-5} M benzapyrene. (From Parker[41].)

excited dimer (equation (16)). Similar results were obtained by Parker[41] with solutions of both pyrene and benz(a)pyrene (see fig. 7) so dilute that the formation of excited dimer by the Förster–Kasper mechanism was negligible. Both Stevens and co-workers and Parker proposed that there intervened at low temperatures an alternative mechanism to equation (16) involving transfer of energy between two triplets at distances greater than that required for excited dimer formation.

All results so far reported in fluid solutions down to a temperature of -110 °C can be satisfactorily explained by what we may term this combined Parker–Stevens mechanism:

$$^3A + {}^3A \begin{cases} \xrightarrow{a} {}^1A_2^* \rightarrow {}^1A^* + {}^1A, & (17a) \\ \xrightarrow{b} {}^1A^* + {}^1A, & (17b) \end{cases}$$

in which (17*a*) predominates at high temperature in fluid solutions and (17*b*) at low temperature in viscous solutions. However, Birks and co-workers[42, 43] have proposed that triplet-triplet interaction yields initially a higher excited state of the dimer which then converts rapidly into either the lower excited state of the dimer or the excited and ground state monomers in a branching ratio of $\alpha:1$:

$$^3A + {}^3A \rightarrow {}^1A_2^{**} \begin{cases} \nearrow {}^1A_2^* \\ \quad \rightleftharpoons \\ \searrow {}^1A^* + {}^1A. \end{cases} \quad (18)$$

Both the Birks mechanism and the Parker–Stevens mechanism give an equally acceptable quantitative interpretation of the results at room temperature. The Parker–Stevens mechanism also gives an acceptable qualitative explanation of why the ratio of dimer to monomer emission should decrease at low temperatures—under these conditions the rate of diffusion is lower, the triplet molecules spend a relatively longer time within the triplet-triplet transfer distance and a greater proportion of them would be expected to interact without the production of excited dimer. Parker[44] criticised the Birks mechanism on the ground that it requires the rate of dissociation of the dimer in its upper excited state to be *greater* at low temperature relative to the competing reaction, and he concluded that on present evidence the Parker–Stevens mechanism is to be preferred.

P-type delayed fluorescence at high rates of light absorption

The square-law dependence of the intensity of *P*-type delayed fluorescence on the intensity of the exciting light, and the exponential decay of the delayed fluorescence, will only apply if the rate of the bimolecular triplet-triplet interaction is small compared with the sum of all first-order triplet decay processes (i.e. k_h where

$$k_h = k_p + k_m + k_e + k_q[q]).$$

In general the rate of triplet decay is given by

$$d[^3A]/dt = k_h[^3A] + k_i[^3A]^2. \quad (19)$$

Thus, as the rate of light absorption is increased $k_i[^3A]^2$ ultimately becomes appreciable compared with $k_h[^3A]$. The efficiency of delayed fluorescence then increases more slowly than predicted by equation (12) and the decay curve of delayed fluorescence contains an increasing proportion of second-order component. Finally, at very high rates of light absorption, $k_i[^3A]^2$ becomes very large, triplet decay becomes second order and the decay of delayed fluorescence becomes non-exponential.

Although at high rates of light absorption the square-law dependence of delayed fluorescence on the *rate of absorption exciting light* is no longer obeyed, the intensity of P-type delayed fluorescence is still proportional to the square of the intensity of *triplet-singlet* emission because the latter is always proportional to $k_p[^3A]$ and the former to $k_i[^3A]^2$. Thus, at high rates of light absorption the intensity of P-type delayed fluorescence increases at the expense of phosphorescence. Ultimately its *intensity* becomes proportional to I_a and the intensity of phosphorescence becomes proportional to $(I_a)^{\frac{1}{2}}$.

The conditions under which the second-order triplet decay becomes appreciable are discussed by Parker[1, 17] in terms of the values of I_a and the relevant rate constants. Moderate proportions of second-order decay produce comparatively little deviation of the decay curve of the delayed fluorescence from the exponential form, but simply increase the slope of the log plot so that an apparent decrease in triplet lifetime is observed with high intensities of exciting light. Such effects were observed in the early measurements with solutions of pyrene.[17] It is therefore important to measure the lifetimes of delayed fluorescence at two intensities of exciting light to confirm that the proportion of second-order triplet decay is not sufficient to introduce an error in the measured lifetime.

P-type delayed fluorescence in rigid medium

As the temperature of the solution is reduced and its viscosity increases the effect of increased triplet lifetime is opposed by the reduced value of k_i (which is often close to the diffusion-controlled value in fluid solution) and it is difficult to predict from equation (12) how the efficiency of delayed fluorescence will vary in the intermediate temperature range. At low temperatures in 'very rigid' solvents the efficiency is expected to be very low, and in fact it is found that rates of light absorption sufficient to give quite high efficiencies in fluid solution give negligible efficiencies in rigid media. If, however, the rate of light absorption is increased by using relatively high concentrations of solute, quite high efficiencies of delayed fluorescence are sometimes observed.

Somewhat conflicting results have been obtained by different workers in rigid media. This is perhaps not surprising because the physical properties of a frozen glass will depend critically on its composition, and it is not always possible to be sure that some of the luminescent solutes are retained completely in solution at low temperature with the high concentrations necessary to observe delayed fluorescence.

Muel[33] found that the intensity of the delayed fluorescence of benz(a)pyrene in ethanol at 77 °K was proportional to the square of the intensity of the exciting light and decayed exponentially with a lifetime equal to half that of the triplet. To explain the great increase in intensity on warming to −140 °C he suggested two possibilities: either the formation of a new long-lived dimeric state, or the successive transfer of energy from molecule to molecule until it is collected by a pair of molecules capable of interacting. He did not apparently consider the possibility that the increased emission could simply be the result of the increased value of k_i in equation (9) as the viscosity of the solvent decreased on warming.

Azumi & McGlynn[45, 46] observed delayed fluorescence from several aromatic and halogenated aromatic hydrocarbons in ether–pentane–ethanol glass at 77 °K and found that although the phosphorescence decayed exponentially (or nearly so) the delayed fluorescence could be represented by the sum of two first-order processes, the faster of which had a lifetime much less than one-half that of the phosphorescence. They assumed that direct resonance interaction between triplets occurred but did not explain what was the rate-controlling process responsible for the lifetime of the delayed fluorescence.

Dupuy[47] observed no delayed fluorescence from chrysene, naphthalene or phenanthrene in certain crystalline saturated hydrocarbon matrices, but found that the presence of a small quantity of benzene was sufficient to cause the emission of delayed fluorescence. He suggested that this was due to dimer formation resulting from concentration of the fluorescent hydrocarbon in the 'islets' of benzene formed by rejection of the benzene from the crystallising matrix. Dupuy's mechanism, like that of Azumi & McGlynn, assumed that diffusional transport in the glassy medium at 77 °K was negligible, and thus accounted for the absence of oxygen quenching. It did not, however, explain what is the rate-controlling process that gives rise to fluorescence emission of long lifetime.

Azumi & McGlynn[48] investigated the delayed fluorescence of pyrene in rigid media. In pure isopentane solvent at 77 °K they observed only phosphorescence. If small amounts of water were introduced into the solvent, the glasses formed were slightly foggy and both phosphorescence and delayed fluorescence were observed. At intermediate concentrations delayed fluorescence showed bands due both to monomer and dimer. The intensity of emission from the monomer was proportional to the square of the intensity of the exciting light while that from the dimer was proportional to the first power of the intensity of the exciting light. Azumi & McGlynn therefore concluded that the dimer emission was not produced by triplet-triplet interaction but was due to the formation of a *long-lived* excited dimer. Their experiments did not, however, rule out the possibility of the formation of micro crystals of pyrene in the wet solvent. The results of Ferguson[49] seem to be related to those of Azumi & McGlynn. Ferguson measured the *prompt* fluorescence emission from crystallised solutions of pyrene and other aromatic hydrocarbons in cyclohexane and observed both monomer and dimer emission. He assumed that the solute molecules took up adjacent sites in the lattice of the solvent crystals so that they were favourably situated for the formation of excited dimers.

It may be significant that the presence of excited dimers in rigid media seems to be associated with the introduction of two phases, e.g. by the addition of benzene in the experiments of Dupuy, or the addition of water in those of Azumi & McGlynn. The possibility has to be considered therefore that the formation of excited dimers in rigid media may be due to the formation of micro-crystals of solute. The *prompt* fluorescence of pyrene crystals is known to consist of dimer emission[50] and delayed fluorescence from crystals is probably associated with the annihilation of triplet excitons. If this were rapid compared with first-order triplet decay, the intensity of delayed fluorescence would be proportional to the first power of the intensity of exciting light as observed by Azumi & McGlynn[48] in wet isopentane glasses and by Parker[51] with deliberately crystallised solutions of phenanthrene.

Parker[51] investigated the delayed fluorescence from solutions of phenanthrene in various solvents at 77 °K. He found that concentrated solutions in clear ether–pentane–ethanol glass gave a delayed fluorescence spectrum identical with that of prompt fluorescence and having an intensity proportional to the square of the rate of light absorption. Fine crystalline suspensions of phenanthrene in iso-pentane at 77 °K also gave delayed fluorescence. The spectrum was identical with that of the prompt fluorescence of the crystals but was different from the spectrum obtained in the clear glass. Unlike that from the clear glass, the intensity was proportional to the first power of the intensity of the exciting light. Parker therefore concluded that the delayed fluorescence from the glass was not due to fine crystals but to phenanthrene held in true solution. In glasses of higher rigidity the intensity of delayed fluorescence was less, and in all glasses its decay was non-exponential (see fig. 8). The initial rate of

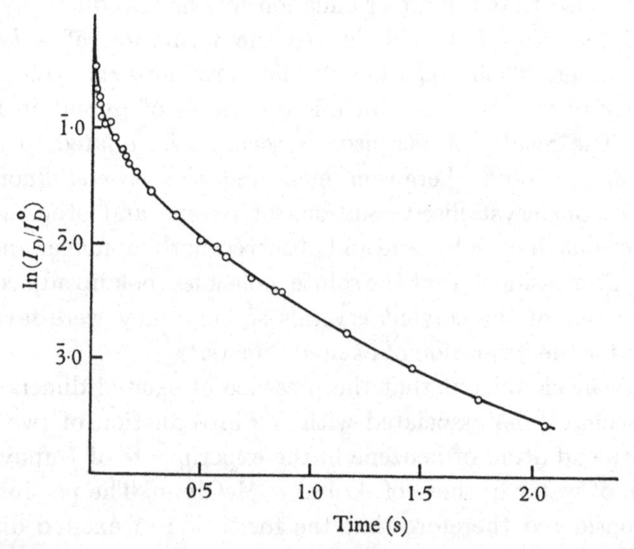

Fig. 8. Decay of delayed fluorescence in E.P.A. at 77 °K; rate of absorption of exciting light was 6×10^{-5} Einstein litre^{-1} s^{-1}. Similar curves were obtained in the more viscous media, but the initial intensity was lower. (From Parker[51].)

decay was much faster than the decay of phosphorescence which was exponential. When the wavelength of the exciting light was varied, the intensity of delayed fluorescence (I_D) was found to obey the relation

$$I_D \propto I_0^2 \epsilon_s, \tag{20}$$

where ϵ_s is the extinction coefficient of the ground state. A mechanism involving direct absorption by the triplet state was thus excluded because it would require

$$I_D \propto I_0^2 \epsilon_T. \tag{21}$$

By measuring the rate of growth of the delayed fluorescence when the exciting light was switched on after varying periods of darkness, Parker showed that the results could be interpreted in terms of a triplet-triplet interaction. He proposed two mechanisms. The first requires slow diffusive transport in the glassy media and the establishment of a non-random distribution of triplet molecules as a result of their interaction. The alternative mechanism attributes the delayed fluorescence to the interaction between those pairs of triplets initially formed at sites within the interaction distance, the delay in emission corresponding to the time taken for rotation into a favourable configuration for interaction. Quantitative treatment indicated that the two mechanisms treated separately would require triplet-triplet interaction distances of at least 30*A* and 15*A* respectively. He suggested that both mechanisms might operate simultaneously. The scheme satisfactorily explained the absence of oxygen quenching and could also be used to account for the results of Dupuy.

Recombination of delayed fluorescence

Before proceeding to consider sensitised delayed fluorescence in fluid solution, it will be as well to describe briefly the third type of delayed fluorescence listed in table 1, because this arises from recombination of photochemical products and could be confused with the *P*-type delayed fluorescence in rigid media described in the previous section.

The first detailed investigation of the photoreaction of organic compounds in rigid media was reported by Lewis & Lipkin[(52)] who showed that three kinds of primary reaction could occur, viz. (*a*) photodissociation, e.g. the formation of triphenylmethyl radicals from hexaphenylethane; (*b*) photo-oxidation, e.g. the formation of the diphenylamine radical ion from diphenylamine by ejection of an electron; and (*c*) photo-ionisation, e.g. dissociation of the diphenylamine radical ion to give a proton and the diphenyl nitrogen radical. Lewis & Kasha[(11)] proposed that such photochemical reactions in rigid media could be brought about by two distinct photochemical processes. With easily oxidisable compounds, absorption of a photon results in the immediate ejection of an electron and the rate of reaction is proportional to the first power of the intensity of exciting light. With other compounds the rate of reaction is proportional to the square of the intensity of exciting light and in this case the ejection of the electron results from the absorption of a photon by the triplet state which can build up to a high concentration in strongly irradiated glasses at low temperature. The occurrence of such biphotonic chemical reactions has recently been investigated in detail[53] and recombination of the fragments can give rise to luminescence. It had not always been clear from the observations of previous workers whether the primary processes involved were mono- or biphotonic.

One of the first investigations of the light emission resulting from recombination was that of Debye & Edwards.[54] They investigated the emission of extremely long lifetime ($> 10^2$ s) from irradiated solutions of easily oxidisable compounds such as phenol and toluidine in rigid glasses at 77 °K. They interpreted the non-exponential decay of luminescence in terms of photo-ionisation and subsequent diffusion of the trapped electrons back to the ionised molecules. They did not determine the spectral distribution of the long-lived emission.

Linschitz, Berry & Schweitzer[55] found that the absorption spectra of solutions of lithium in mixed amine glasses at low temperature showed intense peaks at 600 nm with a less strong background absorption extending into the infra red. Illumination with light of shorter wavelengths caused the intensity of the 600 nm band to decrease and the longer wavelength background to increase. They attributed the 600 nm band to strongly solvated electrons, i.e. to electrons in relatively deep traps, and the longer wavelength absorption to less strongly solvated electrons, i.e. to electrons in shallower traps. They then irradiated solutions of lithium diphenylamide, *N*-lithium carbazole, and other compounds, and were able to identify the spectra of both the radicals or radical ions, and the solvated electrons. Recombination at liquid nitrogen temperature was very slow, but, as the irradiated solution was allowed to warm up, luminescence appeared and the absorption bands of both radical and solvated electron decreased. This provided strong evidence that the luminescence was indeed due to the recombination of ion and electron. Of particular interest is the fact that in the experiments of Linschitz and co-workers, the spectrum of the luminescence was the same as that of triplet-singlet phosphorescence (i.e. it was recombination phosphorescence) and no luminescence corresponding to the singlet-singlet transition was observed. This was not due to a high rate of intersystem crossing since photo-excitation produced intense prompt fluorescence. This observation is qualitatively similar to the results obtained by Albrecht and co-workers[56] with *N*, *N'*-tetramethylparaphenylenediamine in hydrocarbon glasses at 77 °K. When they exposed a photo-ionised glass to infra red radiation they observed the emission of both recombination phosphorescence and recombination delayed fluorescence. The ratio of the intensities of the two recombination emissions was much greater than that observed for ordinary phosphorescence and prompt fluorescence, indicating that the recombination process favoured the direct population of the triplet state.

With acriflavine and related dyestuffs in ether-pentane-ethanol glasses, Lim and co-workers(57) observed the emission of delayed fluorescence at 77 °K lasting for some seconds. They also observed a transient absorption which they attributed to the positive radical ion formed by photo-ionisation. Except in the early stages, the delayed fluorescence decayed exponentially at a rate equal to that of the rate of decay of the radical ion. It was greater with flash excitation than with steady excitation. Because the integrated intensity of the delayed fluorescence, and the initial concentration of radical ion were directly proportional to the intensity of the exciting light, they concluded that the excitation was a one-photon process. The efficiency of the delayed fluorescence was greater with short wavelength excitation. The results were explained in terms of a model similar to that suggested earlier(56) for the luminescence of tetramethylparaphenylene diamine although the latter has now been re-interpreted in terms of a biphotonic process.(58)

A delayed fluorescence dependent on the wavelength of excitation has also been observed by Stevens & Walker(59) with perylene in liquid paraffin solutions at 77 °K. They found that the excitation spectrum of the delayed fluorescence coincided approximately with the triplet-triplet absorption continuum of perylene, and interpreted the results in terms of a biphotonic mechanism. They proposed that absorption of light by the triplet state leads to ionisation and that the slow recombination of positive ion and electron populates both the excited singlet and triplet states. They explained the linear relationship between the intensity of the delayed fluorescence and the intensity of the exciting light by assuming that the rate of unimolecular triplet decay was slow compared with its rate of photo-ionisation.

The observation by Parker & Joyce(60) of artifacts and trivial effects in the measurement of delayed fluorescence are of particular relevance in the investigation of recombination delayed fluorescence. They arise when a long-lived emission from one component of a solution (or even from the container itself) is situated in a spectral region absorbed by a second component. The long-lived emission from the first component will then excite *prompt* fluorescence of the second component and this prompt fluorescence will decay at a rate identical with that of the long-lived emission from the first component. Parker & Joyce observed such artifacts and trivial effects with solutions of perylene in fused quartz vessels that were phosphorescent at 77 °K under short wavelength excitation, and also with solutions of perylene in liquid paraffin containing phosphorescent impurities. Such artifacts and trivial effects are likely to be particularly important at low temperatures and the results of measurements of delayed fluorescence in these conditions must therefore be carefully scrutinised to ensure that artifacts or trivial effects are not responsible.

Porter and co-workers[53] have made a study of biphotonic photochemical processes resulting from the absorption of light by triplet states in rigid media at 77 °K. They observed two kinds of process with solutions of aromatic compounds in rigid aliphatic hydrocarbon glasses, namely: ionisation of the solute and sensitised dissociation of the solvent to give free radicals and hydrogen atoms. The latter then abstract hydrogen atoms from either the solvent or the solute and so produce solute radicals. With many of the irradiated solutions, exposure to infra red light, or gentle warming, resulted in the emission of both fluorescence and phosphorescence which was attributed to the recombination of ions. The phosphorescence/fluorescence ratio was higher than that observed in normal optical excitation and in this respect the results are similar to those of Albrecht and co-workers[56] which have since been re-interpreted by Cadogan & Albrecht[58] in terms of a biphotonic process.

Sensitised *P*-type delayed fluorescence

The process of energy transfer from a donor molecule (D) in the triplet state to an acceptor molecule (A) in the ground state, as represented by the equation

$$^3D + {}^1A \rightarrow {}^1D + {}^3A, \tag{22}$$

was first observed by Terenin & Ermolaev[61] in rigid media. It has since been investigated in fluid solution by various methods[15, 62–65] and it is now generally accepted that if the triplet level of the acceptor lies well below that of the donor, the transfer rate is diffusion-controlled with a rate constant k_c given approximately by

$$k_c = 8RT/3\,000\eta, \tag{23}$$

where η is the viscosity of the solvent. If the acceptor A is a compound that gives P-type delayed fluorescence when excited on its own by direct light absorption, it is to be expected that process (22) will result in the emission of sensitised P-type delayed fluorescence of the acceptor when a mixed solution is irradiated by light absorbed by the donor. Such sensitised delayed fluorescence in solution was first

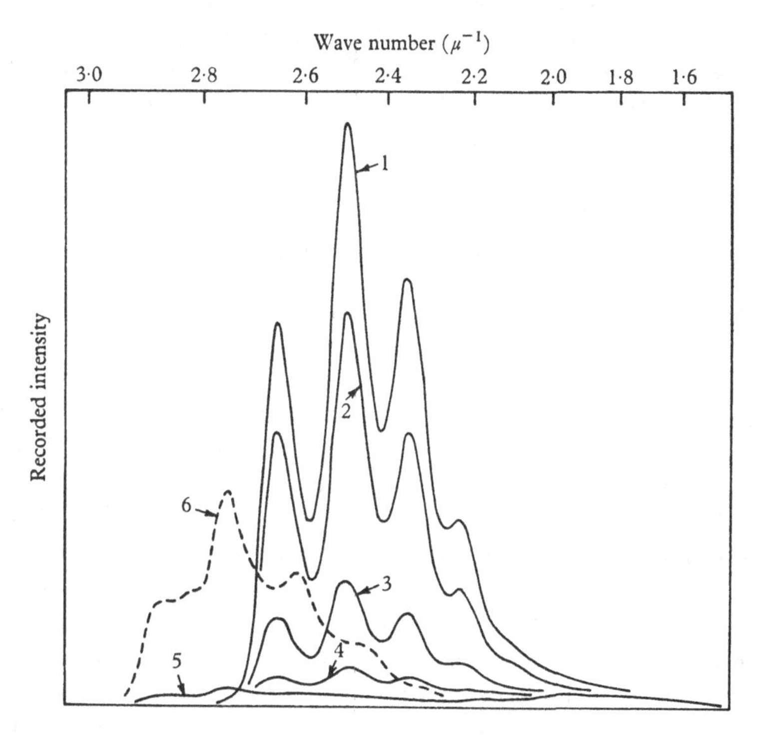

Fig. 9. Sensitised delayed fluorescence spectra of anthracene in 10^{-3} M phenanthrene solution. Intensity of exciting light was approximately $2{\cdot}7 \times 10^{-9}$ Einstein cm^{-2} s^{-1} at $3{\cdot}19\ \mu^{-1}$ (313 mμ). Delayed emission spectra with anthracene concentrations of: (1) 10^{-6} M, (2) 5×10^{-7} M, (3) 10^{-7} M, (4) 10^{-8} M, (5) 10^{-9} M. Curve (6) fluorescence emission spectrum of solution (1) at 260 times less sensitivity. (Not corrected for phosphorimeter factor.) (From Parker & Hatchard[30].)

observed[30] for the system phenanthrene (donor) with anthracene (acceptor) (see fig. 9), and has since been found in many other systems, some of which will be discussed later. It is noted here that with a typical triplet lifetime of $\sim 10^{-2}$ s (in fluid solution) and $k_c \sim 10^{10}$ litre mole^{-1} s^{-1}, the concentration of acceptor required to quench the donor triplet by 50 % is $\sim 10^{-8}$ M. Since the rate of light absorption by the donor can be made large, and about half of the resulting triplet energy is transferred to the acceptor at a concentration of 10^{-8} M, the sensitised delayed fluorescence from the latter is quite strong and its observation provides a sensitive method for detecting the presence of the acceptor at concentrations of 10^{-9} M or less.

In general, there are two mechanisms by which the sensitised delayed fluorescence can be produced. Both require for their initial stages the population of the triplet level of the donor by light absorption, followed by triplet to singlet energy transfer (equation (22)) with the formation of acceptor triplets. The first of the two mechanisms then proceeds by interaction between two acceptor triplets

$$^3A + ^3A \rightarrow ^1A^* + ^1A, \qquad (24)$$

and the second by interaction between unlike triplets

$$^3A + ^3D \rightarrow ^1A^* + ^1D. \qquad (25)$$

There is also evidence[66] that the process

$$^3A + ^3D \rightarrow ^1A + ^1D^* \qquad (26)$$

can also take place in some systems. The occurrence of process (25) was first demonstrated by Parker[67] with the system anthracene (donor) and naphthacene (acceptor). The triplet energy of naphthacene is slightly less than one-half that of its excited singlet and hence process (24) occurs only with great difficulty at room temperature and delayed fluorescence from solutions of naphthacene alone is extremely weak. However, process (25) is still energetically possible and the excitation of anthracene in solutions containing small concentrations of naphthacene causes the emission of sensitised delayed fluorescence of the latter (see fig. 10). As the concentration of naphthacene is increased the intensity of its delayed fluorescence at first increases, passes through a maximum and then decreases (see curve 2 in fig. 11). The decrease is caused by quenching of the anthracene triplet (curve 1 in fig. 11) via process (22) and the latter thus competes with (25) for the anthracene triplets.

By choosing as a donor a compound whose lowest excited singlet state lies below that of the acceptor, but whose triplet state lies above that of the acceptor (see fig. 12) it is possible to arrange for the acceptor to emit delayed fluorescence having a wavelength *shorter* than that of the exciting light. Such *sensitised anti-Stokes delayed fluorescence* was first observed[68] with the systems phenanthrene–naphthalene and proflavine–anthracene. A striking example is shown in fig. 13, which

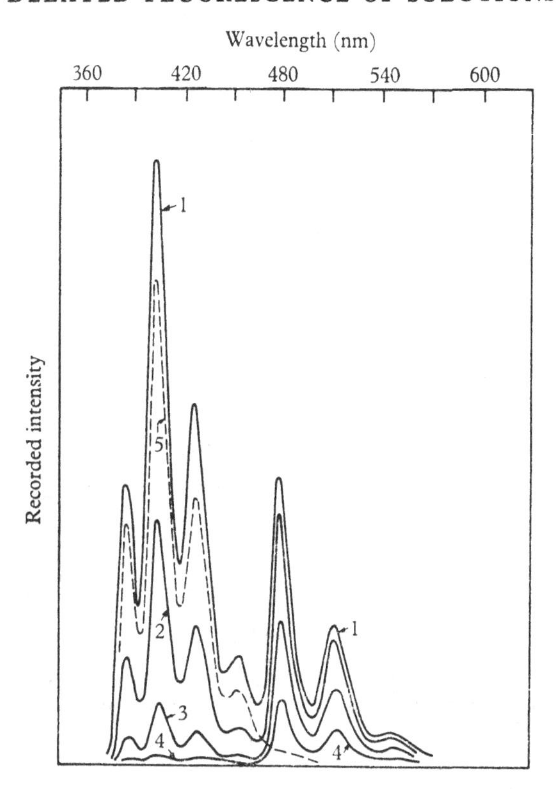

Fig. 10. Delayed fluorescence spectra at 20 °C. All solutions contained 5×10^{-5} M anthracene in ethanol. Rate of light absorption was $0{\cdot}4 \times 10^{-5}$ Einstein litre^{-1} s^{-1} at $2{\cdot}73\ \mu m^{-1}$ (366 nm). Delayed fluorescence at naphthacene concentrations of (1) 4×10^{-8} M, (2) 8×10^{-8} M, (3) $1{\cdot}5 \times 10^{-7}$ M, (4) 3×10^{-7} M. (5) Normal fluorescence of 4 at 100 times lower sensitivity (not corrected for phosphorimeter factor). (From Parker[67].)

refers to the excitation of the donor, eosin, with green light (546 nm) and the emission of sensitised delayed fluorescence of the acceptor, anthracene (curve 2), the shortest wavelengths of which were in the ultraviolet region at ~380 nm. The P-type delayed fluorescence of the eosin at ~550 nm is greater in the presence of anthracene (curve 2) than in its absence (curve 1), although the phosphorescence of the eosin at ~690 nm is quenched by the presence of anthracene. This is an example of *mutual sensitisation* of delayed fluorescence[66] by processes (25) and (26). It should be noted that dyestuffs such as eosin or proflavine[18] show both E-type and P-type delayed fluorescence, the former predominating at high temperatures and the latter at low temperatures, so long as the solution remains fluid.

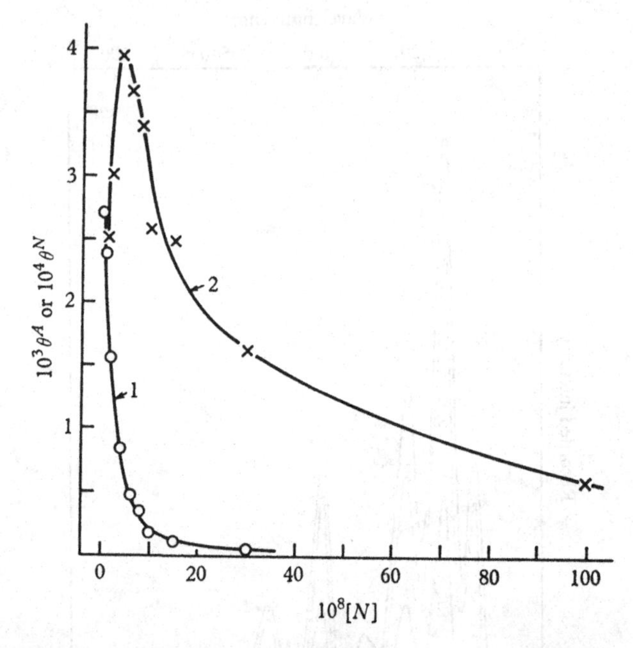

Fig. 11. Variation of delayed fluorescence efficiencies with concentration of naphthacene. Delayed fluorescence efficiency of: (1) anthracene (θ^A), and (2) naphthacene (θ^N). (From Parker[67].)

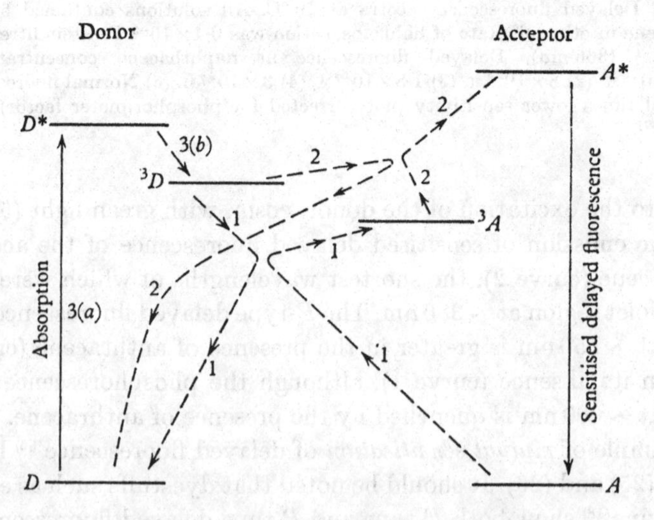

Fig. 12. Diagram of sensitised delayed fluorescence by the mixed-triplet mechanism.

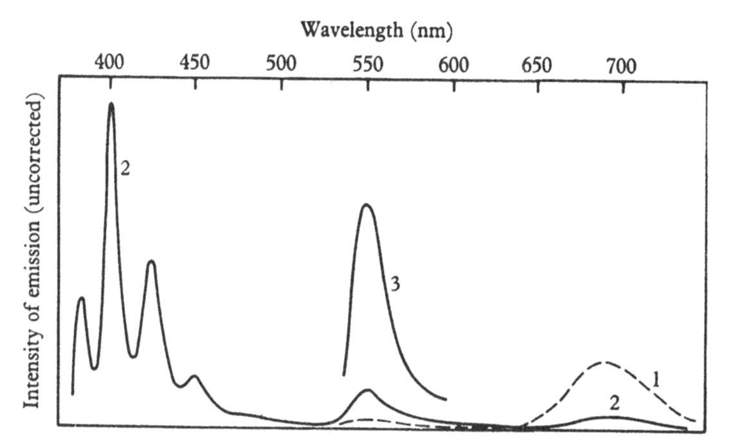

Fig. 13. Mutual sensitisation of delayed fluorescence with eosin as donor. Delayed emission spectra of 5×10^{-6} M eosin di-sodium salt in ethanol at -75 °C with (1) no addition, and (2) 10^{-4} M anthracene. (3) Prompt fluorescence emission spectrum of (1) and (2) at 3000 times lower sensitivity. Rate of light absorption was 1×10^{-5} Einstein litre^{-1} s^{-1}. (From Parker, Hatchard & Joyce[66].)

Applications of delayed fluorescence measurement

(*a*) *Investigation of triplet lifetimes in solutions*

Measurement of the lifetime of either *E*-type or *P*-type delayed fluorescence provides a convenient method for investigating the triplet lifetime in fluid solution under conditions where the phosphorescence is too weak to be observed. It has the advantage over flash absorption spectroscopy that relatively low triplet concentrations can be observed and the difficulties associated with the separation of first-order and second-order triplet decay processes can thus be circumvented. For example, Stevens & Walker[37] have investigated the decay of aromatic hydrocarbon triplets in liquid paraffin by this method, and Parker & Hatchard[69] have applied it to the investigation of the quenching of triplet benz(a)pyrene by cationic polymer and to other problems described in subsections (*c*) and (*d*) below.

(*b*) *Determination of triplet energies*

Both *E*-type and *P*-type delayed fluorescence can be used to determine triplet energies with fluorescent compounds for which phosphorescence cannot be observed directly. Parker & Joyce[19, 20] determined the ΔE value of chlorophylls *a* and *b* by measuring the intensity and lifetime of the *E*-type delayed fluorescence in propylene glycol and applying equation (8). They determined approximate limits for the triplet energy of perylene[70] by the observations that: (i) perylene gives rise to weak P-type delayed fluorescence by direct excitation and hence its triplet energy must be at least one-half of that of the excited singlet, i.e. $\geqslant 1{\cdot}14\,\mu\text{m}^{-1}$, and (ii) its P-type delayed fluorescence is strongly sensitised by anthracene and hence its triplet energy must be $< 1{\cdot}48\,\mu\text{m}^{-1}$. In general, the triplet energy of a compound may be located by reference to two other 'test' compounds with known triplet energies. If the compound under investigation can act as a donor to one of the 'test' compounds and as acceptor to the other, its triplet energy must lie between those of the two test compounds.[66]

(c) *Probability of triplet to triplet energy transfer*

Equation (12) may be written in the form

$$\theta/\phi_f = \tfrac{1}{2}pk_c I_a(\phi_t\tau)^2, \tag{27}$$

where k_c is the diffusion-controlled rate constant (equation (23)) and p is the probability that an encounter between triplets will ultimately give rise to a singlet excited molecule. Thus by measuring θ/ϕ_f and τ at a known value of I_a, the value of $(\sqrt{p})\,\phi_t$ may be calculated.[18] If the value of ϕ_t can be determined by an independent method (see below) the value of p can be derived. Some values of p are shown in table 3. The data in table 3 refer to dilute solutions of compounds for which the formation of excited dimer is small or negligible. Let us therefore re-write equation (9), followed by (10), (11) or (14) as follows:

$$^3A + {}^3A \rightarrow {}^1A + {}^1A, \tag{28}$$

$$\rightarrow {}^1A^* + {}^1A, \tag{29}$$

$$\rightarrow {}^5A + {}^1A. \tag{30}$$

Table 3. *Probability of triplet to triplet energy transfer*

Compound	ϕ_f	ϕ_t	$(\sqrt{p})\phi_t$	p
Naphthalene	0·21	0·71	0·53	0·56
Acenaphthene	0·39	0·45	0·25	0·31
Pyrene	0·72	0·27	0·14	0·27
Anthracene	0·30	0·70	0·22	0·10
1,2-benzanthracene	0·20	0·80†	0·37	0·21†
Benz(a)pyrene	0·42	0·58†	0·09	0·02†
Fluoranthene	0·21	0·79†	0·16	0·04†
Proflavine	0·40	0·60†	0·006	0·0001†
Acridine 0	0·46	0·54†	0·006	0·0001†

† Calculated by assuming $\phi_t = (1-\phi_f)$. Corresponding value of p is therefore a minimum value.

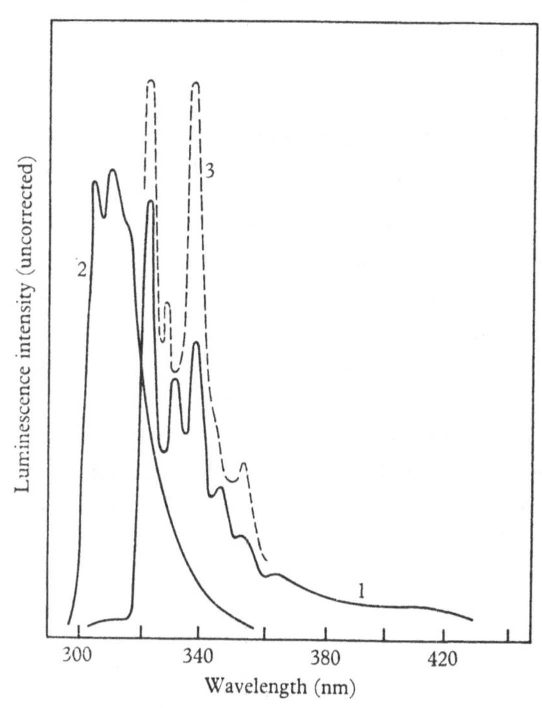

Fig. 14. Curves of sensitised delayed fluorescence of impurity in fluorene. Solution, 5×10^{-5} M fluorene in ethanol at -75 °C. Curve 1, delayed fluorescence excited by 302 nm (3·31 μm^{-1}) radiation at a rate of absorption of light of $0{\cdot}9 \times 10^{-6}$ Einstein litre^{-1} s^{-1}; curve 2, prompt fluorescence at 2000 times lower sensitivity; curve 3, prompt fluorescence of acenaphthene solution. (From Parker, Hatchard & Joyce[73].)

To these should be added the possibility that the two molecules, after interaction, separate again as triplets, viz.

$$^3A + {}^3A \rightarrow {}^3A + {}^3A. \tag{31}$$

Thus, the probability p is given by

$$p = k_{29}/(k_{28} + k_{29} + k_{30} + k_{31}). \tag{32}$$

With some compounds (e.g. naphthalene) p is high. This can be interpreted in two ways. If the process of triplet to triplet energy transfer (equation (29)) does indeed require an actual encounter, the probability that it occurs at each encounter must be high. Alternatively, the high value of p may simply indicate that triplet energy is transferred across distances greater than that required for an encounter as defined by equation (23), and that one or more of the other processes ((28), (30) and (31)) occur with higher probability. It is possible that equation (30) may also contribute to delayed fluorescence by

$$^5A \rightarrow {}^1A^*. \tag{33}$$

(d) Determination of triplet formation efficiencies

The intensity of sensitised delayed fluorescence from a solution containing a sufficient concentration of acceptor to quench the donor triplet completely (by process (22)) is analogous to equation (27), viz.

$$\theta_s/\phi_f = \tfrac{1}{2}pk_c I_a(p_e\phi_t^D\tau)^2, \tag{34}$$

where θ_s and ϕ_f are the efficiencies of sensitised delayed fluorescence and prompt fluorescence of the acceptor, I_a is now the rate of light absorption by the *donor*, ϕ_t^D is the triplet formation efficiency of the *donor*, τ is the lifetime of the *acceptor* triplet and p_e is the proportion of quenching encounters between donor triplet and acceptor ground state that give rise to acceptor triplets by process (22). Hence the relationship between the intensities of sensitised delayed fluorescence (I_{DF}) emitted by two solutions containing the same high concentration of acceptor but different donors, measured at the same rate of light absorption, is:

$$\frac{(I_{DF})_1}{(I_{DF})_2} = \left[\frac{(p_e\phi_t^D\tau)_1}{(p_e\phi_t^D\tau)_2}\right]^2. \tag{35}$$

Thus if it is assumed that p_e is the same for both donors (it is generally assumed to be unity), the ratio $(\phi_t^D)_1/(\phi_t^D)_2$ can be simply determined. Parker & Joyce[71, 72] have used this method to determine the triplet

Table 4. *Triplet formation efficiencies in ethanol at* 21 °C

Compound	ϕ_t	ϕ_f	$(\phi_t+\phi_f)$
Anthracene	[0·70]	[0·30]	—
Naphthalene	0·71	0·21	0·92
Phenanthrene	0·80	0·13	0·93
Triphenylene	0·89	0·09	0·98
Chrysene	0·82	0·17	0·99
Acenaphthene	0·45	0·39	0·84
1-Methoxynaphthalene	0·46	0·53	0·99
Pyrene	0·27	0·72	0·99

formation efficiencies of the compounds shown in table 4 by assuming a value of 0·7 for anthracene. It will be observed that $(\theta_t+\phi_f) \sim 1{\cdot}0$, suggesting that direct internal conversion to the ground state (k_n in fig. 1) is small.

If the compound to be investigated has a low triplet energy, so that it cannot conveniently be used as a donor, a somewhat different procedure may be employed. The value of $(\surd p)\,\phi_t$ is determined as described in (*c*) above by the measurement of directly excited delayed fluorescence. The value of $(\surd p)\,\phi_t^D$ is determined by measuring the intensity of delayed fluorescence sensitised by a donor of known triplet formation efficiency (ϕ_t^D). From these results the value of both p and ϕ_t can be calculated. Parker & Joyce have used this method to determine the triplet formation efficiencies of perylene[70] and chlorophylls a and b.[19, 20]

(e) *Organic trace analysis*

Sensitised delayed fluorescence can be observed with acceptor concentrations as low as 10^{-9} M and the technique thus has considerable potential as a method of trace analysis. It provides a valuable criterion

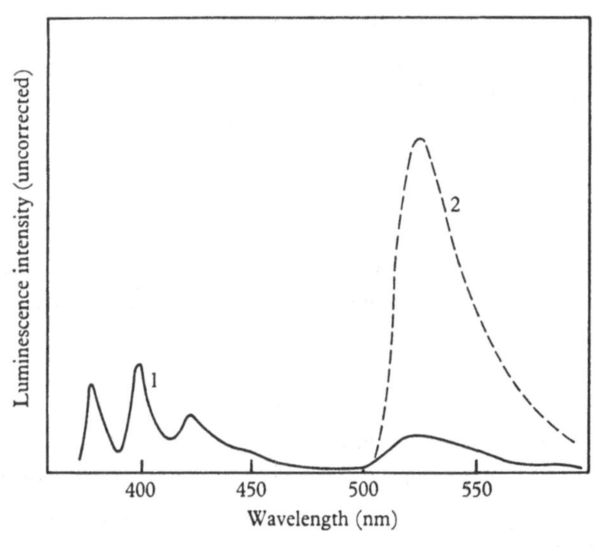

Fig. 15. Curves showing detection of anthracene in impure pyrene. Solution, 10^{-2} M pyrene, 10^{-6} M 1,2-benzanthracene and 10^{-6} M anthracene with $1{\cdot}7 \times 10^{-5}$ M acridine orange hydrochloride as sensitiser at 22 °C. Curve 1, delayed fluorescence excited by 436 nm ($2{\cdot}29\ \mu m^{-1}$) radiation at a rate of absorption of light of 6×10^{-6} Einstein litre^{-1} s^{-1}; curve 2, prompt fluorescence at 4000 times lower sensitivity. (From Parker, Hatchard & Joyce[73].)

of the purity of compounds to be used in photochemical investigations. Thus, if a moderately concentrated solution of the material is found to give a delayed fluorescence spectrum identical with that of its prompt fluorescence, it can be safely assumed that the concentration of impurities capable of quenching the triplet state of the substance is low. The presence of such impurities is often indicated by the appearance of their characteristic sensitised delayed fluorescence, e.g. traces of anthracene in 'pure' phenanthrene or carbazole. An example of a system in which the observation of the delayed fluorescence spectrum indicated the presence of an unsuspected impurity is shown in fig. 14. This refers to a specimen of fluorene that had been recrystallised and subjected to exhaustive zone refining. It gave the prompt fluorescence spectrum of fluorene but a different delayed fluorescence spectrum that closely resembled that of acenaphthene and was probably due to a simple derivative of the latter. Parker, Hatchard & Joyce[73] have shown that the delayed fluorescence of one component of a mixture of fluorescent compounds can be selectively sensitised by the choice of a donor having the appropriate triplet energy. The application of this principle to the detection of 0·01 % of anthracene in an impure specimen of pyrene is illustrated by the spectra shown in fig. 15. Because of the sensitivity of triplet molecules in solution to quenching by impurities, the purity of all solvents and solutes used for investigations of delayed fluorescence have to be carefully checked. This problem has been discussed by Parker.[1] Some solutes are particularly difficult to purify, even by exhaustive zone refining. The measurement of the prompt and delayed fluorescence in solution, and prompt fluorescence and phosphorescence at low temperature in rigid medium, provide sensitive methods for the identification and determination of some impurities.[73, 74, 75]

(*f*) *Measurement of photochemical reaction products*

If the product of a photochemical reaction is fluorescent and has a lower triplet energy than the reactant, the delayed fluorescence of the product will be sensitised by the reactant and the observation of this sensitised delayed fluorescence provides an extremely delicate method of detecting the occurrence of the photochemical reaction. Parker & Hatchard[69] have used this principle to investigate the polymer-catalysed photo-reaction of benz(a)pyrene solutions. Triplet to triplet energy transfer may itself result in photochemical reaction. This seems to be the case in the system anthracene–naphthacene[67] where the consumption of naphthacene may be followed by the observation of its sensitised delayed fluorescence.

References

1 C. A. Parker, in *Advances in Photochemistry*, **2**, ed. W. A. Noyes Jr., G. S. Hammond and J. N. Pitts Jr. (Interscience Publishers Ltd., London, 1964).
2 E. Wiedemann, *Ann. Physik*, **34**, 446 (1888).

3 E. Tiede, *Chem. Ber.* **53**, 2214 (1920).
4 P. Pringsheim, *Fluorescence and Phosphorescence* (Interscience Publishers Ltd., London, 1949).
5 F. Perrin, *Ann. Phys. (Paris)*, **12**, 169 (1929).
6 A. Jablonski, *Nature*, **131**, 839 (1933).
7 A. Jablonski *Z. Physik*, **94**, 38 (1935).
8 W. L. Levshin & L. A. Vinokurov, *Phys. Z. Sow.* **10**, 10 (1936).
9 G. N. Lewis, D. Lipkin & T. T. Magel, *J. Am. Chem. Soc.* **63**, 3005 (1941).
10 P. Pringsheim & H. Vogels, *J. Chim. Phys.* **33**, 345 (1936).
11 G. N. Lewis & M. Kasha, *J. Am. Chem. Soc.* **66**, 2100 (1944).
12 S. Boudin, *J. Chim. Phys.* **27**, 285 (1930).
13 C. A. Parker & C. G. Hatchard, *Trans. Faraday Soc.* **57**, 1894 (1961).
14 H. Kautsky, *Chem. Ber.* **68**, 152 (1935).
15 H. L. J. Bäckström & K. Sandros, *Acta Chem. Scand.* **12**, 823 (1958).
16 C. A. Parker & C. G. Hatchard, *J. Phys. Chem.* **66**, 2506 (1962).
17 C. A. Parker & C. G. Hatchard, *Trans. Faraday Soc.* **59**, 284 (1963).
18 C. A. Parker, C. G. Hatchard & T. A. Joyce, *J. Mol. Spectr.* **14**, 311 (1964).
19 C. A. Parker & T. A. Joyce, *Nature*, **210**, 701 (1966).
20 C. A. Parker & T. A. Joyce (to be published).
21 P. P. Dikun, *Zhur. Eksp. i Teoret. Fiz.* **20**, 193 (1950).
22 R. Williams, *J. Chem. Phys.* **28**, 577 (1958).
23 B. Stevens, E. Hutton & G. Porter, *Nature*, **185**, 917 (1960).
24 Th. Förster & K. Kasper, *Z. Elektrochem.* **59**, 977 (1955).
25 B. Stevens & E. Hutton, *Nature*, **186**, 1045 (1960).
26 B. Stevens & E. Hutton, *Spectrochim. Acta*, **18**, 425 (1962).
27 C. A. Parker & C. G. Hatchard, *Nature*, **190**, 165 (1961).
28 J. B. Birks & I. H. Munro, in *Luminescence of Organic and Inorganic Materials*, H. P. Kallmann and G. M. Spruch, eds. (John Wiley, New York, 1962), p. 230.
29 C. A. Parker & C. G. Hatchard, *Proc. Chem. Soc.* p. 147 (1962).
30 C. A. Parker & C. G. Hatchard, *Proc. Roy. Soc. (London)* A, **269**, 574 (1962).
31 B. Stevens, M. S. Walker & E. Hutton, *Proc. Chem. Soc.* p. 62 (1963).
32 S. Czarnecki, *Bull. Acad. Pol. Sci.* **9**, 561 (1961).
33 B. Muel, *C.R. Acad. Sci. (Paris)* **255**, 3149 (1962).
34 M. A. El-Sayed, M. T. Wauk & G. W. Robinson, *Mol. Physics*, **5**, 205 (1962).
35 R. W. Brandon, R. E. Gerkin & C. A. Hutchison Jr., *J. Chem. Phys.* **37**, 447 (1962).
36 B. Stevens & M. S. Walker, *Proc. Chem. Soc.* p. 181, (1963).
37 B. Stevens & M. S. Walker, *Proc. Roy. Soc. (London)* A, **281**, 420 (1964).
38 C. A. Parker, *Spectrochim. Acta*, **19**, 989 (1963).
39 E. Döller & Th. Förster, *Z. Physik. Chem., Ne.F. (Frankfurt)*, **31**, 274 (1962).

40 C. Tanaka, J. Tanaka, E. Hutton & B. Stevens, *Nature*, **198**, 1192 (1963).
41 C. A. Parker, *Nature*, **200**, 331 (1963).
42 J. B. Birks, *J. Phys. Chem.* **67**, 1299 (1963).
43 J. B. Birks, G. F. Moore & I. H. Munro, *Spectrochim. Acta*, **22**, 323 (1966).
44 C. A. Parker, *Spectrochim. Acta* (in the Press).
45 T. Azumi & S. P. McGlynn, *J. Chem. Phys.* **38**, 2773 (1963).
46 T. Azumi & S. P. McGlynn, *J. Chem. Phys.* **39**, 1186 (1963).
47 F. Dupuy, Thesis (University of Bordeaux, 1964).
48 T. Azumi & S. P. McGlynn, *J. Chem. Phys.* **39**, 3533 (1963).
49 J. Ferguson, *J. Chem. Phys.* **43**, 306 (1965).
50 J. Ferguson, *J. Chem. Phys.* **28**, 765 (1958).
51 C. A. Parker, *Trans. Faraday Soc.* **60**, 1998 (1964).
52 G. N. Lewis & D. Lipkin, *J. Am. Chem. Soc.* **64**, 2801 (1942).
53 B. Brocklehurst, W. A. Gibbons, F. T. Lang, G. Porter & M. I. Savadatti, *Trans. Faraday Soc.* **62**, 1793 (1966).
54 P. Debye & J. O. Edwards, *J. Chem. Phys.* **20**, 236 (1952).
55 H. Linschitz, M. G. Berry & D. Schweitzer, *J. Am. Chem. Soc.* **76**, 5833 (1954).
56 W. C. Meyer & A. C. Albrecht, *J. Phys. Chem.* **66**, 1168 (1962); E. Dolan & A. C. Albrecht, *J. Chem. Phys.* **37**, 1149 (1962); **38**, 567 (1963); W. M. McClain & A. C. Albrecht, *J. Chem. Phys.* **43**, 465 (1965).
57 E. C. Lim & G. W. Swenson, *J. Chem. Phys.* **36**, 118 (1962); **39**, 2768 (1963); E. C. Lim & W.-Y. Wen, *J. Chem. Phys.* **39**, 847 (1963); E. C. Lim, C. P. Lazzara, M. Y. Yang & G. W. Swenson, *J. Chem. Phys.* **43**, 970 (1965).
58 K. D. Cadogan & A. C. Albrecht, *J. Chem. Phys.* **43**, 2550 (1965).
59 B. Stevens & M. S. Walker, *Chem. Comm.* p. 8 (1965).
60 C. A. Parker & T. A. Joyce, *J. Chem. Soc.* A, p. 821 (1966).
61 A. N. Terenin & V. L. Ermolaev, *Trans. Faraday Soc.* **52**, 1042 (1956).
62 H. L. J. Bäckström & K. Sandros, *Acta Chem. Scand.* **14**, 48 (1960).
63 G. Porter & F. Wilkinson, *Proc. Roy. Soc.* (*London*) A, **264**, 1 (1961).
64 H. L. J. Bäckström & K. Sandros, *Acta Chem. Scand.* **16**, 958 (1962).
65 F. Wilkinson, *J. Phys. Chem.* **66**, 2569 (1962).
66 C. A. Parker, C. G. Hatchard & T. A. Joyce, *Nature*, **205**, 1282 (1965).
67 C. A. Parker, *Proc. Roy. Soc.* (*London*) A, **276**, 125 (1963).
68 C. A. Parker & C. G. Hatchard, *Proc. Chem. Soc.* p. 386 (1962).
69 C. A. Parker & C. G. Hatchard, *J. Photochem. and Photobiol.* (in the Press).
70 C. A. Parker & T. A. Joyce, *Chem. Comm.* p. 108 (1966).
71 C. A. Parker & T. A. Joyce, *Chem. Comm.* p. 234 (1966).
72 C. A. Parker & T. A. Joyce, *Trans. Faraday Soc.* (in the Press).
73 C. A. Parker, C. G. Hatchard & T. A. Joyce, *Analyst*, **90**, 1 (1965).
74 C. A. Parker, *Proceedings of the S.A.C. Conference, Nottingham* (W. Heffer and Sons Ltd., Cambridge, 1965), p. 208.
75 C. A. Parker, *Chem. in Britain*, **2**, 160 (1966).

THE KINETICS OF THE EXCITED STATES OF ANTHRACENE AND PHENANTHRENE VAPOR

G. L. POWELL, R. C. JARNIGAN AND M. SILVER

Abstract

The intersystem crossing quantum yield $\phi_{ST} = k_{ST}/k$ has been determined in anthracene and phenanthrene using the delayed fluorescence techniques introduced by Zahlan *et al.*[1] The original analysis has been extended to include the first-order triplet decay (β), non-radiative triplet-triplet annihilation (γ_2), as well as radiative triplet-triplet annihilation (γ_1). The lifetime of the excited singlet ($1/k$) has also been determined using a nanosecond lamp for excitation, a fast photomultiplier and a pulse sampling oscilloscope to observe the fluorescence decay.

The triplet state decays with a lifetime given by $\tau_T = 1/(\beta+\gamma T_0)$ where γ is the observed bimolecular triplet decay constant given by

$$\gamma = (1-\tfrac{1}{2}\phi_{ST})\gamma_1+\gamma_2.$$

From this ϕ_{ST} can be expressed as $\phi_{ST} = Q(1+\gamma_2/\gamma_1)$ where Q is an experimentally determined quantity and is the lower limit of ϕ_{ST}. A Strickler–Berg calculation was used to determine the radiative decay constant, k_f, and from our value of k the fluorescence quantum yield, ϕ_f. Since the sum of all quantum yields must be unity, then

$$1-\phi_f-Q = Q\gamma_2/(\gamma_1)+\phi_{IC}$$

(ϕ_{IC} is the quantum yield for internal conversion of the first excited singlet state directly to the ground state). Thus, upper limits can be placed on γ_2/γ_1 and ϕ_{IC}.

For phenanthrene, excitation in the first singlet band gives $\tau = 50$ ns, $Q = 0{\cdot}88$, $\phi_f = 0{\cdot}07$, $\gamma_2/\gamma_1 < 0{\cdot}05$, $\phi_{IC} < 0{\cdot}05$, and $\beta = 0{\cdot}30\times10^3$ s^{-1}. Excitation in the second singlet band gives $\tau = 45$ ns. All these quantities are independent of pressure. For anthracene at a pressure of one torr, excitation in the first singlet band gives $\tau = 6$ ns, $Q = 0{\cdot}15$, $\phi_f = 0{\cdot}20$, and $\beta = 1{\cdot}0\times10^3$ s^{-1}. At higher pressures, self quenching with unit efficiency upon collision was observed for the first excited singlet state consistent with the results of Ware & Cunningham.[2] Relative fluorescence quantum yields indicated that k_f is not pressure dependent. Since triplet-triplet annihilation and self quenching of the singlet state are energetically similar, it is reasonable to assume that γ_2/γ_1 is appreciable and at low pressures ϕ_{ST} may be as large as $0{\cdot}80$.

References

1 A. B. Zahlan, S. Z. Weisz, R. C. Jarnigan & M. Silver, *J. Chem. Phys.* **42**, 4244 (1965).
2 W. R. Ware & P. T. Cunningham, *J. Chem. Phys.* **43**, 3826 (1965).

OPTICAL INVESTIGATIONS OF THE TRIPLET STATE OF NAPHTHALENE IN DIFFERENT CRYSTALLINE ENVIRONMENTS

H. PORT AND H. C. WOLF

Abstract

The spectra of phosphorescence and delayed fluorescence were measured as a function of temperature for the following systems: naphthalene-h_8 in durene, naphthalene-d_8 in durene, naphthalene in β-methyl-naphthalene, naphthalene containing durene or thionaphthene and naphthalene-h_8 in naphthalene-d_8. There is evidence for emission from disturbed exciton states of the host in some cases. Trap-trap and guest-guest annihilation is responsible for delayed fluorescence. The thermal and the spectroscopic depth are in good agreement. The mechanism of trap to trap energy transfer is discussed.

Introduction

The lowest triplet state of organic molecules like naphthalene is very sensitive to perturbations by the environment. The lifetime changes considerably with temperature, and the concentration dependence of lifetime and intensity is very pronounced. It is well established now, that the actual (non-radiative) lifetime of the lowest triplet state in crystals like anthracene and naphthalene is, by a factor of 100 or 1 000, shorter than the lifetime of the molecules isolated as guest in a similar host crystal.

In order to get a better understanding of the interactions in the solid state which affect the triplet state of naphthalene, we investigated the phosphorescence and the delayed fluorescence spectra of naphthalene in different crystalline environments. We give here a preliminary report on some spectroscopic results. A more detailed paper including the decay time measurements is in preparation.[1]

Experiments

All the measurements were made at low temperatures, between 1·6 and 40 °K, using photo-electric registration of the emission spectra with a spectral resolution of approximately 5 cm^{-1}. In order to separate phosphorescence and delayed fluorescence from prompt fluorescence and reflected exciting light we used sectors rotating with a constant frequency of 50 Hz. Therefore we were unable to measure exact intensity of delayed emission with lifetime shorter than 50 ms. All specimens were single crystals. For excitation we used a mercury lamp (Osram HBO 200) with filters suitable to excite the first singlet state of naphthalene.

Naphthalene as guest in durene

We remeasured the phosphorescence spectrum of naphthalene-h_8 and naphthalene-d_8 in durene. The spectra start with 0·0 at 21 355 cm^{-1} (h_8) and 21 452 cm^{-1} (d_8) respectively. The most intense vibronic lines can be analysed using mostly totally symmetric vibrations of the naphthalene molecule. All the lines have superimposed a very characteristic phonon structure.[2] The measured lifetimes are $\tau_p = 2{\cdot}7$ s (h_8) and $\tau_p = 18{\cdot}5$ s (d_8), both at 4·2 °K.

β-Methyl-naphthalene as guest in naphthalene

Naphthalene crystals containing approximately 0·1 % of β-methyl-naphthalene show the β-methyl-naphthalene emission only with 0·0 at 20 967 cm^{-1}. Using the measured value for the energy of the triplet exciton band in naphthalene crystals[3] the triplet trap depth of the β-methyl-naphthalene guests in the naphthalene host is 240 cm^{-1}.

The temperature dependence of the phosphorescence intensity (I_p) and the delayed fluorescence intensity (I_{DF}) is given in fig. 1. If the delayed fluorescence originates from thermally activated guest-guest triplet annihilation one expects the following temperature dependence:[4]

$$I_{DF} \simeq N_T^2 \exp(-\Delta E/kT).$$

N_T is the concentration of β-methyl-naphthalene in the triplet state. Since I_p is proportional to N_T, an Arrhenius plot of $\log I_{DF}/I_p^2$ *vs* $1/T$ should give the activation energy ΔE. The experimental value (fig. 1) is $\Delta E = 240 \pm 40$ cm^{-1}. Within the limits of error, this value is equal to the spectroscopic trap depth. Therefore we suggest the following mechanism for delayed fluorescence in this system. Guest triplet states are thermally activated into the host exciton band. They are trapped again at other excited guest molecules where they cause fluorescence by the well-known annihilation process (guest-guest triplet annihilation).

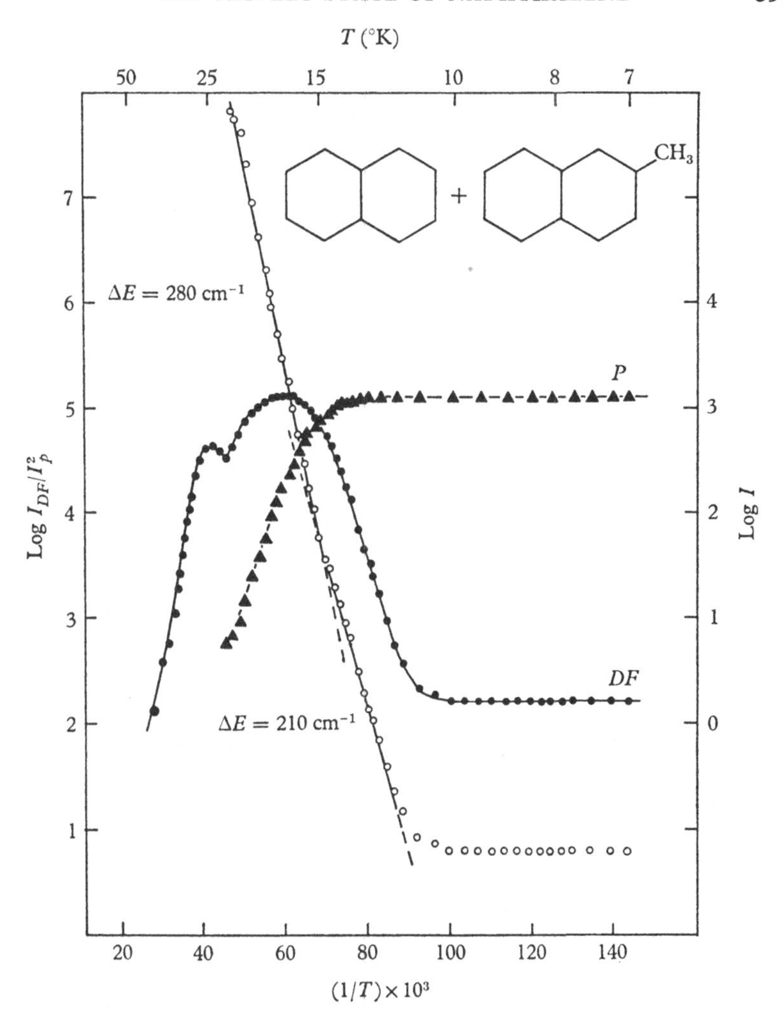

Fig. 1. Intensity of phosphorescence I_p and delayed fluorescence I_{DF} (right ordinate) and ratio I_{DF}/I_p^2 (left ordinate) versus temperature for 0·1% β-methyl-naphthalene in naphthalene.

X-traps in naphthalene

The name X-traps has been suggested for disturbed exciton states of the host crystal.[5] Impurity molecules like thionaphthene cannot act as traps for the energy in the lowest singlet exciton band of naphthalene themselves, but naphthalene molecules near to these impurities are disturbed in such a way that they become shallow traps. Their depth is characteristic for the particular disturbing molecule. Thionaphthene in naphthalene creates a 28 cm^{-1} trap for the singlet state.

Thionaphthene and durene as impurities in naphthalene (concentration $\sim 10^{-4}$) create such X-traps for the triplet state also. At

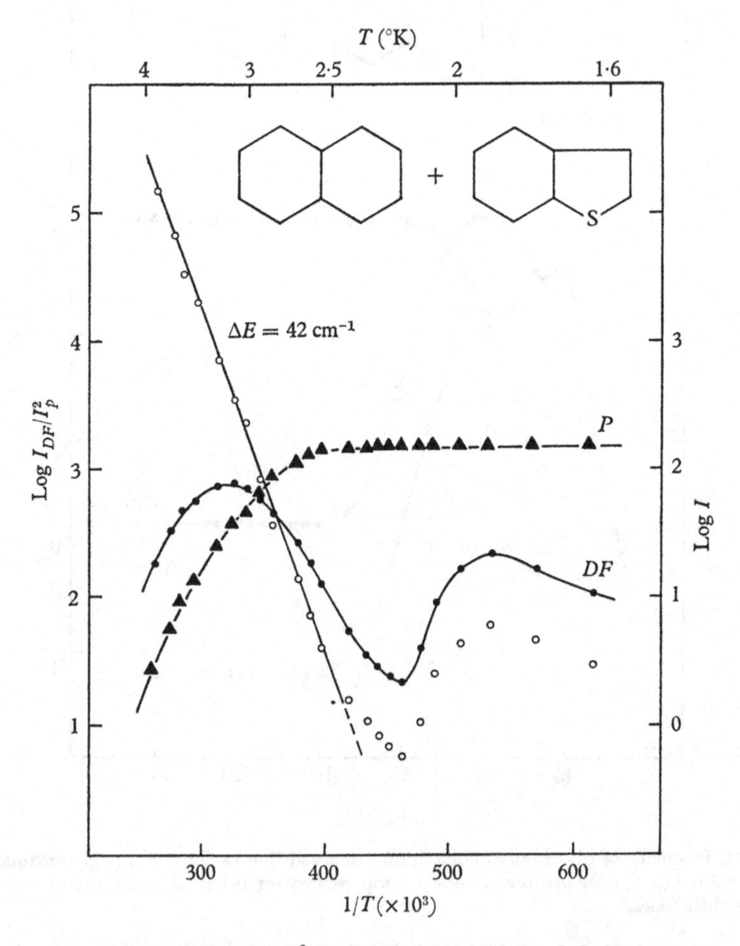

Fig. 2. I_p, I_{DF}, I_{DF}/I_p^2 for durene-X-traps in naphthalene.

1·6 °K one observes a phosphorescence spectrum which is identical to that of naphthalene, but red-shifted. 0·0 is at 21 147 cm^{-1} (durene) and 21 162 cm^{-1} (thionaphthene). The lifetime of phosphorescence at 1·6 °K is $\tau_p = 2{\cdot}7$ s in both cases.

The temperature dependence of I_p and I_{DF} is shown in figs. 2 and 3. The Arrhenius plot for I_{DF}/I_p^2 gives the activation energy for delayed fluorescence as 50 cm^{-1} (durene) and 42 cm^{-1} (thionaphthene). These values are again very similar to the spectroscopic trap depth (60 and 45 cm^{-1} respectively).

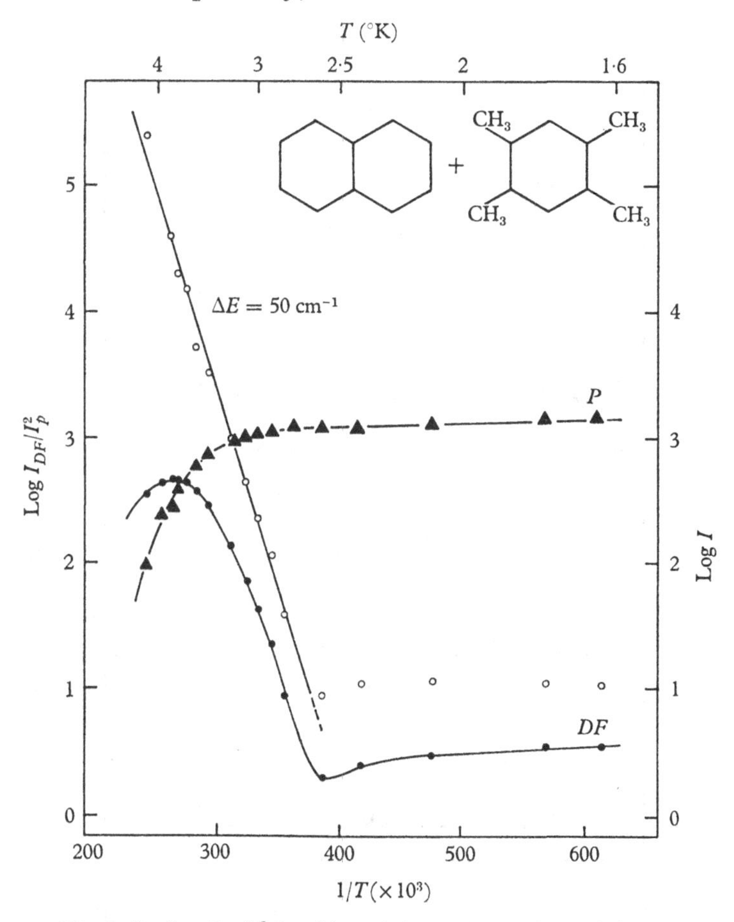

Fig. 3. I_p, I_{DF}, I_{DF}/I_p^2 for thionaphthene-X-traps in naphthalene.

It can be concluded that for these disturbed exciton states *trap-trap triplet annihilation* is the process responsible for delayed fluorescence, where one of the two annihilating traps has to be thermally excited into the host exciton band.

Naphthalene-h_8 as guest in naphthalene-d_8

This system is of special interest, because one is able to grow mixed crystals with high guest concentrations. It has been studied optically in the singlet state[6] in the triplet state[7] and, using electron spin resonance, in the triplet state[8].

In good agreement with El-Sayed, Wauk & Robinson[7] we are

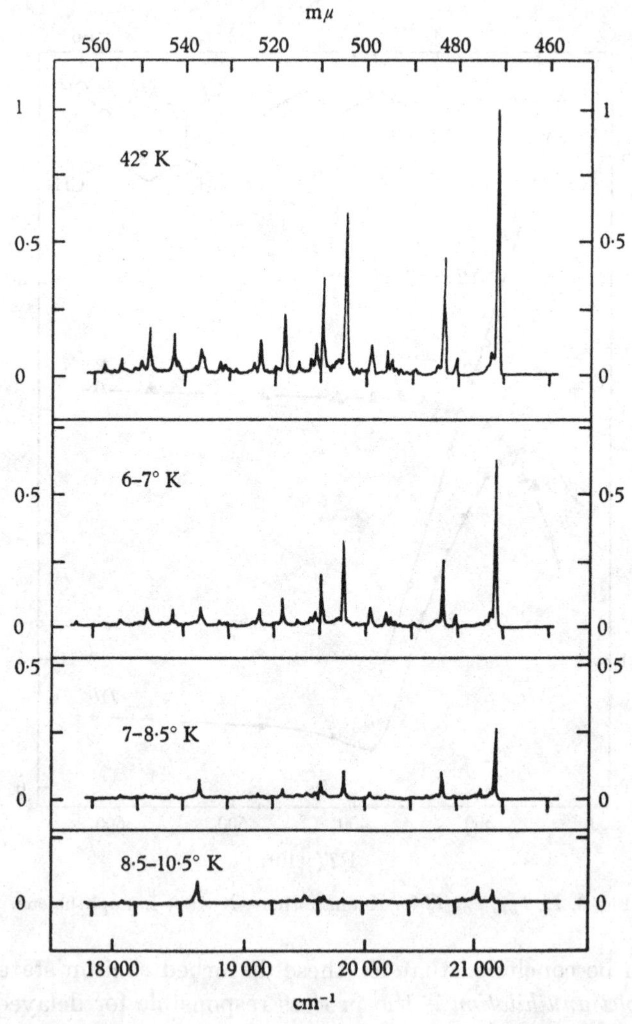

Fig. 4. Phosphorescence spectrum of naphthalene-h_8 (1 %) in naphthalene-d_8 at different temperatures. Unpolarised, spectral resolution 5 cm^{-1}.

able to analyse the phosphorescence spectrum at 4·2 °K using totally symmetric vibrations of the naphthalene molecule with 0·0 at 21 210 cm^{-1}. A part of the spectrum is shown in fig. 4. If one lowers the temperature below 4·2 °K, in the vicinity of 2·5 °K and at lowest

h_8 concentrations, a new additional system of lines appears in the phosphorescence spectrum with 0·0 at 21 270 cm^{-1}. This emission is present also in our purest naphthalene-d_8. It seems very probable that this is the emission of some kind of trap in naphthalene-d_8. At temperatures above 2·5 °K these traps are thermally depopulated in favour of the delayed fluorescence with a measured activation energy of 35 cm^{-1}.

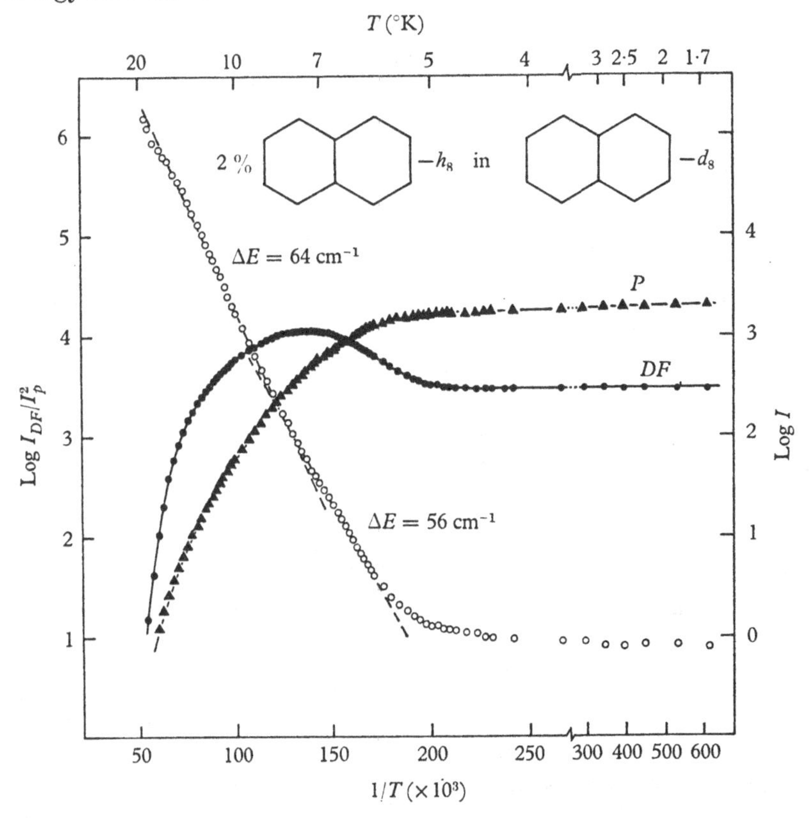

Fig. 5. I_p, I_{DF}, I_{DF}/I_p^2 for naphthalene-h_8 (2 %) as guest in naphthalene-d_8.

The temperature dependence of naphthalene-h_8 phosphorescence and delayed fluorescence is shown in fig. 5. In this case the Arrhenius plot for log $[I_{DF}/I_p^2]$ gives an activation energy 60 cm^{-1}. This energy is not sufficient to excite the traps thermally into the host exciton band (which is expected to be at 21 310 cm^{-1}). The measured activation energy is equal to the distance between the h_8-traps at 21 210 cm^{-1} and the X-traps at 21 270 cm^{-1}. Therefore we suggest that in this system the delayed fluorescence is caused by *guest-guest triplet annihilation with the* $21\,270\,cm^{-1}$ *trap as intermediate step*. One excited guest molecule is thermally excited into the trap level. The temperature needed to excite it further into the host exciton band is lower, therefore it is excited further into the host exciton band. Now it is able to migrate until it is trapped again at a second excited guest molecule. This mechanism has to be studied more in detail by an exact kinetic analysis.

The metastability of the triplet state is responsible for the unusual fact that traps with a spectroscopic depth of $\Delta E = 35\,\text{cm}^{-1}$ are thermally depopulated already at 3 °K, and those with $\Delta E = 60\,\text{cm}^{-1}$ at 8 °K. Therefore already at very low temperatures there exists *an efficient trap-to-trap energy transfer using the exciton state of the host crystal*. The trap-to-trap transfer mechanism by virtual states of the crystal as proposed by Robinson and co-workers[7,9] is at least not the only mechanism by which higher triplet traps are depopulated in favour of lower lying triplet traps.

The lifetime of the phosphorescence in the low temperature–low concentration limit is $\tau_p = 2{\cdot}7$ s. Above 4·5 °K and for concentrations higher than 2 % the decay is faster and non-exponential. This will be discussed elsewhere.[1]

Conclusions

The triplet emission spectra in organic crystals can be analysed by using free exciton states, disturbed exciton states and localised exciton states which are all well known from the singlet state.

The temperature dependence of phosphorescence and delayed fluorescence, at temperatures not very much higher than the temperature of the delayed fluorescence maximum, can be analysed by triplet-triplet annihilation processes. Trap-trap annihilation, guest-guest annihilation and guest-guest annihilation into trap states have been shown to exist.

The sharp decrease in intensity of phosphorescence and delayed fluorescence at higher temperatures are being studied in more detail. The intensities measured so far have to be corrected since the lifetimes become too short for the frequency of our phosphoroscope. A better analysis of this temperature region will give more information on the nature of the radiationless processes which make phosphorescence in pure crystals so extremely weak.

References

1 H. Port, *Diplomarbeit* (Stuttgart, 1967).
2 Paper in preparation.
3 D. M. Hanson & G. W. Robinson, *J. Chem. Phys.* **43**, 4174 (1965).
4 S. P. McGlynn, T. N. Misra, E. F. McCoy, *Int. Lumineszenz Symposium München*, 1965, p. 98. Thiemig Verlag München, 1966.
5 A. Pröpstl & H. C. Wolf, *Z. Naturforsch.* **18***a*, 724 (1963).
6 E. F. Sheka, *Opt. i Spektroskopiya*, **10**, 684 (1961). English translation: *Opt.* and *Spectr.* **10**, 360 (1962).
7 M. A. El-Sayed, M. T. Wauk & G. W. Robinson, *Mol. Physics*, **5**, 205 (1962).
8 M. Schwoerer & H. C. Wolf, paper at this Conference.
9 G. C. Nieman & G. W. Robinson, *J. Chem. Phys.* **37**, 2150 (1962).

EXCITATION OF THE TRIPLET STATES OF ORGANIC MOLECULES

J. B. BIRKS

Three principal methods are currently used for the excitation of the triplet states of organic molecules:

(1) Photo-excitation into the lowest excited singlet state S_1 or higher excited states S_n, leading to intersystem crossing to the triplet manifold.

(2) Photo-excitation into the lowest excited triplet state T_1 or higher excited triplet states T_n. Unless there is reasonably strong internal or external spin-orbit coupling (internal or external heavy atom effect) or magnetic perturbation (paramagnetic effect), the molecular photo-excitation cross-section of T_n is low. Long path lengths, concentrated solutions, condensed phases (liquid or crystal) and/or intense light sources (e.g. ruby lasers) must then be used to achieve a reasonable triplet population.

(3) Excitation by ionising radiation. This gives a high triplet yield, as evidenced by the slow scintillation component observed in aromatic crystals and oxygen-free solutions which is due to $T_1 - T_1$ association producing delayed fluorescence.[(1)] Four processes contribute to the triplet production:[(2)]

(*a*) E_S, direct excitation of S_n by the primary radiation and secondary electrons (δ-rays), leading to intersystem crossing to T_1;

(*b*) E_T, direct excitation of T_n;

(*c*) I_S, ionisation followed by molecular ion-electron recombination into S_n and subsequent intersystem crossing to T_1;

(*d*) I_T, ionisation followed by ion recombination into T_n.

E_T and I_T are the processes with the largest triplet yields, because of the respective multiplicities of T_n and S_n,

$$\sigma(I_T)/\sigma(I_S) \simeq 3,$$

and possibly also $$\sigma(E_T)/\sigma(E_S) \simeq 3,$$

where $\sigma(x)$ refers to the total cross-section of process x.

The S_0–T_n transition, which is spin-forbidden with photon excitation, is spin-allowed with electron (spin 1/2) excitation. This suggests a new, and potentially powerful, experimental approach to the excitation and study of T_n, namely

(4) The use of a mono-energetic beam of low energy electrons for the direct excitation of the triplet states of molecules in the vapour phase.

Work at Manchester University on the design and construction of instruments for the low-energy electron impact spectrometry of atomic and molecular gases and vapours was initiated a few years ago by my colleague, Dr F. H. Read. The experimental problems have proved extremely tough, but not insuperable, and several useful instruments are now in pilot operation. These offer facilities for studying many new atomic and molecular phenomena and parameters. These include: (i) the determination of absolute differential scattering and excitation cross-sections; (ii) the influence of vibronic interactions on these cross-sections;[3] (iii) the determination of the symmetries of excited states of aromatic molecules from the angular and energy distribution of scattered electrons;[4] (iv) the study of the spatial, temporal and energy correlations of scattered electrons and emitted photons; (v) lifetime, spectral and polarisation studies of singlet and triplet emissions from gases and vapours at low pressures, and (vi) prompt and delayed luminescence quantum yields, spectra and lifetimes as a function of pressure for the study of excimer, excimer ion and exciplex formation and pressure quenching in the noble gases and in aromatic hydrocarbons.

To illustrate the use of low-energy electron excitation, a brief account will be given of its application to the observation of the lifetimes of triplet states in helium by Mr K. A. Bridgett and Dr T. A. King, using an experimental method similar to that of Bennett.[5] The triplet electron excitation cross-section rises sharply above the threshold energy T_n, peaks at a few eV above T_n, and then decreases to zero. To avoid complications due to interstate cascading, mono-energetic (to $\pm 0{\cdot}2$ eV) electrons of energy $T_n + 0{\cdot}5$ eV are used for the excitation of T_n. The beam from a special electron gun is modulated by a pulse generator, designed by Mr R. J. Smith-Saville, which yields electron pulses of 100 ns duration, 2 ns cut-off, and 100 kc/s repetition frequency. The lifetime τ of the T_n emission is measured using a modified form of the photon-sampling technique introduced by Bollinger & Thomas.[6] τ is measured as a function of pressure and the data are extrapolated to zero pressure to obtain the radiative lifetime τ_0 of T_n. A fast high-gain photomultiplier P observes the emission after it has passed through a high-resolution grating monochromator, the slits of which are sufficiently narrow that less than 5 % of the light flashes generate a signal from P due to a single photon impinging on its photocathode. The time spectrum of these single-photon signals, relative to a zero-time signal derived from the pulse generator, is observed using a fast-response delayed coincidence detector, a time-to-pulse-amplitude converter and a 256-channel pulse-amplitude analyser. The pulse-amplitude spectrum reproduces the decay function of the T_n emission, a small correction being applied for the instrumental response function when the latter is of similar magnitude to τ. The pulse-amplitude spectrum is recorded by an $X-Y$ recorder and/or by a typewriter read-out for direct data analysis by the Atlas computer which is programmed to yield the most probable value of τ and the standard deviation. To date, τ_0 has been measured for six T_n emission lines in the visible helium spectrum. In all cases, the results agree with the theoretical values within the small experimental error. The measurements will be extended to other emission lines in the near and vacuum ultraviolet, to other noble gases, to aromatic molecules and to higher pressures.

I wish to thank my colleagues, named in the text, for reference to their unpublished work. The research has been supported by the University Grants Committee, the Science Research Council, and the Rank Research Laboratories.

References

1 T. A. King & R. Voltz, *Proc. Roy. Soc.* (*London*) A, **289**, 424 (1966).
2 J. B. Birks, *The Theory and Practice of Scintillation Counting.* (Pergamon Press, Oxford, 1964.)
3 F. H. Read, *Proc. Phys. Soc.* **83**, 619 (1964).
4 F. H. Read & G. L. Whiterod, *Proc. Phys. Soc.* **82**, 434 (1963).
5 W. R. Bennett, Jr. *Appl. Optics*, Suppl. 1, 24 (1962).
6 L. M. Bollinger & G. E. Thomas, *Rev. Sci. Instrum.* **32**, 1044 (1961).

THE DELAYED LUMINESCENCE AND TRIPLET QUANTUM YIELDS OF PYRENE SOLUTIONS

J. B. BIRKS AND G. F. MOORE

Abstract

Expressions are obtained for the quantum yields of triplet formation, delayed monomer and excimer fluorescence, and phosphorescence of aromatic compounds in solution. The three delayed luminescence quantum yields depend on the excitation conditions. The triplet quantum yields ϕ_T of pyrene in cyclohexane (20 to 70 °C) and in ethanol (−80 to −40 °C) are derived from the delayed fluorescence decay curves. The results indicate that $\phi_T = 1 - \phi_F$, where ϕ_F is the normal fluorescence quantum yield.

Introduction

Pyrene, in common with other aromatic hydrocarbons, exhibits five luminescences in solution:

(i) monomer fluorescence (MF) from the excited monomer M^*;

(ii) excimer fluorescence[1] (EF) from the excimer D^* produced by the collisional interaction of M^* and an unexcited monomer M,

$$M^* + M \rightleftarrows D^*; \tag{1}$$

(iii) phosphorescence (P) from the excited triplet T^* produced by intersystem crossing from M^* and D^*;

(iv) delayed monomer fluorescence (DMF); and

(v) delayed excimer fluorescence (DEF), produced by triplet-triplet association,[2,3] yielding M^* and D^* respectively,

$$T^* + T^* \begin{array}{l} \nearrow M^* + M. \\ \quad \Downarrow\!\Uparrow \\ \searrow D^* \end{array} \tag{2}$$

The rate parameters and the thermodynamic properties of process (1) have been determined from the quantum yields[4,5] and from the lifetimes[5,6] of MF and EF as a function of concentration and temperature in various solvents. The branching ratio of process (2) has been evaluated[3,7] from observations[2,7,11] of the relative quantum yields of DMF and DEF.

No experimental data are available on the quantum yield ϕ_T of T^* by intersystem crossing from M^* and D^*, which relates processes (1) and (2) and which influences the quantum yields of P, DMF and DEF. This paper presents theoretical relations for these quantum yields, and it describes an experimental study of ϕ_T for pyrene solutions in cyclohexane and in ethanol as a function of temperature.

Theory

The previous notation,[6,7,8] with slight modifications, is used and extended. The excitation light intensity $I[M]$ is expressed in absorbed photons s^{-1}. $[M]$ is the molar concentration of the solute, and square brackets indicate molar concentrations of excited species. The various rate processes are listed in table 1.

Table 1. *Rate processes*

Process	Description	Rate (s^{-1})	
$M+h\nu \rightarrow M^*$	Excitation of M	$I[M]$	
$M^* \rightarrow M+h\nu_M$	Fluorescence of M^*	$k_{FM}[M^*]$	$k_M[M^*]$
$M^* \rightarrow T^*$	Intersystem crossing from M^* to T^*	$k_{TM}[M^*]$	
$M^* \rightarrow M$	Internal quenching of M^*	$k_{GM}[M^*]$	
$M^*+M \rightarrow D^*$	Formation of D^* from M^*+M	$k_{DM}[M][M^*]$	
$D^* \rightarrow 2M+h\nu_D$	Fluorescence of D^*	$k_{FD}[D^*]$	$k_D[D^*]$
$D^* \rightarrow T^*+M$	Intersystem crossing from D^* to T^*	$k_{TD}[D^*]$	
$D^* \rightarrow 2M$	Internal quenching of D^*	$k_{GD}[D^*]$	
$D^* \rightarrow M^*+M$	Dissociation of D^* into M^*+M	$k_{MD}[D^*]$	
$T^*+T^* \rightarrow M^*+M$	T^*-T^* association yielding M^*	$k_{MT}[T^*]^2$	$k_{TT}[T^*]^2$
$T^*+T^* \rightarrow D^*$	T^*-T^* association yielding D^*	$k_{DT}[T^*]^2$	
$T^* \rightarrow M+h\nu_P$	Phosphorescence of T^*	$k_{PT}[T^*]$	$k_T[T^*]$
$T^* \rightarrow M$	Internal quenching of T^*	$k_{GT}[T^*]$	

The following parameters are defined:

$$k_M = k_{FM}+k_{TM}+k_{GM} = k_{FM}/q_M = 1/\tau_M,$$

$$k_D = k_{FD}+k_{TD}+k_{GD} = k_{FD}/q_D = 1/\tau_D,$$

$$k_T = k_{PT}+k_{GT} = 1/\tau_T,$$

$$k_{TT} = k_{MT}+k_{DT},$$

$$\alpha = k_{DT}/k_{MT},$$

$$K_e = k_{DM}/(k_D+k_{MD}),$$

$$\beta = k_M+k_D K_e[M].$$

The quantum yields of MF and DF are[6] respectively,

$$\phi_M = k_{FM}/\beta, \tag{3}$$

$$\phi_D = k_{FD}K_e[M]/\beta, \tag{4}$$

so that $$\phi_D/\phi_M = (k_{FD}/k_{FM})K_e[M] = K_1[M]. \tag{5}$$

Similarly, the quantum yields of T^* from M^* and D^* are, respectively,

$$\phi_{TM} = k_{TM}/\beta, \tag{6}$$

$$\phi_{TD} = k_{TD}K_e[M]/\beta, \tag{7}$$

so that the total quantum yield of T^* is

$$\phi_T = \phi_{TM} + \phi_{TD} = (k_{TM} + k_{TD}K_e[M])/\beta. \tag{8}$$

Using a phosphoroscope, the system is excited periodically by a light intensity $I\,[M]$ for a time $t_0(\ll \tau_T)$, at intervals of time $t_I > \tau_T$, yielding an initial T^* concentration of

$$[T^*]_0 = \phi_T I[M]t_0, \tag{9}$$

and DMF, DEF and P are observed periodically during the time t ($< t_I$) after t_0. In general the T^* concentration decays as follows:

$$[T^*] = \frac{[T^*]_0 \exp\{-k_T t\}}{1 + [T^*]_0 (k_{TT}/k_T)(1 - \exp\{-k_T t\})}. \tag{10}$$

There are two limiting cases of (10):

(i) $k_{TT}[T^*] \gg k_T$ (large $I[M]t_0$ and/or small t),

$$[T^*] = \frac{[T^*]_0}{1 + k_{TT}[T^*]_0 t}; \tag{10a}$$

(ii) $k_{TT}[T^*] \ll k_T$ (small $I[M]t_0$ and/or large t),

$$[T^*] = [T^*]_0 \exp\{-k_T t\}. \tag{10b}$$

The concentrations of M^* and D^*, produced by $T^* - T^*$ association, are given by[7]

$$[M^*] = R_M[T^*]^2, \tag{11}$$

$$[D^*] = R_D[T^*]^2, \tag{12}$$

where $$R_M = \frac{k_{MT} + k_{DT}k_{MD}/(k_D + k_{MD})}{\beta}, \tag{13}$$

$$R_D = \frac{k_{TT}K_e[M] + k_{DT}k_M/(k_D + k_{MD})}{\beta}. \tag{14}$$

The quantum yields of $P(\phi_P)$, DMF (ϕ_M^d) and DEF (ϕ_D^d) depend on the excitation conditions. For large $I\,[M]t_0$, $k_{TT}[T^*] \gg k_T$ (10a),

$$\phi_P = 0, \tag{15}$$

$$\phi_M^d = \frac{k_{FM}R_M\phi_T}{k_{TT}}, \tag{16}$$

$$\phi_D^d = \frac{k_{FD}R_D\phi_T}{k_{TT}}, \tag{17}$$

so that the intensities of DMF and DEF are proportional to $I[M]t_0$ under these conditions.

For small $I[M]t_0$, $k_{TT}[T^*] \ll k_T$ (10b),

$$\phi_P = \frac{k_{PT}\phi_T}{k_T}, \tag{18}$$

$$\phi_M^d = \frac{k_{FM}R_M\phi_T^2 I[M]t_0}{2k_T}, \tag{19}$$

$$\phi_D^d = \frac{k_{FD}R_D\phi_T^2 I[M]t_0}{2k_T}, \tag{20}$$

so that the intensities of DMF and DEF are proportional to $(I[M]t_0)^2$, and the intensity of P is proportional to $I[M]t_0$, under these conditions.

For all values of $I[M]t_0$ and t, the ratio of the quantum intensities of DEF and DMF is

$$\frac{I_D^d}{I_M^d} = \frac{\phi_D^d}{\phi_M^d} = \frac{k_{FD}R_D}{k_{FM}R_M}$$

$$= \frac{k_{FD}(\alpha k_M + (1+\alpha)k_{DM}[M])}{k_{FM}(k_D + (1+\alpha)k_{MD})}$$

$$= K_2 + K_3[M] \tag{21}$$

a result previously derived[3, 7] for small $I[M]t_0$ (10b) only.

Experimental results and analysis

Measurements were made on the delayed fluorescence of zone-refined pyrene in solution in spectroscopic grade ethanol and cyclohexane at temperatures from −80 to −40 °C and from 20 to 70 °C, respectively. The solutions were contained in cylindrical Pyrex cells, and they were degassed by the cyclic freeze-pump-thaw technique[2, 7] to a pressure of 5×10^{-6} mm Hg prior to sealing.

The delayed fluorescence was excited by filtered and focused radiation of 365 nm wavelength from a 250 W high pressure mercury arc, and observed by a phosphoroscope, monochromator, photomultiplier and oscilloscope, as described previously,[7] under constant excitation ($I[M]t_0$) conditions. Observations were made of the decay of the delayed fluorescence at time $t \ll \tau_T$. The DMF and DEF were observed to decay in the same manner, in agreement with (21).

The conditions of excitation and observation corresponded to

$k_{TT}[T^*] \gg k_T$, since the delayed fluorescence intensity I^d decayed in the manner indicated by (10a), (11) and (12),

$$I^d = \frac{A[T^*]_0^2}{(1+k_{TT}[T^*]_0 t)^2} \tag{22}$$

where $A = BR_M k_{FM}$ and $BR_D k_{FD}$ for DMF and DEF respectively, and B is an instrumental constant. Plots of $(I^d)^{-\frac{1}{2}}$ against t were linear of gradient $A^{-\frac{1}{2}}k_{TT}$, intercept $A^{-\frac{1}{2}}[T^*]_0^{-1}$ and gradient/intercept ratio $U = k_{TT}[T^*]_0$. Typical results for solutions of pyrene in ethanol at three temperatures are plotted in fig. 1.

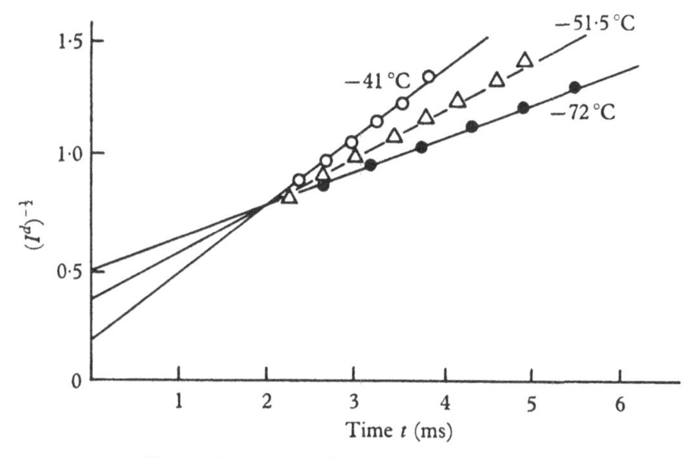

Fig. 1. Pyrene in ethanol at different temperatures. $(I^d)^{-\frac{1}{2}}$ against time t.

The T^*–T^* association process (2) has been shown to be diffusion-controlled,[2] and is thus similar to the excimer formation process (1).[5, 6] Hence k_{TT} is taken as proportional to T/η, where T is the absolute temperature and η is the solvent viscosity, so that

$$\eta U/T \propto [T^*]_0 \propto \phi_T \tag{23}$$

for constant excitation conditions $I[M]\,t_0$ (9).

For pyrene in ethanol or cyclohexane at a given T, q_M and q_D are approximately equal ($= q_F$) and independent of solvent.[5, 6] Hence, from (3) and (4), the total quantum yield of MF and EF,

$$\phi_F = \phi_M + \phi_D = (k_{FM} + k_{FD}K_e[M])/\beta \tag{24}$$

is approximately constant ($= q_F$), and independent of $[M]$ at a given T. Figures 2 and 3 plot the experimental values[5, 6] of $(1-\phi_F)$ for degassed solutions of pyrene in cyclohexane and in ethanol, respectively, as a function of temperature.

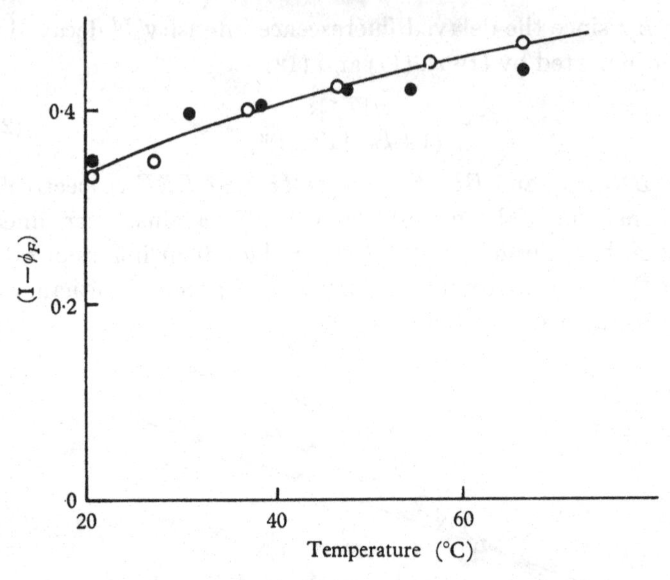

Fig. 2. Temperature dependence of pyrene in cyclohexane. —○—, $1-\phi_F$; ●, ϕ_T (normalised at 39 °C).

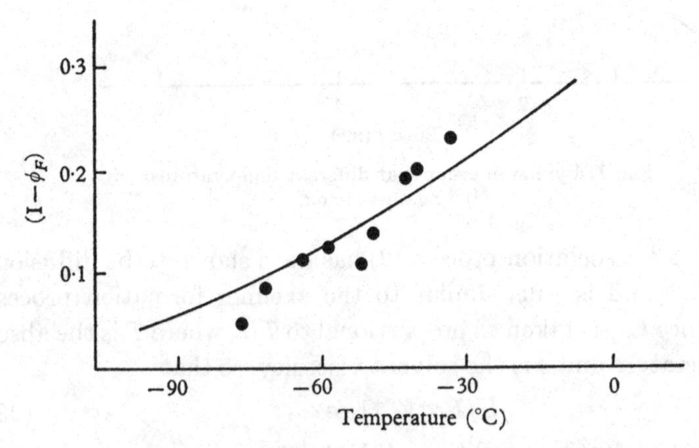

Fig. 3. Temperature dependence of pyrene in ethanol. ——, $1-\phi_F$; ●, ϕ_T (normalised at −58 °C).

In general, for any compound

$$1-\phi_F = \phi_T + \phi_G, \tag{25}$$

where

$$\phi_G = (k_{GM} + k_{GD} K_e[M])/\beta \tag{26}$$

is the total quantum yield of internal quenching of M^* and D^*. Recent studies[9,10] have indicated that $\phi_G \simeq 0$ for some aromatic hydrocarbons (e.g. anthracene and its 9-phenyl, 9-methyl and 9,10-diphenyl derivatives) in solution at normal temperatures. To test whether $\phi_G \simeq 0$ for pyrene, the values of $\eta U/T$ ($\propto \phi_T$) for the cyclohexane and ethanol solutions were normalised to $(1-\phi_F)$ at 39 °C and −58 °C, respectively, and they are plotted in figs. 2 and 3. The close

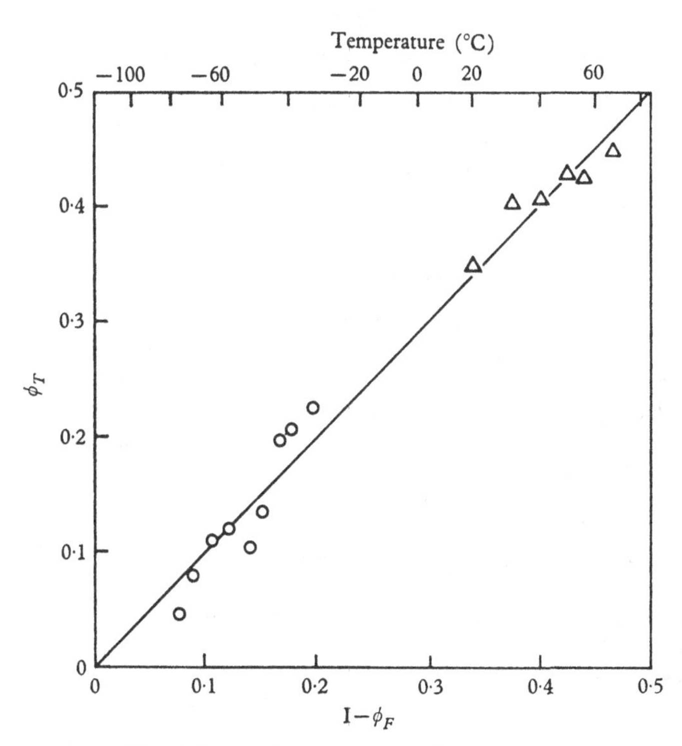

Fig. 4. Pyrene in cyclohexane (△) and ethanol (○). ϕ_T (figs. 2 and 3) against $(1-\phi_F)$.

correspondence, within the experimental error, indicates that either (i) $\phi_G = 0$, or (ii) ϕ_T and ϕ_G vary in the same manner over a wide temperature range in two different solvents. The latter is considered very improbable, because of the different nature of the intersystem crossing (ϕ_T) and internal quenching (ϕ_G) processes. Moreover, for pyrene in cyclohexane and in ethanol, the temperature dependences of the individual monomer and excimer fluorescence quenching rate parameters, $k_{IM}(=k_{TM}+k_{GM})$ and $k_{ID}(=k_{TD}+k_{GD})$, have each been shown[5,6] to be of the simple form

$$k_I = k'_I \exp(-W_I/\mathbf{k}T), \tag{27}$$

indicating that only one fluorescence quenching process is operative in each case. Since intersystem crossing certainly occurs as evidenced by *DMF* and *DEF*, it is concluded that $k_{TM} = k_{IM}$, $k_{TD} = k_{ID}$, $k_{GM} = 0$, $k_{GD} = 0$, $\phi_G = 0$ over the whole temperature range. The data from figs. 2 and 3 are replotted in fig. 4 to test and illustrate the simple relation

$$\phi_T = 1 - \phi_F, \tag{28}$$

which has been deduced from the results. From (28) it is concluded that the triplet quantum yield ϕ_T varies with temperature in the manner shown in figs. 2 and 3. From (27) it is concluded that intersystem crossing in pyrene is a thermally activated process, which is probably induced by solvent collisions.

References

1 T. Förster & K. Kasper, *Z. Elektrochem.* **59**, 977 (1955).
2 C. A. Parker & C. G. Hatchard, *Trans. Faraday Soc.* **57**, 1894 (1961).
3 J. B. Birks, *J. Phys. Chem.* **67**, 2199 (1963); **68**, 439 (1964).
4 E. Döller & T. Förster, *Z. Phys. Chem.* NF, **34**, 132 (1962).
5 J. B. Birks, M. D. Lumb & I. H. Munro, *Proc. Roy. Soc.* (*London*) A, **280**, 289 (1964).
6 J. B. Birks, D. J. Dyson & I. H. Munro, *Proc. Roy. Soc.* (*London*) A, **275**, 575 (1963).
7 J. B. Birks, G. F. Moore & I. H. Munro, *Spectrochim. Acta*, **22**, 323 (1966).
8 J. B. Birks, C. L. Braga & M. D. Lumb, *Proc. Roy. Soc.* (*London*) A, **283**, 83 (1965).
9 T. Medinger & F. Wilkinson, *Trans. Faraday Soc.* **61**, 620 (1965).
10 J. D. Laposa, E. C. Lim & R. E. Kellogg, *J. Chem. Phys.* **42**, 3025 (1965).
11 G. F. Moore & I. H. Munro, *Spectrochim. Acta* (in the Press).

TRIPLET STATE STUDIES OF SOME POLYPHENYLS IN RIGID GLASSES

I. A. RAMSAY AND I. H. MUNRO

Abstract

Triplet lifetimes and absolute triplet-triplet extinction coefficients of diphenyl, ortho-, meta- and para-terphenyl were measured in n-butanol at liquid air temperatures. The relative triplet-triplet absorption spectrum of para-quaterphenyl was also measured. Similar results for the absolute values of triplet-triplet extinction coefficients were obtained using three independent techniques. The fluorescence and phosphorescence spectrum of each compound was measured at the temperature of liquid air. From the ratio of integrated fluorescence and phosphorescence intensities and a knowledge of fluorescence quantum efficiencies, approximate values of radiative triplet lifetimes were obtained. All polyphenyl spectra, whether of fluorescence, phosphorescence or triplet-triplet absorption, were found to have a characteristic shape. The maximum extinction coefficient of para-terphenyl was much greater than that found for ortho- or meta-terphenyl.

Introduction

The properties of the triplet states of aromatic molecules may be observed conveniently in inert rigid matrices, for example, in solutions of n-butanol maintained at liquid air temperatures. Diffusional processes are negligible under such conditions and the lifetime of the metastable triplet level may be several seconds. This means that at the front of the sample almost complete depopulation of the singlet levels may take place at relatively low light intensities ($\sim 10^{18}$ photons/cm^2/s).

The rate equations which describe the populations of the lowest excited singlet (S_1) and triplet (T_1) states cannot be solved explicitly unless the excitation rates are independent of the concentrations of molecules in these states. Only two situations exist in which the excitation rates are independent of the concentrations of excited states: (*a*) when depopulation of the ground (S_0) state is negligible; (*b*) when the ground to excited singlet state extinction coefficient, $\epsilon(\lambda)$, is comparable with the triplet-triplet extinction coefficient, $\sigma(\lambda)$, throughout the excitation spectrum.

In either case, it can be shown[1] that

$$D(t,\lambda) = \sigma(\lambda)c\left[\int_0^l \frac{Z(\Lambda)}{k_1+Z(\Lambda)}\,d\Lambda - e^{-k_1 t}\int_0^l \frac{Z(\Lambda)\,e^{-Z(\Lambda)t}}{k_1+Z(\Lambda)}\,d\Lambda\right], \quad (1)$$

where $D(t,\lambda)$ is the triplet-triplet optical density at wavelength λ, t seconds after the start of excitation, c is the molar solute concentration, l cm is the length of the cell, $k_1\,s^{-1}$ is the triplet first-order decay rate and $Z(\Lambda)\,s^{-1}$ is the triplet excitation rate in the path of the exciting beam Λ cm behind the front face of the cell.

If the exciting beam intensity is made sufficiently low for ground state depletion to be small, then $I_1(Z_0)$ can be defined from (1) such that

$$I_1(Z_0) \equiv \int_0^l \frac{Z(\Lambda)}{Z_0}\,e^{-\tau_f Z(\Lambda)}\,d\Lambda$$

$$= \frac{e^{k_1\tau_f}}{2{\cdot}303\bar{\epsilon}c}\,\frac{1}{A}\ln\left(1+\frac{e^{A}-1}{e}\right), \quad (2)$$

where $Z_0\,s^{-1}$ is the triplet excitation rate at the front face of the cell, τ_f is the $1/e$ fall time of the transmitted light at a triplet-triplet absorption peak, $\bar{\epsilon}$ is the average value of the solute extinction coefficient with respect to the spectrum of exciting light absorbed by the sample and $A(\lambda) \equiv 2{\cdot}303D(\infty,\lambda)$. Also from (1) we define $I_2\,(Z_0)$ such that

$$I_2(Z_0) \equiv \int_0^l \frac{Z(\Lambda)}{k_1+Z(\Lambda)}\,d\Lambda = \frac{D(\infty,\lambda)}{\sigma(\lambda)c}. \quad (3)$$

The parameters $\bar{\epsilon}$, c, A, k_1, τ_f and $D(\infty,\lambda)$ were measured directly. $I_1(Z_0)$ and $I_2(Z_0)$ were then calculated using an Atlas computer. In this way, $\sigma(\lambda)$, the absolute triplet-triplet extinction coefficient, could be determined (method (*a*)).

Since Z_0 is proportional to the exciting intensity, it follows from (3) that $\sigma(\lambda)c$ can be obtained also from the ratio of the intercepts derived from plots of $1/I_2$ against $1/Z_0$ and of $1/D(\infty,\lambda)$ against reciprocal relative exciting light intensity, provided that the range of Z_0 in the calculations corresponds to the range of exciting intensities used in the measurements. This second analysis (method (*b*)) for $\sigma(\lambda)$ is similar to that used by McClure[2] except that head-on, not broadside, excitation is used.

If γ is the fraction of non-fluorescent deactivation which goes from S_1 to S_0 via T_1, q_f is the fluorescence quantum efficiency, P is the number of exciting photons absorbed per second by the sample and

Δ cm^2 is the area of the exciting beam, then it has been shown[1, 3] that

$$Z_0 = \frac{2{\cdot}303\gamma}{N}(1-q_f)\,\bar{\epsilon}\,\frac{P}{\Delta}, \tag{4}$$

where $\gamma(1-q_f)$ represents the quantum yield of triplet formation. If $\gamma = 1$, i.e. intersystem crossing is the only important non-radiative process, then q_f may be determined from Z_0 when P and Δ have been measured. The value of q_f calculated in this way is not precise because of the errors accumulated in measurements of Z_0, $\bar{\epsilon}$, P and Δ. Nevertheless, when a sensible value of q_f is obtained (between 0 and 1) it provides verification of the values of Z_0 and $\sigma(\lambda)$ and hence a third independent check on the results.

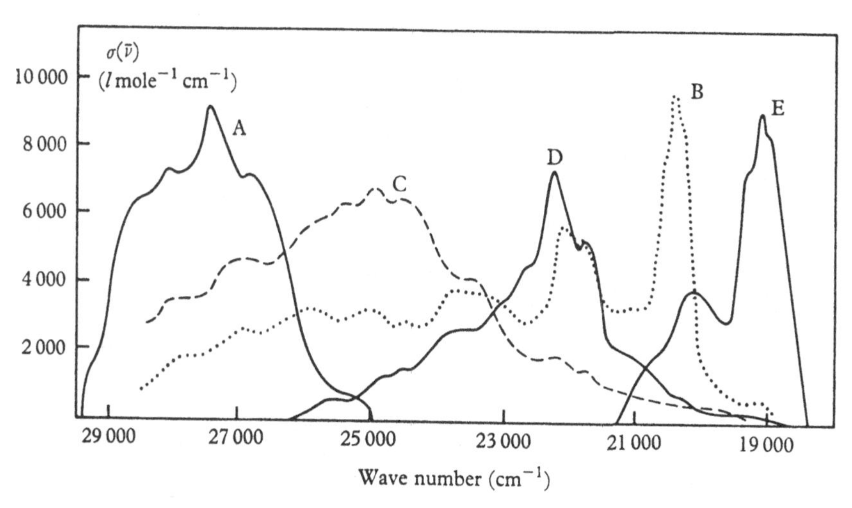

Fig. 1. Triplet-triplet absorption spectra of some polyphenyls on an absolute scale of extinction coefficient.

A, diphenyl: scale $\times 30$. $\sigma(\lambda)_{\text{max.}} = 275$ at 363 nm.
B, ortho-terphenyl: scale $\times 100$. $\sigma(\lambda)_{\text{max.}} = 90$ at 488 nm.
C, meta-terphenyl: scale $\times 100$. $\sigma(\lambda)_{\text{max.}} = 67$ at 401 nm.
D, para-terphenyl: scale $\times 1$. $\sigma(\lambda)_{\text{max.}} = 7400$ at 449 nm.
E, para-quaterphenyl: ordinate scale is relative absorption only.

Experimental and results

Using apparatus described previously,[1] triplet-triplet absorption spectra were measured on an absolute scale of extinction coefficient for solutions of diphenyl, ortho-, meta- and para-terphenyl in n-butanol glasses at liquid air temperatures. The spectra are shown in fig. 1. Because of the low solubility of para-quaterphenyl in n-butanol absolute measurements were not possible and so the spectrum shown in fig. 1 indicates the triplet-triplet absorption on a relative scale only. The chemicals used were of commercial grade purity, and it is clear from the spectra, for example for the terphenyls, that each isomer contains a proportion of the other two, present as impurities. This conclusion was verified by lifetime measurements (see table 1).

The extractions of Z_0 for diphenyl, ortho-, meta- and para-terphenyl, necessary for determinations of $\sigma(\lambda)$ by method (*a*) and q_f, are shown graphically in fig. 2. The corresponding values of I_2 were obtained from

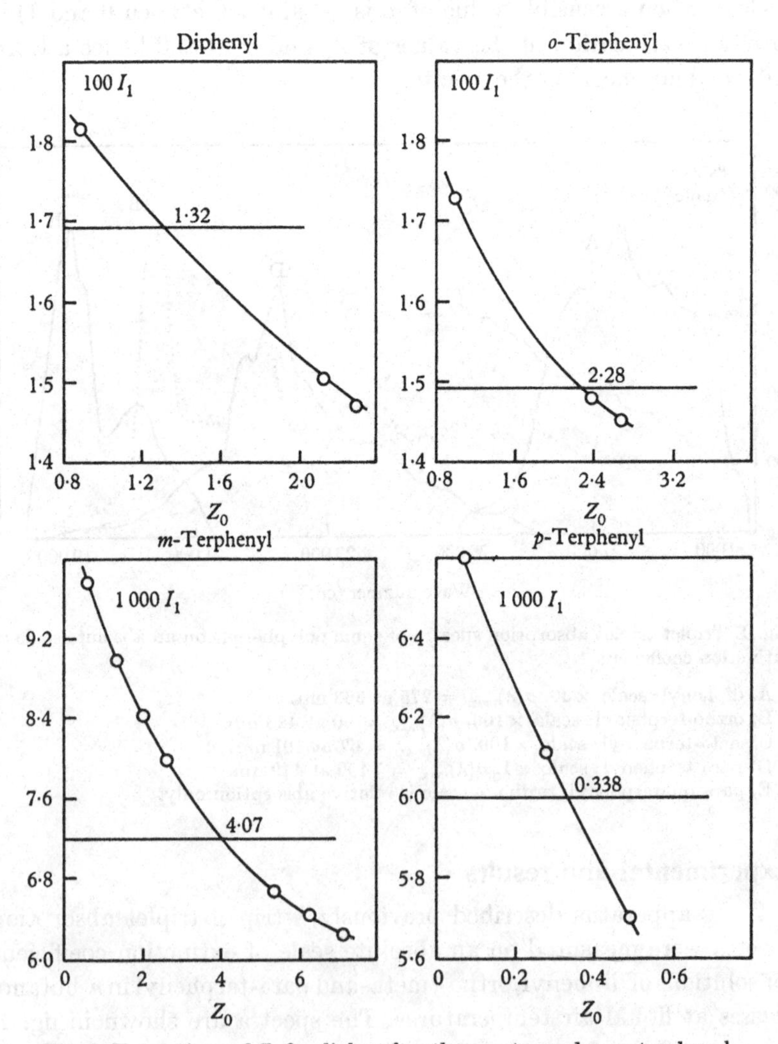

Fig. 2. Extractions of Z_0 for diphenyl, ortho-, meta- and para-terphenyl.

Table 1

Sample	λ(nm)	σ (litre mole^{-1} cm^{-1}) (*a*)	(*b*)	Average	q_f	$1/k_1$ (s)	Assignment	I_p/I_f	$1/k_p$ (s)
Diphenyl	363	300	250	275	0·60	$3{\cdot}93 \pm 0{\cdot}1$	Diphenyl	0·41	$9{\cdot}6\left(\frac{1-q_f}{q_f}\right)$
o-Terphenyl	488	80	100	90	0·52	$3{\cdot}45 \pm 0{\cdot}1$	*o*-Terphenyl	—	—
	401	—	—	—	—	$4{\cdot}41 \pm 0{\cdot}1$	*m*-Terphenyl	—	—
	449	—	—	—	—	$2{\cdot}02 \pm 0{\cdot}1$	*p*-Terphenyl	—	—
m-Terphenyl	401	72	63	67	0·58	$4{\cdot}65 \pm 0{\cdot}2$	*m*-Terphenyl	0·11	$43\left(\frac{1-q_f}{q_f}\right)$
	449	—	—	—	—	$2{\cdot}00 \pm 0{\cdot}06$	*p*-Terphenyl	—	—
p-Terphenyl	449	4880	9920	7400	0·89	$1{\cdot}93 \pm 0{\cdot}06$	*p*-Terphenyl	0·17	$11{\cdot}3\left(\frac{1-q_f}{q_f}\right)$

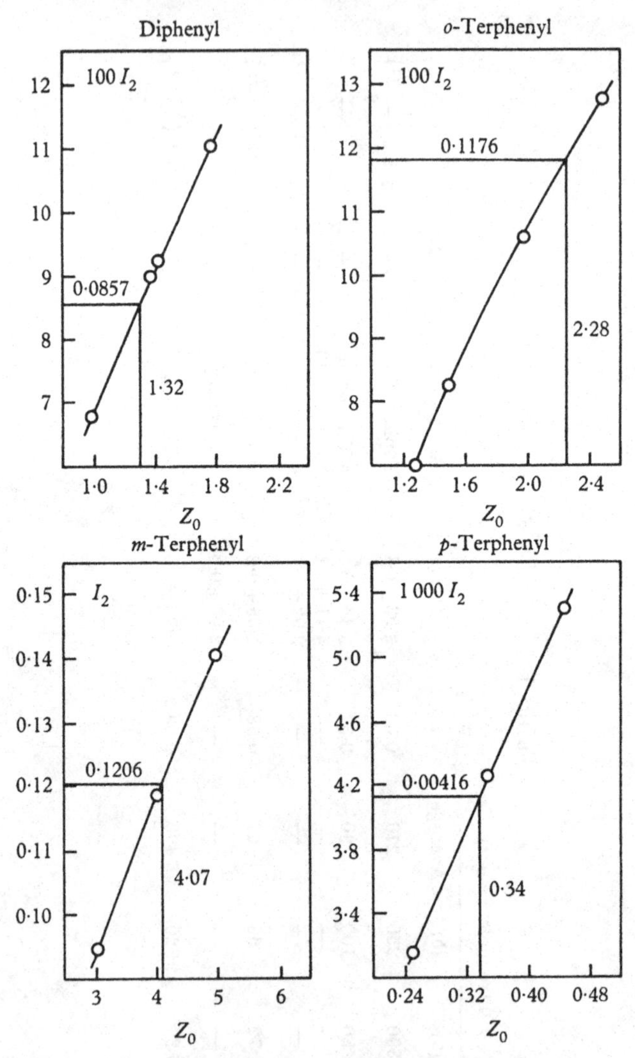

Fig. 3. Derivations of I_2 from Z_0.

the graphs in fig. 3. The use of a series of wire gauze attenuators to change the exciting light intensity (P_{relative}) provided results for a series of graphs of $1/D(\infty, \lambda)$ against $1/P_{\text{relative}}$, and calculated plots of $1/I_2$ against $1/Z_0$ and these are shown in fig. 4.

Maximum values of the absolute triplet-triplet extinction coefficient measured using methods (*a*) and (*b*) are shown in table 1 together with measured first-order triplet lifetimes.

P and Δ were measured using a chemical actinometer and photo-

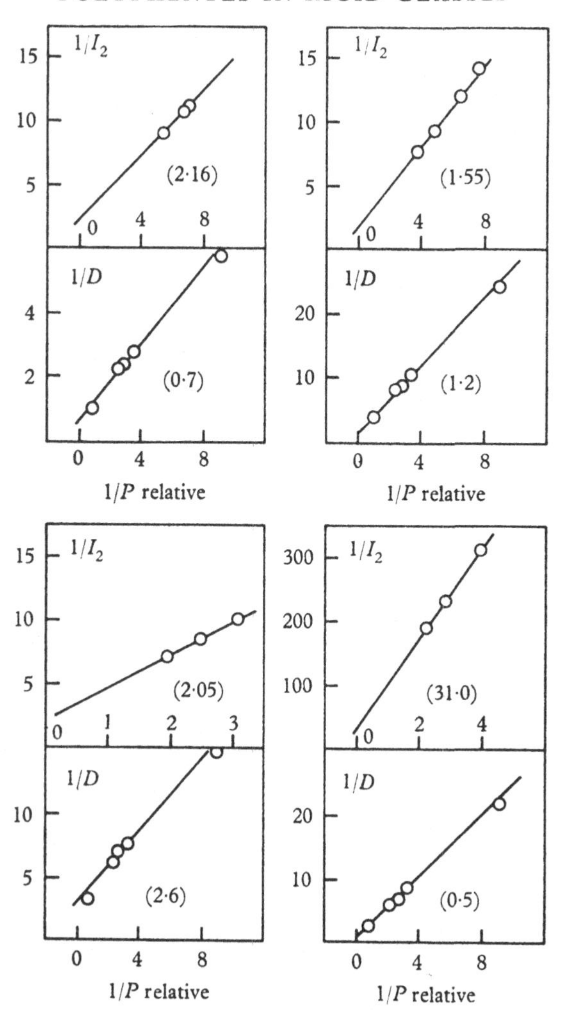

Fig. 4. Calculated ($1/I_2$ against $1/P_{\text{relative}}$) and measured ($1/D$ against $1/P_{\text{relative}}$) McClure plots for the polyphenyls. Ordinate intercept values are shown in brackets.

sensitive plates respectively and the value of q_f calculated from these values using (4) for each compound is also shown in table 1. The values of q_f are uncertain by a factor of 3.

The fluorescence and phosphorescence emission spectra of each compound were measured at liquid air temperatures in n-butanol glass and are shown in fig. 5. From the ratio of the integrated relative intensities of fluorescence and phosphorescence, approximate values were calculated of the radiative triplet lifetime $1/k_p$ and these also are given in table 1.

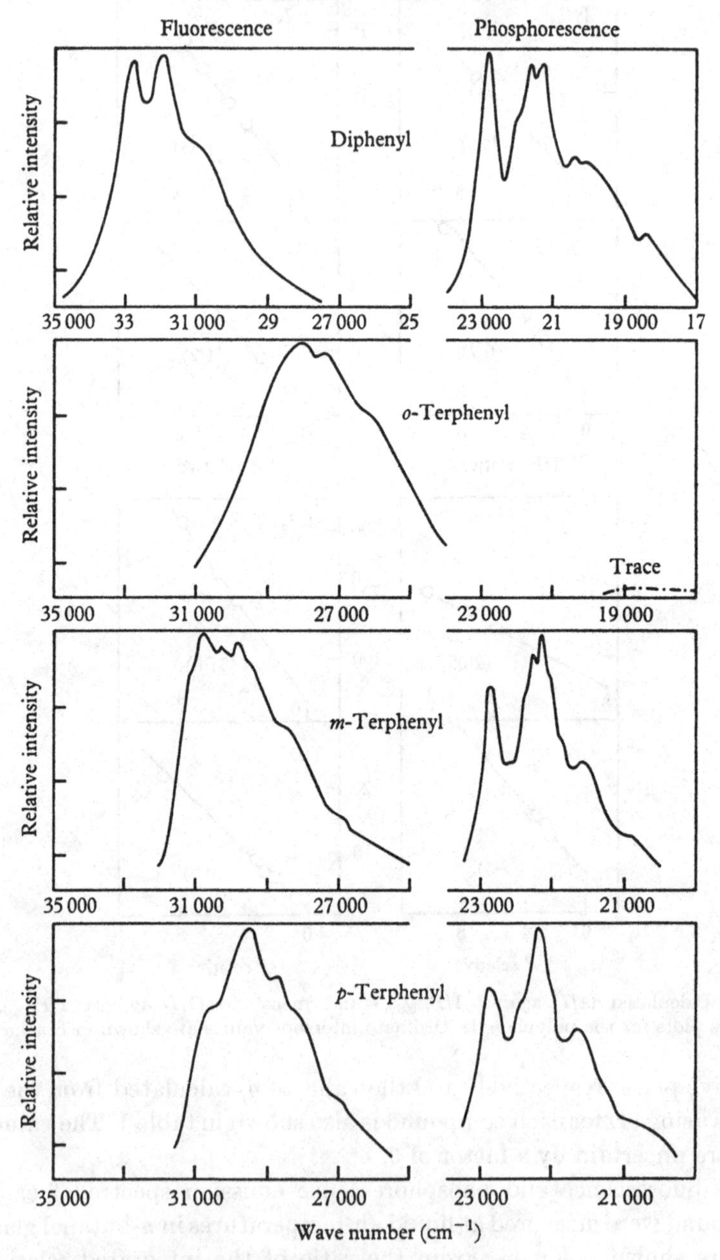

Fig. 5. Fluorescence and phosphorescence spectra (uncorrected) of diphenyl, ortho-, meta- and para-terphenyl in rigid *n*-butanol glasses.

Discussion

The triplet-triplet absorption spectra in fig. 1, obtained using a monochromator bandwidth of 1·75 nm, are plotted on an absolute extinction coefficient scale defined, independently, by two methods of analysis. The values of the maximum absolute extinction coefficient of biphenyl, ortho-, meta- and para-terphenyl are consistent with the measured values of fluorescence quantum efficiency for these compounds.

References

1 I. H. Munro & I. A. Ramsay, *International Conference on Luminescence*, Budapest, 1966. *Acta Phys. Hungarica*, to be published.
2 D. S. McClure, *J. Chem. Phys.* **19**, 670 (1951).
3 R. A. Keller & S. G. Hadley, *J. Chem. Phys.* **42**, 2382 (1965).

DECAY TIME OF DELAYED FLUORESCENCE OF ANTHRACENE AS A FUNCTION OF TEMPERATURE (2–300 °K)†

F. R. LIPSETT AND D. H. GOODE

The story starts in 1963, with the invention of the ruby laser. When this was done it occurred to two groups—Singh & Stoicheff[2] at NRC in Ottawa and Peticolas, Goldsborough & Rieckhoff[3] at IBM in San Jose, California—to irradiate an anthracene crystal with red light from a ruby laser (fig. 1). I (F.R.L.) am not sure what they expected to happen, but what did happen was that blue fluorescence in the usual wavelength region of anthracene fluorescence appeared. This of course

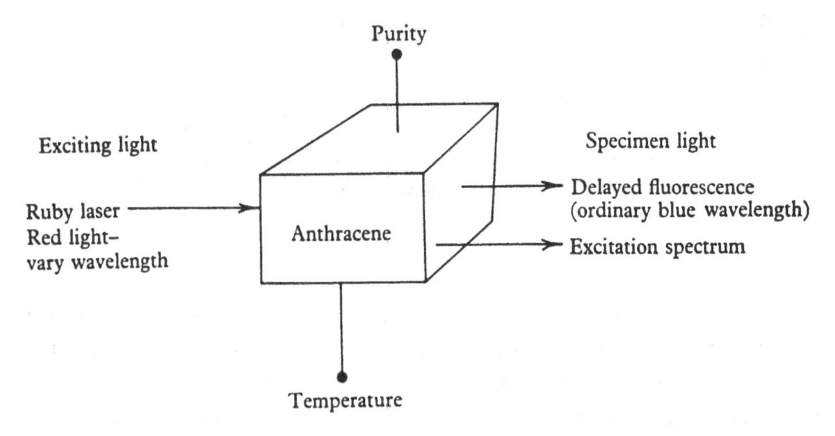

Fig. 1.

was quite unexpected from the point of view that more energy was apparently coming out of the crystal than went in, and I can remember the amazement with which we at NRC learned of this result.

There was of course a hot debate at the time as to how this extraordinary effect could come about, which I do not propose to discuss. It was fairly soon resolved that both two-photon absorption and

† This paper was originally to have been entitled 'Triplet Emission spectra at 4·2 of undoped and β-methyl-naphthalene doped naphthalene crystal.' The results on naphthalene were published before the Symposium took place.[1]

triplet-triplet annihilation could and did play a part. The two-photon absorption is responsible for rapid emission of the anthracene blue fluorescence (to a few microseconds following the flash) and triplet-triplet annihilation for the longer-lasting delayed fluorescence with which we will be concerned from now on.

Well, now that shining red light from a ruby laser (rather than ultraviolet light) on anthracene had been found to produce fluorescence in the ordinary blue region it was a natural thing to try red light from a conventional light source—for example a xenon lamp plus monochromator—rather than the laser. Then the wavelength of the exciting light could be varied. This was done by Avakian and colleagues[(4)] at Du Pont in Wilmington, Delaware.

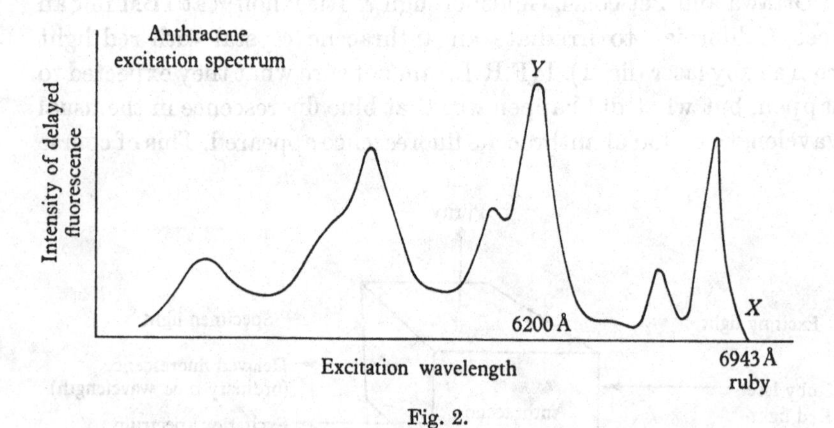

Fig. 2.

They found an excitation spectrum (which of course is closely related to, and sometimes identical with, the absorption spectrum) which corresponded to the triplet absorption spectrum of anthracene in other phases. Here was clear evidence of the important role of the triplet state in the delayed fluorescence. They also found (fig. 2) that the ruby laser was really quite inefficient for producing delayed fluorescence—its wavelength falling here (point *X* in fig. 2) where the absorption coefficient is extremely small—while light of wavelength *Y* was much better. It was only because the ruby laser is so fantastically bright that any fluorescence had been obtained at all. Of course it seems never to have occurred to anybody to use red light for exciting anthracene until the ruby laser was first used.

It was at this stage that I was brought into the story. At NRC Singh was in a very productive phase and was using all suitable optical and electrical equipment he could lay his hands on. I happened to have a xenon lamp and monochromator, liquid helium cryostat, photomultipliers and associated gadgetry and so it was natural that Singh should start using it and to sweep me along in his progress.

We did two things—tried anthracene crystals of varying purity and varied the temperature of the crystals. The impure crystals showed only feeble—if any—delayed fluorescence. In fact later on it became apparent that the longer the decay time of delayed fluorescence the purer the crystal. In at least two laboratories—NRC and Du Pont—the decay time is used as an indication of the purity of a crystal on a routine basis, and we speak of a 5 ms crystal or a 20 ms crystal, and so on.

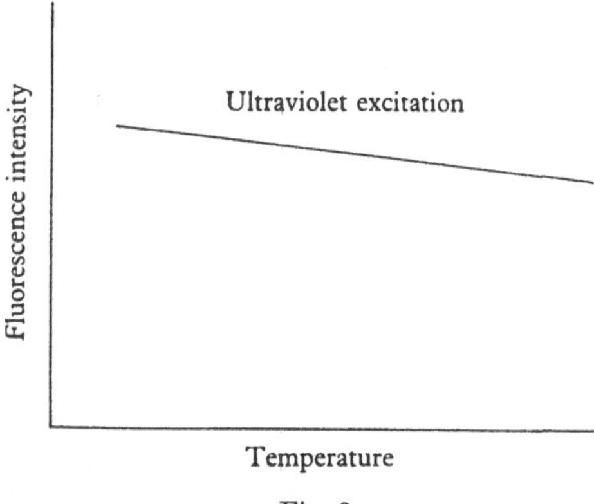

Fig. 3.

The effect of temperature was quite different. When anthracene is excited by ordinary ultraviolet light there is only a slight change in fluorescence intensity as the temperature is changed (fig. 3). The change is monotonic, with the fluorescence a little brighter at low temperatures than at high, and the change itself may be due to an experimental artifact rather than some intrinsic process.

When the anthracene is excited by red light, however, a very unusual temperature dependence is found (fig. 4). We now have several well-pronounced maxima, at approximately 7·5, 35 and 190 °K. We soon found that these were not due to experimental error but were quite real, but could think of no appropriate explanation, and in due course published a description of these experiments[5] without a proper explanation.

After some months, however, Willem Siebrand at NRC was able to explain the temperature effect on the basis of sets of traps.[6] A particular set of traps would have a certain trapping time and trapping temperature, and would thus become effective at a certain temperature. Siebrand was able to produce a curve which fitted the experimental curve very well indeed. His theory does not tell anything about the detailed nature of the traps, but we are fairly certain that they are physical defects in the crystal lattice (rather than chemical impurities). Some of our other experiments at NRC support this view.

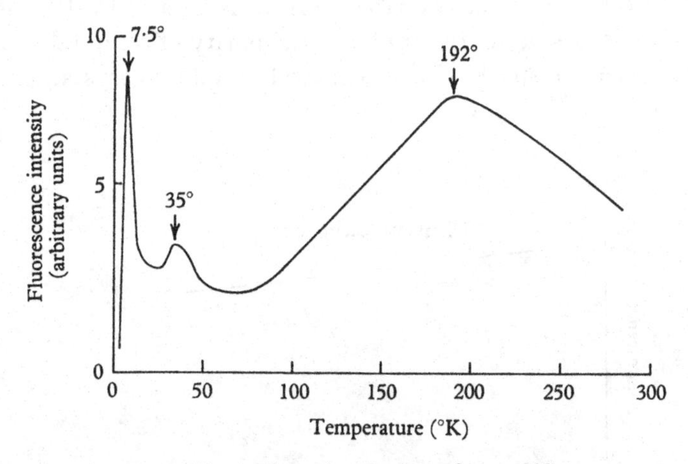

Fig. 4. Natural anthracene, $\lambda_{ex.} = 622$ mμ.

Turning back a bit, the delayed fluorescence can be discussed in terms of the following equation:

$$dn/dt = \alpha I - \beta n - \gamma n^2,$$

where n is the number or concentration of triplet excitons; t is the time; α is the absorption coefficient for the exciting light; I is the intensity of the exciting light; β is the monomolecular decay constant of the triplet state; and γ is the bimolecular interaction rate constant for triplet-triplet annihilation. We assume that since we are dealing with relatively feeble exciting light that n will be small and that the right-hand term—γn^2—can be neglected. In the steady state we can then have β or n change as the temperature changes.

It was therefore quite an obvious experiment to determine whether the decay time β varied as the temperature varied in a manner analogous to the intensity of delayed fluorescence. Although it was obvious and we set out to do it several years ago, the experiment is unfortunately a difficult one and we were plagued with a series of experimental pitfalls, accidents and other sorts of bad luck which would reduce you all to tears if I were to describe them in detail.

While we were going through all this Kepler probably did the whole experiment, as he gave a talk on this subject about a year ago.[7] However, we have not yet seen any publication of his detailed results.

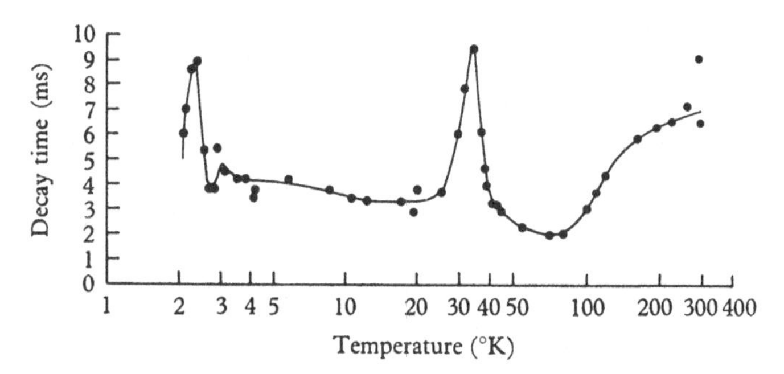

Fig. 5. Anthracene, decay time of delayed fluorescence.

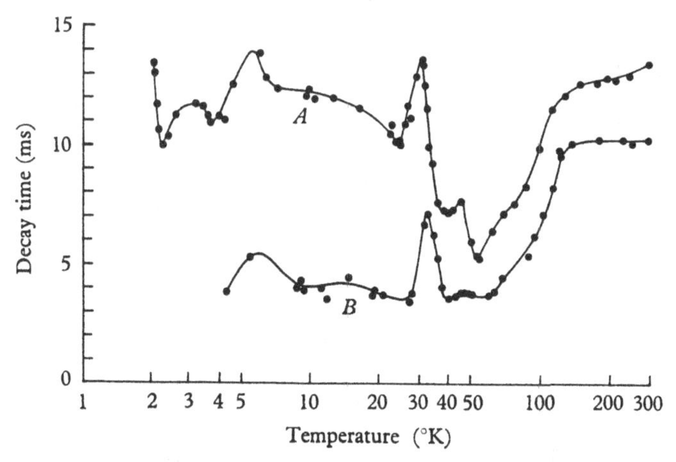

Fig. 6. Anthracene, sample no. 2, decay time of delayed fluorescence, $\lambda_{ex.}$ 6200 Å. *A*, Observed at 15–35 ms after stop of excitation. Also with blue filter (4–72) in front of PM. *B*, Observed under same conditions as sample no. 1, 7–15 ms after stop of excitation.

In any case we finally got everything working simultaneously just a few weeks ago, and the first results are shown in fig. 5. Some more accurate results obtained with an improved phosphoroscope are shown in fig. 6.

You can see at once that the decay time varies in a complex way similar to the delayed fluorescence intensity. There are at least four maxima, and the temperatures of some—at about 7 °K and 35 °K—correspond to those found in intensity. (I should mention that the decay time is the experimental value, and the triplet decay time is given by multiplying these values by two.) You can deduce from the slide that the decay is non-exponential. So far we have only worked with two crystals, and there are distinct differences between them.

That brings the story more or less up to the minute. As yet we do not have any detailed explanation of the results. However, it looks as though physical defects acting as traps are responsible for at least some of the phenomena. While on the subject of physical defects I would like to end by mentioning some results recently obtained by two other research workers—G. Sloan and V. Ern of Du Pont —which point out the importance of physical defects and the complexity of the effects they may produce. (These results have not yet been published.)

Sloan and Ern sealed some anthracene in a container. In the container they were able to grow a crystal from the melt or from the vapour. When grown from the melt the crystal had a delayed fluorescence decay time of about 24 ms. When grown from the vapour, however, the decay time was only 5 ms. The crystal could be grown in the two modes repeatedly, and the result was reproducible. Since the chemical purity must have been the same for the two forms the large difference in decay times must have resulted from differences in the numbers of forms of physical defects. There evidently remains much to be done in this field.

References

1 F. R. Lipsett & G. Macpherson, *Can. J. Phys.* **44**, 1485 (1966).
2 S. Singh & B. P. Stoicheff, *J. Chem. Phys.* **38**, 2032 (1963).
3 W. L. Peticolas, J. P. Goldsborough & K. E. Rieckhoff, *Phys. Rev. Letters*, **10**, 43 (1963).
4 P. Avakian, E. Abramson, R. G. Kepler & J. C. Caris, *J. Chem. Phys.* **39**, 1127 (1963).
5 S. Singh & F. R. Lipsett, *J. Chem. Phys.* **41**, 1163 (1964).
6 W. Siebrand, *J. Chem. Phys.* **42**, 3951 (1965).
7 R. G. Kepler, *Bull. Am. Phys. Soc.* **11**, 269 (1966).

ENERGY TRANSFER BETWEEN BENZENE AND BIACETYL AND THE LIFETIME OF TRIPLET BENZENE IN THE GAS PHASE

C. S. PARMENTER AND B. L. RING

Abstract

The intensity of biacetyl phosphorescence excited by energy transfer from triplet benzene in mixtures of benzene and biacetyl vapours exposed to a flash discharge has been studied as a function of time. 2500–2600 Å light from a 3 μs discharge populates the lowest excited singlet state ($^1B_{2u}$) of benzene in mixtures of 20 torr of benzene and 0·004–0·045 torr of biacetyl. The excited benzene decays by fluorescence to the ground state and by intersystem crossing to a triplet state within the duration of the flash. Energy transfer from triplet benzene then excites triplet biacetyl which can be detected by the biacetyl phosphorescence in the region 5000–6000 Å. This emission rises from zero intensity to its maximum in about 75 μs after the flash. The time required to reach this maximum is a function of the lifetime of triplet benzene in the absence of biacetyl, the rate of energy transfer to biacetyl and the lifetime of triplet biacetyl. The biacetyl emission ultimately displays the exponential decay characteristic of the biacetyl lifetime of $1{\cdot}7 \times 10^{-3}$ s. Observation of the time required for maximum phosphorescence intensity from mixtures with varying biacetyl concentrations indicates that the energy transfer rate constant at 300 °K is $5{\cdot}7 \times 10^{-11}$ cm^3 molecule^{-1} s^{-1} and that the lifetime of $^3B_{1u}$ benzene in the absence of biacetyl is $2{\cdot}6 \times 10^{-5}$ s. We have established that this lifetime is not determined by triplet-triplet interactions or by impurity quenching. The lifetime of this state in rigid media at 77 °K is about six orders of magnitude longer. Some processes which may account for the short gas phase lifetime are discussed.

CHARGE TRANSFER TRIPLET STATE OF MOLECULAR COMPLEXES

SUEHIRO IWATA, JIRO TANAKA AND SABURO NAGAKURA

The phosphorescence spectra of molecular complexes were first found by Reid[1] with the trinitrobenzene and aromatic hydrocarbon system and further investigations have been made by Czekalla *et al.*[3] and McGlynn and his co-workers.[2] In the course of the study on the emission spectra of molecular complexes, we have found the long-lived emissions different from those of the component molecules.[4] In the present paper we will report further investigations along this line. The acceptors used in the present study are 1,2,4,5-tetracyanobenzene (TCNB), tetrachlorophthalic anhydride (TCPA) and phthalic anhydride (PA), and the donors employed are hexamethylbenzene (HMB), durene and mesitylene. The aim of the present study is to find the mechanism of the long-lived emissions, and to correlate the energy levels of the complex system with those of component molecules. It will be seen that the charge transfer interaction is involved in the excited states of complexes, and the phosphorescence state is certainly regarded as due to the triplet state with charge transfer character.

Experimental results

Phosphorescence and fluorescence spectra were measured at 77 °K with a grating monochromator of JASCO CT-50 attached with a IP 28 photomultiplier tube as a detector. A mixed solvent of ethyl ether and isopentane (1 : 1) (abbreviated hereafter to EP) was used as a solvent.

The phosphorescence spectra of molecular complexes of TCNB (acceptor) with HMB, durene and mesitylene (donors) are shown in fig. 1, together with that of TCNB itself. The fluorescence spectra of these systems are shown in fig. 2 for the purpose of comparison. The phosphorescence and fluorescence spectra of the other systems are shown in figs. 3 and 4. The observed maximum wave numbers are tabulated in table 1. The lifetime of the phosphorescence was measured using the shutter and synchroscope. The results are shown in table 2.

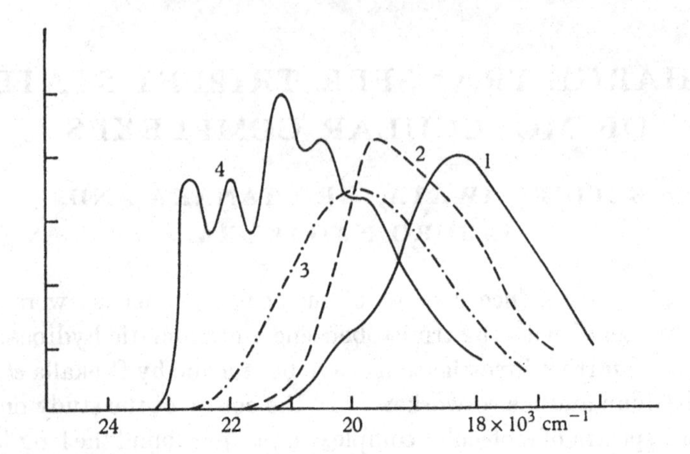

Fig. 1. The phosphorescence spectra of TCNB complexes in EP at 77 °K: curve (1) TCNB–HMB, (2) TCNB–durene, (3) TCNB–mesitylene, and (4) TCNB only.

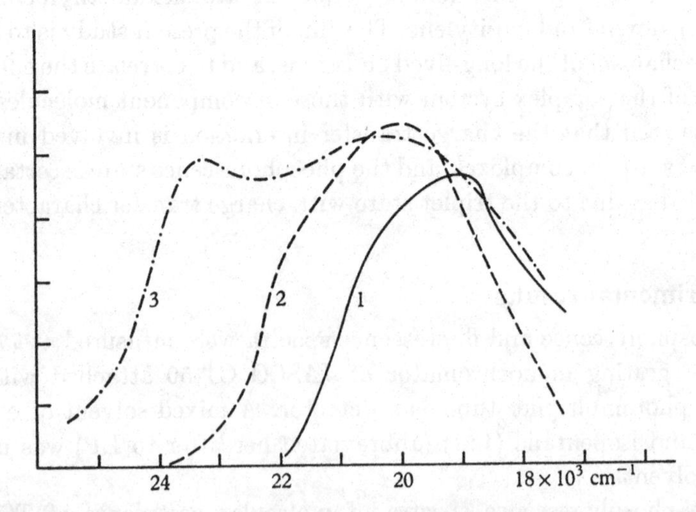

Fig. 2. The total emission (fluorescence plus phosphorescence) spectra of TCNB complexes in EP solution at 77 °K: curve (1) TCNB–HMB, (2) TCNB–durene, and (3) TCNB–mesitylene. The phosphorescence of TCNB–mesitylene complex which appears at 20 000 cm^{-1} is very much stronger than that of other complexes. The band at 23 000 cm^{-1} is fluorescence.

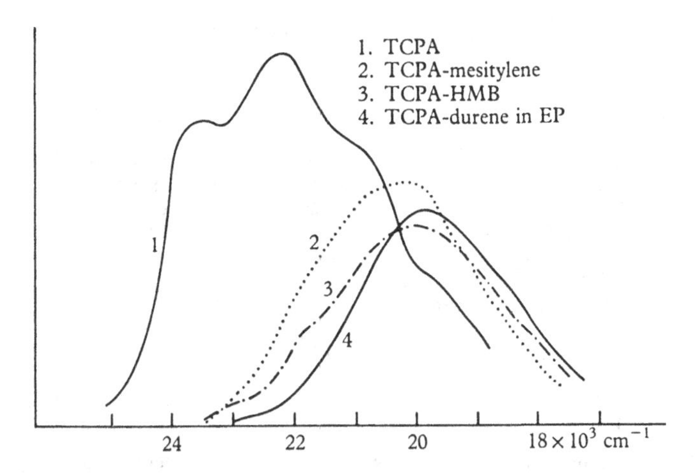

Fig. 3. The emission spectra of TCPA and its HMB complex at 77 °K: curve (1) phosphorescence spectra of TCPA in EP solution, (2) total emission (fluorescence plus phosphorescence) spectra of TCPA–HMB complex crystal, and (3) phosphorescence spectra of TCPA–HMB complex crystal.

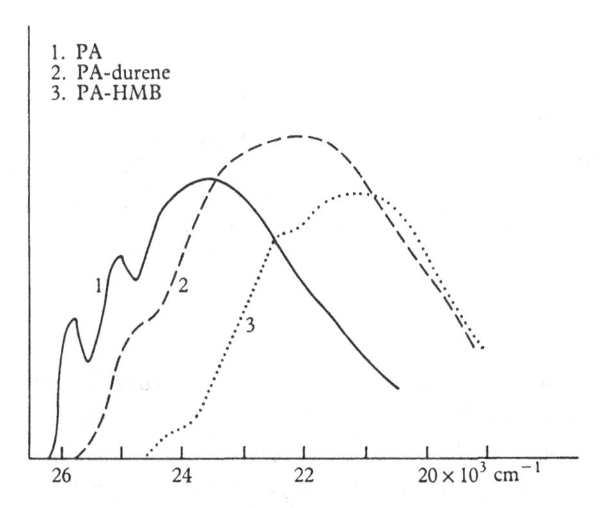

Fig. 4. The emission spectra of PA and its HMB complex in EP solution at 77 °K: curve (1) phosphorescence spectra of PA, (2) fluorescence spectra, and (3) phosphorescence spectra of PA–HMB complex.

Table 1. *Position of the fluorescence and the phosphorescence maxima in EP solution at* 77 °K

	Fluorescence (cm^{-1})	Phosphorescence (cm^{-1})
TCNB–Mesitylene	$23{\cdot}3 \times 10^3$	$19{\cdot}5 \times 10^3$
TCNB–Durene	20·0	18·5
TCNB–HMB	19·5	17·5
TCPA–Durene	21	19·8
TCPA–HMB	20·2	—†
TCPA–Durene (crystal)	—	19·6
TCPA–HMB (crystal)	20	19·5
PA–Durene	21·7‡	22·2
PA–HMB	21·7	21·0
TCNB	—	20·8, 22·5 (0–0)
TCPA	—	22·2, 23·5 (0–0)
PA	—	23·5, 25·8 (0–0)

† The maximum position was not determined because of the low solubility of the complex and of the strong free acceptor phosphorescence.

‡ After the measurement the solution was coloured.

Table 2. *The lifetime of the phosphorescence in the degassed EP solution at* 77 °K†

	Acceptors		
Donors	TCNB (s)	TCPA (s)	PA (s)
Acceptor only	3·2	0·21	1·4
Mesitylene	3·3	—	—
Durene	1·9	0·25 (0·25)‡	0·78
HMB	0·41	0·004 (0·008)‡	0·3

† The detected emissions were only CT phosphorescences (the influence of the acceptor phosphorescence on the lifetime was strictly avoided).

‡ The values in isopentane and *n*-propyl ether (1:1) mixed solution by Czekalla *et al.*[3]

Discussion

The experimental results indicate the following features: (1) The maximum wave numbers of phosphorescence bands decrease with the increasing donor abilities for a series of complexes with the same acceptor, namely ν(phosphorescence) in table 1 decrease in the order of mesitylene, durene and hexamethylbenzene complexes. The same tendency was observed with the CT fluorescence band which satisfies the mirror image relationship to the charge transfer absorption. (2) The phosphorescence bands are closely related to the charge transfer fluorescence bands rather than to the triplet states of acceptors. (3) The phosphorescence states of the acceptor are themselves higher in energy than the charge transfer fluorescence states of the complexes. This means that the charge transfer triplet states are the lowest excited states of the complexes, as is shown in fig. 5. In this figure are given the excited states of the HMB complexes with TCNB, TCPA and PA.

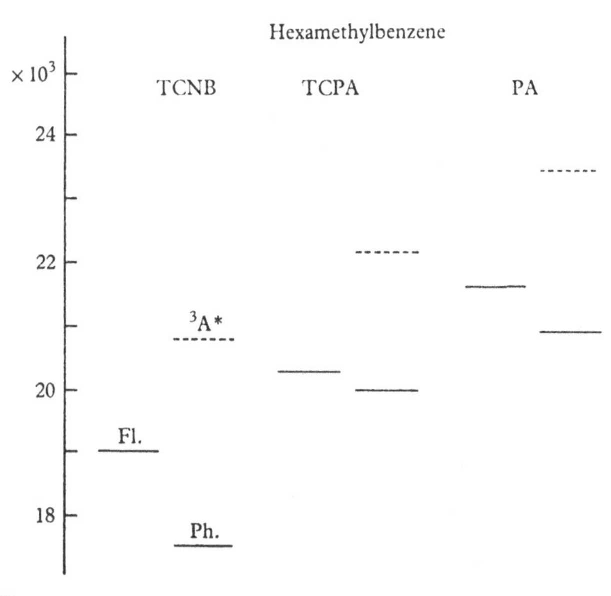

Fig. 5. The energy diagrams of the CT singlet states, CT triplet states and the acceptor lowest triplet states of TCNB–HMB, TCPA–HMB and PA–HMB complexes. Each level on the diagram corresponds to the maximum position of the observed emission spectrum.

From these findings it is most plausible that the phosphorescence state is mainly composed of the charge transfer triplet state.

In the earlier studies of the systems[1, 2, 3] including aromatic hydrocarbons as donors, the phosphorescence states were ascribed to donors' triplet states. This interpretation seems to be reasonable, because the inspection of the energy level diagrams of the complexes shows that their lowest triplet states are those of aromatic hydrocarbons (naphthalene, anthracene and coronene). In the case of the durene and HMB complexes with derivatives of phthalic anhydride, however, Czekalla & Mager[3] could not interpret the observed red shift of the phosphorescence spectrum on this basis.

The triplet level of charge transfer complex has first been discussed by Mulliken.[5] He argued that the location of the level might depend on many complicated factors, such as the height of the charge transfer level and the strength of intermolecular interaction. According to the valence-bond treatment for the charge transfer state, the triplet level will be more unstable by $2K$ than the singlet level, where K means the exchange integral in valence-bond term.[6] However, as was pointed out by Mulliken, the singlet level would be lifted through the charge transfer interaction with the ground state, hence the triplet state might be slightly higher or nearly the same in energy with the singlet charge transfer state.

The mixing of the singlet charge transfer configuration with the singlet locally excited configurations of donor and acceptor was first discussed by Murrell.[7] The interaction between the triplet configurations can be treated in the same manner. In the present case, the triplet charge transfer configuration and the triplet locally excited configuration within acceptor are so close that their mixing should occur considerably. The magnitude of off-diagonal matrix element for this mixing is difficult to determine. We tentatively assume the value to be $0{\cdot}2 \sim 0{\cdot}5$ eV for the interplanar separation of 3·4 Å. Then, the triplet state is considered to be lower in energy by about $0{\cdot}1 \sim 0{\cdot}5$ eV than the charge transfer singlet state.

On the basis of this mechanism, we could succeed in explaining characteristics of the phosphorescence of the molecular complexes under consideration. It may be concluded that the emission occurs from the charge transfer triplet state which was strongly perturbed and mixed by the acceptor's triplet state.

The lifetime of phosphorescence tabulated in table 2 shows a remarkable shortening in the case of HMB complexes. It may imply that an enhancement of intersystem crossing will occur in this case, where the highest occupied and lowest vacant molecular orbitals of donor are doubly degenerate. A further investigation will be required to clear up this point.

References

1 C. Reid, *J. Chem. Phys.* **20**, 1212 (1952).
2 S. P. McGlynn, J. D. Boggus & E. Elder, *J. Chem. Phys.* **32**, 357 (1960).
3 J. Czekalla, G. Briegleb, W. Herre & H. J. Vahlensieck, *Z. Elektrochem.* **63**, 715 (1959); J. Czekalla & K. J. Mager, *ibid.* **66**, 65 (1962).
4 S. Iwata, J. Tanaka & S. Nagakura. A paper presented at the *International Symposium on Photochemistry*, Tokyo, 1965.
5 R. S. Mulliken, *J. chim. phys.* **61**, 20 (1963).
6 H. Eyring, J. Walter & G. E. Kimball, *Quantum Chemistry*, (John Wiley & Sons, New York, 1944), p. 213.
7 J. N. Murrell, *J. Am. Chem. Soc.* **79**, 4839 (1957).

FLASH-PHOTOLYTIC DETECTION OF TRIPLET ACRIDINE FORMED BY ENERGY TRANSFER FROM BIACETYL

A. KELLMANN AND L. LINDQVIST

Abstract

The lowest triplet states of acridine and biacetyl in benzene solution were studied by flash photolysis. The spectrum of the acridine triplet was determined in detail, and first- and second-order rate constants of its decay were obtained ($100\ s^{-1}$ and $3{\cdot}4 \times 10^9\ M^{-1}\ s^{-1}$, respectively). The triplet state of biacetyl was found also to be deactivated by a second-order rate process, with consumption of the biacetyl, in addition to the ordinary phosphorescence process. Flash photolysis of solutions containing both compounds, with excitation limited to the absorption band at 430–450 mμ of biacetyl, established the appearance of triplet acridine formed by energy transfer from triplet biacetyl. It was found that the acridine triplet reacts chemically with ground state and triplet biacetyl.

The photochemical importance of the triplet state is due to a large extent to its long lifetime compared to that of excited singlet states. There are also specific factors determining the reactivities of these states, and one observes cases where only the excited singlet state is active. Thus, according to studies by Kellmann & Dubois,[1] and recently by Kira, Ikeda & Koizumi,[2] the photoreduction of acridine in hydrogen-donating solvents does not seem to be possible via the lowest triplet state (a π–π^* state) but only via electronic states of higher energy. The former authors arrived at this conclusion from studies of the quenching of biacetyl phosphorescence by acridine: they did not observe any reduction of acridine by the solvent (ethanol) in this process, although the π–π^* triplet acridine (the only excited electronic state lying below triplet biacetyl in energy) is presumably formed in the quenching reaction.

The present study using the flash photolysis technique was undertaken in an attempt to confirm the above findings by establishing the actual formation of the acridine triplet in the quenching process.

Experimental

The flash photolysis apparatus consists of a lamp assembly (six oxygen-filled, 20 cm long discharge tubes) producing flashes of a maximum discharge energy of 3750 joule; flash duration time is 4 μs.[3] A jacketed cell (20 cm long) contained the solution to be studied. A saturated copper sulphate solution (cut-off 310 mμ) or alternatively a saturated sodium nitrite solution (cut-off 405 mμ) in the cell jacket were used as filters for the photolytic light. A d.c. xenon lamp, a grating monochromator (band width 1 mμ), a photomultiplier tube (EMI 9558 BQ) and an oscilloscope were used in the usual way to measure the transient changes in optical density of the flash-exposed solutions.

The solutions were degassed in a flask directly attached to the reaction cell, by a repeated freeze-thaw procedure. Oxygen-free argon was introduced above the solution between each degassing cycle. Benzene (Merck's, chromatographically pure) was recrystallised 3 times in ethanol. Biacetyl (Fluka) was used without purification. In a few runs the biacetyl was redistilled, and stored in the dark at low temperature.

Results and discussion

Before studying the interaction between biacetyl and acridine these reagents were investigated separately using the flash technique. Both compounds have been studied previously by this method; however, the conditions were not the same as those of the present study, and complementary information was therefore required.

Acridine. The absorption spectrum of the lowest triplet state of acridine in benzene has been reported;[4] a pronounced absorption maximum at approximately 440 mμ was observed. We obtained the same band which appeared transiently on flashing 10^{-6} M acridine in benzene solution using copper sulphate as filter. Fig. 1 shows the spectrum in detail. A striking feature of the spectrum is the similarity between the ground states and triplet acridine spectra in the range 330–390 mμ. In addition, weak bands are observed at 435 and 520 mμ. The intensity of the transient absorption increased with increasing flash light output reaching an approximately constant level at high light output (2000 joule discharge energy). The saturation effect was assumed to be due to complete conversion of the solute to the triplet state. A value for the extinction coefficient of $2{\cdot}5 \times 10^4$ M^{-1} cm^{-1} was obtained on this basis.

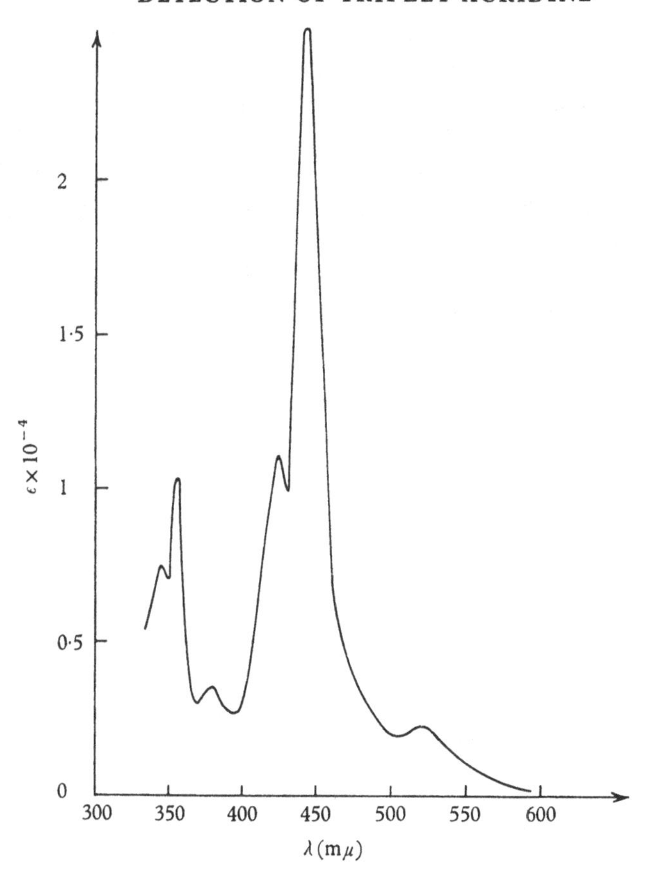

Fig. 1. π–π triplet acridine spectrum.

The decay of the triplet state (A_T) of a molecule to the ground state (A) is assumed to be due, in general, to the following processes:

$$A_T \rightarrow A + h\nu \qquad \text{(rate constant } k_1\text{)}, \tag{1}$$

$$2A_T \rightarrow 2A \qquad (k_2), \tag{2}$$

$$A_T + A \rightarrow 2A \qquad (k_3). \tag{3}$$

Flash studies on $(0{\cdot}5-10)\times 10^{-6}$ M acridine solutions were performed at varying light intensities to determine the relative importance of the above processes. The results did not show any contribution from reaction [3] in the present case. The decay rate of the transient absorption after flashing may then be expressed by the relation

$$d/dt\,(\log \Delta D) = k_1 + k_2 \Delta D/\epsilon l,$$

where ΔD is the difference between the optical density at a time t after the flash and that before flashing. At wavelengths where acridine does not absorb, ϵ is the extinction coefficient of the triplet state; l is the cell length. Fig. 2 shows that this relation is obeyed, and one obtains $k_1 = 100\,s^{-1}$, $k_2 = 3{\cdot}4 \times 10^9\,M^{-1}\,s^{-1}$, $k_3 < 2 \times 10^7\,M^{-1}\,s^{-1}$. The maximum value of k_1 is orders of magnitude lower than the values previously reported;[4] this probably reflects a lower impurity content in the solutions in the present study. The value of k_2 corresponds closely to that for a diffusion-controlled solution.

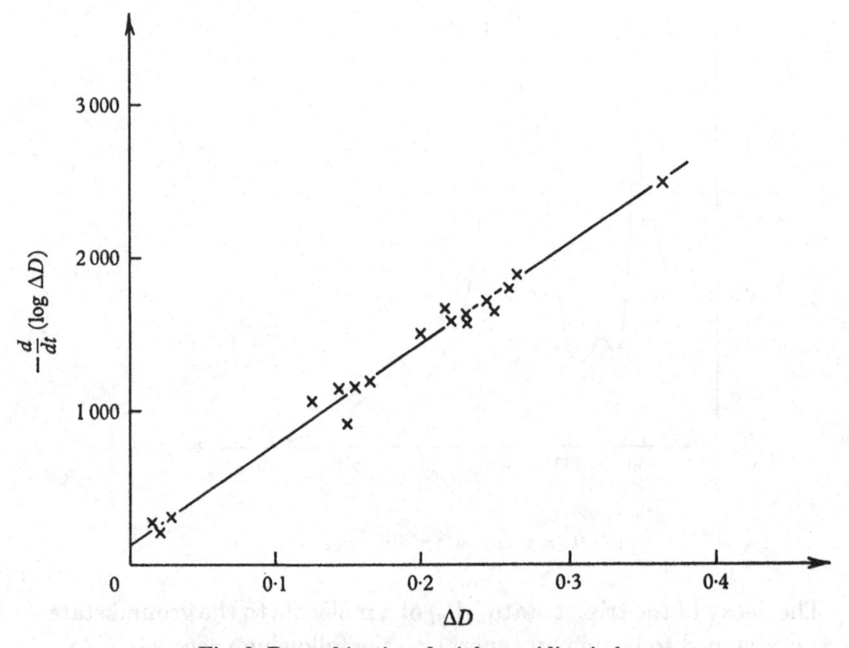

Fig. 2. Decay kinetics of triplet acridine in benzene.

The kinetic study revealed an unexpected increase in decay rate of the triplet with the number of flashes ($\sim$ 10 % increase per flash). This quenching effect was found to be due to the formation of a product which was detected spectrophotometrically. The product showed a monotonically decreasing absorption from 280 to 420 mμ corresponding to the spectrum of fulvene. Flash runs on benzene alone gave the same absorption which is apparently due to photolysis of benzene occurring in spite of the use of the copper sulphate filter.

Biacetyl. Porter & Windsor [5] found on flashing biacetyl a transient absorption with a maximum at 317 mμ that they attributed to the triplet state. We have studied $(4–40) \times 10^{-5}$ M biacetyl in benzene solution using a sodium nitrite filter to isolate the visible range, and found a spectrum similar to that given by these authors; however, a permanent consumption of 1–2% biacetyl per flash was also observed. In this study it was necessary to use relatively high concentrations of biacetyl to obtain a measurable transient absorption, and it was not possible to saturate the solutions with light as was done in the acridine study.

Since the transient absorption was weak it was felt that it might possibly be due to primary decomposition products of the biacetyl. An attempt was therefore made to confirm the triplet assignment by studying the biacetyl phosphorescence after the flash. Using the flash apparatus set for study at 525 mμ (biacetyl phosphorescence peak wavelength) without the analysing xenon light source and at high photomultiplier gain it was possible to observe the decay of the phosphorescence after the flash. Although these measurements, due to instrumental difficulties, were only semi-quantitative, they showed that the phosphorescence and absorption decays were directly related to each other. This confirms that the transient spectrum is due to triplet biacetyl.

The rate of decay of the 320 mμ transient absorption indicated an important contribution by a bimolecular triplet biacetyl (B_T) process. This bimolecular reaction leads to the formation of products or directly to the ground state. (B):

$$2B_T \rightarrow 2B \text{ or products} \quad (k_4), \tag{4}$$

in addition to the decay due to phosphorescence

$$B_T \rightarrow B + h\nu. \tag{5}$$

A relation between k_4 and the unknown ϵ of biacetyl triplet at 320 mμ was obtained: $k_4 = 10^6 \times \epsilon$. Assuming a diffusion-controlled reaction an approximate value of $\epsilon = 5 \times 10^3 \text{M}^{-1} \text{cm}^{-1}$ was obtained and the population of the triplet state during the flash would then amount to approximately 1% for a 4×10^{-4} M solution of biacetyl.

Acridine and biacetyl. Kellmann & Dubois[1] demonstrated the quenching of biacetyl phosphorescence by acridine. The following reaction was assumed to occur:

$$B_T + A \rightarrow B + A_T. \tag{6}$$

The present flash study was done on solutions containing acridine (10^{-6} M) and biacetyl ($(4–40) \times 10^{-5}$ M). A sodium nitrite filter was used to limit the excitation to the biacetyl. The formation of triplet acridine was followed at 440 mμ. Due to stray light the observations could be started only 50 μs after the flash; by this time a high percentage of the acridine was converted to the triplet state. This finding clearly confirms the occurrence of reaction (6). The triplet concentration decreased at first relatively slowly (during approximately 200 μs) and then more rapidly—the rate in this later stage was faster than that of triplet acridine in the absence of biacetyl and at the same concentration. The first slow decay is apparently due to continued activation of acridine by triplet biacetyl present in excess, competing with the deactivating processes. The fast decay is supposed to be due to the quenching reactions

$$A_T + B_T \rightarrow A + B \tag{7}$$

and

$$A_T + B_T \rightarrow \text{products} \tag{8}$$

in addition to reactions (1)–(3).

The occurrence of reaction (8) would explain the high irreversible acridine consumption which amounted to 15 % per flash. This consumption made a quantitative analysis of the results difficult. The quenching of the triplet acridine could possibly also be due to biacetyl in the ground state

$$A_T + B \rightarrow A + B \text{ or products.} \tag{9}$$

The relative importance of reactions (7)–(9) was not determined; a very detailed study would be required to separate these reactions.

The formation of triplet acridine was also studied as a function of the flash output. At high outputs a constant value of 40 % conversion to the triplet state was observed. This steady state established during flashing appears to confirm the occurrence of the deactivating reaction (7), competing with the activation reaction (6).

Conclusions

It is seen from this study that the photochemical interactions of biacetyl with acridine are complex, and more work needs to be done to obtain a complete picture of the reaction scheme. However, the present results show definitely that acridine is formed in the lowest triplet state by energy transfer from triplet biacetyl and confirm thus the results of Kellmann & Dubois which indicated that acridine is not reduced via its lowest triplet state. However, the present work has pointed out that quenching by biacetyl (reactions (7)–(9)) present in the solution may compete with a reduction reaction that might otherwise be occurring.

References

1 A. Kellmann & J. T. Dubois, *J. Chem. Phys.* **42**, 2518 (1965).
2 A. Kira, Y. Ikeda & M. Koizumi, *Bull. Chem. Soc. Japan*, **29**, 1673 (1966).
3 L. Lindqvist (to be published).
4 A. V. Buettner, *Dissertation Abstr.* **23,** 4125 (1963).
5 G. Porter & M. W. Windsor, *Proc. Roy. Soc.* (*London*) A, **245**, 238 (1958).

EXTINCTION COEFFICIENTS OF TRIPLET-TRIPLET TRANSITIONS BETWEEN 3000 AND 8800 Å IN ANTHRACENE

R. ASTIER AND Y. H. MEYER

Abstract

The complete triplet-triplet anthracene absorption spectrum has been obtained in solution between 3000 and 8800 Å. New weak bands are observed in the visible range. Assignments for the different electronic levels observed are made by comparison with Pariser's and Kearns' calculations.

The direct measurement of the absolute value of the maximum extinction coefficient for the anthracene 4250 Å band gives: 90000 (litre mole^{-1} cm^{-1}) at 113 °K. By way of comparison the value of the 4130 Å triplet-triplet naphthalene band has been measured: 45000 (litre mole^{-1} cm^{-1}). Other previously reported values are discussed.

Since the higher triplet states were identified by Lewis, Lipkin & Magel,[1] McClure,[2] Porter & Windsor,[3] the transient absorption of several[4, 5, 6] molecules has been studied. However, only strong transient absorption bands were reported. Only recently was Kellogg[7] able to identify new much weaker anthracene bands in the near infra red.

We have been looking for other weak bands corresponding to parity forbidden transitions for this molecule in the as yet unexplored visible range, as the calculations of Pariser[8] and of Kearns[9] led to expect.

In so far as bands already known are concerned, extinction coefficients given by the different authors are rare and divergent. Thus we have tried to make a *direct* measurement in the best possible conditions.

The case of naphthalene, on which more results are available, was re-examined by way of comparison.

Experimental

We used the standard technique known as 'flash photolysis'. However the sensitivity and precision were increased thanks to the simultaneous use of: (1) very high power pumping and a large excitation surface; (2) low concentrations and temperatures which reduce radiationless deactivation; (3) an absorption cell of optical quality even at low temperatures, in order to have a straight and known light path (length 21 cm).

The solvent used was a mixture of spectrograde ethanol and methanol (3 : 1), and the concentration varied from 10^{-4} to 10^{-7} M/l. At the temperature used (– 160 °C) the mixture is highly viscous. The solution is contained in a transparent silica tube, 10 mm I.D., closed at each end by a ground silica piston. The grinding allows the piston to follow the liquid contraction (15 %) during cooling. The cell is cooled by a nitrogen flow in a transparent quartz dewar and is lit by two xenon flash tubes VQ X 621 mounted in a bi-elliptical reflector. An energy of 180 joules per flash is delivered in 30 μs. The white light used for spectrum analysis is furnished by another xenon flash tube VQ X 65. The parallel beam crosses the cell and is received in a Hilger and Watts 'Medium' spectrograph. A chopper in front of the slit stops any stray excitation light or fluorescence from entering. By means of a delay line, excitation can be triggered some time before observation (between 200 μs and several ms). We used the Kodak II L plate for which the γ was determined by densitometry.

The product used was Eastman 'X 480' and 'H 480 synthetic' anthracene. No difference was observed in the transient spectra.

Degassing the solution proved to be unnecessary. We verified that degassing by repeated cooling in liquid nitrogen and thawing under a vacuum of 5×10^{-6} torr followed by filling of the cell in a nitrogen atmosphere does not modify the transient spectrum obtained at – 160 °C in any way.

Repeated irradiations diminish the absorption band intensity, while a permanent absorption band continuum appears in the ultraviolet. The decay rate of the compound was measured and taken into account in the calculation of the ϵ.

Results

(1) *Triplet-triplet anthracene spectrum*

The spectrum we obtained (fig. 1) includes the two band groups found respectively by Porter [6] and Kellogg [7] and also shows new bands in between these groups as well as a new band at 3810 Å.

This weak band seems to correspond to the band found at 3700 Å in naphthalene.[5, 6] In the case of anthracene it could be observed only thanks to the total depletion of the ground state. We shall assign it to the vibrational level $2 \times 1370\,cm^{-1}$ of the T_5 state rather than to the $^3A_{1g}^+$ state as Pariser suggested for the corresponding naphthalene band. For the $^3A_{1g}^+$ state appears more logically on our anthracene spectrum at 5400 Å (T_3).

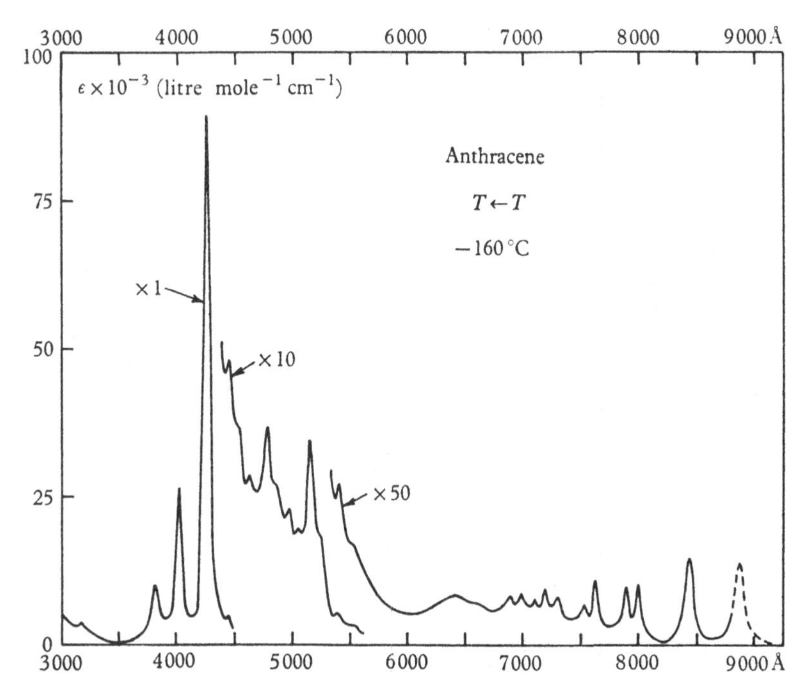

Fig. 1. Transient absorption spectrum of anthracene in solution (ethanol: 3, methanol: 1) at 113 °K. The band at about 8900 Å is taken from Kellogg.[7]

The weak bands between 8000 and 8800 Å exactly confirm Kellogg's result. On our spectrum the first band was eliminated by the sensitivity limit of the plate. On the other hand, more bands appear towards higher energies and correspond to higher vibrations, dominated by the frequency $600\,cm^{-1}$. This level was attributed by Kellogg to the forbidden transition predicted by Pariser (fig. 2).

In the 4500–6000 Å region the intermediate intensity bands can be assigned to either the $^3A_{1g}^+$ or $^3B_{3u}^-$ state, since their energy corresponds to Pariser's calculations. They are one order of magnitude stronger than the infra red bands. If one admits that the transitions towards the $^3B_{1g}^+$ and $^3A_{1g}^+$ states have the same oscillator strength (since they are equally forbidden) then one may deduce that the stronger T_4 bands are due to the $^3B_{3u}^-$ state. The vibrational analysis

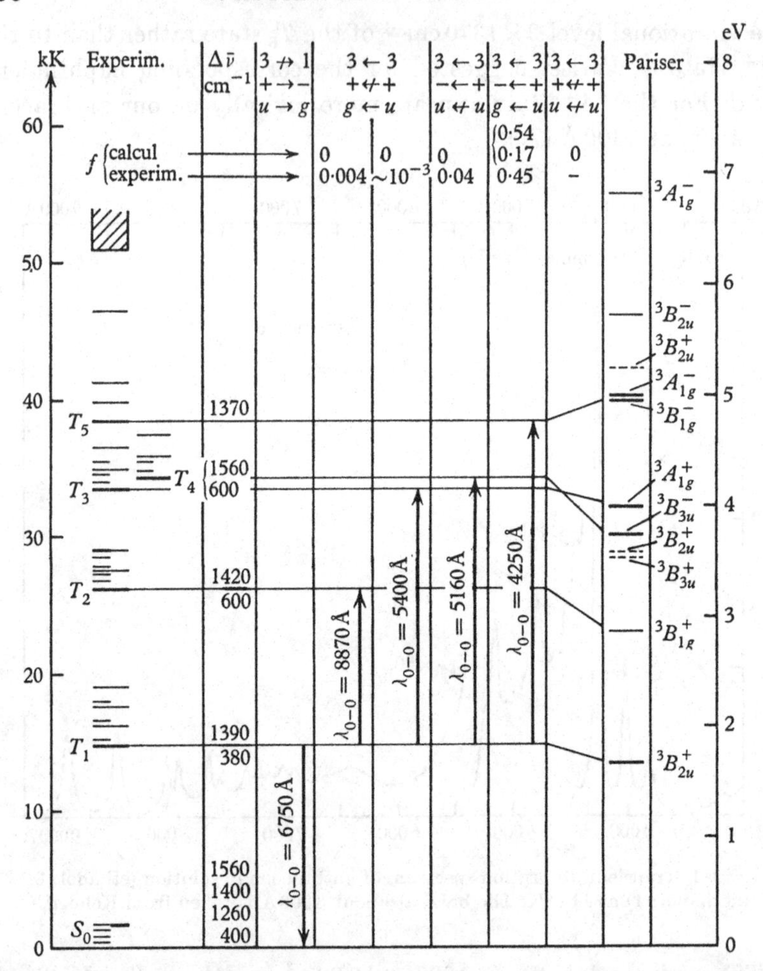

Fig. 2. Diagram of anthracene triplet levels. Frequencies of the T_1 state are taken from ref. (11). The position of the T_1 state and ground state vibrations are from ref. (12). Oscillator strengths were calculated by means of the well-known but approximate formula: $f = 4{\cdot}3\ 10^{-9} \int \epsilon\, d\bar{\nu}$.

does seem to confirm the existence of two superimposed electronic levels (table 1).

From these assignments, one can now choose without ambiguity between the six possible states discussed by Kearns.[9] The Moffitt 'perimeter model' method which Kearns applied to the calculation of the triplet state level energies does indeed give for his assignment *I* energy values which correspond within about 1 000 cm^{-1} to the figures we obtained (table 1).

Table 1. *Frequencies observed for triplet-triplet anthracene absorption*

(Comparison of experimental and calculated values for electronic levels)

$\bar{\nu}$ observed (cm^{-1})	Assignment $\Delta\bar{\nu}$ (cm^{-1})		$\bar{\nu}$ calculated Pariser (cm^{-1})	$\bar{\nu}$ calculated perimeter model Kearns (cm^{-1})
(11270[7])	T_1–T_2 0–0	—	13400 ($^3B_{1g} \leftarrow {}^3B_{2u}$)	12400 ($^3B_b \leftarrow {}^3L_a$)
11870	600	—	—	—
12520	2 × 600	—	—	—
12700	1420	—	—	—
13100	3 × 600	—	—	—
13290	1420 + 600	—	—	—
13710	4 × 600	—	—	—
13900	1420 + 2 × 600	—	—	—
14090	2 × 1420	—	—	—
14310	5 × 600	—	—	—
14510	1420 + 3 × 600	—	—	—
18200	?	—	—	—
18500	$T_1 - T_3$ 0–0	—	18700 ($^3A_{1g} \leftarrow {}^3B_{2u}$)	19000 ($^3L_b \leftarrow {}^3L_a$)
19080	600	—	—	—
19380	—	$T_1 - T_4$ 0–0	16600 ($^3B_{3u} \leftarrow {}^3B_{2u}$)	~20000 ($^3B_a \leftarrow {}^3L_a$)
19780	2 × 600	—	—	—
20100	1560	600	—	—
20640	1560 + 600	2 × 600	—	—
20930	—	1560	—	—
21660	2 × 1560	1560 + 600?	—	—
22150	—	1560 + 2 × 600	—	—
22510	—	2 × 1560	—	—
23510	$T_1 - T_5$ 0–0	—	{26400 ($^3B_{1g} \leftarrow {}^3B_{2u}$) 26800 ($^3A_{1g} \leftarrow {}^3B_{2u}$)	— —
24880	1370	—	—	—
26250	2 × 1370	—	—	—
31600	?	—	—	—

(2) *Extinction coefficients*

The few reported measurements of the absolute value of anthracene are not in agreement with each other (table 2). There is the same disagreement for naphthalene.

Table 2. *Maximum triplet-triplet extinction coefficients*

	McClure (1951)	Craig & Ross (1954)	Porter & Windsor (1958)	Keller & Hadley (1965)	Ramsay & Munro (1966)	Present work
	77 °K	77 °K	295 °K	77 °K	77 °K	113 °K
	(EPA)	(EPA)	(Paraffin)	(butane, isopentane)	(*n*-butanol)	(*Et*OH, *Me*OH)
Anthracene λ = 4250 Å	2800	> 45000	71500	—	—	90000
Naphthalene λ = 4130 Å	480	> 10000	10000	14000	710	45000

When we took our spectroscopic measurement with a concentration of 10^{-6} the singlet-singlet anthracene absorption bands had completely disappeared. It can thus be deduced that a minimum of 95% of the molecules were in the triplet state. For the 4253 Å band, one is then able to measure *directly* a molar triplet-triplet extinction coefficient: $\epsilon_T = 90\,000 \pm 10\,000$ (litre mole^{-1} cm^{-1}). This value tallies with that of Porter and Windsor, taking into account the temperature band narrowing. The increase of the maximum ϵ between 295 and 113 °K was measured in the singlet-singlet bands to be (140 ± 10)% for both anthracene and naphthalene.

For the 8428 Å infra red band we found an ϵ in good agreement with Kellogg's value.

For naphthalene we were unable to obtain total depletion of the ground state, because the absorption matched poorly with the flash emission. Depopulation of the ground state was found to be 50%. Taking into account not only photolysis but also the transient absorption continuum which lies from 4400 Å to shorter wavelengths, we found $\epsilon = 45\,000 \pm 5\,000$ (litre mole^{-1} cm^{-1}) at 4130 Å and 113 °K. This value is well above those already given, the difference probably arising from an overestimation of the true triplet concentration in previous works.

In conclusion, table 2 results can be divided into two categories:

(1) Those obtained by indirect methods involving such other factors as radiationless deactivation and yield (McClure; Keller & Hadley; Ramsay & Munro).

(2) Those obtained by direct methods (Craig & Ross; Porter & Windsor; present work).

We note that indirect methods generally lead to much lower values for ϵ. If greater confidence is to be given to direct methods, this would seem to indicate that certain data involved in molecular processes and in particular the radiationless deactivation rate, though widely admitted, should be reconsidered.

Acknowledgement

We should like to thank Mr O. de Witte for the design of the electronic equipment and Mr Boisleve for its construction.

References

1 G. N. Lewis, D. Lipkin & T. Magel, *J. Am. Chem. Soc.* **63**, 3005 (1941).
2 D. S. McClure, *J. Chem. Phys.* **19**, 670 (1951).
3 G. Porter & M. W. Windsor, *Disc. Faraday Soc.* **17**, 178 (1954).
4 D. S. McClure & P. L. Hanst, *J. Chem. Phys.* **23**, 1772 (1955).
5 D. P. Craig & I. G. Ross, *J. Chem. Soc.*, p. 1589 (1954).
6 G. Porter & M. W. Windsor, *Proc. Roy. Soc.* (*London*) A, **245**, 238 (1958).
7 R. E. Kellogg, *J. Chem. Phys.* **44**, 411 (1966).
8 R. Pariser, *J. Chem. Phys.* **24**, 250 (1956).
9 D. R. Kearns, *J. Chem. Phys.* **36**, 1608 (1962).
10 R. A. Keller & S. G. Hadley, *J. Chem. Phys.* **42**, 2382 (1965).
11 P. Avakian, E. Abramson, R. G. Kepler & J. C. Caris, *J. Chem. Phys.* **39**, 1127 (1963).
12 M. R. Padhye, S. P. McGlynn & M. Kasha, *J. Chem. Phys.* **24**, 588 (1956).
13 I. A. Ramsay & I. H. Munro, *International Conference on Luminescence*, Budapest, 1966.

ANTHRACENE TRIPLET-TRIPLET ANNIHILATION RATE CONSTANT

GERHARD FINGER, O. ZAMANI-KHAMIRI,
JOHN OLMSTED III, AND A. B. ZAHLAN

Abstract

A new experimental method is described for the determination of (γ/c), where γ is the triplet-triplet bimolecular annihilation rate constant and c the fractional yield of excited singlets: $2T \xrightarrow{\gamma}$ products, $2T \xrightarrow{c\gamma} S^* + S_0$. The method has been applied to anthracene in the vapour phase in the temperature range 430–553 °K and vapour pressure range 0·5–1·5 torr. (γ/c) is found to decrease from 21×10^{-10} at 430 °K to $7{\cdot}2 \times 10^{-10}$ cm^3 s^{-1} at 553 °K; and from this it is estimated that c varies from 0·3 to 1·0.

Introduction

Porter & West[1] have studied the kinetics of bimolecular decay of the triplet state of anthracene in the vapour phase using flash photolysis and kinetic spectrophotometry

$$2T \xrightarrow{\gamma} \text{products}. \tag{1}$$

However, accurate knowledge is not available of the rate of the annihilation reaction of two triplets to give an excited singlet and a ground state singlet,

$$2T \xrightarrow{c\gamma} S^* + S_0, \tag{2}$$

where c is the ratio of triplets annihilating via reaction 2 to those annihilating via reaction 1. Reaction 1 includes 2.

Triplet-triplet collisions may also be elastic. Thus, γ can be written as

$$\gamma = f\sigma v. \tag{3}$$

Here f is the fraction of collisions leading to a reaction, σ is an appropriate molecular collision cross-section, and v is the molecular thermal velocity. If one assumes $f = 1$ and substitutes 'reasonable' values for σ and v, one estimates that γ is of the order of 10^{-10} cm^3 s^{-1}.

We have modified the experimental flash lamp technique utilised earlier[2] to measure the ratio of singlet-triplet rate constant k_{ST} to the overall singlet decay rate constant k. This modification, under certain appropriate assumptions, permits the determination of the ratio (γ/c) from the experimental data obtained. The technique does not require independent knowledge of other physical quantities, such as the absorption coefficient. Moreover, when combined with measurements of the type made by Porter and West, it permits the evaluation of both γ and c.

Theory and procedure

We assume that the following set of reactions adequately describes the processes occurring in a cell containing pure hydrocarbon vapour in the presence of exciting radiation:

$$S_0 + h\nu \longrightarrow S^*, \tag{4}$$

$$S^* \xrightarrow{k_{ST}} T, \tag{5}$$

$$S^* \xrightarrow{k_S} S_0 + h\nu \text{ or heat}, \tag{6}$$

$$T + T \xrightarrow{c\gamma} S^* + S_0, \tag{2}$$

$$T + T \xrightarrow{(1-c)\gamma} 2S_0 + \text{heat}, \tag{7}$$

$$T \xrightarrow{\beta} S_0 + \text{heat}. \tag{8}$$

Under these conditions, one may write the rate of production of triplet states T:

$$dn_T/dt = k_{ST}n_S - 2\gamma n_T^2 - \beta n_T. \tag{9}$$

Consider a system excited by a 2 μs flash. In the time interval immediately after termination of the flash, the rate of production of excited singlets is

$$dn_S/dt = c\gamma n_T^2 - kn_S, \tag{10}$$

where $k = k_{ST} + 1/\tau$ and $1/\tau = k_S$.

The primary observable quantity in such a system is the emitted photons, which can be observed in the form of the fluorescent signal, $F_1(t)$, generated at the output of a photomultiplier. This signal can be related directly to the excited singlet state concentration:

$$F_1(t) = VBQkn_S(t), \tag{11}$$

where Q is the quantum yield of the process

$$S^* \rightarrow S_0 + h\nu, \tag{12}$$

B is an instrumental constant, and V is the emitting volume.

In the time interval immediately following the termination of the flash, the concentration of singlets can be found by invoking the steady-state approximation:

$$dn_S/dt = c\gamma n_T^2 - kn_S = 0; \quad n_S = c\gamma n_T^2/k. \tag{13}$$

If this is substituted into the differential equation for n_T and the integration performed, one finds

$$n_T = \frac{n_{T_0}}{1+(\beta+\alpha n_{T_0})t}, \tag{14}$$

and therefore $$F_1(t) = VBQ\gamma c\left[\frac{n_{T_0}}{1+(\beta+\alpha n_{T_0})t}\right]^2. \tag{15}$$

Here n_{T_0} is the number of triplet states immediately after termination of the flash, and $\alpha = \gamma(2-ck_{ST}/k)$.

This can be cast into a form convenient for graphical display by inverting and taking the square root. Furthermore, for reasonable initial concentrations of triplet states, the β term is negligible, yielding

$$F_1(t)^{-\frac{1}{2}} = (BQV\gamma c)^{-\frac{1}{2}}[1/n_{T_0}+\gamma(2-ck_{ST}/k)t]. \tag{16}$$

Thus, the inverse square root of the observed fluorescent decay signal plotted as a function of time should give a straight line with slope,

$$S = \left(\frac{\gamma}{VBQc}\right)^{\frac{1}{2}}(2-ck_{ST}/k)\,\text{volt}^{-\frac{1}{2}}\,\text{s}^{-1}. \tag{17}$$

Measured values for the ratio k_{ST}/k are in the literature.[2] If a method for measuring BQ is available, the ratio γ/c can be determined. Such a method is described below.

The apparatus used for the flash lamp experiments has been described previously.[2] Briefly, it consists of a double aluminum block oven into which a fused silica cylindrical cell is placed, in such a way that the lower portion of the oven controls the vapour pressure, and the upper portion controls the temperature within the cell. This cell is irradiated with the output of a GE BH 6 mercury arc flash lamp, and the resulting delayed fluorescence signal monitored by an IP 28 photomultiplier and displayed on the screen of a Tektronix 555 double-beam oscilloscope. The lamp output is filtered through Corning 7–60 + 0–54 filters and the fluorescence signal through 2 × 3–73 + 1 × 5–58 filters.

For the steady-state experiments, the flash lamp is replaced by a 1 kW GE BH 6 lamp operated continuously. Its output is passed through a Carl Leiss double monochromator and the 3663 Å mercury line selected to fall on the cell. The power of the transmitted beam is determined by a calibrated Eppley thermopile connected to an L and N Type K-3 potentiometer and an L and N model 9834 DC null detector. The same photomultiplier and oven are used as in the flash lamp experiment. The procedure used in the set of experiments was as follows: an 18 mm I.D., 41 mm optical path fused silica cell, previously scrubbed with hot 1 : 1 H_2SO_4–HNO_3 followed by complete rinsing and baking, was filled with excess anthracene by sublimation under vacuum. This cell was introduced into the aluminum oven, and initial measurements were made to obtain P_0 and F_{S0}, using steady-state illumination with the vaporisation oven at room temperature to maintain vapour pressure less than 10^{-2} torr. The transmitted power P_1 and fluorescent signal F_S were then measured for a variety of pressure and temperature settings. The transmitted power readings were corrected for the effect of thermal radiation reaching the thermopile from the oven by subtracting the power readings obtained at the appropriate temperatures with the exciting light beam off. This was deemed more accurate than the use of an infrared filter because it eliminated the need for correction of the power reading for transmission and reflectance losses in the filter. Under these conditions it can be shown that

$$BQ = F_S h\nu/(P_0 - P_1), \tag{18}$$

where ν is the frequency of exciting radiation.

Subsequent to these measurements, the Eppley thermopile was removed and the steady-state light source replaced with the flash lamp arrangement, without disturbing the oven-photomultiplier geometry. The temperature and pressure settings utilised in the earlier runs were then duplicated and the delayed fluorescence signal measured in the time domain 200–700 μs after the flash.

Results and discussion

When one combines equations (17) and (18), one finds for (γ/c) the expression

$$\frac{\gamma}{c} = \frac{S^2(h\nu)\, F_S\, V D}{(2 - c k_{ST}/k)^2\, (P_0 - P_1)}. \tag{19}$$

The constant D is the calibration factor for the Eppley thermopile.

For the apparatus used for anthracene, the following constants were determined: $V = 10{\cdot}2\ \text{cm}^3$, $k_{ST}/k = 0{\cdot}3 \pm 0{\cdot}1$, $h\nu = 5{\cdot}43 \times 10^{-12}$ erg, $D = 6{\cdot}0 \times 10^{-11}$ volt s erg^{-1}, $F_S/(P_0 - P_1) = (4{\cdot}7 \pm 0{\cdot}6) \times 10^4$. The anthracene vapour pressure was in the range 0·5–1·5 torr. Slope values and the calculated values of (γ/c) at different vapour temperatures are recorded in table 1. For these calculations it was assumed that $c = 0{\cdot}5$; varying c from 0 to 1 on the right-hand side of equation (19) alters (γ/c) by $\pm 20\,\%$.

Table 1

T(°K)	$S \times 10^3$	$(\gamma/c) \times 10^{10}$ ($cm^3 s^{-1}$)
433	6·7	21
456	7·0	23
472	4·8	11
474	4·6	10
522	4·4	9·1
548	4·4	9·1
553	4·0	7·2

The major sources of error arise from:

Stability of steady state light source—15 %.
Scattered light corrections—5 %.
Accuracy of Eppley readings—2 %.
Accuracy of determination of scope trace average—10 %.
Uniformity of excited state distribution—5 %.

From the above, we estimate that our values for (γ/c) are accurate to about 35 %.

It is clear from table 1 that (γ/c) is temperature dependent outside the range of error. The velocity, which enters into γ, varies as the square root of the temperature. This predicts an increase by a factor of 1·13 in γ over the 433–553 °K temperature range. The observed trend in (γ/c) is in the opposite direction. This could be due to changes in f, σ, and/or c. It is reasonable to attribute this variation to c, which implies that vibrationally excited triplet molecules have a greater probability of leading to an excited singlet. Porter & West, find for anthracene, $\gamma = 6{\cdot}4 \times 10^{-10}\,cm^3\,s^{-1}$ at a vapour temperature of 413 °K. This, in conjunction with the above data, leads to $c = 0{\cdot}3$ at 433 °K and $c = 1{\cdot}0$ at 553 °K. Porter & West used solution triplet-triplet extinction coefficients in their determination of γ, and therefore these c values may involve considerably greater error than the 35 % estimated for the (γ/c) ratios.

Acknowledgements

Acknowledgement is made to the donors of the Petroleum Research Fund, administered by the American Chemical Society, for partial support of this work. We also wish to thank Dr R. C. Jarnigan for the sample of zone refined anthracene.

References

1 G. Porter & P. West, *Proc. Roy. Soc.* (*London*) A, **279**, 302 (1964).
2 A. B. Zahlan, S. Z. Weisz, R. C. Jarnigan & M. Silver, *J. Chem. Phys.* **42**, 4244 (1965).

DISCUSSION

PARKER

Kellogg to Parker: What distances do you find for the triplet-triplet interaction?

Parker to Kellogg: 15 or 30 Å according to whether the rate is controlled by rotational movements alone, or by slow diffusive transport in the rigid medium—but the values are only very approximate. This is an awkward range to explain. There are two possible mechanisms: (1) annihilation by an exchange interaction which appears to occur in the range of 10–15 Å; (2) Förster-type transfer from one triplet to the T–T absorption band of the other giving a higher triplet. This could occur over distances of 30 Å. To yield S_1 the latter mechanism would have to be followed by feed-back into the singlet manifold, which does not seem likely due to the probability of rapid internal conversion to T_1.

Avery to Parker: I think that there are some theoretical reasons for believing that transfer of triplet excitation energy can take place over distance as large as 15–30 Å. When the Breit interaction is expanded in a Taylor series, terms result which are very small in magnitude but which can nevertheless be effective in causing energy transfer during the extremely long lifetime of the triplet state. For example, I believe that we must consider spin-spin coupling.

Robinson to Parker: What is wrong with Kellogg's dipole-dipole mechanism for long-range transfer via the triplet-triplet absorption band? I would think that high up on the energy level diagram, the multiplicity does not matter so much, but rather it is the exact nature of the energy surfaces which count. Thus even though the final state for such long range transfer is a 'triplet' it could happen that this is an efficient mechanism for population of lower singlets. Certainly one could design experiments to check this point, or are there experimental data already that bear on this?

Parker to Robinson: (No comment.)

Rice to Parker: I believe that too much emphasis is being placed on the precise value of the 'range' of interaction (15–30 Å); because of the approximations made in the analysis of the data, I believe the figures quoted may be very different—perhaps as small as 5 Å. This belief is based on the observation that the diffusion kinetics use macroscopic parameters to describe microscopic events. Collision (i.e. reaction) diameters, are unknown. While they remain valuable as a general guide, I doubt that kinetic analyses of this type can give precise information on the interaction range.

Stevens to Parker: When unimolecular triplet state relaxation predominates, no triplet state concentration gradient is set up; it is therefore doubtful whether the Debye equation, based essentially on the existence of this gradient, and which you use to estimate the encounter annihilation probability, has any validity.

Parker to Stevens: What we measure is a bimolecular rate constant for $T+T \rightarrow S^*+S$. We express this rate constant as the product (pk_c) where k_c is the value calculated from the Debye equation. I agree that the actual rate of $T+T \rightarrow$ products may be higher than k_c.

Stevens to Parker: P-type and E-type delayed fluorescence are essentially associated with aromatic hydrocarbons and dye molecules respectively. Have you any evidence for a dye exhibiting P-type delayed fluorescence or for an aromatic hydrocarbon exhibiting the P-type emission?

Parker to Stevens: We have seen weak P-type delayed fluorescence from proflavine, acridine orange and eosin (very weak) by working in ethanol at about −80° to −100 °C where E-type delayed fluorescence is small. We have observed both P-type delayed fluorescence and E-type delayed fluorescence in solutions of chlorophylls *a* and *b* under appropriate conditions.

We have not observed E-type delayed fluorescence from any of the aromatic hydrocarbons that we have investigated.

Windsor to Parker: Dr Dawson and Dr Kropp in my laboratory have observed a delayed fluorescence from coronene in a plastic medium. The fluorescence intensity is a function of temperature. We believe this delayed fluorescence to be of Parker's E-type; i.e. it occurs by thermal population of S_1 from T_1. As in eosin the $S_1 - T_1$ gap in coronene is relatively small (4 500 cm^{-1}) and therefore the E-type process is favored.

Kazzaz to Parker: With reference to your results on the delayed fluorescence of pyrene solutions, containing anthracene as an im-

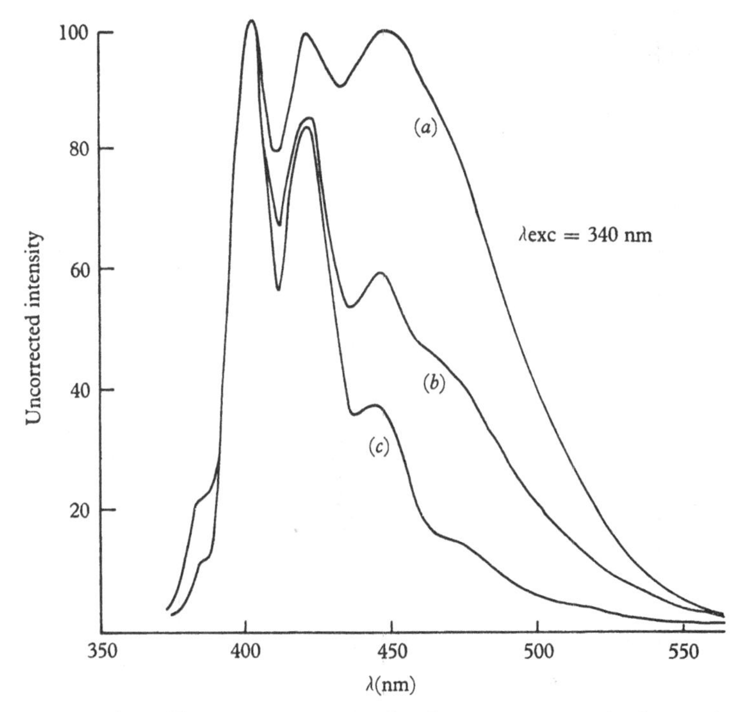

Fig. 1. Fluorescence spectra of anthracene–pyrene mixed crystal. (*a*) 10^{-4} mole/mole; (*b*) 10^{-3} mole/mole; (*c*) 10^{-2} mole/mole.

purity, in which anthracene emission is detected, I wish to report the observation of the effect of anthracene impurity on the prompt fluorescence of pyrene in the crystalline phase.

Fig. 1 shows the fluorescence spectra from three pyrene crystals containing controlled concentrations of anthracene of 10^{-4}, 10^{-3} and 10^{-2} mole/mole. As the anthracene concentration increases, the pyrene fluorescence yield is quenched and that of the anthracene emission increases.

Excitation spectra of the anthracene fluorescence correspond to the absorption spectrum of a pure pyrene crystal. This demonstrates that the anthracene molecules are excited indirectly via energy transfer from the pyrene.

Parker to Kazzaz: In your crystal experiments it seems that sensitisation of the anthracene occurs via the excited singlet states. All our experiments on *delayed* fluorescence refer to transfer of energy from

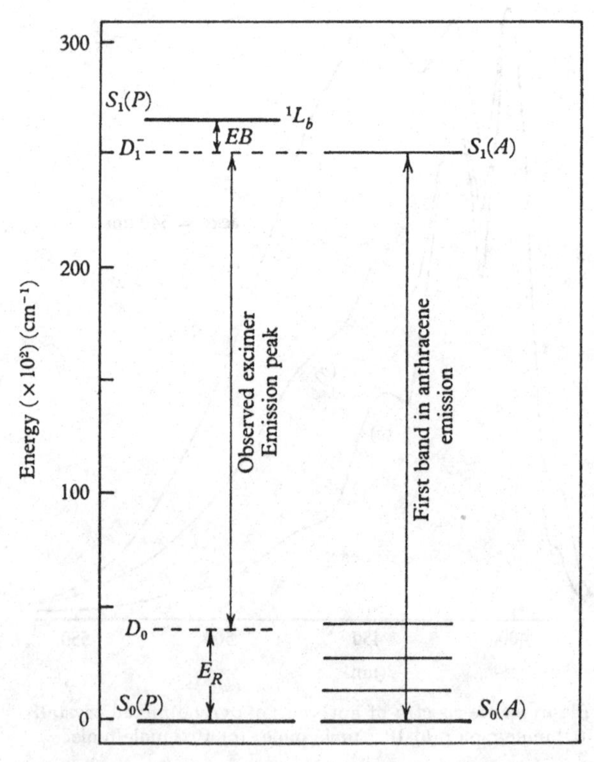

Fig. 2. Energy diagram for anthracene (A) in pyrene (P) system. D_1, lowest excited singlet state of the excimer; D_0, is the ground state of the excimer; E_R, is the repulsive energy of the pyrene excimer ground state.

triplet states. In one particular experiment the solution contained a high concentration of pyrene and much lower concentrations of 1,2-benzanthracene and anthracene. Acridine orange was added as a triplet sensitiser—its triplet lies *above* that of anthracene but below that of pyrene and 1,2-benzanthracene. We irradiated with 436 nm which was absorbed *only* by the acridine orange and hence only the triplet state of anthracene was populated. The singlet state of the pyrene was never populated so the question of pyrene dimer does not arise.

Meyer to Parker: What was the concentration of triplets obtained when using cooled solutions? Is it possible that simple triplet-triplet reabsorption can then interfere?

Parker to Meyer: Down to $-100\,°C$ the concentration of triplets was not sufficient for T–T absorption to be appreciable.

Birks to Parker: The triplet-triplet energy transfer process suggested by Parker, to explain the delayed fluorescence behaviour of pyrene solutions at low temperatures, is improbable and unnecessary. A long-range process can transfer energy, but it is most unlikely to transfer angular momentum (spin). Hence, triplet-triplet energy transfer produces a molecule in an excited triplet state T_n of twice the energy of the lowest excited triplet state T_1. T_n is then internally converted through the vibrational levels of the triplet manifold to T_1, so that the net result is triplet quenching. Kellogg has shown that Förster theory satisfactorily explains triplet-triplet energy transfer. Unless intersystem crossing from the triplet manifold in the region of T_n to the singlet manifold is much more efficient than the internal conversion through the triplet manifold, T^*-T^* energy transfer cannot produce a molecule in the lowest excited singlet state S_1 to yield delayed fluorescence. I know of no experimental or theoretical evidence for the assumption required by Parker's hypothesis, that intersystem crossing can be much more efficient than internal conversion.

Parker's hypothesis was introduced to explain the temperature dependence of ϕ_D^d/ϕ_M^d, the ratio of the quantum yields of delayed excimer and monomer fluorescence. Parker, and independently Tanaka, Tanaka, Stevens and Hutton, consider the low-temperature behaviour of ϕ_D^d/ϕ_M^d for pyrene solutions in ethanol to be anomalous and they have invoked long-range T^*–T^* interaction to explain it. The reaction kinetic analysis (Birks and Moore, this Conference), which is independent of the exact nature of the T^*–T^* association process, shows that ϕ_D^d/ϕ_M^d (equation (21) *loc. cit.*) depends on eight rate parameters, six of which (k_M, k_D, k_{DM}, k_{MD}, k_{TM}, k_{TD}) are temperature dependent. The temperature dependences of the six corresponding rate processes all contribute to the temperature dependence of ϕ_D^d/ϕ_M^d and it is incorrect to ascribe the latter only to the temperature dependence of k_{TD} and k_{TM} or of their ratio α.

Birks, Moore & Munro (ref. 7 *loc. cit.*), and Moore & Munro (ref. 11, *loc. cit.*) have measured α and various rate parameters for several solution systems. For pyrene in cyclohexane or ethanol, $\alpha = 2{\cdot}0$ at 20 °C, and it retains this value (within the experimental error) at higher temperatures. As originally suggested by Birks (ref. 3, *loc. cit.*), the collisional diffusion-controlled triplet-triplet association of rate constant k_{TT} initially yields an excited excimer D^{**}. A fraction $\alpha/(1+\alpha)$ of D^{**} are internally converted to the lowest excimer singlet state D^*. The remaining fraction $1/(1+\alpha)$ dissociates into $M^{**}+M$, and M^{**} then internally converts into the lowest monomer singlet state M^*. Hence, for $\alpha = 2$, 6/9 of D^{**} yield D^*, 3/9 of D^{**} yield M^*+M. These values can be explained on a simple multiplicity argument, originally suggested by Dr B. Stevens (private communication). There are nine possible states of D^{**}, one singlet ${}^1D^{**}$, three triplets ${}^3D^{**}$, and five quintets ${}^5D^{**}$, and it is assumed that these states are formed with equal probability. ${}^3D^{**}$ dissociates, so that at least 3/9 of D^{**} yields M^*+M. ${}^1D^{**}$ converts to D^*, and if ${}^5D^{**}$ also converts to D^*, the remaining 6/9 of D^{**} yields D^*. Hence, $\alpha = 2$.

A reduction in temperature is unlikely to affect the behaviour of ${}^3D^{**}$ or ${}^1D^{**}$. By Hund's rule ${}^5D^*$ the lowest quintet state is lower in energy than ${}^1D^*(\equiv D^*)$, the lowest singlet state of the excimer. Let us assume that conversion from ${}^5D^{**}$ to ${}^1D^*$ proceeds by the following steps: internal conversion to ${}^5D^*$ through the quintet manifold, followed by thermal activation through an energy gap

$$\Delta_{15}(={}^1D^*-{}^5D^*)$$

and intersystem crossing to ${}^1D^*$. This process is in competition with the internal quenching and fluorescence of ${}^5D^*$, but the latter may be assumed negligible because of the identity of the prompt (${}^1D^*$) and delayed excimer fluorescence spectra. If the quantum yield of ${}^1D^*$ from ${}^5D^*$ at absolute temperature T is $\phi_{15}(T)$, then

$$\alpha = \frac{1+5\phi_{15}(T)}{3}, \qquad (1)$$

Moore and Munro (ref. 11, *loc. cit.*) have observed a decrease in α with T for pyrene in ethanol consistent with (1). The results have been analysed (Birks, *Physics Letters*, **24** *A*, 479 (1967) submitted for publication) to evaluate $\phi_{15}(T)$ and estimate Δ_{15}. Equation (1) predicts that α varies between 0·33 (low temperature limit) and 2·0 (high temperature limit). All the present data on α (refs. 3, 7, 11, *loc. cit.*) are consistent with this.

The reaction kinetics require a slight modification to allow for the intermediate state D^{**}, which was not included in the previous analysis (table 1, *loc. cit.*). Instead of the conventional description of the diffusion-controlled D^{**} formation process as *unidirectional*, of rate constant

$$k_{TT} = 8\mathbf{R}Tp/3\,000\eta, \tag{2}$$

where $p(\leqslant 1)$ is the reaction probability per encounter, it should be described as *reversible* ($T^* + T^* \rightleftharpoons D^{**}$) with a forward rate of $k'_{TT}[T]$ and a backward rate of k_{TTD}, where

$$k'_{TT} = 8\mathbf{R}T/3\,000\eta. \tag{3}$$

These equations, which apply only to collisions between similar molecules, have been verified experimentally for pyrene (refs. 5, 6, *loc. cit.*). Comparison of the unidirectional and reversible reactions, and consideration of the reaction thermodynamics (ref. 8, *loc. cit.*) shows that

$$p = k_{TT}/k'_{TT} = 1 - k_{TTD}/k'_{TT}, \tag{4}$$

$$= 1 - \exp(-\Delta S/\mathbf{R})\exp(\Delta H/\mathbf{R}T), \tag{5}$$

where ΔH and ΔS are the enthalpy and entropy of the reaction, respectively. Equations (1)–(5) provide an adequate description of triplet-triplet association and delayed fluorescence, without the introduction of Parker's hypothesis. The previous analysis (*loc. cit.*) is based on the assumption that p is independent of T, which from (5) can only occur when $p = 1$. The linear relation between ϕ_T and $1 - \phi_F$ (fig. 4, *loc. cit.*) which results from this assumption, provides experimental evidence for the validity of (3) and (2), with $p = 1$, for pyrene at normal temperatures. Equation (5) predicts a decrease of p below unity at low temperatures. If the ϕ_T data in figs. 3 and 4 for ethanolic solutions are renormalised to $(1 - \phi_F)$ at the highest temperature of measurement, where $p \simeq 1{\cdot}0$, the deviation from linearity at lower temperatures appears consistent with the temperature variation of p predicted by (5).

D^{**} is a special example of an exciplex (excited complex dissociated in the ground state) in which the two components are identical. The general reaction kinetics of the formation of exciplexes, which are applicable to excimers, π–π exciplexes, n–π excipleses and excifluors (fluorescent exciplexes) and to concentration and collisional impurity quenching, have been considered elsewhere (Birks, *Nature*, submitted for publication). When the exciplex components differ, the equations equivalent to (4) and (5) remain valid, but those equivalent to (2) and (3) require multiplication by a geometrical factor A to allow for the difference in the van der Waals (Stokes) and collision radii of the reactants. It is common practice to apply (2) directly to collisions between dissimilar molecules. This procedure yields an apparent value of p, which we may designate p'', which is no longer the reaction probability per encounter, since $p'' = Ap$.

SILVER

Windsor to Silver: Dr Dawson and I have measured the triplet yield of anthracene (actually $C_{14}D_{10}$) in EPA (*Molecular Crystals*, March–April 1967, in the Press). We obtain a value of $0{\cdot}53 \pm 0{\cdot}10$. Since the second triplet of anthracene lies below the first excited singlet state, the triplet yield should be independent of temperature and I would expect ϕ_T in the vapour to be similar to the value we obtain. Your value for ϕ_T depends very much on the value of γ_2/γ_1, used. What do you believe to be the correct value of γ_2/γ_1, and what value of ϕ_T does this give?

The case of phenanthrene is probably very similar. Although an excited triplet lying below S_1 has not yet been detected experimentally it has been predicted theoretically (Pariser, see also Kellogg's comment on my paper). Therefore, the value of 0·78 for ϕ_T reported by Lamola and Hammond (*J. Chem. Phys.* **43**, 2129 (1965)) can probably be transferred to the gas phase without risk of great error. It would then agree fairly well with your measurement of 0·88 for phenanthrene vapour.

Silver to Windsor: Just from the data, and adding up $\phi_{ST}+\phi_F+\phi_{IC}$, one is left with a value of γ_2/γ_1 of about 4 or 5 assuming that $\phi_{IC} \ll 1$. Our value is of course a lower limit on ϕ_{ST} since we do not know γ_2/γ_1 and ϕ_{IC}. I agree with your comment on the phenanthrene.

Stevens to Silver: The independence of the excited singlet state lifetime of phenanthrene of vapour pressure is consistent with the absence of fluorescence self-quenching in solutions of this compound at room temperature, attributed to dominant dissociative relaxation of the excimer under these conditions. However, in the case of anthracene vapour, the fluorescence self-quenching probability has a negative temperature coefficient, ascribed to a temperature-dependent dissociative excimer relaxation. Under these conditions, self-quenching may involve intersystem crossing of singlet to triplet excimer which then dissociates to ground and triplet states, as suggested by Wilkinson for pyrene. This would lead to an alternative pressure-dependent path of molecular intersystem crossing—have you taken this into account?

Silver to Stevens: I believe that this comment on anthracene is more relevant to the paper by Finger *et al.* because they have clearly done much more than we have done.

WOLF

Robinson to Wolf: In the benzene-h_6: benzene-d_6 system which we have studied extensively in the region 1·6–4·2 °K, we still believe there is a long-range T–T annihilation process that is not thermally activated. However, we now believe that the mechanism may not involve the virtual states of the host but could be long-range dipole-dipole through a strong $T \rightarrow T^*$ absorption in the acceptor molecule, as discussed by Kellogg. However, our temperature-dependent measurements have not been nearly as complete as yours so we cannot really rule out the thermally activated process in this case, on the basis of any experiment. Since the energy gap is 200 cm^{-1} and the temperature is equal to or below 4·2 °K, thermal activation does seem unlikely. Both the thermal and non-thermal processes could occur with domination of one over the other depending on the nature of the system, temperature, concentration, etc.

We used to see the 60 (or 35) cm^{-1} traps that you mention for the naphthalene-h_8: naphthalene-d_8 system, but lately we have not been able to see them even though we have tried to produce them. Thus I think whatever causes these traps can be removed, but I cannot say how, since we do not know what change was made in our purification and crystal growing procedure that got rid of these traps.

Wolf to Robinson: There is no question that the trap with 0·0 at 21 270 cm^{-1} can be removed by still higher purification.

Parker to Wolf: I have a comment relevant to that of Professor Robinson. We have tried a double-beam experiment in which we produced a high population of phenanthrene triplets in EPA at 77 °K by excitation at 313 nm and then excited the triplets by simultaneous irradiation in their visible absorption band. We found no increases in the intensity of the phenanthrene delayed fluorescence when the visible beam was switched on. Of course we may not have populated the same upper triplet level as that populated by triplet-to-triplet energy transfer.

Stevens to Wolf: From your slides it appears that, relative to the phosphorescence intensity, the temperature-independent delayed fluorescence I°_{DF} at low temperatures is much higher in the naphthalene-h_8 : naphthalene-d_8 system than in any of the others. Perhaps if this were subtracted from the temperature-dependent intensity, i.e. the plotted function is amended to $\log[(I_{DF} - I^{\circ}_{DF})/I^2_p]$, the thermal activation energy would approach the spectroscopic value as in the other systems. Have you tried this?

Wolf to Stevens: Thank you for the suggestion. We did not try these corrections, because the delayed fluorescence at lowest temperatures is certainly due to still more shallow traps which we do not know in detail. If we apply the correction you suggest, the activation energy in the low temperature region increases, but only very little. The overall picture remains unchanged so far as I see it today.

Sharnoff to Wolf: How sharp is the zero-phonon phosphorescence line of naphthalene-d_8 in durene? Are the zero-phonon lines in the other systems you have studied of similar sharpness?

Wolf to Sharnoff: I cannot give an exact value for the linewidth. So far the sharpest lines we have seen are of the order of 2 cm^{-1}. If the trap depth is lower the line shape and linewidth change and become more like a free exciton line.

Van der Waals to Wolf: Do you see durene phosphorescence from the system naphthalene in durene at very low temperature?

Wolf to van der Waals: I see a very intense phosphorescence which partially overlaps that of naphthalene and is due to the matrix. But this is the emission of the impurity mentioned by Dr Siebrand in an earlier discussion.

Hutchison to Wolf: The paramagnetic resonance measurements of triplet state energy transfer rates made by Dr Ashford and myself and mentioned in my paper, give a value of $300 \pm 40\,cm^{-1}$ for the energy of thermal activation of the transfer process in the case of β-methylnaphthalene in a naphthalene host crystal. This is in agreement with the value reported by Dr Wolf.

BIRKS

Hochstrasser to Birks: Can you utilise the electron impact method to detect higher energy triplet states that are hidden under singlet states?

Birks to Hochstrasser: The electron excitation cross-sections of triplet states are relatively sharp functions of the electron energy, while those of singlet states are broad. It should, therefore, be quite easy to resolve the sharp triplet excitation spectrum against the broad singlet excitation spectrum. Moreover, if the fluorescence photons originating from singlet excitation are detected and any signals coincident in time with these are rejected, the singlet background can be completely removed.

Cundall to Birks: I should like to support Dr Birks remarks concerning excitation of triplet states by ionising radiation. Pulse radiolysis in particular is very well adapted to studying processes such as triplet quenching and energy transfer.

It can be used to measure triplet-triplet extinction coefficients. Dr Land of the Christie Hospital, Manchester, has measured values at the absolute maximum for anthracene ($57\,200\,M^{-1}\,cm^{-1}$), naphthalene ($22\,600\,M^{-1}\,cm^{-1}$), phenanthrene ($22\,000\,M^{-1}\,cm^{-1}$), 1,2-benzanthracene ($25\,100\,M^{-1}\,cm^{-1}$), all relative to a maximum extinction coefficient for benzophenone ketyl around $540\,m\mu$, determined to be $3\,220\,M^{-1}\,cm^{-1}$. A figure of $\sim 17\,700\,M^{-1}\,cm^{-1}$ was obtained for the extinction coefficient of the benzophenone triplet from similar energy transfer experiments in benzene. The value for naphthalene differs from that obtained by Moore and Munro, referred to by Dr Birks.

BIRKS [AND MOORE]

Silver to Birks: Regarding the use of multiplicity, the best data on this are from Helfrich of NRC from a study of charge carrier recombination. However, even here, a very different path involving trapping in charge transfer states and thermal detrapping gives almost the same yields for prompt and delayed recombination radiation.

LIPSETT

Silver to Lipsett: The evidence for many but not all traps being physical is increasing. Annealing experiments in naphthalene and anthracene have indicated this. Many other workers have come to similar conclusions.

Lipsett to Silver: Certainly. As you know, we have carried out some of these experiments at the National Research Council. However, because of the (sometimes) enormous effects of trace quantities of chemical impurities on the spectra of organic compounds, I still think one must be cautious about drawing *firm* conclusions. In particular, Craig and Santi-Faroni have shown that anthracene dimerises even in the dark in a sealed container, so one must bear in mind that these dimers *may* be influencing the decay times.

Dexter to Lipsett: Does anthracene, like graphite and many other substances, ever grow from the vapour phase in the form of whiskers, thus suggesting the presence of screw dislocations?

Lipsett to Dexter: In the vapour phase, anthracene ordinarily grows in the form of hexagonal platelets. In some solutions, it may grow in the form of needles. Other organic materials do grow in the form of needles or whiskers in the vapour phase.

Kommandeur to Dexter: In answer to Dr Dexter's question with regard to the presence of dislocations in anthracene, I can remark the following: by evaporating the surface off bent naphthalene and anthracene crystals, we observed regular patterns of etch-pits, presumably arising from the presence of dislocation. At room temperature, however, the patterns soon became rather arbitrary, and lost their orientation with respect to the crystal axes, indicating rapid polygonisation.

Stevens to Lipsett: Are you justified in neglecting triplet exciton annihilation, even under low intensity irradiation, when the delayed fluorescence exhibits a non-exponential decay?

Lipsett to Stevens: Possibly not. However, the experiments have only been carried out during the past few weeks, and we have as yet not had enough time to carry out an analysis of the results.

PARMENTER

Windsor to Parmenter: For benzene in solution, the sum $(\phi_T + \phi_F)$ is equal to 0·30 which is appreciably less than unity. For naphthalene the corresponding sum is 0·59; this sum does not change on deuteration. We thus have, in addition to fluorescence emission and intersystem crossing, a decay mode of the first excited singlet state that consumes an appreciable fraction of the excited singlets, but which shows no isotope effect despite the large $S_1 S_0$ energy gap. To my mind, this is consistent with a hypothesis of decay by a pathway involving chemical isomerisation as you suggest for benzene in the gas phase.

Robinson to Parmenter: For the reasons I stated the other day after Dr Cundall's paper, I have an uneasy feeling about these short triplet lifetimes in the gas phase. Triplets from intersystem crossing are being produced in high vibrational states and, in the gas phase, dissipation of this vibrational energy may either not occur (low total pressure) or take place by different routes than in condensed phases. This could be the difference if there really is one.

Hochstrasser to Parmenter: I detect a note of generality in this discussion which is about shortened triplet state lifetimes for gas-phase molecules compared with the same ones in the solid phase, yet in your system there is a molecule, namely biacetyl, for which no shortening occurs. There is, therefore, no generality to the gas-phase solid-phase lifetime anomaly.

I suggest we consider what are the basic differences between aromatic hydrocarbons and ketones in regard to electronic properties and so on. Possibly the degree of localisation of the excitation reduces the probability of intermolecular quenching by impurities? Or possibly this localisation means that only a few of the vibrational modes are being involved, thereby influencing the probability of gross isomerisations, and radiationless transitions.

Cundall to Parmenter: 1. The possibility of some form of chemical quenching may be responsible for your second-order quenching process. This may or may not yield an identifiable chemical product. Such a second-order quenching process could be responsible for the very short lifetime of the benzene triplet in the liquid state.

2. I assume that you can exclude effects arising from decomposition of biacetyl.

3. Impurity quenching cannot surely be that of the type normally understood. The impurity must have an effect specifically on benzene. This is difficult to understand: as pointed out elsewhere in the discussion in the isomerisation work, the amount of impurity needed to bring the triplet lifetime in the vapour into line with a radiative lifetime of the benzene triplet of 10 s would be *very* large. I think we would accept that rapidity of deactivation involves a unique process. Would you agree?

Parmenter to Cundall: 1. There is at present no evidence for or against chemical quenching of the benzene triplet by ground state molecules. We cannot yet even be sure that the observed second-order quenching is caused by ground state molecules, rather than by impurities introduced with benzene.

2. The behaviour of the system does not change after the mixture has been exposed to many irradiations, and thus a stable decomposition product that interacts with triplet benzene is not formed. Formation of a transient photochemical product, which quenches benzene triplets, is ruled out because of the absence of intensity effects on the kinetics of the mixture. Quenching by such a product would appear kinetically very much like triplet-triplet annihilation.

3. Impurities are certainly not establishing the first-order lifetime, since their effects are all contained in the observed second-order quenching. I think that it is quite likely that a new triplet relaxation path is in effect in the gas phase.

Kommandeur to Parmenter: Would 'chemical quenching' by an isomer be effective? A Dewar type structure is a vibronically excited state; i.e. one point of Professor Robinson's bedspring. Maybe other relaxation processes should be considered for the gas-phase triplets, such as coupling to the rotation via vibronic and vibration rotation interaction.

Parmenter to Kommandeur: Relaxation of excited states by isomer formation is commonly observed and is a decay path which must be considered for this triplet state. This, and vibrational rotational interactions, may be more significant in the gas phase, but we really need to know whether this fast relaxation results from removal of benzene from a rigid environment or results from the temperature increase.

KELLMANN

Stevens to Kellmann: Is it safe to use argon for outgassing purposes in a fluid system in which triplet states are being generated, in view of the possibility of external spin-orbit perturbation of the triplet state by this fairly heavy atom?

Kellogg to Stevens: On the question of whether residual argon could give enough spin-orbit coupling to quench the triplet, the answer is probably no. The work of Wilkinson shows that inert gases (xenon) are much less effective, externally, than the equivalent halogen (bromine). Since chlorinated solvents are poor spin-orbit inducers and since only one atmosphere argon pressure was present, the above mentioned effect is probably negligible.

Cundall to Kellmann: Dr E. J. Land and I have recently measured the extinction coefficient of the biacetyl triplet in benzene solution. $\epsilon_{3150\text{Å}}$ values of 6300 (relative to anthracene $\epsilon_{4288\text{Å}} = 57200$) and 6100 (relative to 1,2 benzanthracene $\epsilon_{4875\text{Å}} = 25100$). Units $\text{M}^{-1}\,\text{cm}^{-1}$.

MEYER

Windsor to Meyer: To your list of values of extinction coefficients for triplet anthracene we can add the value 115000 ± 15000 measured in EPA at 77 °K by singlet depletion studies by Dawson and Windsor (*Molecular Crystals*, April 1967). This is in good agreement with your value of 90000.

We also observed new bands at 5200 and 4800 Å for anthracene in epoxy plastic at room temperature and confirmed their assignment to the T–T absorption spectrum of anthracene from their decay rate; they had the same lifetime as the well-known intense bands at 4200 and 4000 Å. Regarding naphthalene, I want to point out that the value of 10000 for ϵ_T, which you ascribed to Porter and Windsor, was actually a *lower limit* since in those experiments we were unable to detect any singlet depletion. It is not therefore inconsistent with the higher value of 45000 which you now report. Finally, let me compliment you on your T–T absorption spectrum of anthracene which shows more structural details than any published previously.

Meyer to Windsor: Yes, if one can extrapolate the temperature band narrowing, we then find a value of 108000 at 77 °K for ϵ_T.

Kellogg to Meyer: When measuring weak bands, the possibility of impurities must be carefully ruled out. Since you did not measure lifetimes, how did you check for this?

Meyer to Kellogg: First, the total concentration used to obtain the weak bands was never higher than 10^{-4} M. The impurity concentration in our purest anthracene is smaller than 10^{-4} M. The impurity concentration in the solution is then smaller than 10^{-8} M. To be of apparent intensity (say for $\epsilon \sim 100$) the impurity ϵ should be as high as 10^6. Such a value is very improbable at this wavelength.

Secondly, we observed exactly the same spectrum with samples of different origins (purified or 'synthetic').

Thirdly, the rate of photolysis was identical when observed with S–S bands and with any T–T bands.

OLMSTED

Parker to Olmsted: I understand that your estimate for c, the fractional yield of excited singlet from triplet-triplet annihilation, may be as high as unity under some conditions, i.e. higher than would be expected on statistical grounds alone? We found high values of our factor p (in solution) for some compounds (e.g. naphthalene $\sim 0{\cdot}6$), although for anthracene we found only 0·1. Our factor p represents the probability that an 'encounter' between triplets in solutions will give rise to an excited singlet, the encounter rate being defined by $8\mathbf{R}T/3000\eta$. Of course, the high values we find may simply mean that triplet-to-triplet energy transfer is taking place over distances greater than that corresponding to the 'encounter' distance.

Olmsted to Parker: Our value of 1·0 is a preliminary value only, but we can definitely set a lower limit of $c = 0{\cdot}3$ at 553 °K, which is higher than that expected on statistical grounds alone. Our c is somewhat different from your p, being the fraction of *reactive* encounters that yield an excited singlet, and therefore long-distance 'encounters' cannot be used to explain this higher value.

Kellogg to Olmsted: It should be possible to get annihilation of two triplets to give one triplet back. Would this effect change your diffusion rate?

Olmsted to Kellogg: Triplet-triplet annihilation giving back one triplet is included in the general rate expressions, and therefore the overall kinetics are not affected by this possibility. However, since this is a second-order rate process by which triplets disappear only half as rapidly as they do in other second-order annihilation reactions, explicit inclusion would result in a different interpretation of c.

7 TRIPLET STATE RELATED TO BIOLOGY

ESR AND OPTICAL STUDIES OF SOME TRIPLET STATES OF BIOLOGICAL INTEREST†

J. M. LHOSTE, C. HELENE AND M. PTAK

Abstract

The electronic structure of the lowest triplet state has been investigated by ESR in the following molecules: phenoxazine and phenothiazine diluted in a single crystal of diphenyl, free base, di-cation and magnesium-complexed porphyrines as well as flavines in solution in rigid glasses. It is shown that most conjugated biomolecules have different electronic configurations in their lowest excited singlet and triplet states. This could explain the differences observed in their photochemical and physicochemical behaviour.

The role of the triplet state in energy transfer in macromolecules, such as proteins and nucleic acids, is discussed. The mechanisms of interaction of the excited aromatic residues depend on the macromolecular structure as well as on their intrinsic electronic properties.

Due to their long lifetime, triplet biomolecules may absorb a second photon before deactivation. These biphotonic processes provide high energy levels at which ionisation reactions or bond splitting in sensitised molecules may occur.

In conclusion, it is emphasised that both the electronic properties of the individual chromophores and the supramolecular organisation govern the fate of electronic excitation in biological systems.

Introduction

Metastable triplet states of conjugated biomolecules have been proposed and, in some instances, have been shown to be important intermediates in photochemical reactions of biological importance.[1] Apart from bioluminescence phenomena which are known to involve mostly excited states of singlet multiplicity,[2] electronically excited

† Supported in part by a contract between the Division of Biology and Medicine, U.S. Atomic Energy Commission, and the Florida State University.

states can be populated only via electronic transitions induced by external excitation. The absorption of excitation energy in a given molecule may be direct or may occur through energy transfer mechanisms. Such reactions may occur in natural processes like photosynthesis or vision, or during processes that involve damage in biological metabolism or genetic information by ionising or ultraviolet radiations.

Before assuming a specific role for the triplet state in the primary steps of these processes or in simplified photochemical reactions studied *in vitro*, a good knowledge of the physical properties of the excited states of the chromophore molecules involved seems desirable. These considerations have led us to investigate the electronic structure and the role of the lowest triplet state in transfer and accumulation of excitation energy in the following systems: porphyrins and natural or artificial (flavins and azine dyes) sensitising agents, proteins (aromatic amino acids) and nucleic acid constituents.

Electronic structure and fundamental properties of biological molecules in the lowest triplet state

Experimental investigation of electron configuration in triplet states of biological interest

Conjugated biomolecules have no, or few, symmetry elements and a large number of heteroatoms. This introduces serious difficulties in both experimental investigation and theoretical prediction on the electronic structure of excited states. Even the assignment of a π, π^* or an n, π^* configuration for the lowest triplet may be difficult. The electronic structure of metastable triplet states in conjugated molecules may be derived from optical investigations and mostly from techniques using the paramagnetic properties associated with the spin $S = 1$.

The best description of these electronic properties is obtained from the paramagnetic resonance studies of photo-excited molecules diluted and oriented by isomorphous substitution in an organic single crystal, following the techniques developed by Hutchison & Mangum.[3] From these experiments, the zero-field splitting (ZFS) energies due to the dipolar interaction of the elementary spins are measured, and the principal axes of the spin-spin coupling tensor are related to the geometry of the molecule. Moreover, the spin distribution over the molecular ring may be derived from the analysis of the anisotropic hyperfine interactions of the electron spin with the nuclei. This method has been so far applied only to a small number of conjugated hydrocarbons and N-heterocyclics because of the difficulties of obtaining suitable mixed crystals and the limited possibility of observing and interpreting resolved hyperfine structures in the ESR spectra. Generally one has to refer to the ESR spectra obtained from photo-excited rigid solutions of randomly oriented molecules.[4, 5] These easier experiments place no restriction on the complexity of the molecule, but provide only a measure of the ZFS energies. In some cases, the study of photo-selected triplet molecules[6, 7] gives some information about the orientation of the principal axes of the spin-spin coupling tensor with respect to the molecular frame. For many bio-molecules a population high enough for the detection of the $\Delta m = \pm 1$ transitions cannot be reached and the only magnetic parameter accessible is the mean-square ZFS derived from the position of the strong $H_{\text{min.}}(\Delta m = \pm 2)$ transitions.[4] This may be sufficient to estimate the extent of delocalisation of the excited state and the nature of the molecular orbitals involved.[8]

Optical studies of the triplet state may provide some complementary information from the characteristics of the energy, vibrational structure, lifetime, quantum yield and polarisation of the phosphorescence emission or from spectral data on triplet-triplet absorption.

An empirical approach may help in the understanding of the electronic structure of triplet states of biological molecules, using model compounds (e.g. toluene, phenol, indole for the aromatic amino acids phenylalanine, tyrosine and tryptophane respectively), or investigating the effects on the ZFS energies of weakly perturbing substitutions of the parent molecules. One can take advantage of the natural versatility of biomolecules; most have reduced or oxidised, protonated or deprotonated, as well as tautomeric forms. These various forms may have different electronic structures in the triplet state.

The following three examples describe investigations carried out on molecules of increasing complexity.

(*a*) *Phenoxazine and phenothiazine.* The lowest triplet state and cation radical of phenoxazine and phenothiazine have been investigated in order to obtain a better understanding of the electronic properties of these molecules and their derivatives. They are common rings for a large family of photosensitising dyes and for major pharmacological derivatives. Phenoxazine and phenothiazine are tricyclic molecules of C_{2v} symmetry which can be incorporated in diphenyl host crystals. These mixed crystals are suitable for paramagnetic resonance experiments[9] for the following reasons: (1) the substitution of guest molecules is isomorphous with two magnetically distinct types of molecules corresponding to the two molecules of the diphenyl crystal unit cell, (2) the guest molecules have a triplet state of lowest energy which can trap the triplet excitation, and (3) the crystallography of the host crystal is known. Experimental conditions and results concerning preliminary investigations of phenoxazine have already been reported.[6]

Table 1. *Experimental spin Hamiltonian parameters*

	Phenoxazine		Phenothiazine	
Table 1*a*	Diphenyl crystal†	Ethanol glass	Diphenyl crystal	Ethanol glass
D/hc (cm^{-1})	$\pm 0{\cdot}0990$ (3)	0·1249	$\pm 0{\cdot}1230$ (5)	0·1237
E/hc	$\mp 0{\cdot}0154$ (5)	0·0119	$\pm 0{\cdot}0142$ (5)	0·0135
X/hc	$\pm 0{\cdot}0482$	0·0535	$\pm 0{\cdot}0270$	0·0278
Y/hc	$\pm 0{\cdot}0175$	0·0297	$\pm 0{\cdot}0553$	0·0547
Z/hc	$\mp 0{\cdot}0659$	0·0833	$\mp 0{\cdot}0820$	0·0825
g_{xx}	2·0037 (4)	—	2·0042 (4)	—
g_{yy}	2·0037 (4)	—	2·0045 (4)	—
g_{zz}	2·0020 (4)	—	2·0014 (6)	—
τ (s)	2·4	2·2	$\leqslant 0{\cdot}1$	$\leqslant 0{\cdot}1$
$\lambda_{ph.}$ (0–0) (mμ)	472	475	516	512

	Pyrene		Quinoxaline	
Table 1*b*	Fluorene crystal‡	Ethanol glass	Durene crystal§	Ethanol glass
D/hc	$\pm 0{\cdot}0657$	0·0683	$\pm 0{\cdot}1007$	0·0959
E/hc	$\mp 0{\cdot}0316$	0·0345	$\mp 0{\cdot}0182$	0·0183

† Numbers in parentheses are the uncertainties in the last quoted figures.
‡ Reference (10). § Reference (11).

The experimental spin Hamiltonian parameters of the lowest triplet state of the two molecules in diphenyl and rigid ethanol at 77 °K are reported in table 1*a*. With phenoxazine, a noticeable difference in ZFS energies appears for diphenyl and ethanol matrices. We observed small variations in ZFS energies for *N*-heterocyclic (quinoxaline) or hydrocarbon (pyrene) triplet states in ethanol compared to the values found in fluorene[10] and durene[11] crystals respectively (table 1*b*). Such variations may be due to the polarity of the solvent, hydrogen bonding or small changes in geometry as suggested by the polarisation of phosphorescence.[12] In diphenyl crystals the molecules may be forced to a planar configuration while they may be slightly folded in rigid ethanol. But phenothiazine has more tendency toward folding than phenoxazine[13] and its triplet state exhibits the same ZFS in both matrices. The ZFS are characteristic of π, π^* triplet states of aromatic molecules with the D parameter probably positive.[14] The anisotropy of the g tensor is in good agreement with the theory developed by Stone[15] for aromatic molecules. The low anisotropy of the g tensor, the long lifetime of the excited state and the orientation of the principal axes of the spin-spin coupling tensor with respect to the host crystal unit cell favor a planar configuration for the two molecules in the triplet state. With phenothiazine, probably because of the size of the sulphur atom, the direction of these principal axes differs by a few degrees from the direction of the symmetry axes of the diphenyl molecules in the crystal unit cell.

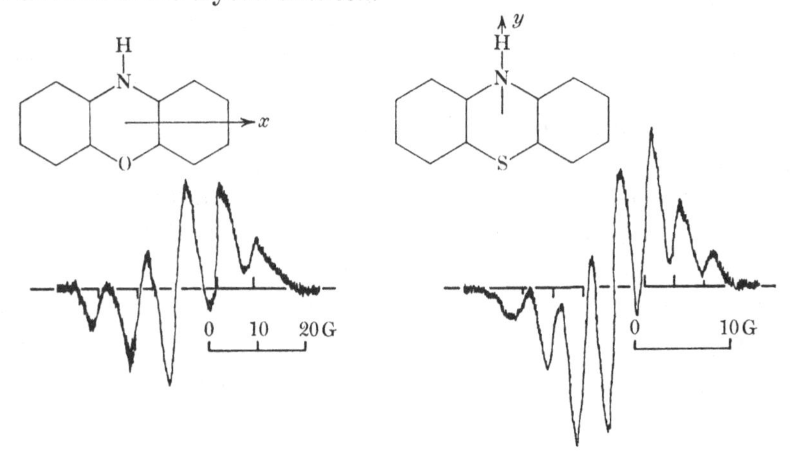

Fig. 1. Hyperfine structure of the $H \| x$ transitions in the phenoxazine triplet state and $H \| y$ transition (high field only) in the phenothiazine triplet state.

The partial resolution of the hyperfine structure provides more details about the electronic structure of the two triplet states (fig. 1):

(1) In phenoxazine the $H \| x$ transitions present a hyperfine structure corresponding to a partly resolved 1 : 4 : 6 : 4 : 1 intensity ratio quintet with a splitting of 7·8 G between adjacent lines. This structure can arise only from the interaction of the electron spin with four equivalent α protons. The principal values of the C–H hyperfine tensor[16, 17] give normalised spin densities of 0·22 on the four α-C's. The linewidths of the $H \| y$ and $H \| z$ transitions suggest a low spin density on the β-C's. Because of the high z component of the N hyperfine tensor[11] any noticeable spin density on the central nitrogen atom

should contribute significantly to the line shape of the $H \parallel z$ transitions. However, the four α-C's spin densities give the main contribution to the gaussian lineshape of the $H \parallel z$ transitions, and thus the spin density on the central nitrogen must be small. The equivalence of the four α-C's suggests a low spin density on the oxygen atom and a node in the triplet state wavefunction through the y axis. In an ethanol glass the apparent symmetry of the triplet wavefunction with respect to the x axis is probably destroyed considering the changes in ZFS energies.

(2) In phenothiazine, even in diphenyl crystals, the four α-C–H are no longer equivalent for the hyperfine interactions. The hyperfine structure of the $H \parallel x$ transitions is not resolved, but the $H \parallel y$ high field transition is a well-resolved septet with a splitting of 3·15 G between adjacent lines. The intensity ratio of these lines agrees with a binomial distribution corresponding to the interaction of the electron spin with six equivalent (in that direction) protons. The linewidths of the unresolved $H \parallel x$ and $H \parallel z$ transitions compared with the principal values of the C–H hyperfine tensor suggest that the $H \parallel y$ structure arises from a normalised spin density of 0·29 on two α-C's and 0·11 on the four β-C's. The spin density on the central nitrogen must be as small as for phenoxazine. The change in spin distribution compared to phenoxazine (fig. 2) is reflected in the change in the values of the ZFS energies and in the sign of the E parameter. Such a difference has already been reported for the phenoxazine and phenothiazine cations.[18]

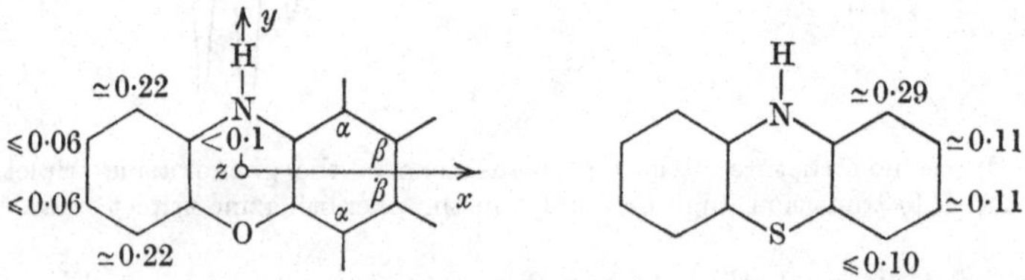

Fig. 2. Normalised spin densities in the phenoxazine and phenothiazine lowest triplet state as derived from the analysis of the hyperfine structure of the ESR spectra using the principal values of the C—H and N hyperfine tensors of refs. (17) and (11). (The spin distribution in phenothiazine is tentative).

The spin distributions observed for the phosphorescent state of phenoxazine and phenothiazine indicate that the wavefunction of the lowest triplet state has probably a node through the line joining the two heteroatoms.† This configuration differs from that of the lowest

† In the C_{2v} symmetry point group, the lowest triplet belongs to the B_2 representation, whereas the lowest excited singlet belongs to the A_1 representation as indicated by the magnetophotoselection experiments of ref. (6).

triplet state of anthracene where the highest spin density is predicted on the two mesocarbons.[19, 20] This difference, similar to the difference in spin distribution between the triplet states and the cation radicals,[18] may be explained by the non-pairing of bonding and antibonding orbitals in the tricyclic azines due to the presence of the heteroatoms and the number of π electrons ($\neq 4n+2$), and a possible orbital degeneracy and high mixing of configurations for the lowest triplet state suggested by preliminary calculations using Hückel molecular orbitals.[18]

Differences in the electronic structure of the excited states of the heterocyclic molecules compared to anthracene appear in the photochemical behaviour as well as in the effect of substituents on the electronic properties of these molecules, and may explain the high efficiency of the heterocyclic dyes in photosensitised reactions such as photodynamical effects in living organisms.

(*b*) *Porphyrins*. The first experimental evidence for a metastable triplet state in tetrapyrrolic molecules was the observation by Calvin & Dorough[21] of the phosphorescence of chlorophyll *b* at low temperature. This observation was followed by a detailed investigation of the luminescence properties of chlorophylls[22, 23] and porphyrins[24] in rigid solvents. Phosphorescence of chlorophyll was thought to originate from an n, π^* or a π, π^* triplet state, depending on the environment. The evidence for dimerisation of chlorophylls in non-polar solvents,[25, 26] triplet-triplet absorption spectra of porphyrins,[27, 28] and finally the observation of a $\Delta m = \pm 2$ ESR transition for the triplet state of porphyrins[29] and chlorophyll *b*[30] eliminated the possibility of a low-lying n, π^* triplet state for these molecules. This fact has not weakened the interest for a role of triplet states of chlorophylls in photosynthesis.† The ESR spectra of some porphyrins (fig. 3) photo-excited in rigid ethanol allow one to go further in the description of the electronic configuration of the lowest π, π^* triplet states in porphyrins.

The experimental results may be summarised as follows (fig. 4):

(1) The free base porphyrins (D_{2h}), as well as the di-cations or magnesium-complexed porphyrins (D_{4h}), have a triplet state with a spin-spin coupling tensor of cylindrical symmetry ($E = 0$).

† For example, the observation of chlorophyll delayed fluorescence (C. A. Parker & T. A. Joyce, *Nature*, **210**, 710, 1966) and chlorophyll triplet-triplet absorption (G. Porter & G. Strauss, *Proc. Roy. Soc.* (*London*) A, **295**, 1, 1966) in green leaves, strongly support the existence of high yield of chlorophyll triplet excitation in these systems upon illumination.

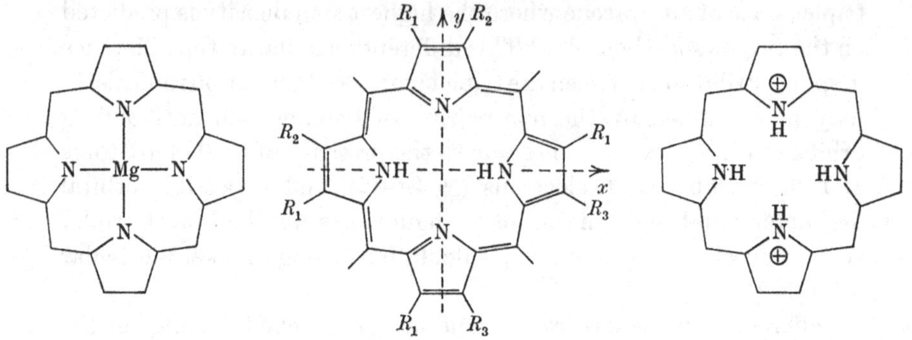

II I III

Porphyrins IX	Hemato-	Proto-	Deutero-	Etio-
$R_1 =$	$-CH_3$	$-CH_3$	$-CH_3$	$-CH_3$
$R_2 =$	$-CHOHCH_3$	$-CH{=}CH_2$	$-H$	$-CH_2CH_3$
$R_3 =$	$-CH_2CH_2COOH$	$-CH_2CH_2COOH$	$-CH_2CH_2COOH$	$-CH_2CH_3$

Fig. 3. Porphyrins IX: I, Free base; II, magnesium complex; III, diprotic acid (di-cation).

(2) The two species of D_{4h} symmetry have very similar ZFS. The zero-field D parameter of Mg-etioporphyrin has been calculated from the spectrum published by Azizova *et al.*[31] which was the only $\Delta m = \pm 1$ spectrum of porphyrin triplet state previously published and is the only one presently available for metal porphyrins.† A direct comparison between porphyrins having different lateral substituents seems reasonable since for hematoporphyrin IX, protoporphyrin IX dimethyl ester ($\Delta m = \pm 2$ transition only), and deuteroporphyrin IX dimethyl ester, we observed the same spectra for the free bases as well as for the di-cations. Furthermore, the asymmetric substitutions of the isomers IX (fig. 3) do not perturb the cylindrical symmetry of the spin-spin coupling tensor in the lowest triplet states.

(3) The H_0 peaks of the di-cation triplet spectra are broadened, but the H_z peaks remain essentially unchanged.

(4) The D-parameter of the free base triplet is much higher than that of the di-cation or magnesium complex triplets. This difference is too high to be considered as a simple perturbation of a unique electronic configuration and suggests two orbitally different configurations.

† Recently we measured for the Mg-deuteroporphyrin IX diamethylester and Mg-protoporphyrin IX dimethylester triplet states, $D/hc = 0{\cdot}035\ \text{cm}^{-1}$, $E = 0$, supporting better the similarity in electronic structure of the cationic and magnesium complexed triplets.

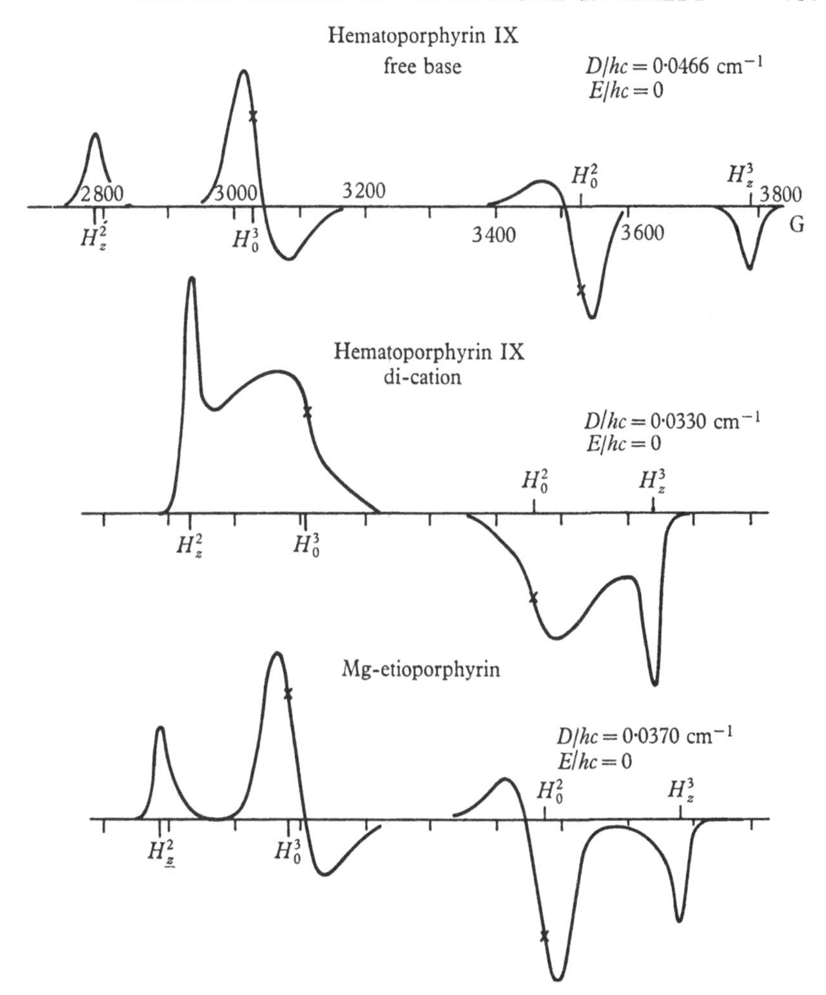

Fig. 4. $\Delta m = 1$ ESR spectra of the lowest triplet state of porphyrins photo-excited ($\lambda > 380\ m\mu$) in rigid glass solution at 77° K (ethanol/pyridine 5%; ethanol/HCl 0·1 N; EPA, respectively). The zero-field parameters are calculated assuming $g = 2{\cdot}003$. The spectrum of Mg-etioporphyrin is that published by Azizova *et al.* (ref. 31) recalculated for the microwave frequency ($\nu = 9220$ MHz) at which the other spectra have been recorded. In the center of the spectra is a free radical signal which has been subtracted from the total spectra.

(5) Despite the absence or great weakness of phosphorescence, the free base and di-cation porphyrins may have a triplet population high enough for ESR measurements. The fluorescence yield is still high at low temperature, but the rate of intersystem crossing to the triplet state must be of the same order of magnitude and the lowest triplet state must be deactivated mostly by non-radiative decay. Metal

porphyrins have a stronger phosphorescence,[24] a much higher steady-state triplet population and a higher lifetime[31] due to the lowering of the rate of non-radiative decay.

A theoretical interpretation of the electronic properties of porphyrins has been given by Gouterman and collaborators[32, 33] following the ideas originated by Simpson[34] and Platt.[35] These calculations show that the ground-state absorption spectra can be interpreted using a four-orbital model with two nearly degenerate lower orbitals of a_{1u} and a_{2u} symmetry and a higher doubly degenerate orbital of e_g symmetry. The validity of this model has been confirmed by a self-consistent molecular orbital (Parr–Pariser–Pople) treatment of porphyrin π system.[33] An extended configuration interaction treatment of singly and doubly excited configurations for the lowest and higher triplet states shows that the four-orbital model is too simple for explaining the triplet-triplet absorption spectrum, but gives a 92% 'pure' configuration for the lowest triplet state of metal porphyrins (D_{4h}). Following these calculations,[33] in D_{4h} porphyrins the energy of the pair of degenerate $^3E_u(3a_{2u}, 4e_g)$ states is found to be 0·2–0·4 eV lower than that of the degenerate $^3E_u(1a_{1u}, 4e_g)$ states. In D_{2h} porphyrins (free base) the degeneracy of the states is removed. The splitting of the $^3E_u(1a_{1u}, 4e_g)$ configurations is much higher than the splitting of the $^3E_u(3a_{2u}, 4e_g)$ configurations. The energy of the y-component of the configurations including an a_{1u} orbital is now at least 0·5 eV lower than that of any of the three other configurations. Even if these calculations have to be considered qualitative, the interchange of an a_{1u} orbital, with a node through the nitrogen and methene carbon atoms, and an a_{2u} orbital, with a node through the α- and β-carbons of the pyrrole rings, in the dominant configuration of the lowest triplet state of free base and metal or di-cation porphyrins may account for the differences in ZFS energy observed. Furthermore, one must admit that the symmetry perturbation introduced in the free base, although being high enough for removing the degeneracy of the configurations, is not sufficient to destroy the cylindrical symmetry of the spin-spin coupling tensor, as observed for the effect of lateral substituents. A small difference in ZFS energy may be expected for the di-cation and the magnesium complex triplets due to the effect of charges on the conjugation through the nitrogen $p\pi$ orbitals and a possibility of conjugation of a $p\pi$ orbital of the magnesium atom with molecular orbitals of proper symmetry. Complexing diamagnetic ions of different electronegativities will test the latter effect.

The anomalous lineshape of the ESR spectra of the phosphorescent triplet state of odd aromatic ions observed by de Groot *et al.*[36] has been proposed by these authors as evidence for Jahn–Teller instability in aromatic systems for which theory predicts orbitally degenerate triplet states. They proposed that the spectra arise from a range of distorted configurations strongly stabilised by interaction with the polar solvent. The spectra should exhibit a broadening of the H_0 lines due to the displacement of π electron charges in the molecular plane when the degeneracy of the triplet state is removed by distortion of the molecular frame, while the Z zero-field parameter associated with the quantisation of the electron spin in the molecular plane, and therefore the H_z stationary fields, remains essentially unaffected in first order because of the symmetry requirement of vibronic interactions. The anomalous lineshape of the ESR spectrum of the triplet state of porphyrin di-cations (fig. 4) can be interpreted in the same way since theory predicts an orbital degeneracy for the lowest triplet state (^3E_u). Furthermore, the experimental observation of a Jahn–Teller instability in the di-cations is favored compared to the metal complexes because of the higher flexibility of the porphyrin ring and the possibility of stabilisation of the distorted configurations by strong interactions between the doubly charged molecules and the polar solvent glass.

In chlorophylls the chlorin ring is complexed with a magnesium atom, but one of the pyrrole rings is hydrogenated and excluded from the conjugated π orbitals. The x and y directions are no longer equivalent. The $\Delta m = \pm 2$ ESR transition observed for the chlorophyll b triplet state[30] is characteristic of a highly delocalised π, π^* state similar to the porphyrin one, but no $\Delta m = \pm 1$ transitions have been so far observed. This hinders the prediction of the symmetry of the spin-spin coupling tensor or the molecular orbital configuration of the lowest triplet state.

(*c*) *Flavins and other conjugated biomolecules.* For other phosphorescent biomolecules a precise description of the electronic structure of the lowest triplet state in terms of spin densities or symmetry and nodal properties of the electron wave function is not available. For aromatic amino acids[37, 38] and nucleic acid bases[39, 40] only $\Delta m = \pm 2$ transitions have been so far observed except for adenine and guanine nucleotides, for which $\Delta m = \pm 1$ transitions have been recorded.[40] $\Delta m = \pm 1$ transitions have been observed in model compounds like indole[41] for tryptophane, and unsubstituted purine[42] and orotic acid[40] for the purine and pyrimidine nucleic acid derivatives. Experimental and theoretical studies of these chromophore models may provide detailed information about the electronic structure of the parent molecules. Tyrosine has been shown to possess different triplet states in the ionised and neutral forms.[43] A detailed study of luminescence and ESR properties of nucleic acid components carried out by Shulman and collaborators[40] emphasises the effect of protonation on the electronic structure of the excited states of nucleic bases. Furthermore, these authors propose[44] that the phosphorescent emission and ESR signal of photo-excited DNA is that of the triplet state of ionised thymine. This point will be discussed in the following section. These ESR experiments leave no doubt that a π, π^* configuration is involved for the lowest triplet state of all phosphorescent biomolecules investigated thus far.

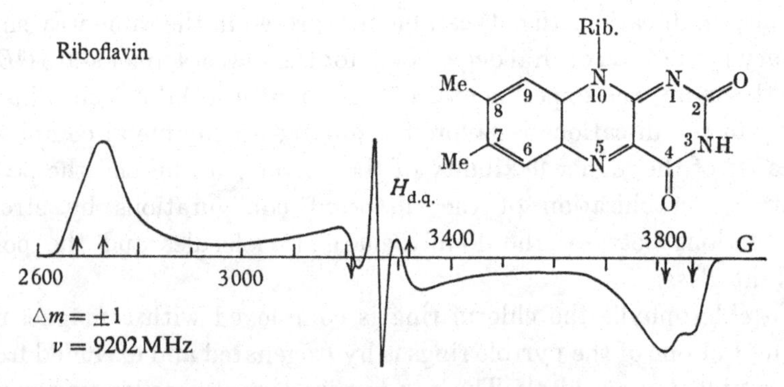

Fig. 5. $\Delta m = 1$ ESR spectrum of the riboflavin lowest triplet state in rigid ethanol. The narrow line in the center is the double quantum transition. The free radical signal around $g = 2$ has been subtracted from the total spectrum.

A detailed investigation of the electron configuration of the flavin and pteridine triplet state has been previously reported.[45] More information about the electronic properties of these molecules seemed desirable since flavosemiquinones are electron carriers in biological oxido-reduction and the triplet state of flavoquinones has been found responsible for several photochemical reactions[46] and was suggested as a possible intermediate in photosynthesis.[47] Using the assumption that the effect of weak perturbations, such as hyperconjugation through methyl groups or charge effect in protonation or deprotonation, on the ZFS energies derived from ESR experiments in rigid glass solutions (fig. 5) should reflect the electron distribution in the photo-excited state, a qualitative description of the electron configuration of the triplet state was depicted[45] for these rather complex molecules. The ZFS parameters measured in substituted, tautomeric, protonated and reduced forms of flavins and pteridines indicate that the spin distribution in the lowest triplet state of flavoquinones is qualitatively similar to that in the flavosemiquinone radicals.[48] In spite of the similarity of flavin and pteridine triplet states, the spin distribution in the former molecules is mainly localised in the quinoxaline part. A low spin density on the N 3 atom which is used in the binding with the apoenzyme proteins, high spin densities on the N 10–N 1 and probably N 5 ring nitrogens, and a non-equivalence of the methyl groups in positions 7 and 8 are the most significant data to compare with the biological properties of flavins. This configuration for both the triplet and semiquinone states is similar to the electron properties of anthracene-like molecules rather than to the electron configuration described previously for the heteroazines.

The chemical properties of the lowest triplet state of conjugated biomolecules

The main conclusion which arises from the ESR studies carried out on the lowest triplet state of conjugated biological molecules is that all have delocalised π, π^* orbital configurations. For molecules with an n, π^* configuration for the lowest excited singlet state, e.g. unsubstituted purine, the singlet-triplet splitting in that type of electron configuration is not large enough compared to that in low-lying π, π^* configurations, and the lowest triplet state is thus a π, π^* state. When the lowest excited states of singlet and triplet multiplicity have both a π, π^* orbital configuration, they do not necessarily correspond to the same orbital excitation as observed for benzene and naphthalene. Finally, even when the two states correspond to the same pure orbital excitation, the triplet state is expected to possess a charge distribution different from that of the singlet excited state because of differences in electron correlation energy.

The reasons for a specific chemical reactivity of triplet states in photochemistry and photobiology have been extensively discussed and it has been often emphasised that triplet states must be considered as different chemical species with their own physical and chemical properties (see, for example, several discussions in reference (1)). Some specific properties of triplet states that have been put forward to explain their chemical behaviour are a lifetime of milliseconds to seconds, a high electron polarisability, or spin conservation in primary reactions involving triplet substrates or ground triplet oxygen. But usually the difference in electronic structure can explain the specific chemical reactivity of triplet and singlet excited states. Little is known about the charge densities in excited singlet states, and in triplet states the spin densities measured from the paramagnetic properties do not correspond directly to the charge densities, but the investigations described in this paper support the existence of different electronic structures in the singlet and triplet states of most photochemically important biomolecules.

Some differences in physico-chemical properties of electronic states of different multiplicity are easily observable. For example, flavins are better electron acceptors in the triplet state than in the ground state[47] and have a higher basicity in the first excited singlet state than in the lowest triplet state or in the ground state.[45] Furthermore, the electronic structure found for the triplet state of flavoquinones favours a metal affinity much higher than in the ground state and comparable to that of the flavosemiquinone radical.[49] Such a variety of redox properties or proton or metal affinities is characteristic of many biological molecules and must be carefully considered in interpreting the behaviour of these molecules in photobiological processes.

Excitation energy transfer and primary photochemical reactions involving the triplet states of biomolecules

It has been shown in the previous section that differences in electronic structure could explain the specific role of the lowest triplet state of biomolecules as a photochemical precursor. But triplet states do not necessarily react chemically and may transfer their excitation energy by physical processes before that energy could activate photochemical reactions. The following mechanisms of that type have been investigated in our laboratory: (*a*) the physical mechanism of excitation energy transfer between chromophores, with particular attention to nucleic acids and proteins, and (*b*) photosensitisation of aliphatic molecules in biphotonic reactions involving low energy triplet states of conjugated biomolecules.

Excitation energy transfer in nucleotides and aromatic aminoacids

Luminescence and paramagnetic resonance studies of covalently bound dinucleotides or dipeptides and of equimolecular mixtures of different mononucleotides or aromatic amino acids in organic and aqueous frozen solutions at 77 °K were carried out in order to provide more information about the interactions of photo-excited molecules in nucleic acids and proteins.

(*a*) *Excitation energy transfer in dinucleotides.* Strong interactions occur in dinucleotides photo-excited in rigid solvents with a possibility of energy transfer between the two stacked bases.[50, 51, 52] An interaction in the singlet excited states is indicated by the red shift and broadening of fluorescence spectra as compared to equimolecular mixtures. This has been interpreted as an excimer interaction.[52, 53] This kind of interaction makes the interpretation of experimental evidence of triplet-triplet energy transfer difficult since all naturally occurring nucleic bases absorb in the same wavelength range. For dinucleotides containing an adenine moiety and another naturally occurring nucleotide (G, U, C), Guéron *et al.*[52] propose that the energy transfer to the adenine part occurs at the singlet level, through strong excimer interaction followed by singlet-triplet conversion in the adenine part, although the lowest excited singlet is higher in adenine. A systematic study of fourteen dinucleotides[53] containing guanine, adenine, uracil and cytosine confirms the idea that adenine in its triplet state is the final excitation energy acceptor in all dinucleotides containing adenine (a previous assignment[51] of transfer to cytosine or uracil in ApC or ApU was due to partial hydrolysis and paramagnetic impurities), and suggests that guanine is the energy donor when it is associated with a pyrimidine. Comparing the quantum yield of luminescence in the dinucleotides and in equimolecular mixtures of mononucleotides[53] shows that energy transfer mechanisms may be complex in some nucleotide pairs. With substituted bases one can show unambiguously an energy transfer at the triplet level. For example, in AppAcC, acetylcytosine may be selectively excited at wavelengths higher than 300 mμ and the excitation energy may be partly transferred to the adenine moiety via the triplet states (fig. 6).

(*b*) *Excitation energy transfer in mixed aggregates in ice.*[53, 54, 55] In hydroxylic rigid glass solutions, solute molecules are dispersed. In dinucleotides the two bases are stacked one over the other. At the same concentrations (*ca.* 10^{-3} M) in aqueous solutions at 77 °K, solute molecules form aggregates. The luminescence of aromatic amino acids or purine nucleotides is strongly quenched and is comparable to that observed in microcrystalline powders. Adding an inorganic salt (NaCl) before freezing increases the luminescence yield up to a rapidly reached maximum (NaCl $\geqslant$ 0·2 M for purines 10^{-3} M or NaCl $\geqslant$ 2 M for trypto-

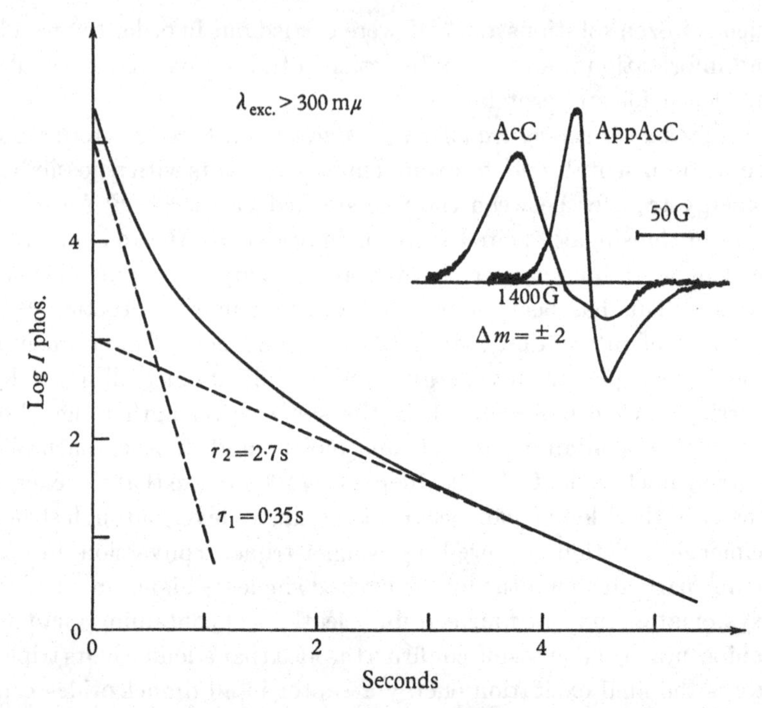

Fig. 6. The phosphorescence decay and ESR triplet signal of AppAcC dinucleotide in rigid ethanol. With an excitation light above 300 mμ the acetylcytosine moiety is selectively excited. The long component τ_2 in the complex phosphorescence decay and the ESR signal of AppAcC are those observed for pure adenosine solution excited at 250 mμ.

phane $5{\cdot}10^{-3}$ M). With purine nucleotides 10^{-3} M at a concentration of 0·3 M in NaCl the stacking of the molecules is not destroyed and the optical properties of mixed solutions are similar to those in dinucleotides with excimer interaction and a possibility of energy transfer between molecules. On the other hand, adding hydroxylic solvents (e.g. ethanol or propyleneglycol $\geqslant 2\,\%$) completely destroys the aggregates and the luminescence is similar to that observed in pure hydroxylic solvents.

Energy transfer between stacked nucleic bases may occur via the triplet states. Among these reactions, the role of thymine is of particular interest. For example, in ice at 77 °K the phosphorescence of acetyl-cytidine (AcC) is quenched by thymidine (T) without quenching of the fluorescence (fig. 7). Similar results have been observed with 3-methylthymidine or 1,3-dimethylthymine. The triplet-triplet energy transfer is unambiguous in that case and does not require a deprotonation of thymine as proposed by Rahn *et al.*[44] for the triplet state observed in DNA. In any equimolecular solid mixture in ice of adenosine (A), guanosine (G), or cytidine (C) with thymidine, the same phosphorescence is observed ($\lambda_{max.} = 470\,m\mu$, $\tau = 0{\cdot}22\,s$) and may be attributed to thymidine.

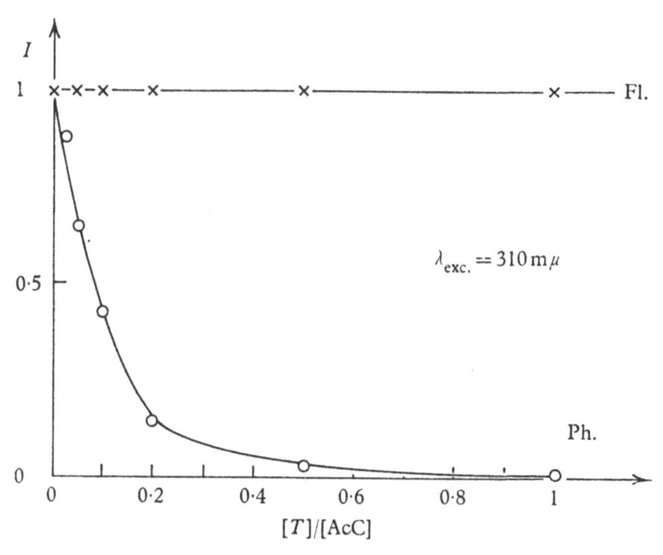

Fig. 7. Normalised fluorescence and phosphorescence intensities of acetylcytidine (10^{-3} M) in frozen aqueous solution containing increasing concentration of thymidine. Acetylcytidine is selectively excited at 310 mμ. A similar quenching of AcC phosphorescence without extinction of fluorescence is observed with 3-methylthymidine and 1, 3-dimethylthymine.

In mixtures of nucleosides that form base pairs (e.g. AcC + G) one observes a quenching of luminescence as with double or triple strand polynucleotides.[56] It is probable that in the base aggregates in ice the luminescence quenching is due mostly to hydrogen bonds which may be partly destroyed by addition of inorganic salts.

These experiments in dinucleotides and mixtures of mononucleotides clearly indicate a possibility of energy transfer between nucleic acid bases at the triplet level and/or at the singlet level depending on the nature of the bases involved. Furthermore, they suggest that in ice the energy may be transferred to the thymine triplet state without ionisation of that molecule. These conclusions may impose some limitations to the mechanism proposed by Rahn *et al.*[44] to explain the origin of the triplet state observed in polynucleotides and DNA. They propose that the ESR signal and the phosphorescence observed in these polymers correspond to the triplet state of ionised thymine and suggest that a proton transfer occurs through the adenine–thymine hydrogen bond in the excited state, which quenches adenine and allows thymine to phosphoresce. Yet, a possibility of direct energy transfer to thymine without proton transfer cannot be excluded in DNA. Further investigations on thymine triplet state in methylated polynucleotides may clarify this problem in the near future.†

(c) *Excitation energy transfer in dipeptides and proteins.* Energy transfer between aromatic amino acids in proteins can be easily detected by optical and ESR spectroscopy. The latter is more convenient than phosphorescence measurements to identify triplet acceptors because it provides a better resolution than optical spectra. Singlet-singlet energy transfer between aromatic amino acids in proteins is known from the fluorescence studies of Weber & Teale[57] and occurs at normal and low temperature.

At low temperature, part of the transferred energy can be converted to a long lifetime triplet state of the acceptor residue. In aromatic dipeptides including a tryptophan residue the fluorescence at 293 °K[58] or the phosphorescence and ESR triplet signal at 77 °K[59] are those of tryptophane. In polypeptides and proteins the transfer from tyrosine to tryptophane is not complete. Only an ESR signal of the tryptophane triplet state has been observed[38, 60] but the phosphorescence and the fluorescence of tyrosine are only partly quenched.[61] In the latter case, the nature of the transfer mechanism is not ambiguous since aromatic residues in proteins are diluted among aliphatic residues and the lack of direct overlap between the chromophores prevents transfer at the triplet level. In dipeptides or molecular aggregates of aromatic amino acids, the relative yield of fluorescence and phosphorescence indicates that the energy transfer still occurs via the singlet state followed by partial singlet-triplet conversion in tryptophane molecules. However, at high pH values, when the OH group of tyrosine is ionised, the direction of the singlet-singlet energy transfer in the tryptophyl-tyrosine dipeptide is reversed,[58] but at low temperature the excitation energy may be transferred backward to the tryptophane residue at the triplet level.

† The diminution of the rate of intersystem crossing S_1–T_1 in thymine and uracil upon methylation (A. A. Lamola & J. P. Mittal, *Science*, **154**, 1560, (1966)) could explain the role of ionisation of phosphorescence yield in case of direct excitation, but would not affect the triplet population of pyrimidines if the lowest triplet state is populated via intermolecular triplet energy transfer mechanism.

Biphotonic processes involving the triplet states of biological molecules

In frozen solutions at 77 °K aromatic biomolecules initiate photo-reactions involving excitation energies higher than the quantum energy of the exciting light. The observations made in our laboratory[37, 45, 59, 62, 63, 64] and by Russian authors (see refs. (31), (65) and references therein) lead to distinguish at least three types of reactions.

(*a*) *Photo-ionisation reactions.* Photo-ionisation of purine nucleosides and aromatic amino acids has been observed in different media: in pure ice[67, 62] or in ice including NaCl, NaOH or HCl (unpublished results) and in boric acid glass.[64] Upon photo-excitation an ESR signal corresponding to the photo-oxidised radical ions and trapped electrons appears. In aqueous solutions 8 M in NaOH the absorption of the radical ions and of solvated electrons ($\lambda = 590\,\mathrm{m}\mu$) is observable. In acidic media the ejected electrons recombine with protons and H atoms appear in the ESR spectra.[64] Simultaneously, the triplet signal of the purine or the amino acid is quenched and a delayed luminescence which reaches rapidly a steady intensity is emitted.[66] This luminescence which arises from recombination of the ions and the electrons can be enhanced by heating or by the effect of visible light (R. Guermonprez, to be published). It has the spectral characteristics of phosphorescence of the parent molecule.

(*b*) *Photo-sensitisation reactions.* At low intensity of irradiation the photo-sensitisation of hydroxylic glassy solvents or of a second aliphatic solute in ice by purines or aromatic amino acids may occur without ionisation or reduction of the sensitising molecules. The latter reaction (e.g. deamination of glycine or alanine[31, 63]) is observed only for concentrations of the sensitised molecules at least 100 times higher than that of the sensitising solute. At lower relative concentration one observes simultaneously a photo-ionisation of the phosphorescent molecules. This suggests that photo-ionisation and photo-sensitisation are two competitive phenomena although Azizova *et al.*[31] propose that deamination of amino acids and polypeptides occurs by reaction of protonated amino groups with free electrons. Sensitised bond splitting without chemical modification of the sensitising molecules has been observed with many non-biological aromatic molecules including dyes.[65]

(*c*) *Photo-reduction by hydrogen abstraction from solvent molecules.* Hydrogen abstraction in hydroxylic solvents may be carried out with reduction of the phosphorescent molecule. With porphyrins[29, 31] or flavins[45] the photo-reduction is indicated by the appearance in the ESR spectrum at $g = 2$ of a singlet signal of the solute molecules superimposed to the signal of hydroxylic radicals and by a new band in the absorption spectrum.

The quadratic dependence of the photochemical damage on light intensity[68] and the kinetics of the free radical production and triplet

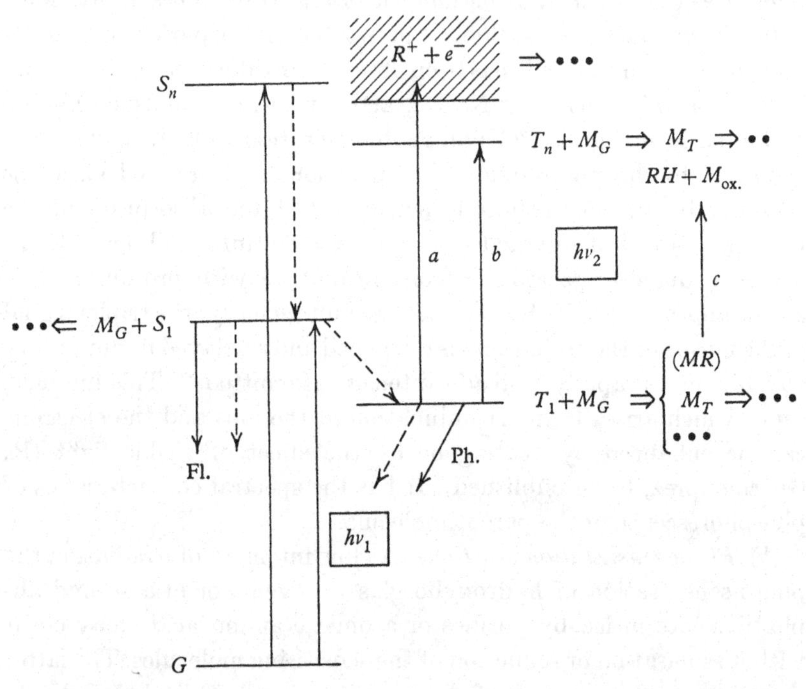

Fig. 8. Primary physical processes in photochemical reactions involving triplet states of a conjugated molecule *R*. *M* is a sensitised molecule. Dots represent subsequent chemical steps.

population quenching,[69] as well as the use of alternate[31] or double[70, 71] monochromatic light beams and the quenching of the reactions by paramagnetic ions (Cu^{2+}) indicate that the three types of reactions occur via the lowest triplet state of the phosphorescent molecules by biphotonic processes which satisfy the energy requirement in each individual reaction when the irradiation is carried out with light quanta of low energy. These reactions differ in the nature of the second light step (fig. 8). The lack of a light polarisation effect[72] and of spectral requirements,[73] as far as a minimum energy is provided for the photo-ionisation of aromatic molecules suggests that reactions of type (*a*) occur by direct ionisation of the lowest triplet state of the phosphorescent molecule. Reactions of type (*b*) may occur via triplet-triplet absorption followed by energy transfer to the sensitised molecules as shown unambiguously by Terenin and collaborators in the photo-splitting of alkyl iodides and tertiary butanol by aromatic molecules[65] including porphyrins.[71] Triplet-triplet energy transfer to the sensitised molecules explains the high yield of the reactions due to the conservation of the total spin in the system. The photo-excitation and splitting of a solvent-triplet solute complex of charge transfer character has been proposed by several authors[74, 75] for reactions of type (*c*). Such a mechanism may be favored for solute molecules like flavins which have been shown to be good electron acceptors in the triplet state. Nevertheless, not one of these mechanisms seems to be proved as unique and in many heterocyclic molecules several types of reactions may occur simultaneously. This makes the detailed analysis of the mechanisms difficult.

Whatever the mechanism of the second light step could be, these biphotonic processes provide excitation energies at a level corresponding to high yield for many photochemical reactions even when the excitation light is in the visible or near ultraviolet wavelength range. Accumulation of energy from triplet states of aromatic molecules could occur by triplet-triplet annihilation processes similar to those observed in organic crystals.[76] Yet, the strict requirement in geometry and homogeneity of molecules for migration of the triplet excitation over large distances makes a process of that type rather unprobable in biological materials compared to the monomolecular processes.

Conclusion

Photo-excited biomolecules must be considered as different chemical species and, among the excited states, the lowest triplet state may have a specific role in the chemical conversion, the transfer or the accumulation of light quanta energy in photo-synthetic or in radiation damage processes (fig. 8). This specificity arises from the high singlet-triplet conversion rate, the long lifetimes and the electronic structure observed for the lowest triplet state of most conjugated biomolecules. These properties have been observed mostly in experiments at low temperatures in frozen rigid media at high intensities of irradiation. Such conditions apparently differ from those found in living organisms which contain a high amount of water as well as paramagnetic ions and oxygen quenchers and are generally exposed to much lower light intensities. Flash photolysis techniques in liquids, as well as the use of rigid matrices at normal temperature, show that the behaviour of triplet states is not qualitatively very dependent upon these environmental factors. Furthermore, the structural organisation of biological materials, like the helical structure of nucleic acids or the compact folding of protein chains, does not differ fundamentally from the molecular arrangement in the low-temperature experiments. Nevertheless, lower yields are expected for triplet excitation transfer and biphotonic processes compared to chemical reactions in the lowest triplet and excited singlet states.

Triplet states have not been so far detected *in vivo*, but the detailed study of the chemical steps of photo-reactions *in vitro* as well as *in vivo* has yet brought strong evidence for the role of triplet states of conjugated molecules in the absorption, redistribution and chemical conversion of light quanta in photo-biological processes. A better knowledge of the electronic structure and physical properties of photoexcited biomolecules will undoubtedly bring more insight into these processes.

Acknowledgements

The authors are pleased to acknowledge helpful co-operation at various stages of these investigations with Drs P. Douzou, R. Guermonprez, A. Haug, A. M. Michelson and R. Santus. One of them (J. M. L.) expresses his appreciation for the hospitality, encouragement and criticism of Dr M. Kasha, and for helpful discussions with Dr J. H. van der Waals during the preparation of the manuscript.

References

1 See, for example, the *Proceedings of a Conference on Molecular Mechanisms in Photobiology*, in *Photochem. Photobiol.* **3**, 269–580 (1964).

2 W. D. McElroy & H. H. Seliger, in *Light and Life*. W. D. McElroy and B. Glass, eds. (The Johns Hopkins Press, 1961.)

3 C. A. Hutchison Jr. & B. W. Mangum, *J. Chem. Phys.* **34**, 908 (1961).

4 J. H. van der Waals & M. S. de Groot, *Mol. Physics*, **2**, 333 (1959); **3**, 190 (1960).

5 E. Wasserman, L. C. Snyder & W. A. Yager, *J. Chem. Phys.* **41**, 1763 (1964).

6 J. M. Lhoste, A. Haug & M. Ptak, *J. Chem. Phys.* **44**, 648, 654 (1966).
7 M. A. El-Sayed & S. Siegel, *J. Chem. Phys.* **44**, 1416 (1966).
8 H. Sternlicht, *J. Chem. Phys.* **38**, 2316 (1963).
9 R. W. Brandon, R. E. Gerkin & C. A. Hutchison, Jr., *J. Chem. Phys.* **41**, 3717 (1964).
10 S. W. Charles, P. H. H. Fischer & C. A. McDowell, *Mol. Physics*, **9**, 517 (1965).
11 J. S. Vincent & A. H. Maki, *J. Chem. Phys.* **39**, 3088 (1963).
12 N. K. Chaudhuri & M. A. El-Sayed, *J. Chem. Phys.* **44**, 3728 (1966).
13 S. Hosoya, *Acta Cryst.* **16**, 310 (1963).
14 A. D. McLachlan, *Mol. Physics*, **5**, 51 (1962).
15 A. J. Stone, *Mol. Phys.* **7**, 311 (1964).
16 H. M. McConnell & J. Strathdee, *Mol. Physics*, **2**, 129 (1959).
17 T. Cole, C. Heller & H. M. McConnell, *Proc. Natl. Acad. Sci. U.S.* **45**, 525 (1959).
18 J. M. Lhoste & F. Tonnard, *J. chim. phys.* **63**, 678 (1966).
19 A. T. Amos, *Mol. Physics*, **5**, 91 (1962).
20 J. H. van der Waals & G. ter Maten, *Mol. Physics*, **8**, 301 (1964).
21 M. Calvin & G. D. Dorough, *J. Am. Chem. Soc.* **70**, 699 (1948).
22 J. Fernandez & R. S. Becker, *J. Chem. Phys.* **31**, 467 (1959).
23 I. S. Singh & R. S. Becker, *J. Am. Chem. Soc.* **82**, 2083 (1960).
24 J. B. Allison & R. S. Becker, *J. Chem. Phys.* **32**, 1410 (1960); *J. Chem. Phys.* **67**, 2662, (1963).
25 S. Aronoff, *Arch. Biochem. Biophys.* **98**, 344 (1962).
26 J. J. Katz, G. L. Closs, F. C. Pennington, M. R. Thomas & H. H. Strain, *J. Am. Chem. Soc.* **85**, 3801 (1963).
27 R. Livingston, *J. Am. Chem. Soc.* **77**, 2179 (1955).
28 L. Pekkarinen & H. Linschitz, *J. Am. Chem. Soc.* **82**, 2407 (1960).
29 Z. P. Gribova, R. P. Yevstigneva, A. F. Mironova, L. P. Kayushin, V. N. Luzina & A. K. Piskunov, *Biofizika*, **8**, 550 (1963).
30 G. T. Rikhireva, Z. P. Gribova, L. P. Kayushin, A. V. Umrikhina & A. A. Krasnovskii, *Dokl. Akad. Nauk SSSR*, **159**, 196 (1964).
31 O. A. Azizova, Z. P. Gribova, L. P. Kayushin & M. K. Pulatova, *Photochem. Photobiol.* **5**, 763 (1966).
32 M. Gouterman, *J. Mol. Spectr.* **6**, 138 (1961).
33 C. Weiss, H. Kobayashi & M. Gouterman, *J. Mol. Spectr.* **16**, 415 (1965).
34 W. T. Simpson, *J. Chem. Phys.* **17**, 1218 (1949).
35 J. R. Platt in *Radiation Biology*, A. Hollaender, ed., vol. III, chap. 2 (McGraw Hill, New York, 1956).
36 M. S. de Groot, I. A. M. Hesselmann & J. H. van der Waals. *Mol. Phys.* **10**, 241 (1965).
37 M. Ptak & P. Douzou, *Nature*, **199**, 1092 (1963).
38 T. Shiga & L. H. Piette, *Photochem. Photobiol.* **3**, 223 (1964).
39 R. O. Rahn, J. W. Longworth, J. Eisinger & R. G. Shulman, *Proc. Natl Acad. Sci. U.S.* **51**, 1299 (1964).

40 R. G. Shulman & R. O. Rahn, *J. Chem. Phys.* **45**, 2940 (1966).
41 B. Smaller, *Adv. Biol. Med. Phys.* **9**, 225 (1963).
42 R. Santus & C. Hélène, unpublished results.
43 J. E. Maling, K. Rosenheck & M. Weissbluth, *Photochem. Photobiol.* **4**, 241 (1965).
44 R. O. Rahn, R. G. Shulman & J. W. Longworth, *J. Chem. Phys.* **45**, 2955 (1966).
45 J. M. Lhoste, A. Haug & P. Hemmerich, *Biochem.* **5**, 3290 (1966).
46 B. Holmström, *Arkiv Kemi* **22**, 329 (1964).
47 G. K. Radda & M. Calvin, *Biochem.* **3**, 384 (1964).
48 A. Ehrenberg, G. Eriksson & P. Hemmerich, in *Oxidases and Related Redox Systems*, J. E. King, S. Mason and M. Morrison, eds., vol. I (John Wiley and Sons, New York, 1965), p. 179.
49 P. Hemmerich, F. Muller & A. Ehrenberg, *ibid.* vol. I, p. 157.
50 C. Hélène, P. Douzou & A. M. Michelson, *Biochim. Biophys. Acta*, **109**, 261 (1965).
51 C. Hélène, P. Douzou & A. M. Michelson, *Proc. Natl Acad. Sci. U.S.* **55**, 376 (1966).
52 M. Guéron, R. G. Shulman & J. Eisinger, *Proc. Natl Acad. Sci. U.S.* **56**, 814 (1966).
53 C. Hélène & A. M. Michelson, *Biochim. Biophys. Acta*, 1967 (in press).
54 C. Hélène, *Biochem. Biophys. Res. Com.* **22**, 237 (1966).
55 C. Hélène, R. Santus & M. Ptak, *C.R. Acad. Sci., Paris*, **262**C, 1349 (1966).
56 R. O. Rahn, T. Yamane, J. Eisinger, J. W. Longworth & R. G. Shulman, *J. Chem. Phys.* **45**, 2947 (1966).
57 G. Weber, in *Light and Life*, W. D. McElroy and B. Glass, eds. (The Johns Hopkins Press, 1961).
58 R. W. Cowgill, *Biochim. Biophys. Acta* **75**, 272 (1963).
59 P. Douzou & M. Ptak, *J. chim. phys.* **61**, 1681 (1964).
60 T. Shiga, S. Mason & C. Simo, *Biochem.* **5**, 1877 (1966).
61 E. Yeargers, F. R. Bishai & L. Augenstein, *Biochem. Biophys. Res. Com.* **23**, 570 (1966).
62 C. Hélène, R. Santus & P. Douzou, *Photochem. Photobiol.* **5**, 127 (1966).
63 R. Santus, C. Hélène & M. Ptak, *C.R. Acad. Sci., Paris*, **262**D, 2077 (1966).
64 R. Santus, R. Guermonprez & M. Ptak, *C.R. Acad. Sci., Paris*, **261**, 117 (1965).
65 A. Terenin, V. Rylkov & V. Kholmogorov, *Photochem. Photobiol.* **5**, 543 (1966).
66 P. Debye & J. O. Edwards, *J. Chem. Phys.* **20**, 236 (1952).
67 O. A. Azizova, *Biofizika*, **9**, 745 (1964).
68 K. S. Bagdasaryan, Z. A. Sinitsyna & V. I. Muromtzev, *Dokl. Akad. Nauk SSSR*, **152**, 349 (1963).
69 S. Siegel & K. Eisenthal, *J. Chem. Phys.* **42**, 2494 (1965).
70 Z. P. Gribova & L. P. Kayushin, *Biofizika*, **9**, 627 (1964).

71 V. Rylkov, V. Kholmogorov & A. Terenin, *Dokl. Akad. Nauk. SSSR*, **169**, 1373 (1966).
72 K. D. Cadogan & A. C. Albrecht, *J. Chem. Phys.* **43**, 2550 (1965).
73 G. E. Johnson & A. C. Albrecht, *J. Chem. Phys.* **44**, 3179 (1966).
74 B. Smaller, *Nature*, **195**, 593 (1962).
75 E. C. Lim & G. W. Swenson, *J. Chem. Phys.* **39**, 2768 (1963).
76 H. Sternlicht, G. C. Nieman & G. W. Robinson, *J. Chem. Phys.* **38**, 1326 (1963).

THE TRIPLET STATE OF DNA

M. GUERON, R. G. SHULMAN AND J. EISINGER

Abstract

DNA dissolved in ethylene glycol–water glass at 77 °K exhibits only one excited triplet. Optical and ESR experiments on DNA of various A:T content show that the triplet is localised on thymine. In poly-dAT, an excimer, to which lights absorbed by A and T contribute, is shown to be a precursor, possibly the direct precursor, of the thymine triplet. The question of proton transfer from thymine to adenine is discussed. On the basis of similar quenching by paramagnetic metal ions, the triplet is identified as a precursor to the thymine free radical.

Introduction

Several photochemical products of irradiated DNA are now known, first among them being the cyclobutane-type thymine dimer.[1] A thymine free radical, which is not long-lived under biological conditions, has also been identified and studied.[2] In our laboratory we have been investigating the steps which lead to the chemical damage, starting from the time that energy is first absorbed by the DNA polymer. Our main tools have been absorption and emission spectroscopy, and electron spin resonance (ESR) of the triplet state. In the course of this work it has become clear that the excited states in DNA and some polynucleotides are quite different from those of the component nucleotides, because of interactions between the latter. Excitation quenching, energy transfer, excimer formation, triplet-triplet transfer and possibly proton transfer across a hydrogen-bond are some effects of the interactions, which have been observed in different systems.

We recall that DNA is a double-stranded helical polymer. Each strand has a sugar-phosphate backbone, with one base attached to each sugar. There are four bases: two purine derivatives adenine (A) and guanine (G), and two pyrimidine derivatives thymine (T) and cytosine (C). Each base from one strand is hydrogen-bonded to another coplanar base which is attached to the other strand. The hydrogen bonds are specific, allowing only pairing of A to T and of G to C (fig. 1). The planes of the bases are perpendicular to the helical axis, the distance between successive planes being 3·36 Å. Model systems have also been used in the investigations we will discuss:

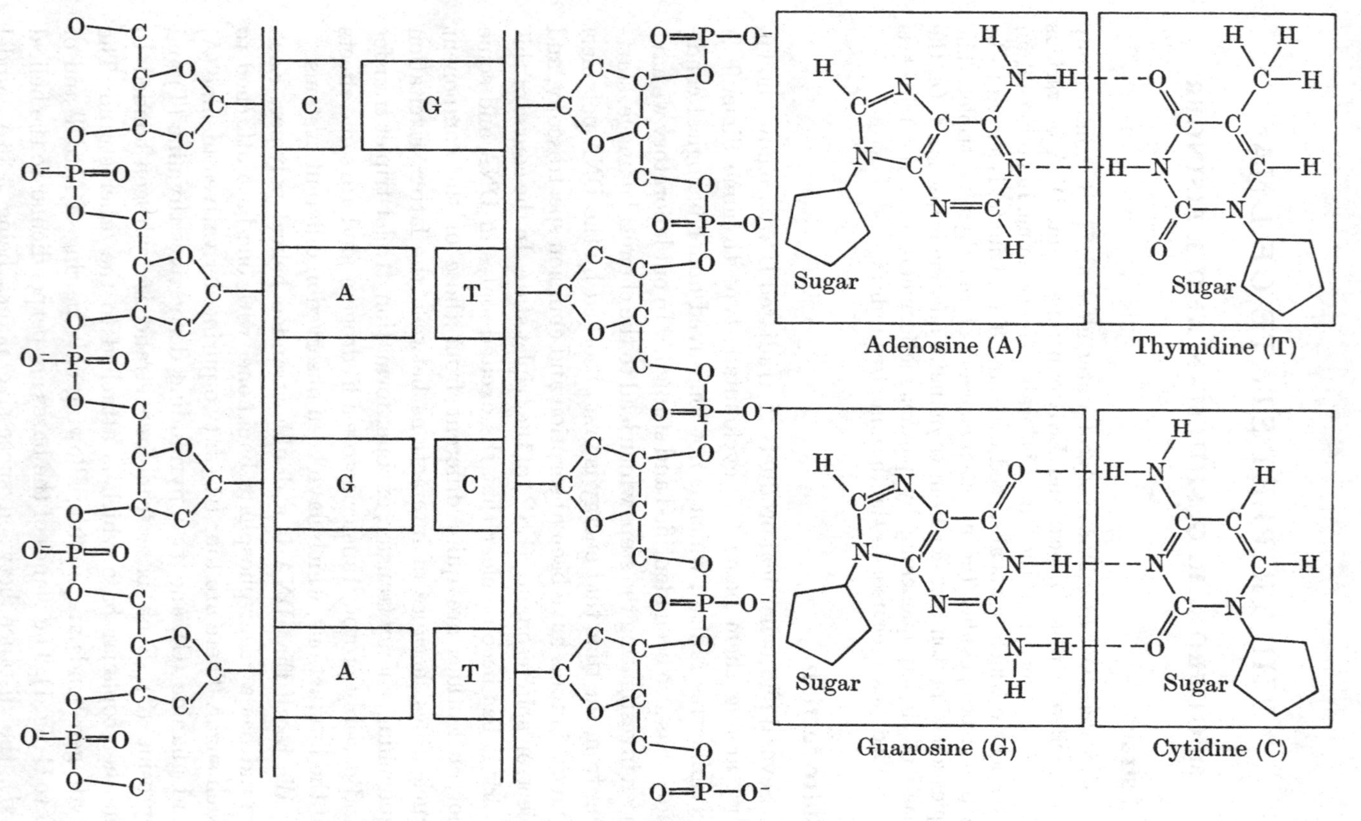

Fig. 1. Schematic structure and base pairing in DNA.

mononucleotides, dinucleotides (a single strand carrying two bases) and various synthetic polynucleotides, both single- and double-stranded.

Until now our work has been done mostly at liquid nitrogen temperature in polar glasses. In this way by paying the price of moving away from biological conditions, one is enabled to study fluorescence, phosphorescence and the triplet-state electron spin resonance, none of which have been observed at room temperature in DNA. In this paper, we describe the identification of the DNA triplet, discuss the possible means of its formation and the evidence that it is the precursor of the thymine radical previously identified.

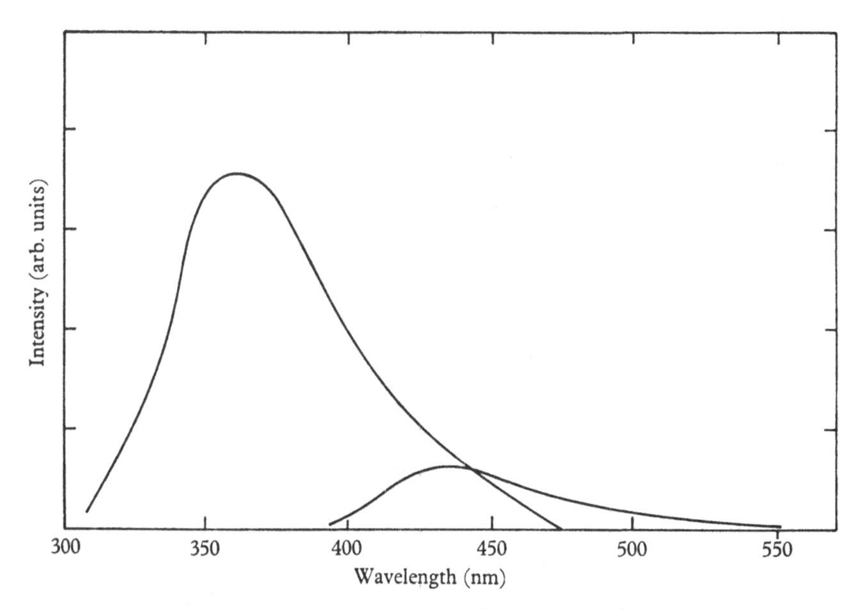

Fig. 2. Fluorescence and phosphorescence of calf-thymus DNA. The data are corrected for variations of sensitivity with wavelength. The yields are determined by comparison with the fluorescence of para-terphenyl. For unknown reasons their values, and the ratio of fluorescence to phosphorescence, vary appreciably with sample. Fluorescence Q.Y. = $0{\cdot}7 \times 10^{-2}$; phosphorescence Q.Y. = $0{\cdot}1 \times 10^{-2}$; phosphorescence decay time = 0·3 s; T = 80 °K. EG: H_2O glass.

Identification of the DNA triplet

Upon ultraviolet excitation at 77 °K in EG: H_2O glass, purified DNA exhibits fluorescence and phosphorescence (fig. 2). Each emission is broad and structureless, with a quantum yield of less than 1 %.

Comparison with the monomer emissions[4–7] indicates a general similarity but also pronounced differences. The emissions of DNA and the monomers occur in the same spectral regions, and the phosphorescence decay times are all of the order of 1 s, indicating π, π^* triplets. However, neither fluorescence nor phosphorescence of DNA can be matched with the corresponding emission of any one of the monomers (fig. 3). It has since been shown by comparison with dinucleotides that the fluorescence of DNA comes from excimers.[8]

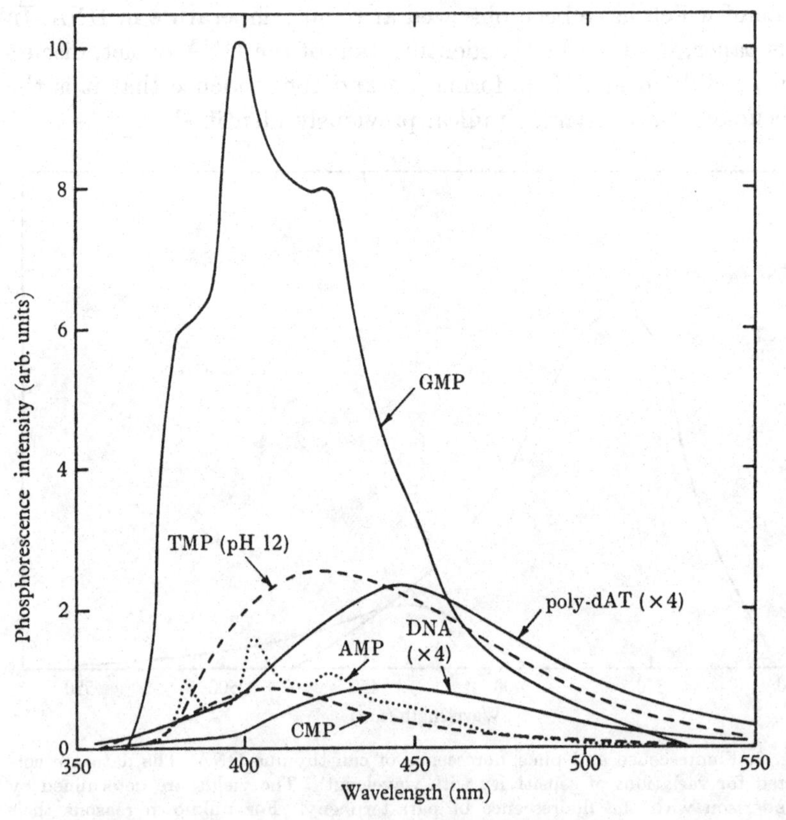

Fig. 3. Phosphorescence of monomers, calf-thymus DNA and poly-dAT.

The first attempts to observe the ESR signal of the DNA triplet were unsuccessful. At neutral pH, strong signals were observed from adenosine and guanosine monophosphates (AMP and GMP); there was only a weak phosphorescence from CMP; and TMP exhibited no phosphorescence or ESR. However, in DNA no signal was found at fields where the triplets of the mononucleotides were observed. Subsequently, the ESR of TMP at pH 12 (where it is ionised by the loss of the N_3 proton) was observed,[3] with a decay time of 0·45 s which was close to that of 0·3 s observed for the DNA phosphorescence. More importantly the DNA triplet ESR was found at almost the same magnetic field as the high pH TMP resonance (table 1). The picture of the DNA triplet which emerged was: the triplet is localised on a thymine residue which has become ionised by transferring its labile proton at N_3 across the hydrogen bond to the adenine with which it is paired. The proposed proton transfer explains at one step why a triplet is observed from T, why the A triplet is not observed, and why the ESR and emission properties of DNA are so close to those of TMP^-. Evidence for the localisation of the triplet on thymine follows. The possibility of proton transfer is discussed below.

Table 1. ESR *properties of mono- and poly-nucleotides*

	$(D^2+3E^2)^{\frac{1}{2}}$ cm^{-1}	τ(s)
AMP	0·126	2·5
GMP	0·145	1·25
TMP^-	0·198	0·45
DNA	0·200	0·3
dAT	0·201	0·3

The main spectroscopic argument for the localisation of the triplet in thymine lies in the near identity of the zero-field splitting

$$D' = [D^2+3E^2]^{\frac{1}{2}}$$

as mentioned above. The very small difference, $D' = 0{\cdot}200\ \mathrm{cm}^{-1}$ for DNA versus $D' = 0{\cdot}198\ \mathrm{cm}^{-1}$ for TMP^-, is perhaps ascribable to the different environments. Similarly, the differences in the decay times and the slight shifts in the phosphorescence spectra suggest only minor differences between the triplets of TMP^- and DNA, and do not put in question the localisation of the triplet on thymine.

Another strong argument for this localisation is given by our study of the triplet in various DNA's and model polymers. No triplet population is observed (in phosphorescence or ESR) from synthetic poly-dG : dC or poly-rG : rC. On the other hand, both poly-dAT (alternating double stranded ATAT...) and poly-dA:dT (strands of A hydrogen-bonded to strands of T) exhibit a triplet nearly identical to that of calf thymus DNA, as judged from phosphorescence spectra, ESR values of D' and decay times. By contrast, a dilute mixture of GMP and CMP exhibits the strong GMP triplet, while a dilute mixture of AMP and TMP shows exclusively the AMP triplet. We have also studied the intensities of both phosphorescence and ESR for various DNA's, prepared by T. Yamane, as a function of A–T content. Both intensities grow with the A–T content, in qualitative agreement with our model (fig. 4).

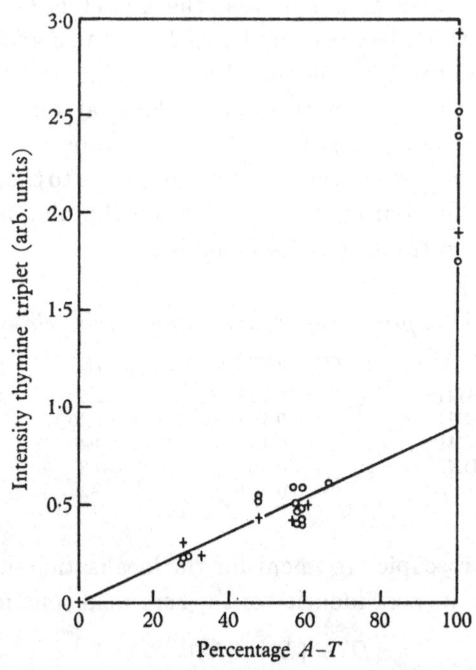

Fig. 4. Phosphorescence (○) and ESR (+) intensities of various DNA's as a function of A–T content.

In the course of the studies described above we found no indication of other triplet states in native DNA, although small signals from A or G triplet have been observed in denatured or degraded DNA samples. A triplet with a phosphorescent yield larger than 10^{-3} would certainly have been detected in emission; while with a lifetime of approximately 1 s, and intersystem crossing yield of a few per cent, it would have been detected in ESR. It must therefore be explained why the C, G and A triplets observed in the monomers are not present in DNA. Are they not populated, or are they populated and then quenched, or are they populated and then transferred[10] rapidly to the thymine triplet? For G and C, the singlet and triplet quenching observed in poly G : C indicates that the triplets are not populated. This is also the case of A,[3] in poly-rA : rU, in which the A fluorescence and phosphorescence are both quenched.

Paths of excitation

We now consider the question of the possible paths of excitation and products of the thymine triplet in DNA. Because the absorption spectrum of DNA does not show large wavelength shifts from that of a mononucleotide mixture of the same composition, we infer that there are no strong interactions between the bases in their ground states so that absorption takes place in the localised chromophores; this means that we must deal with four distinct absorption processes. After a chromophore is excited, the following processes must be considered: fluorescence, intersystem crossing, non-radiative decay to the ground state, energy transfer to other chromophores, excimer formation and photochemical reactions.

Because no fluorescence characteristic of any of the four bases is observed, we conclude that, in the polymer, the monomer-like singlet states formed upon absorption have lifetimes much shorter than in the isolated monomers. Hence intersystem crossing from these states is excluded because in the monomers it occurs at rates comparable to fluorescence. Singlet quenching has been demonstrated[(8)] for the hydrogen-bonded G–C pairs, and also for A in A–U base pairs (in which U resembles T). A theoretical study of energy transfer between nucleotides[(7)] predicts singlet energy transfer between most bases in DNA. It cannot be decided at present whether the range of transfer is so limited that the fate of the singlet depends on the site of energy absorption, or not. In the former case, one would have to consider a number of different singlets. Excimer formation, which may or may not be preceded by energy transfer, has been demonstrated. Consideration of the competing rates of intersystem crossing, excimer formation and fluorescence shows that excimers, the only observed singlet states, are capable of undergoing intersystem crossing.[(9)] They may also lead to damaged products although this has not been proven, and in fact damage from the singlet state may be a competing path.

Turning now to the thymine triplet, we ask the following questions:

(*a*) Is there direct evidence that an excimer is its singlet precursor and how is the excimer populated?

(*b*) It has been proposed[(10)] that the thymine triplet is populated by triplet transfer from the other bases. However, we have given[(3)] rather strong evidence against this possibility. We therefore assume that the thymine triplet has a singlet as its *direct* precursor, and is therefore populated by intersystem crossing. But we have seen that intersystem crossing does not occur in neutral TMP. Is it allowed in DNA because thymine has transferred its N_3 proton to adenine, which we have shown[3] previously would explain the A quenching and the D' values of the DNA ESR signal? Or because interactions between the stacked bases increase the probability of intersystem crossing in neutral thymine?

(*c*) Is the thymine triplet the precursor of any specific damage product?

As regards question (*a*), we present an analysis of a naturally occurring DNA, poly-dAT, containing only alternating A:T base pairs, which should be simpler to understand than a DNA containing all four bases. As in other DNA's, the fluorescence is of the excimer type, broad, structureless, red-shifted from those of the monomers; the triplet is the thymine triplet.[7] We have taken excitation spectra of both fluorescence and phosphorescence. We first observed that the shape of the fluorescence was independent of excitation wavelength, implying that only one excimer is present. The shape of the excimer excitation spectrum shows that it is generated by light absorbed by both T and A with relative yields of approximately 1 and 0·7. The fact that there is no monomer emission suggests that these numbers represent the absolute yields for populating the excimer from light absorbed by T and A.

From the observation that the ratio of fluorescence to phosphorescence is independent of excitation wavelength, we conclude that the excimer is a precursor of the thymine triplet in poly-dAT.

If, as was discussed in question (*b*), triplet transfer is excluded, the excimer becomes a good candidate for the (singlet) direct precursor of the thymine triplet. This implies that the excimer is at least partially located on thymine.

We cannot resolve question (*b*) at this time. The evidence in favour of T^- as the origin of the DNA triplet has been presented and depends mainly upon the A quenching[3] and the similar ESR positions. However, we have accumulated considerable evidence that shows hydrogen-bonding is not required for this triplet but that aggregation of the chromophores (presumably by stacking) is sufficient.

The first indication of this came from neutral solutions of thymine, thymidine, aza-thymine and *N*-methylthymine, all of which showed[13] thymine-type ESR resonances at high concentrations (10^{-2} M). All except thymine lost the resonance at low concentrations (10^{-3} M) in contrast to the pH 12 samples where the resonances persisted at low concentrations. Another indication was the negligible reduction in the thymine resonance in DNA after denaturation.[3] Still another indication that hydrogen-bonding might not be required was the rather strong thymine resonances observed[7] in ϕX–174 DNA which is a single stranded DNA and in poly-dT. In all these cases it seemed quite possible that in our low-temperature glasses considerable aggregation involving hydrogen-bonding had occurred. This was notably suspected in ϕX–174 DNA because no G or A triplet was observed. Recently, however, Lamola[14] has looked at the ESR of 1,3-dimethylthymine at high concentrations ($> 10^{-3}$ M) in EG : H_2O glasses and observed an ESR signal which is not seen at lower concentrations. This ESR signal with $D' \approx 0{\cdot}187\ \text{cm}^{-1}$ is close to that observed in *N*-methylthymine and it appears under conditions where the molecules are aggregated but cannot, of course, be hydrogen-bonded since the dimethylthymine molecule has no aza hydrogens. Also, energy transfer at the triplet level cannot be responsible for these triplets because no triplet donor is present. Thus under conditions where both hydrogen-bonding and triplet-triplet transfer are impossible, a triplet appears with a value of D' which is reasonably close to that of T^- and very close to that of *N*-methylthymine[13] at pH 12. Hence either the intersystem crossing rate has been increased by the aggregation or the rate of triplet quenching decreased. Disentangling these various possibilities will require additional experiments, some of which are in progress.[17]

Question (*c*) concerning photochemical reactions of the thymine triplet has been studied in preliminary experiments by quenching the DNA triplet with divalent transition metal ions.[3] It had been previously shown[15] that these ions bind to the phosphate groups. The comparison of emission from samples with and without metal ions shows that the fluorescence is unperturbed, while as Bersohn & Isenberg[16] have shown the T triplet is quenched. The thymine free radical in DNA is quenched at the same rate as the thymine triplet which is therefore implicated as a precursor. On the other hand, it is not known how the triplet gives rise to the free radical, which has gained a proton at the C_6 position, or what governs the free radical quantum yield, which[2] is $\sim 10^{-3}$, as compared to the triplet yield, which, measured by ESR, is[7] equal to 2×10^{-2}. Lastly, the quenching range[16] of the metal ion is of the order of six base pairs. This would imply triplet transfer (the fluorescence is not quenched) and is surprising in view of the energy differences between the different triplets and of the fact that no other triplets are observed.

Conclusion

The following picture of the DNA triplet is emerging. The triplet is located on the thymine moiety. The poly-dAT experiments show that light absorbed by adenine and thymine contributes to it via a singlet excimer. It seems likely, but not yet certain, that the excimer is the triplet's immediate precursor and is located on thymine. Metal ion triplet quenching shows that the triplet is a precursor of the thymine free radical. We hope that it is also clear from this report that a number of questions are still unanswered.

Acknowledgements

As indicated in the references, our early work on this subject was done with Drs R. O. Rahn and J. W. Longworth. In addition, we have been helped considerably at various phases by the samples supplied by Dr T. Yamane and the assistance of Mr B. J. Wyluda.

References

1 R. Beukers & W. Berends, *Biochim. Biophys. Acta*, **49**, 181 (1961).
2 P. S. Pershan, R. G. Shulman, B. J. Wyluda & J. Eisinger, *Physics*, **1**, 163 (1964).
3 R. O. Rahn, R. G. Shulman & J. W. Longworth, *Proc. Natl Acad. Sci. U.S.* **53**, 893 (1965); *J. Chem. Phys.* **45**, 2955 (1966).
4 P. Douzou, J. Francq, M. Hanss & M. Ptak, *J. Chim. Phys.* **58**, 926 (1961).
5 J. W. Longworth, *Biochem. J.* **81**, 23 (1961).
6 J. W. Longworth, R. O. Rahn & R. G. Shulman, *J. Chem. Phys.* **45**, 2930 (1966).
7 M. Guéron, J. Eisinger & R. G. Shulman (to be published).
8 J. Eisinger, M. Guéron, R. G. Shulman & T. Yamane, *Proc. Natl. Acad. Sci. U.S.* **55**, 1015 (1966).
9 M. Guéron, R. G. Shulman & J. Eisinger, *Proc. Natl. Acad. Sci. U.S.* **56**, 814 (1966).
10 C. Hélène, *Biochem. Biophys. Res. Commun.* **22**, 237 (1966).
11 R. O. Rahn, T. Yamane, J. Eisinger, J. W. Longworth & R. G. Shulman, *J. Chem. Phys.* **45**, 2947 (1966).
12 J. Eisinger & R. G. Shulman, *Proc. Natl. Acad. Sci. U.S.* **55**, 1387 (1966).
13 R. G. Shulman & R. O. Rahn, *J. Chem. Phys.* **45**, 2940 (1966).
14 A. A. Lamola (private communication).
15 R. G. Shulman, H. Sternlicht & B. J. Wyluda, *J. Chem. Phys.* **43**, 3116 (1965).
16 R. Bersohn & I. Isenberg, *J. Chem. Phys.* **40**, 3175 (1964).
17 A. A. Lamola, M. Gueron, T. Yamane, J. Eisinger & R. G. Shulman (to be published).

SOME CHARACTERISTICS OF THE TRIPLET STATES OF THE NUCLEIC BASES†

ALBERTE PULLMAN

Abstract

Calculations on the first excited triplet states of the nucleic purines and pyrimidines and of related compounds have been performed in different approximations of the method of molecular orbitals. The results concerning the positions of the triplet states and their electronic structure (electron distribution, spin densities and zero-field splitting parameters) are discussed in terms of the approximations involved and compared with the available experimental data.

Introduction

In the last two years, our research group has undertaken a systematic study of the structure of the nucleic bases and their analogs by refined theoretical procedures.[1–6] This has included both the study of the ground-state characteristics of the bases and of their oxygenated derivatives, as well as the calculation of the position and electronic structure of their excited singlet and triplet states.

The method used was fundamentally the self-consistent approach in the framework of the Pariser–Parr–Pople approximations. Our study differs however from similar calculations recently[7] published by: (*a*) its systematic character; (*b*) a careful choice of the semi-empirical integrals of the Pariser–Parr–Pople scheme so as to reproduce as many properties as possible for a large set of reference molecules; (*c*) the inclusion of an approximate description of the σ-framework.

Three descriptions of the excited states have been studied: *the single-configuration* representation using the virtual orbitals of the ground state calculations (SCF), *the state* representation obtained by mixing all the singly-excited configurations (SCFCM), and the *unrestricted Hartree–Fock* procedure using the same set of approximations as in the closed-shell computation (UHF). Details on the calculation procedures can be found in our earlier work.[1,3,6]

† This work was supported by grant no. GM 12289-02 of the United States Public Health Service (National Institute of General Medical Sciences).

Table 1. *Charges and bond orders in the ground and first triplet states*

[(*a*) SCF; (*b*) SCFCM; (*c*) UHF.]

	Charges					Bond orders			
	S_0	T_1			Bond	S_0	T_1		
Atom		(*a*)	(*b*)	(*c*)			(*a*)	(*b*)	(*c*)
					Uracil				
N_1	1·730	1·572	1·691	1·687	N_1C_2	0·369	0·399	0·364	0·348
C_2	0·806	0·817	0·806	0·826	C_2N_3	0·397	0·373	0·396	0·368
N_3	1·740	1·773	1·747	1·786	N_3C_4	0·373	0·305	0·317	0·310
C_4	0·815	0·947	1·005	0·944	C_4C_5	0·332	0·561	0·518	0·650
C_5	1·089	0·986	0·979	0·997	C_5C_6	0·873	0·377	0·449	0·209
C_6	0·985	1·246	1·145	1·184	N_1C_6	0·404	0·431	0·396	0·392
O_7	1·451	1·397	1·445	1·398	C_2O_7	0·807	0·795	0·808	0·835
O_8	1·384	1·263	1·182	1·177	C_4O_8	0·835	0·643	0·500	0·618
					Cytosine				
N_1	1·690	1·677	1·721	1·734	N_1C_2	0·385	0·389	0·287	0·402
C_2	0·793	0·792	0·993	0·829	C_2N_3	0·393	0·524	0·527	0·665
N_3	1·297	1·273	1·130	1·185	N_3C_4	0·751	0·448	0·500	0·288
C_4	0·872	1·150	1·120	1·151	C_4C_5	0·369	0·529	0·391	0·506
C_5	1·134	1·104	1·103	1·055	C_5C_6	0·853	0·695	0·826	0·775
C_6	0·952	1·187	1·003	1·098	N_1C_6	0·441	0·339	0·394	0·315
O_7	1·461	1·006	1·141	1·139	C_2O_7	0·794	0·668	0·475	0·518
N_8	1·801	1·810	1·789	1·808	C_4N_8	0·469	0·363	0·391	0·365

					Adenine				
N_1	1·265	1·219	1·237	1·213	N_1C_2	0·613	0·720	0·691	0·804
C_2	0·836	0·904	0·867	0·887	C_2N_3	0·700	0·482	0·497	0·290
N_3	1·273	1·249	1·233	1·220	N_3C_4	0·539	0·571	0·546	0·556
C_4	0·971	0·951	0·953	0·994	C_4C_5	0·622	0·574	0·584	0·570
C_5	1·071	0·971	0·994	0·939	C_5C_6	0·514	0·382	0·380	0·262
C_6	0·854	1·056	1·037	1·052	C_5N_7	0·442	0·568	0·549	0·617
N_7	1·323	1·353	1·271	1·294	N_7C_8	0·764	0·521	0·460	0·614
C_8	0·937	1·102	1·043	0·980	C_8N_9	0·537	0·489	0·500	0·551
N_9	1·632	1·641	1·644	1·617	C_4C_9	0·420	0·411	0·397	0·415
N_{10}	1·838	1·553	1·723	1·803	C_6N_{10}	0·425	0·453	0·426	0·317
					N_1C_6	0·644	0·509	0·486	0·451
					Guanine				
N_1	1·707	1·738	1·723	1·725	N_1C_2	0·458	0·321	0·418	0·316
C_2	0·883	1·269	0·939	1·248	C_2N_3	0·721	0·410	0·667	0·269
N_3	1·386	1·342	1·316	1·257	N_3C_4	0·408	0·635	0·477	0·712
C_4	1·031	0·952	1·082	0·940	C_4C_5	0·686	0·451	0·490	0·458
C_5	1·068	0·997	1·005	1·002	C_5C_6	0·338	0·342	0·473	0·364
C_6	0·785	0·785	1·065	0·805	C_5N_7	0·505	0·595	0·483	0·521
N_7	1·273	1·236	1·294	1·277	N_7C_8	0·750	0·669	0·737	0·723
C_8	0·973	0·916	0·926	0·932	C_8N_9	0·539	0·495	0·539	0·509
N_9	1·614	1·667	1·637	1·664	C_4N_9	0·451	0·365	0·387	0·375
N_{10}	1·852	1·854	1·849	1·862	C_2N_{10}	0·405	0·298	0·394	0·297
O_{11}	1·428	1·242	1·165	1·286	C_6O_{11}	0·821	0·816	0·407	0·739
					N_1C_6	0·362	0·366	0·258	0·420

This report deals with some results concerning the electronic structure of the triplets.

π-electron distribution and spin densities in the triplet state

As far as electron distribution is concerned, the SCF representation does not distinguish between a given singlet and the corresponding triplet (built with the same virtual orbitals). In this sense it is similar to the Hückel approximation and intrinsically it does not give a satisfactory representation of the triplet structure.

From that point of view, the SCFCM approximation is more satisfactory. However, restricting the mixing to singly-excited configurations precludes the possibility of obtaining negative spin densities. Such negative spin densities can result only from the inclusion of doubly-excited configurations involving four singly-occupied orbitals.[6,8] This shortcoming does not exist in the UHF representation which can be shown to be equivalent to mixing essentially doubly-excited configurations and some singly-excited ones with the lowest SCF triplet.[4,8] But then, however, most of the singly-excited configurations are excluded afterwards by the spin-projection method[8] so that the doubly-excited ones become probably overweighted. It seems as if the correct structure lies somewhere between the SCFCM and the UHF representations. It is thus interesting to carry out both calculations.

Table 1 gives the distribution of charges and bond orders in the first triplet state in the three approximations, as compared with the corresponding quantities of the ground state for the four bases uracil, cytosine, adenine and guanine. The spin densities in the first triplet are given in table 2.

Keeping the previous remarks in mind, the examination of the results shows the following features (for the numbering of the atoms, see fig. 1).

(*a*) *Pyrimidines.* The ground and the triplet states of uracil appear to differ strongly, essentially in the region of the C_5C_6 bond which undergoes an inversion of polarity and a striking decrease of its double-bond character, whereas all the rest of the molecular picture remains qualitatively (and almost quantitatively) unmodified (except for a discharge of O_8). Interestingly enough, a similar effect can be seen in the lowest excited singlet.

Table 3 shows the situation of the C_5C_6 bond in uracil and cytosine for both excited states. According to our previous discussion, the

Table 2. *Spin densities in the lowest triplets in different approximations*

Atom	SCF	SCFCM	UHF	Atom	SCF	SCFCM	UHF
	Uracil				Cytosine		
N_1	0·266	0·337	0·188	N_1	0·202	0·243	0·098
C_2	0·011	0·021	−0·005	C_2	0·044	0·273	−0·084
N_3	0·057	0·117	0·015	N_3	0·384	0·221	0·404
C_4	0·155	0·233	−0·059	C_4	0·325	0·297	0·492
C_5	0·509	0·416	0·529	C_5	0·123	0·145	−0·069
C_6	0·548	0·417	0·735	C_6	0·294	0·134	0·305
O_7	0·069	0·188	0·018	O_7	0·502	0·548	0·695
O_8	0·385	0·270	0·579	N_8	0·125	0·140	0·157
	Adenine				Guanine		
N_1	0·088	0·129	−0·011	N_1	0·132	0·131	0·083
C_1	0·169	0·143	0·257	C_2	0·502	0·150	0·636
N_3	0·301	0·262	0·528	N_3	0·257	0·166	0·336
C_4	0·043	0·069	−0·010	C_4	0·203	0·258	0·017
C_5	0·121	0·137	0·028	C_5	0·274	0·211	0·361
C_6	0·318	0·297	0·245	C_6	0·000	0·329	−0·110
N_7	0·184	0·121	0·049	N_7	0·038	0·209	−0·069
C_8	0·330	0·267	0·187	C_8	0·222	0·168	0·203
N_9	0·067	0·089	0·550	N_9	0·063	0·094	0·036
N_{10}	0·380	0·485	0·177	N_{10}	0·123	0·157	0·130
				O_{11}	0·186	0·126	0·376

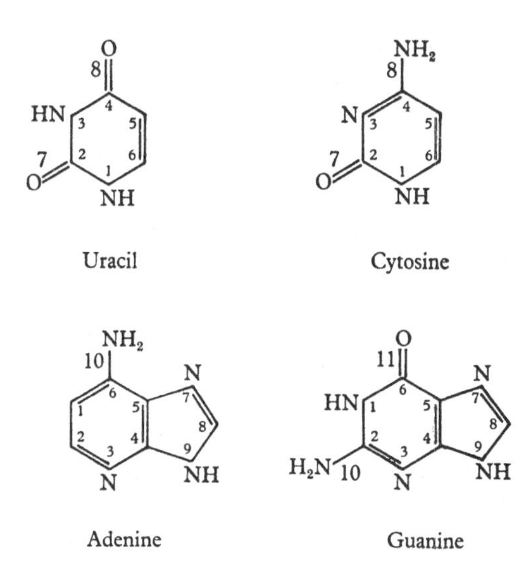

Fig. 1. The numbering of the atoms in the bases.

mean value of the indices corresponding to SCFCM and UHF has been taken for the triplet and singlet. In uracil the strong delocalisation of the double bond upon excitation appears clearly, with a slight piling up of negative charge on both ends in the singlet, on C_6 only in the triplet. The free valences on C_5 and C_6 are strongly increased in both excited states with respect to their ground-state values. Moreover, in the triplet, roughly one half of a spin is localised on each end of the C_5–C_6 bond. A very similar situation is found in thymine.[5]

Table 3. *Charges, bond orders and spin densities* (ρ) *on the* C_5C_6 *bond of uracil and cytosine*

[The values in column 2, 3 and 4 are averages between the approximations SCFCM and UHF.]

		Charges and bond order			
		Ground state	Singlet	Triplet	ρ
U	C_5	1·09	0·99	1·14	0·47
	C_6	0·98	1·16	1·14	0·57
	C_5 C_6	0·87	0·32	0·38	
C	C_5	1·13	1·09	1·07	0·03
	C_6	0·95	1·09	1·17	0·22
	C_5 C_6	0·85	0·80	0·65	

A rather different situation prevails in cytosine. The double-bond character of the C_5C_6 bond is comparable to that of uracil in the ground state, but it does not show any striking decrease in the triplet (the decrease is slightly larger in the singlet state). Carbon C_5 is only slightly affected and the polarity of C_6 is reversed as in uracil. Free valence on C_5 and C_6 is much smaller than in excited uracil. The spin density is appreciable on C_6 but very small on C_5 (on which it appears as negative in the UHF procedure). The spin distribution is displaced towards N_3, C_4 and O_7.

The contrast between uracil and cytosine, as well as the localized character of the excitation in uracil has been noted previously on the basis of Hückel calculations.[9] The fact that refined calculations in different approximations yield the same general conclusions is rather comforting and strengthens the interpretation of the photochemical reactions of the pyrimidine bases, in particular the relative abilities of uracil and cytosine to undergo photodimerisation.[9] It is interesting to mention that the selective photoreduction of uridine in polynucleotides by $NaBH_4$ was discovered on the basis of the theoretical conclusions.[10]

(*b*) *Purines.* In adenine, the positions which are the most affected by excitation to the triplet are C_5 and C_6, which undergo an inversion of polarity, and C_8. The bond order of the C_5—C_6 bond is the most affected by the excitation. The spin-density distribution however does not show a particular localisation of odd spins on a given bond as in uracil. Adenine does not photodimerise and is fairly stable to photodamage in general.[11]

On the whole, for this molecule, the relative order of the spin distribution agrees in the three representations with a more contrasted aspect in the UHF results. The main difference appears on N_9. Calculations of the zero-field splitting parameters in adenine might show which representation is more satisfactory.[12]

Guanine offers an example of the caution which must be exercised when comparing different procedures: rather different distributions of spins appear after and before configuration mixing. This comes from the fact that the two lowest SCF triplets are quasi-degenerate in this molecule, so that their mixing is very strong and the weight of the second configuration is the larger of the two in the first triplet after mixing. However, the UHF calculation gives as the lowest triplet a state whose structure corresponds to the second triplet obtained after mixing. It would be interesting to have more experimental knowledge about the triplet of guanine in order to decide between the two possibilities. Calculations of the zero-field splitting parameters corresponding to the different descriptions are in progress.[12]

The zero-field splitting parameters

The knowledge of the wavefunction Ψ of the triplet state of a molecule allows the calculation of the zero-field splitting parameters D and E which can be determined experimentally by analysis of the electron spin resonance experiments.[13] The first-order perturbation theory of the spin-spin interaction yields the theoretical expression for D and E:

$$D = \frac{3g^2\beta^2}{2} \langle\Psi| \sum_{i<j} \frac{r_{ij}^2 - 3z_{ij}^2}{r_{ij}^5} (3s_{iz}s_{jz} - \mathscr{S}_i\mathscr{S}_j)|\Psi\rangle \tag{1}$$

$$E = \frac{3g^2\beta^2}{2} \langle\Psi| \sum_{i<j} \frac{y_{ij}^2 - x_{ij}^2}{r_{ij}^5} (3s_{iz}s_{jz} - \mathscr{S}_i\mathscr{S}_j)|\Psi\rangle, \tag{2}$$

where r_{ij} is the distance from electron i to electron j; x, y, z being the principal axes for the molecule (z is chosen perpendicular to the molecular plane) while $\mathscr{S}_i$ and $\mathscr{S}_j$ are the electron spin operators.[14] When the unperturbed wavefunction of the triplet state is written as a linear combination of Slater determinants, constructed on individual spin-orbitals, the summation in the expectation bracket reduces to the uncoupled spins,[15,16] and, in particular, in the case of a single configuration wavefunction, the zero-field splitting parameters reduce to[15]

$$\begin{pmatrix} D \\ E \end{pmatrix} = \tfrac{3}{4}g^2\beta^2 \langle \Psi(1,2)| \begin{pmatrix} \mathscr{D}_{\text{op.}} \\ \mathscr{E}_{\text{op.}} \end{pmatrix} |\Psi(1,2)\rangle \tag{3}$$

where $\Psi(1, 2)$ is the antisymmetric orbital part relating to the uncoupled electrons occupying the orbitals ϕ_i and ϕ_j:

$$\Psi(1,2) = \frac{1}{\sqrt{2}}[\phi_i(1)\,\phi_j(2) - \phi_j(1)\,\phi_i(2)], \tag{4}$$

with

$$\mathscr{D}_{\text{op.}} = \frac{r_{12}^2 - 3z_{12}^2}{r_{12}^5}, \tag{5}$$

$$\mathscr{E}_{\text{op.}} = \frac{y_{12}^2 - x_{12}^2}{r_{12}^5}. \tag{6}$$

When configuration mixing is introduced in the triplet wavefunction, D and E are obtained by appropriate combination of elements of this and similar kind.[15,19]

In the usual LCAO approximation of the molecular orbitals

$$\phi_i = \sum_r c_{ir}\,\chi_r. \tag{7}$$

Equation (3) becomes

$$\begin{pmatrix} D \\ E \end{pmatrix} = \tfrac{3}{4}\,g^2\beta^2 \sum_{p,q,r,s} c_{ip}\,c_{jq}\,c_{ir}\,c_{js}[\langle pr|qs\rangle - \langle ps|rq\rangle], \tag{8}$$

where

$$\langle pr|qs\rangle = \langle \chi_p(1)\,\chi_r(1)| \begin{pmatrix} \mathscr{D}_{\text{op.}} \\ \mathscr{E}_{\text{op.}} \end{pmatrix} \chi_q(2)\,\chi_s(2)\rangle. \tag{9}$$

In a very detailed examination of the approximations involved, Godfrey, Kern & Karplus[20] have shown recently that the parameter E was extremely sensitive to the inclusion or neglect of multi-center and exchange integrals, whereas the neglect of all but two-center integrals did not introduce an error superior to 10 % in the D-value. Starting with these premises, we have recently investigated the possibilities of a *semi-empirical* evaluation of D and we have shown that reasonable values could be obtained by equation (8), limited to the coulomb terms *in the single-determinant approximation*, namely,

$$D = \tfrac{3}{4}g^2\beta^2 \sum_{p<q} (c_{ip}\,c_{jq} - c_{iq}\,c_{jp})^2 \langle pp|qq\rangle, \tag{10}$$

provided that the $\langle pp|qq \rangle$ integrals be replaced by an *ad hoc* empirically determined r_{pq} function, and that the correct expression in terms of the orbital coefficients be used.[21] This simplified procedure has been utilised to evaluate D in the nucleic bases. The results obtained are shown in table 4 and compared with the empirical data when available. A few other characteristic compounds are given for comparison. Both the old Hückel coefficients[11] and the SCF coefficients[1] in the molecular orbitals have been used. It is seen that the order of magnitude of D is satisfactorily obtained as is also its relative order in the different compounds. In agreement with experiment, the D value of the thymine anion appears as very high and guanine has a larger D than adenine whereas purine should be intermediate. The Hückel and the SCF coefficients yield the same order although the values are slightly higher in the SCF approximation. The fact that D for guanine appears to be of the correct order of magnitude seems to indicate that its triplet structure is rather well approximated by the lowest SCF configuration (see section II). The identity of the D values for thymine and its anion is interesting and seems confirmed by recent measurements.[26]

Table 4. *Comparison of calculated and experimental D-values*

	$D_{\text{calc.}}$		$D_{\text{exp.}}$
	Hückel	SCF	
Thymine	0·180	—	—
Thymine ion	0·180	—	0·198(D^*)†
Guanine	0·135	0·151	0·141[22]
Purine	0·125	0·132	—
Adenine	0·118	0·121	0·121[22]
Indole	0·117	0·129	0·114[23]
Alloxazine	0·077	—	0·069[24]
Isoalloxazine	0·070	—	0·055[24]
Naphthalene	0·103	—	0·099[25]
Phenanthrene	0·098	—	0·101[25]
Anthracene	0·067	—	0·072[18]

† D^* is $(D^2+3E^2)^{\frac{1}{2}}$; the value D is not yet available for this molecule.[22]

Refined calculations of the D-values are in progress using exact values of the integrals and our SCFCM and UHF wavefunctions and a detailed discussion of the different approximations and of their interconnections will be given elsewhere.[12]

References

1 H. Berthod, C. Giessner-Prettre & A. Pullman, *Theor. Chim. Acta*, **5**, 53 (1966).
2 H. Berthod, C. Giessner-Prettre & A. Pullman, *C.R. Acad. Sci. Paris*, **262**, 2657 (1966).
3 H. Berthod, C. Giessner-Prettre & A. Pullman, *Int. J. Quantum Chem.* (in the Press).
4 J. P. Malrieu, *C.R. Acad. Sci., Paris* (in the Press).
5 A. Denis & A. Pullman, *Theor. Chim. Acta* (in the Press).
6 A. Pullman & B. Pullman, *Advances in Quantum Chemistry*, vol. v (in the Press).
7 A survey of the literature can be found in ref. (6).
8 T. Amos & L. Snyder, *J. Chem. Phys.* **43**, 2146 (1965).
9 M.-J. Mantione & B. Pullman, *Biochim. Biophys. Acta*, **91**, 387 (1964); **95**, 668 (1965).
10 P. Cerutti, K. Ikedo & B. Witkop, *J. Am. Chem. Soc.* **87**, 2505 (1965).
11 B. Pullman & A. Pullman, *Quantum Biochemistry* (Wiley-Interscience, N.Y., 1963).
12 A. Pullman & E. Kochansky (in preparation).
13 For a review, see M. Weissbluth in *Molecular Biophysics*, B. Pullman & M. Weissbluth, eds., (Acad. Press, N.Y. 1965), pp. 206–238.
14 J. H. Van Vleck, *Rev. Mod. Phys.* **23**, 213 (1951).
15 M. Gouterman, *J. Chem. Phys.* **30**, 1369 (1959).
16 H. M. McConnell, *Proc. Natl. Acad. Sci. U.S.* **45**, 172 (1959).
17 A. D. McLachlan, *Mol. Physics*, **5**, 51 (1962).
18 A. van der Waals & G. Ter Maten, *Mol. Physics*, **8**, 301 (1964).
19 S. A. Boorstein & M. Gouterman, *J. Chem. Phys.* **39**, 2443 (1963).
20 M. Godfrey, C. W. Kern & M. Karplus, *J. Chem. Phys.* **44**, 4459 (1966).
21 E. Kochansky & A. Pullman, *J. Chem. Phys.* (to be published).
22 R. G. Shulman & R. O. Rahn, *J. Chem. Phys.* **45**, 2940 (1966).
23 C. Cailly & R. Boukhors, *C.R. Acad. Sci., Paris* (in the Press).
24 J. M. Lhoste, A. Haug & P. Hemmerich, *Biochemistry*, **5**, 3290 (1966).
25 R. W. Brandon, R. E. Gerkin & C. A. Hutchison Jr., *J. Chem. Phys.* **37**, 447 (1962).
26 M. Guéron (private communication).

DISCUSSION

LHOSTE

Parker to Lhoste: I should like to draw attention to the fact that the behaviour of triplet molecules in solution can be quite different in the presence of polymeric molecules. For example, the benz(a)pyrene triplet is quite stable in ethanolic solution but reacts rapidly with the solvent in the presence of small quantities of certain polymers.

Wolf to Lhoste: Is it impossible to incorporate large molecules like porphyrine into organic single crystals like biphenyl? Do you know the upper limit of the concentration which one can get?

Lhoste to Wolf: Up to this time, all our attempts to incorporate porphyrines in organic crystalline materials have been unsuccessful. With large and heteroatomic molecules like these, it should be very difficult to observe the hyperfine structure of the ESR spectra.

Stevens to Lhoste: Triplet-triplet annihilation would be expected to produce adjacent singlet excited and ground state molecules which could photo-associate and form the photodimer; thus both the triplet and the excited singlet states might be regarded as the thymine dimer precursor.

The red-shift of the excimer band, observed from base pairs, appears to be about 3000 cm^{-1} compared with that of 6000 cm^{-1} exhibited by aromatic hydrocarbons. Is this because the components of the base pair are chemically different (mixed dimers) or do you observe the same shift for excimers of identical molecules?

Hélène to Stevens: The red-shift of the fluorescence band observed in dinucleotides containing two different bases (such as *ApC*, *ApU*) is roughly 3500 cm^{-1}. The red-shift is slightly larger for identical bases (such as *CpC*), but anyway it is much smaller than in the case of aromatic hydrocarbons.

Siegel to Lhoste: From the energy sum in the triplet manifold

$$E_T + h\nu \rightarrow E_T^*,$$

necessary to dissociate ice into H (and OH) by energy transfer, can you determine the acceptor energy level in the ice? Using different solutes might allow the determination of the triplet energy levels in ice.

Lhoste to Siegel: In our experimental conditions, we observe photoionisation of the solute molecules. In an acid medium, the electrons are trapped by protons to give H atoms. In a neutral medium, the yield of H atoms is much lower, but they are still detectable by ESR. It is not yet established whether these H atoms originate from sensitised ice, from solute molecules or from recombination with protons.

Guéron to Lhoste: Regarding the experiments on *AppAcC* if triplet-triplet transfer is much faster than the triplet donor lifetime, the acceptor decay should be purely exponential; only a very limited range of values of the transfer interaction, those giving a rate comparable to the decay rate of the donor, would result in a non-exponential decay-rate of emission (as presented), if the donor-acceptor geometry is well defined. Is there other evidence for the triplet transfer mechanism?

Hélène to Guéron: It is quite likely that there is not one single fixed configuration, particularly for nucleotides linked by a pyrophosphate group (as in *AppAcC*). Thus, triplet transfer must occur only in the dinucleotide molecules which have stacked bases and overlapping electron clouds. The phosphorescence spectra of *AcC* and *A* are overlapping and the decay was recorded at a wavelength where these two molecules are able to emit. This can explain why the decay is non-exponential for *AppAcC*.

GUERON

Pullman to Guéron: I was very interested by your statement on the similarity of the triplets in T and T^-. We have calculated the zero-field splitting parameter D for both thymine and its anion and have found the same value, 0·180 cm^{-1} which, on the other hand, agrees very well with your experimental value. However, since this evaluation of D was based on the Hückel coefficients we are looking forward to the results of more elaborate calculations that are being performed in our laboratory.

Stevens to Guéron: Your suggestion, that the excimer is the triplet thymine precursor, is interesting in view of the accumulating evidence for intersystem crossing in aromatic hydrocarbon excimers followed by triplet excimer dissociation to ground and triplet state molecules. In a mixed excimer, this overall process would presumably populate the molecular triplet state of lowest energy although this molecule is not necessarily that excited initially in the singlet state.

Rice to Guéron: 1. Do the deoxynucleotides in water solution phosphoresce?

2. Why is this phosphorescence not quenched by hydrogen-bonding to the water, in line with your postulated quenching mechanism in the H-bonded polymer?

Hélène to Guéron: If the thymine triplet is obtained by energy transfer from adenine, its population should be proportional to the probability of A and T to be one over another. May this explain why poly-dAT, in which A and T are alternated, has a higher thymine triplet population than is to be expected from the measurements on DNA's of different A–T contents?

Guéron to Hélène: Such an explanation would also require that the intensity of the T triplet be a quadratic function of A–T content for low values of A–T content, and this is apparently not observed. We cannot presently draw detailed conclusions from the triplet yield versus A–T content-curve.

Guéron to Rice: 1. The deoxynucleotides GMP, AMP, CMP phosphoresce in the ethylene-glycol-water glass.

2. We definitely observe quenching in the hydrogen-bonded polymers poly-rG: rC, poly-dG:dC and poly-rA:rU, but we are not postulating any one mechanism for explaining this.

PULLMAN

Tanaka to Pullman: I think that you have obtained the energy levels of both singlet and triplet excited states. Do they fit well with experiment?

Pullman to Tanaka: We indeed have obtained the energy levels of all singlet and triplets in these compounds by all three procedures that I have mentioned. The agreement with the experimental data is extremely satisfactory for singlets in the SCFCM representation. Much less so in the UHF calculation for the lowest singlet. Details concerning this point are given in references (1) to (5) of my paper.

As to the locations of the triplets, they are, as a rule, less satisfactory. This seems to be a general feature of the Pariser–Parr–Pople type of approximation, where the integral values used are fitted without reference to triplet location, which was the case in our calculations. In fact we are studying this particular problem in more detail at the present moment in a careful investigation of the benzene molecule and other simple reference compounds.

Van der Waals to Pullman: How do you determine your integrals for the spin dipolar interaction between atomic orbitals, in order to increase the Hückel value of D relative to that obtained with *a priori* calculated values of the atomic orbital integrals?

Pullman to van der Waals: When exact values of the coefficient of each atomic orbital integral are used together with *a priori* calculated integrals, the D values are too small in the simple determinant representation and configuration mixing is necessary. Roughly speaking, one may consider that by mixing configurations, one smears out the odd-electron distribution in the triplet. This, on an average, increases the close-interaction terms in D (by close-interaction, I mean not only adjacent but at least 1–3 and 1–4-type interactions). This made us think that the same effect could be obtained in the simple determinant representation by *artificially* increasing the close-interaction integral values. Thus, we have adopted for $\langle pp|qq\rangle$ a law of the form:

$$\frac{1}{2}\left[\frac{1}{r^3}+\frac{1}{(r^2+d^2)^{\frac{3}{2}}}\right]$$

which satisfies the above condition, where r is the distance between atoms p and q and $d \approx 1{\cdot}4$ Å. The reasons for this choice are explored in a paper to be published in *J. Chem. Phys.*

Lhoste to Pullman: 1. Could you calculate the orientation of the main axes of the spin-spin coupling tensor in the triplet pyrimidines? This is a very sensitive means of testing the precision of calculations on triplet states of low-symmetry molecules and it can be checked experimentally.

2. The similarity in electronic structure of the lowest triplet state of the pyrimidines (uracil, thymine), pteridines (lumazine) and flavins (isoalloxazine and alloxazine) is surprising. Could you comment on the role of the pyrimidinoid ring in the large molecules?

Pullman to Lhoste: 1. We are doing this in the calculation based on the refined wavefunctions (SCFCM and UHF). We calculated them also in the crude approximation that is mentioned in my paper, but merely as a subproduct of the calculation that we did not consider worth commenting on since this is much more sensitive to the approximations made than the D value (see the reference to the paper of Godfrey, Kerr and Karplus in my communication).

2. Looking at it from a superficial point of view, one could be tempted to say that it is due to the

```
O       O
‖       ‖
C—NH—C—NH
```

grouping, which seems to keep a strong individuality, but it would be wiser to compare the calculated electronic structures of these compounds. In fact, we intend to do this.

Hutchison to Pullman: I am interested in your comment that the UHF method, using the usual annihilation techniques to generate eigenstates of **S**, overemphasizes doubly excited configurations which in turn leads to an exaggeration of the magnitudes of the negative spin densities. There are only two cases, of which I am aware, in which spin densities are known for all the C positions in a triplet state organic molecule. These are naphthalene, and the fluorenylidine case which I discussed in my lecture. The UHF method in the naphthalene case gives a negative spin density in good agreement with the observed value, and in the fluorenylidene case seems to give negative spin densities which are much too small in magnitude, not too large as you suggest. The calculation is actually for a doublet state of a fluorenylidene-like molecule as discussed in my lecture, but I believe this makes no essential difference with respect to the point you were making. Do you have any comments on this situation?

Pullman to Hutchison: The statement concerning the importance of certain doubly excited configurations which are the origin of the negative spin densities was made by T. Amos and L. C. Snyder (*J. Chem. Phys.* **43**, 2146 (1965)), and their demonstration was made for triplets. I do not think that a doublet is similar.

NAME INDEX

The numbers in italics are those of pages listing bibliographical references

SUBJECT INDEX